ADVANCED MICROPROCESSORS AND MICROCONTROLLERS

ADVANCED MICROPROCESSORS AND MICROCONTROLLERS

(Third Edition)

B P Singh

Professor
Deptt. of Electronics & Communication Engineering
Madan Mohan Malvia Engineering College
Gorakhpur, U.P.

Renu Singh

Former Guest Lecturer
Madan Mohan Malvia Engineering College
Gorakhpur, U.P.

Publishing Globally

NEW AGE INTERNATIONAL (P) LIMITED, PUBLISHERS

New Delhi • Bangalore • Chennai • Cochin • Guwahati
Hyderabad • Kolkata • Lucknow • Mumbai
Visit us at www.newagepublishers.com

Published by New Age International (P) Ltd., Publishers
First Edition: 2002
Third Edition: 2008
Reprint: 2013

BRANCHES

- **Bangalore** 37/10, 8th Cross (Near Hanuman Temple), Azad Nagar, Chamarajpet, Bangalore-560 018 Tel.: (080) 26756823, Telefax: 26756820, **E-mail: bangalore@newagepublishers.com**
- **Chennai** 26, Damodaran Street, T. Nagar, Chennai-600 017, Tel.: (044) 24353401 Telefax: 24351463, **E-mail: chennai@newagepublishers.com**
- **Cochin** CC-39/1016, Carrier Station Road, Ernakulam South, Cochin-682 016 Tel.: (0484) 2377303, Telefax: 4051303, **E-mail: cochin@newagepublishers.com**
- **Guwahati** Hemsen Complex, Mohd. Shah Road, Paltan Bazar, Near Starline Hotel, Guwahati-781 008 Tel.: (0361) 2513881, Telefax: 2543669, **E-mail: guwahati@newagepublishers.com**
- **Hyderabad** 105, 1st Floor, Madhiray Kaveri Tower, 3-2-19, Azam Jahi Road, Near Kumar Theater Nimboliadda, Kachiguda, Hyderabad-500 027, Tel.: (040) 24652456, Telefax: 24652457 **E-mail: hyderabad@newagepublishers.com**
- **Kolkata** RDB Chambers (Formerly Lotus Cinema) 106A, 1st Floor, S.N. Banerjee Road, Kolkata-700 014 Tel.: (033) 22273773, Telefax: 22275247, **E-mail: kolkata@newagepublishers.com**
- **Lucknow** 16-A, Jopling Road, Lucknow-226 001, Tel.: (0522) 2209578, 4045297, Telefax: 2204098 **E-mail:lucknow@newagepublishers.com**
- **Mumbai** 142C, Victor House, Ground Floor, N.M. Joshi Marg, Lower Parel, Mumbai-400 013 Tel.: (022) 24927869, Telefax: 24915415, **E-mail: mumbai@newagepublishers.com**
- **New Delhi** 22, Golden House, Daryaganj, New Delhi-110 002, Tel.: (011) 23262368, 23262370 Telefax: 43551305, **E-mail: sales@newagepublishers.com**

ISBN: 978-81-224-2285-6

₹ 360.00

C-12-08-6518

Printed in India at Ramprintograph, Delhi.
Typeset at Kalyani Computer Services, Delhi.

PUBLISHING GLOBALLY
NEW AGE INTERNATIONAL (P) LIMITED, PUBLISHERS
7/30 A, Daryaganj, New Delhi-110002
Visit us at **www.newagepublishers.com**

PREFACE TO THE THIRD EDITION

The third edition contains many organizational changes. The important among them are addition of many articles to enrich the contents of the book such as:

- Chapter–1 includes historical development of microprocessors and dual core processor.
- Chapter–2 and 3 have been renovated.
- Chapter–7 has been given a new look to suit the embeded system approach.

PREFACE TO THE FIRST EDITION

The wealth of the natural sciences
no longer consists in the abundance of facts,
but in the way they are linked together.
ALEXANDER VON HUMBOLDT

There is no dearth of books on **Advanced Microprocessors and Microcontrollers.** Why then write yet another?

A student very often finds certain books more helpful than others. I too believe in having a variety of presentations from which to choose the most appropriate. The variety also allows students to appreciate the diversity within the scope of the field.

The quantum jump from 8-bit to 16-bit microprocessor has completely changed the architectural philosophy from single microprocessor to two and then to many. The term 'advanced microprocessor' has been used in this book to include the family. A comprehensive coverage of 8086, 80286, 80386, 80486, Pentium, Motorola microprocessors, PowerPC, and Microcontrollers have been taken up in this book. Sufficient details have been provided on the floating-point processors also.

The goal of the book was to introduce the necessary hardware and software concepts in an elementary, systematic, and integrated manner and to build upon these concepts logically. It is well known that the study of one book does not make one a master of the subject, yet I expect that this book will provide the reader solid foundation for the development of skills in designing the microprocessor based system.

Every effort has been made to treat the subject so as to ensure an easy flow of understanding of readers. Fundamental concepts have been taken up through illustrative physical examples. A larger number of solved examples have been introduced throughout the text. Short question-answers at the end of chapters have been included to facilitate students to absorb the concepts.

Inspite of the best efforts, it is possible that errors of omission and commission might have crept in. I will thankfully acknowledge any mistakes pointed out and any suggestions to further enrich the contents of the book.

—B.P.SINGH

ACKNOWLEDGEMENT

We are extremely thankful to Dr. B.S. Rai, Deptt. of Electronics & Communication, M.M.M. Engg. College, Gorakhpur for going through the manuscript and providing suggestions for addition and deletion wherever needed for the easy flow of the matter to the readers and to Mrs. Meena Singh, M.Tech. student, for helping in proof reading.

Expressing gratitude publicly to wife is almost absent in Hindu society. But the coauthorship of wife is tacitly assumed as wife and husband of Hindu society is considered as one inseparable unit. I owe deep sense of appreciation to my wife Smt. Bimla Singh and children Upendra & Rita, Rita & Sanjay, Rekha & Rajesh and grand children Kaushik, Kinshuk, Honey, Rishan, Rusheel and Rachit for sparing me from day to day work for preparation of the manuscript of the book, even during their short stay with me.

I thankfully acknowledge the contributions of various authors of different books, magazines, data manuals, journals etc. from where materials have been collected to enrich the contents of the book.

(B.P. SINGH)

• • •

I am extremely thankful to my husband Mr. Ashok Kumar for sparing me from household work for concentrating to prepare the text of the book.

(RENU SINGH)

CONTENTS

1

EVOLUTION OF MICROPROCESSORS

1.1 INTRODUCTION

A microprocessor is the chip containing some control and logic circuits that is capable of making arithmetic and logical decisions based on input data and produces the corresponding arithmetic or logical output. The word 'processor' is the derivative of the word 'process' that means to carry out systematic operations on data. The computer we are using to write this page of the manuscript uses a **microprocessor** to do its work. The microprocessor is the **heart** of any computer, whether it is a **desktop** machine, a **server or a laptop.** The microprocessor we are using might be a **Pentium, a K6, a PowerPC, a Sparc** or any of the many other brands and types of microprocessors, but they all do approximately the same thing in approximately the same way. No logically enabled device can do any thing without it. The microprocessor not only forms the very basis of computers, but also many other devices such as cell phones, satellites, and many other hand held devices. They are also present in modern day cars in the form of microcontrollers.

A microprocessor is also known as a **CPU** or central processing unit, which is a complete computational engine that is fabricated on a single chip. Here we will discuss the history of the 80x86 CPU family and the major improvements occurring along the line. The historical background will help us to better understand the design compromises they made as well as to understand the legacy issues surrounding the CPU's design. We are discussing the major advances in computer architecture that Intel employed while improving the x86.

1.2 HISTORICAL DEVELOPMENT OF THE MICROPROCESSORS

Intel developed and delivered the first commercially viable microprocessor way back in the early 1970's: the 4004 and 4040 devices. The 4004 was not very powerful and all it could do was **add** and **subtract** with 4-bit data only at a time. But it was amazing those days that everything was on one chip. Prior to the 4004, engineers built computers either from collections of chips or from discrete components (Transistor wired one at a time). The machines then, were not portable and were very bulky, and power hungry. The 4004 changed the scene with all its circuitry on a single chip. The 4004 powered one of the first portable electronic calculators named **'Busicom'.** These 4-bit microprocessors, intended for use in calculators, required very little power Nevertheless, they demonstrated the future potential of the microprocessor - an entire CPU on a single piece of silicon. Intel rapidly followed their 4-bit offerings with their 8008 and 8080 eight-bit CPUs. A small outfit in Santa Fe, New Mexico, incorporated the 8080 CPU into a box they called the Altair 8800. Although this was not the world's first "personal computer" (there were some limited distribution machines built around the 8008 prior to this), the Altair was the device that sparked the imaginations of hobbyists of the world and the personal computer revolution was born.

The trends in processor design had impact of historical development of microprocessors from different manufacturers. Intel started facing competition from Motorola, MOS Technology, and an upstart company formed by disgruntled Intel employees, Zilog. To compete, Intel produced the 8085 microprocessor. To the software engineer, the 8085 was essentially the same as the 8080. However, the 8085 had lots of hardware improvements that made it easier to design into a circuit. Unfortunately, from software perspective the other manufacturer's offerings were better. Motorola's 6800 series was easier to program, MOS Technologies' 65xx family was also easier to program but very inexpensive, and Zilog's Z80 chip was upward compatible with the 8080 with lots of additional instructions and other features. By 1978 most personal computers were using the 6502 or Z80 chips, not the Intel offerings.

The first microprocessor to make a real splash in the market was the Intel 8088, introduced in 1979 and incorporated into the IBM PC (which appeared around 1982 for the first time). If we are familiar with the PC market and its history, we know that the PC market moved from the 8088 to the 80286 to the 80386 to the 80486 to the Pentium to the Pentium II to the Pentium III to the Pentium 4. Intel makes all of these microprocessors and all of them are improvements of design base of the 8088. The Pentium 4 can execute any piece of code that ran on the original 8088, but it does it about 5,000 times faster!

Sometime between 1976 and 1978 Intel decided that they needed to *leap-frog* the competition and produced a 16-bit microprocessor that offered substantially more power than their competitor's eight-bit offerings. This initiative led to the design of the 8086 microprocessor. The 8086 microprocessor was not the world's first 16-bit microprocessor (there were some oddball 16-bit microprocessors prior to this point) but it was certainly the highest performance single-chip 16-bit microprocessor when it was first introduced.

During the design timeframe of the 8086 *memory* was very *expensive.* Sixteen Kilobytes of RAM was selling above $200 at the time. One problem with a 16-bit CPU is that programs tend to consume more memory than their counterparts on an 8-bit CPU. Intel, ever cogniscent of the fact that designers would reject their CPU if the total system cost was too high, made a special effort to design an instruction set that had a high memory density (that is, packed as many instructions into as little RAM as possible). Intel achieved their design goal and programs written for the 8086 were comparable in size to code running on 8-bit microprocessors. However, those design decisions still haunt us today.

At the time Intel designed the 8086 CPU the average lifetime of a CPU was only a couple of years. Their experiences with the 4004, 4040, 8008, 8080, and 8085 taught them that designers would quickly ditch the old technology in favour of the new technology as long as the new stuff was radically better. So Intel designed the 8086 assuming that whatever compromises they made in order to achieve a high instruction density would be fixed in newer chips. Based on their experience, this was a reasonable assumption.

Intel's competitors were not standing still. Zilog created their own 16-bit processor that they called it Z8000, Motorola created the 68000, their own 16-bit processor, and National Semiconductor introduced the 16032 device (later to be renamed the 32016). The designers of these chips had different design goals than Intel. Primarily, they were more interested in providing a reasonable instruction set for programmers even if their code density was not anywhere near as high as the 8086. The Motorola and National offers even provided 32-bit integer registers, making programming the chips even easier. All in all, these chips were much better (from a software development standpoint) than the Intel chip.

Intel was not resting on its laurels with the 8086. Immediately after the release of the 8086 they created an eight-bit version, the 8088. The purpose of this chip was to reduce system cost (a minimal system was possible with half the memory chips and cheaper peripherals since the 8088 had an 8-bit data bus). In the very early 1980's, Intel also began work on their intended successor to the 8086—the iAPX432 CPU.

In 1980 a small group at IBM got the go-ahead to create a "personal computer" along the lines of the Apple II and TRS-80 computers (the most popular PCs at the time). IBM's engineers probably evaluated lots of different CPUs and system designs. Ultimately, they settled on the 8088 chip. Most likely they chose this chip because they could create a minimal system with only 16 kilobytes of RAM and a set of cheap eight-bit peripheral devices. So Intel's design goals of creating CPUs that worked well in low-cost systems landed them a very big "design win" from IBM.

Intel was still hard at work on the (ill-fated) iAPX432 project, but a funny thing happened - IBM PCs started selling far better than anyone had ever dreamed. As the popularity of the IBM PCs increased (and as people began "cloning" the PC), lots of software developers began writing software for the 8088 (and 8086) CPU, mostly in assembly language. In the meantime, Intel was pushing their iAPX432 with the Ada programming language (which was supposed to be the next big thing after Pascal, a popular language at that time). Unfortunately for Intel, no one was interested in the '432. Their PC software, written mostly in assembly language would not run on the '432 and the '432 was notoriously slow. It took a while, but the iAPX432 project eventually died off completely and remains a black spot on Intel's record to this day.

Intel was not sitting pretty on the 8086 and 8088 CPUs, however. In the late 1970's and early 1980's they developed the 80186 and 80188 CPUs. These CPUs, unlike their previous CPU offerings, were fully upwards compatible with the 8086 and 8088 CPUs. In the past, whenever Intel produced a new CPU it did not necessarily run the programs written for the previous processors. For example, the 8086 did not run 8080 software and the 8080 did not run 4040 software. Intel, recognizing that there was a tremendous investment in 8086 software, decided to create an upgrade to the 8086 that was superior (both in terms of hardware capability and with respect to the software it would execute). Although the 80186 did not find its way into many PCs, it was a very popular chip in embedded applications (i.e., non-computer devices that use a CPU to control their functions). Indeed, variants of the 80186 are in common use even today.

The unexpected popularity of the IBM PC created a problem for Intel. This popularity obliterated the assumption that designers would be willing to switch to a better chip when such a chip arrived, even if it meant rewriting their software. Unfortunately, IBM and tens of thousands of software developers were not willing to do this to make life easy for Intel. They wanted not only to stick with the 8086 software they had written but they also wanted something a little better than the 8086. If they were going to be forced into jumping ship to a new CPU, the Motorola, Zilog, and National offerings were starting to look pretty good. So Intel did something that saved their bacon and has infuriated computer architects ever since: they started creating upwards compatible CPUs that continued to execute programs written for previous members of their growing CPU family while adding new features.

As noted earlier, memory was very expensive when Intel first designed the 8086 CPU. At that time, computer systems with a megabyte of memory usually cost megabucks. Intel was expecting a typical computer system employing the 8086 to have somewhere between 4 kilobytes to 64 kilobytes of memory. So when they designed in a one-megabyte limitation, they feared no one would ever install that much memory in a system. Of course, by 1983 people were still using 8086 and 8088 CPUs in their systems and memory prices had dropped to the point where it was very common to install 640 kilobytes of memory on a PC (the IBM PC design effectively limited the amount of RAM to 640 kilobytes even though the 8086 was capable of addressing one megabyte). By this time software developers were starting to write more sophisticated programs and users were starting to use these programs in more sophisticated ways. The bottom line was that everyone was bumping up against the one-megabyte limit of the 8086. Despite the investment in existing software, Intel was about to lose their cash cow if they did not do something about the memory addressing limitations of their 8086 family (the 68000 and 32016 CPUs could address up to 16 megabytes at the time and many system designers [e.g., Apple] were defecting to these other chips). So Intel introduced the 80286 that was a big improvement over the previous CPUs. The 80286 added lots of new instructions to make programming a whole lot easier and they added a new "protected" mode of operation that allowed access to as much as 16 megabytes of memory. They also improved the internal operation of the CPU and bumped up the clock frequency so that the 80286 ran about 10 times faster than the 8088 in IBM PC systems.

IBM introduced the 80286 in their IBM PC/AT (AT = "advanced technology"). This change proved enormously popular. PC/AT clones based on the 80286 started appearing everywhere and Intel's financial future was assured.

Realizing that the 80x86 (x = "1", "2", "3" or "4") family was a big money maker, Intel immediately began the process of designing new chips that continued to execute the old code while improving performance and adding new features. Intel was still playing catch-up with their competitors in the CPU arena with respect to features, but they were definitely the king of the hill with respect to CPUs installed in PCs. One significant difference between Intel's chips and many of their competitors was that their competitors (notably Motorola and National) had a 32-bit internal architecture while the 80x86 family was stuck at 16-bits. Again, concerned that people would eventually switch to the 32-bit devices their competitors offered, Intel upgraded the 80x86 family to 32-bits by adding the 80386 to the product line.

The 80386 was truly a remarkable chip. It maintained almost complete compatibility with the previous 16-bit CPUs while fixing most of the real complaints people had with those older chips. In addition to supporting 32-bit computing, the 80386 also bumped up the maximum addressability to four gigabytes as well as solving some problems with the "segmented" organization of the previous chips (a big complaint by software developers at the time). The 80386 also represented the most radical change to ever occur in the 80x86 family. Intel more than doubled the total number of instructions, added new memory management facilities, added hardware debugging support for software, and introduced many other features. Continuing the trend they set with the 80286, the 80386 executed instructions faster than previous generation chips, even when running at the same clock speed plus the new chip ran at a higher clock speed than the previous generation chips. Therefore, it ran existing 8088 and 80286 programs faster on these older chips. Unfortunately, while people adopted the new chip for its higher performance, they did not write new software to take advantage of the chip's new features.

Although the 80386 represented the most radical change in the 80x86 architecture from the programmer's view, Intel was not done wringing all the performance out of the 80x86 family. By the time the 80386 appeared, computer architects were making a big noise about the so-called RISC (Reduced Instruction Set Computer) CPUs. While there were several advantages to these new RISC chips, an important advantage of these chips is that they purported to execute one instruction every clock cycle. The 80386 instructions required a widely varying number of cycles to execute ranging from a few cycles per instruction to well over a hundred. Although comparing RISC processors directly with the 80386 was dangerous (because many 80386 instructions actually did the work of two or more RISC instructions), there was a general perception that, at the same clock speed, the 80386 was slower since it executed fewer instructions in a given amount of time.

The 80486 CPU introduced 2-major advances in the 80x86 design. First, the 80486 integrated the *floating-point unit* (or FPU) directly onto the CPU die. *Prior* to this point Intel supplied a separate, *external,* chip to provide floating-point calculations (these were the 8087, 80287, and 80387 devices). By incorporating the FPU with the CPU, Intel was able to speed up floating-point operations and provided this capability at a lower cost (at least on systems that required floating-point arithmetic). The second major architectural advance was the use of *pipelined instruction execution.* This feature (which we will discuss in detail a little later) allowed Intel to overlap the execution of two or more instructions. The end result of pipelining is that they effectively reduced the number of cycles each instruction required for execution. With pipelining, many of the simpler instructions had an aggregate throughput of one instruction per clock cycle (under ideal conditions) so the 80486 was able to compete with RISC chips in terms of clocks per instruction cycle.

While Intel was busy adding pipelining to their 80x86 family, the companies building RISC CPUs were not standing still. To create ever faster and faster-CPU offerings, RISC designers began creating *superscalar* CPUs that could actually execute more than one instruction per clock cycle. Once again, Intel's CPUs were perceived as following the leaders in terms of CPU performance. Another problem with Intel's CPU is that the integrated FPU, though faster than the earlier models, was significantly slower than the FPUs on the RISC chips. As a result, those designing high-end engineering workstations (that typically require good floating-point hardware support) began using the RISC chips because they were faster than Intel's offerings.

From the programmer's perspective, there was very little difference between an 80386 with an 80387 FPU and an 80486 CPU. There were only a handful of new instructions (most of which had very little utility in standard applications) and not much in the way of other architectural features that software could use. The 80486, from the software engineer's point of view, was just a really fast 80386/80387 combination.

So Intel went back to their CAD tools and began work on their next CPU. This new CPU featured a superscalar design with vastly improved floating-point performance. Finally, Intel was closing in on the performance of the RISC chips. Like the 80486 before it, this new CPU added only a small number of new instructions and most of those were intended for use by operating systems, not application software.

Intel did not designate this new chip the 80586. Instead, they called it the *Pentium Processor*. The reason they discontinued referring to processors by number and started naming them was because of confusion in the market place. Intel was not the only company producing 80x86 compatible CPUs. AMD, Cyrix, and a host of others were also building and selling these chips in direct competition with Intel. Until the 80486 came along, the internal design of the CPUs were relatively simple and even small companies could faithfully reproduce the functionality of Intel's CPUs. The 80486 was a different story altogether. This chip was quite complex and taxed the design capabilities of the smaller companies. Some companies, like AMD, actually licensed Intel's design and they were able to produce chips that were compatible with Intel's (since they were, effectively, Intel's chips). Other companies attempted to create their own version of the 80486 and fell short of the goal. Perhaps they did not integrate an FPU or the new instructions on the 80486. Many did not support pipelining. Some chips lacked other features found on the 80486. In fact, most of the (non-Intel) chips were really 80386 devices with some very slight improvements. Nevertheless, they called these chips 80486 CPUs.

This created massive confusion in the market place. Prior to this, if we would purchase a computer with an 80386 chip we knew the capabilities of the CPU. All 80386 chips were equivalent. However, when the 80486 came along and we purchased a computer system with an 80486, we did not know if we were getting an actual 80486 or a remarked 80386 CPU. To counter this, Intel began their enormously successful "Intel Inside" campaign to let people know that there was a difference between Intel CPUs and CPUs from other vendors. This marketing campaign was so successful that people began specifying Intel CPUs even though some other vendor's chips (i.e., AMD) were completely compatible.

Not wanting to repeat this problem with the 80586 generation, Intel ditched the numeric designation of their chips. They created the term "Pentium Processor" to describe their new CPU so they could trademark the name and prevent other manufacturers from using the same designation for their chip. Initially, of course, savvy computer users griped about Intel's strong-arm tactics but the average user benefited quite a bit from Intel's marketing strategy. Other manufacturers release their own 80586 chips (some even used the "586" designation), but they could not use the Pentium Processor name on their parts so when someone purchased a system with a Pentium in it, they knew it was going to have all the capabilities of Intel's chip since it had to be Intel's chip. This was a good thing because most of the other '586 class chips that people produced at that time were, not as powerful as the Pentium.

The Pentium cemented Intel's position as champion of the personal computer. It had near RISC performance and ran tons of existing software. Only the Apple Macintosh and high-end

UNIX workstations and servers went the RISC route. Together, these other machines comprised less than 10% of the total desktop computer market. Intel still was not satisfied. They wanted to control the server market as well. So they developed the Pentium Pro CPU. The Pentium Pro had a couple of features that made it ideal for servers. Intel improved the 32-bit performance of the CPU (at the expense of its 16-bit performance), they added better support for multiprocessing to allow multiple CPUs in a system (high-end servers usually have 2 or more processors), and they added a handful of new instructions to improve the performance of certain instruction sequences on the pipelined architecture. Unfortunately, most application software written at the time of the Pentium Pro's release was 16-bit software that actually ran slower on the Pentium Pro than it did on a Pentium at equivalent clock frequencies. So although the Pentium Pro did wind up in a few server machines, it was never as popular as the other chips in the Intel line.

The Pentium Pro had another big strike against it: shortly after the introduction of the Pentium Pro, Intel's engineers introduced an upgrade to the standard Pentium chip, the MMX (multimedia extension) instruction set. These new instructions (nearly 60 in all) gave the Pentium additional power to handle computer video and audio applications. These extensions became popular overnight, putting the last nail in the Pentium Pro's coffin. The Pentium Pro was slower than the standard Pentium chip and slower than high-end RISC chips, so it did not see much of Sunrise.

Intel corrected the 16-bit performance in the Pentium Pro, added the MMX extensions and called the result the Pentium II. The Pentium II demonstrated an interesting point. Computers had reached a point where they were powerful enough for most people's everyday activities. Prior to the introduction of the Pentium II, Intel (and most industry pundits) had assumed that people would always want more power out of their computer systems. Even if they did not need the machines to run faster, surely the software developers would write larger (and slower) systems requiring more and more CPU power. The Pentium II proved this idea wrong. The average user needed e-mail, word processing, Internet access, multimedia support, simple graphics editing capabilities, and a spreadsheet now and then. Most of these applications, at least as home users employed them, were fast enough on existing CPUs. The applications that were slow (e.g., Internet access) were generally beyond the control of the CPU (i.e., the modem was the bottleneck not the CPU). As a result, when Intel introduced their pricey Pentium II CPUs, they discovered that system manufacturers started buying other people's 80x86 chips because they were far less expensive and quite suitable for their customer's applications. This nearly stunned Intel since it contradicted their experience up to that point.

Realizing that the competition was capturing the low-end market and stealing sales away, Intel devised a low-cost (lower performance) version of the Pentium II that they named *Celeron.* The initial Celerons consisted of a Pentium II CPU without the on-board level two caches. Without the cache, the chip ran only a little bit better than half the speed of the Pentium II part. Nevertheless, the performance was comparable to other low-cost parts so Intel's fortunes improved once more.

While designing the low-end Celeron, Intel had not lost sight of the fact that they wanted to capture a chunk of the high-end workstation and server market as well. So they created a third version of the Pentium II, the Xeon Processor with improved cache and the capability of multiprocessor more than two CPUs. The Pentium II supports a two CPU multiprocessor system but it is not easy to expand it beyond this number; the Xeon processor corrected this limitation. With the introduction of the Xeon processor (plus special versions of Unix and Windows NT), Intel finally started to make some serious inroads into the server and high-end workstation markets.

We can probably imagine what followed the Pentium II. The Pentium III introduced the SIMD (pronounced SIM-DEE) extensions to the instruction set. These new instructions provided high performance floating point operations for certain types of computations that allow the Pentium III to compete with high-end RISC CPUs. The Pentium III also introduced another handful of integer instructions to aid certain applications.

With the introduction of the Pentium III, nearly all-serious claims about RISC chips offering better performance were fading away. In fact, for most applications, the Intel chips were actually faster than the RISC chips available at the time. Next, of course, Intel introduced the Pentium 4 chip (it was running at 2 GHz as this was being written, a much higher clock frequency than its RISC contemporaries). An interesting issue concerning the Pentium 4 is that it does not execute code faster than the Pentium III when running at the same clock frequency (it runs slower, in fact). The Pentium IV makes up for this problem by executing at a much higher clock frequency than was possible with the Pentium III. One would think that Intel would soon own it all. Surely by the time of the Pentium V, the RISC competition would not be a factor anymore.

There is one problem with this theory: even Intel is admitting that they have pushed the 80x86 architecture about as far as they can. For nearly 20 years, computer architects have blasted Intel's architecture as being gross and bloated having to support code written for the 8086 processor way back in 1978. Indeed, Intel's design decisions (like high instruction density) that seemed so important in 1978 are holding back the CPU today. So-called "clean" designs, that do not have to support legacy applications, allow CPU designers to create high-performance CPUs with far less effort than Intel's. Worse, those decisions Intel made in the 1976-1978 time frame are beginning to catch up with them and will eventually stall further development of the CPU. Computer architects have been warning everyone about this problem for twenty years; it is a testament to Intel's design effort (and willingness to put money into R&D) that they have taken the CPU as far as they have.

The biggest problem on the horizon is that most RISC manufacturers were extending their architectures to 64-bits. This had 2-important impacts on computer systems. First, arithmetic calculations will be somewhat faster as will many internal operations and second, the CPUs will be able to directly address more than 4-gigabytes of main memory. This last factor is probably the most important for server and workstation systems. Already, high-end servers have more than 4-gigabytes installed. In future, the ability to address more than 4-gigabytes of physical RAM will become essential for servers and high-end workstations. As the price of a gigabyte or more of memory drops below $100, we will see low-end personal computers with more than 4-gigabytes installed. To effectively handle this kind of memory, Intel needs a 64-bit processor to compete with the RISC chips.

Perhaps Intel has seen the light and decided to give up on the 80x86 architecture. Towards the middle to end of the 1990's Intel announced that they in partnership with Hewlet-Packard are going to create a new 64-bit processor based around HP's PA-RISC architecture. This new

64-bit chip would execute 80x86 code in a special "emulation" mode and run native 64-bit code using a new instruction set. It was too early to tell if Intel would be successful with this strategy, but there were some major risks (pardon the pun) with this approach. The first such CPUs (just becoming available as this is being written) ran 32-bit code far slower than the Pentium III and IV chips. Not only does the emulation of the 80x86 instruction set slow things down, but the clock speeds of the early CPUs were half the speed of the Pentium IVs. This is roughly the same situation Intel had with the Pentium Pro running 16-bit code slower than the Pentium. Second, the 64-bit CPUs (the IA64 family) rely heavily on compiler technology and are using a commercially untested architecture. This is similar to the situation with the iAPX432 project that failed quite miserably. Hopefully Intel knows what they are doing and ten years from then we will all be using IA64 processors and wondering why anyone ever stuck with the IA32.

Intel is betting that people will move to the IA64 when they need 64-bit computing capabilities. AMD, on the other hand, is betting that people would rather have a 64-bit 80x86 processor. Although the details are sketchy, AMD has announced that they will extend the 80x86 architecture to 64-bits in much the same way that Intel extends the 8086 and 80286 to 32-bits with the introduction of the 80386 microprocessor. Only time will tell if Intel or AMD (or both) are successful with their visions.

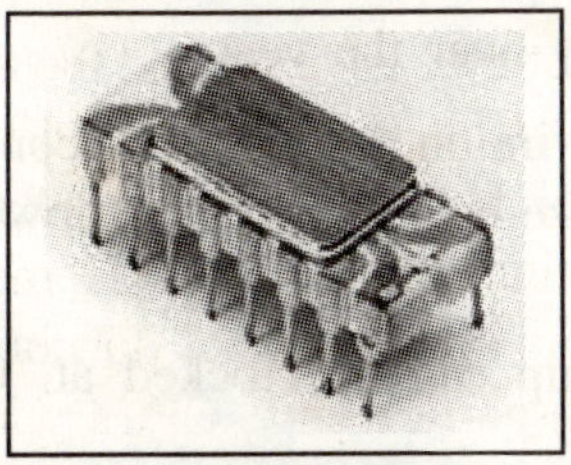

Intel 4004 chip

Intel 8080

Intel Pentium 4 processor

1.2.1 Microprocessor Progress: Intel

Table 1.2.1 helps us to understand the differences between the different processors that Intel has introduced over the years.

Table 1.2.1 Comparison of different microprocessors

Name	*Date*	*Transistors used*	*Microns*	*Max clock speed*	*Address lines*	*Data width*	*MIPS*	*Max addressable memory*
8080	1974	6 K	6	2 MHz	16-bit	8-bit	0.64	64KB, no cache
8086	1978	29 K	6	8 MHz	20-bit	16-bit	0.80	1MB, no cache
8088	1979	29 K	3	5 MHz	20-bit	8-bit bus/ 16-bit	0.33	1MB, no cache
80286	1982	134 K	1.5	12.5 MHz	24-bit	16-bit	2.7	16MB, no cache
80386	1985	275 K	1.5	20 MHz	32-bit	32 bit	6	4GB, no cache
80486	1989	1.2 M	1	25 MHz	32-bit	32 bit	20	4GB, 8K level 1

Pentium	1993	3.1M	0.8	100 MHz	32-bit	64-bit/32-bit	100	4GB, 16K level 1
Pentium Pro	1995	5.5M	0.8	440 MHz	32-bit	64-bit/32-bit	440	64GB, 16K level 1 256K/512 level 2
Pentium II	1997	7.5M	0.35	266 MHz	32-bit	64-bit/32-bit	446	64GB, 32K level 1 256K/512 level 2
Pentium III	1999	9.5M	0.25	500 MHz	32-bit	64-bit/32-bit	1000	64GB, 32K level 1 256K/512 level 2
Pentium 4	2000	42M	0.18	1.5 GHz	32-bit	64-bit/32-bit	1,700	64GB, 32K level 1 256K/512 level 2
Pentium 4 Prescott	2004	125M	0.09	3.6 GHz	32-bit	64-bit/32-bit	7,000	64GB, 32K level 1 256K/512 level 2

Information about this table

- **The date** is the year that the processor was 1st introduced. Many processors are re-introduced at higher clock speeds for many years after the original release date.
- **Transistors mean** the number of transistors used on the chip. We can see that the number of transistors on a single chip has risen steadily over the years.
- **Microns** indicate the width, in microns, of the thinnest wire on the chip. For comparison, a human hair is 100 microns thick. As the feature size on the chip goes down, the number of transistors rises.
- **Clock speed** stands for the maximum rate that the chip can be clocked at. Clock speed will make more sense later.
- **Data Width** is the width of the ALU in bits. A 8-bit ALU can add/ subtract/ multiply/ etc. two 8-bit numbers, while a 32-bit ALU can manipulate 32-bit numbers. A 8-bit ALU would have to execute 4-instructions to add two 32 -bit numbers, while a 32-bit ALU can do it in one instruction. In many cases, the external data bus is of the same width as the ALU, but not always. The 8088 had a 16-bit ALU and a 8-bit external data bus, while the modern Pentiums fetch data 64-bits at a time for their 32-bit ALUs.
- **MIPS** stands for 'millions of instructions per second' and is a rough measure of the performance of a CPU. Modern CPUs can do so many different things that MIPS ratings lose a lot of their meaning, but we can get a general sense of the relative power of the CPU from this column.

From Table 1.2.1 we can infer that, in general, there is a relationship between clock speed and MIPS. The maximum clock speed is a function of the manufacturing process and delays within the chip. There is also a relationship between the number of transistors and MIPS. For example, the 8088 clocked at 5MHz but only executed at 0.33 MIPS (about one instruction per 5/0.33 = 15 clock cycles). Modern processors can often execute at a rate of 3.6/7 = 0.5 clocks/ instruction i.e., 2-instructions per clock cycle. That improvement is directly related to the number of transistors on the chip and will make more sense later.

1.3 HOW MICROPROCESSOR WORKS

To understand how a microprocessor works, it is helpful to look inside and learn about the logic used to create one. In the process we can also learn about **assembly language** → the native language of a microprocessor → and many of the things that engineers can do to boost the speed of a processor.

A microprocessor executes a collection of machine instructions that tell the processor what to do. Based on the instructions, a microprocessor does 3-basic things:

- Using its ALU (Arithmetic Logic Unit), a microprocessor can perform mathematical operations like addition, subtraction, multiplication and division. Modern microprocessors contain complete floating-point processors that can perform extremely sophisticated operations on large floating-point numbers.
- A microprocessor can move data from one memory location to another.
- A microprocessor can make decisions and jump to a new set of instructions based on those decisions.

There may be very sophisticated things that a microprocessor does, but those are its 3-basic activities. The diagram shown in Fig.1.3 is an extremely simple representation of the microprocessor capable of doing those 3-things.

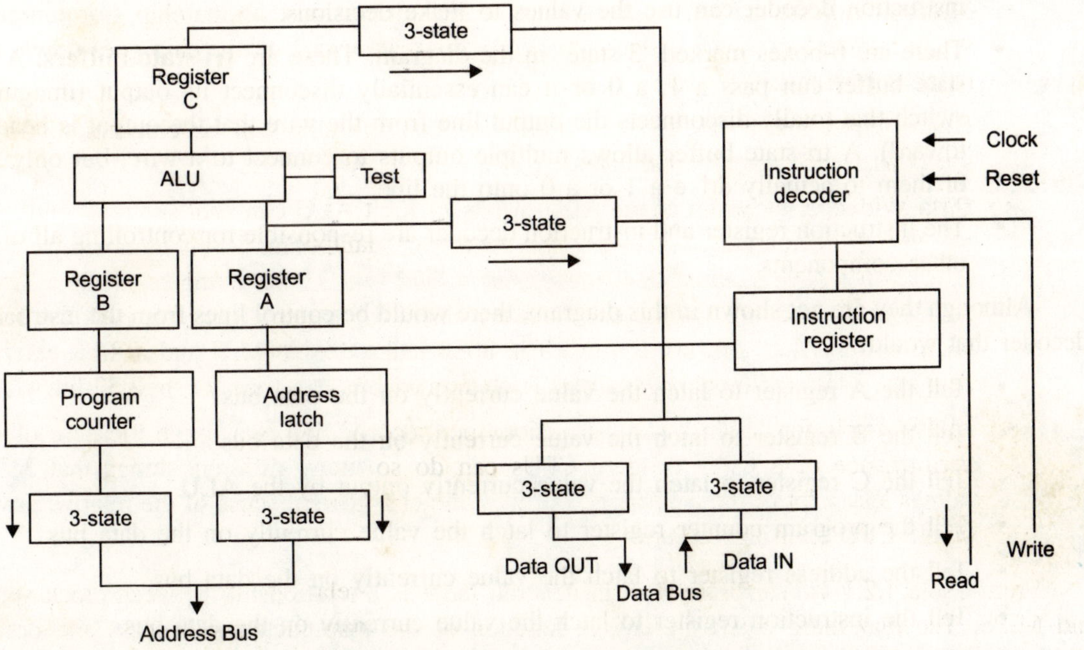

Fig. 1.3 Block diagram of microprocessor

This is about as simple as a microprocessor gets. This microprocessor has:

- **An address bus** (it may be 8, 16, 32 or 64-bits wide) that sends an address to memory or I/O devices.

- **A data bus** (it may be 8, 16, 32 or 64-bits wide)'that can send data to memory or receive data from memory.
- An $\overline{RD}$ (read bar) and $\overline{WR}$ (write bar) line to tell the memory whether it wants to receive or deliver to it.
- **A clock line** that sequences the processor clock pulses.
- **A reset line** that resets the program counter to zero (or whatever) and restarts execution.

Let us assume that both the address and data buses are 8-bits wide in this example. Here are the components of this simple microprocessor:

- Registers A, B and C are simply registers (latches) made out of 8-flip-flops.
- The address registers (latch) is just like registers A, B and C.
- The program counter is a register with the extra ability to increment by 1 and also to reset to zero when told to do so.
- The ALU could be as simple as an 8-bit adder, or it might be able to add, subtract, multiply and divide 8-bit data.
- The test register is a special latch that can hold values from comparisons performed in the ALU. A ALU can normally compare 2-numbers and determine if they are equal, if one is greater than the other, etc. The test register can also normally hold a carry bit from the last stage of the adder. It stores these values in flip-flops and then the instruction decoder can use the values to make decisions.
- There are 6-boxes marked '3-state' in the diagram. These are **tri-state buffers.** A tri-state buffer can pass a 1, a 0 or it can essentially disconnect its output (imagine a switch that totally disconnects the output line from the wire that the output is heading toward). A tri-state buffer allows multiple outputs to connect to a wire, but only one of them to actually drive a 1 or a 0 onto the line.
- The instruction register and instruction decoder are responsible for controlling all of the other components.

Although they are not shown in this diagram, there would be control lines from the instruction decoder that would:

- Tell the A register to latch the value currently on the data bus
- Tell the B register to latch the value currently on the data bus
- Tell the C register to latch the value currently output by the ALU
- Tell the program counter register to latch the value currently on the data bus
- Tell the address register to latch the value currently on the data bus
- Tell the instruction register to latch the value currently on the data bus
- Tell the program counter to increment
- Tell the program counter to reset to zero
- Activate any of the 6-tri-state buffers (6-separate lines)
- Tell the ALU what operation to perform

- Tell the test register to latch the ALU's test bits
- Activate the $\overline{RD}$ line
- Activate the $\overline{WR}$ line

Coming into the instruction decoder are the bits from the test register and the clock line, as well as the bits from the instruction register.

1.3.1 ROM and RAM

The previous section talked about the address and data buses, as well as the $\overline{RD}$ and $\overline{WR}$ lines. These buses and lines connect either to RAM or ROM, generally both. In our sample microprocessor, we have an address bus 8-bit wide and data bus 8-bit wide. That means that the microprocessor can address (2^8) 256 bytes of memory, and it can read or write 8-bit of the memory at a time. Let us assume that this simple microprocessor has 128D location of ROM starting at address 0D (0000H) to 127D(007FH) locations and RAM starting at address 128D (0080H) to 257D (00FFH) locations. Typical look of the ROM and RAM chips is shown in Fig.1.3.1.

By the way, nearly all computers contain some amount of ROM (it is possible to create a simple computer that contains no RAM. Many microcontrollers do this by placing a handful of RAM bytes on the processor chip itself, but generally impossible to create one that contains no ROM). On a PC, the ROM is called the BIOS (Basic Input/Output System). When the microprocessor starts, it begins executing instructions found in the BIOS. The BIOS instructions do things like **test** the hardware in the machine, and then it goes to the **hard disk** to fetch the **boot sector.** This boot sector is another small program, and the BIOS **stores** it in RAM after reading it off the disk. The microprocessor then begins executing the boot sector's instructions from RAM. The boot sector program will tell the microprocessor to fetch something else from the hard disk into RAM that the microprocessor then executes, and so on. This is how the microprocessor loads and executes the entire operating system.

ROM chip

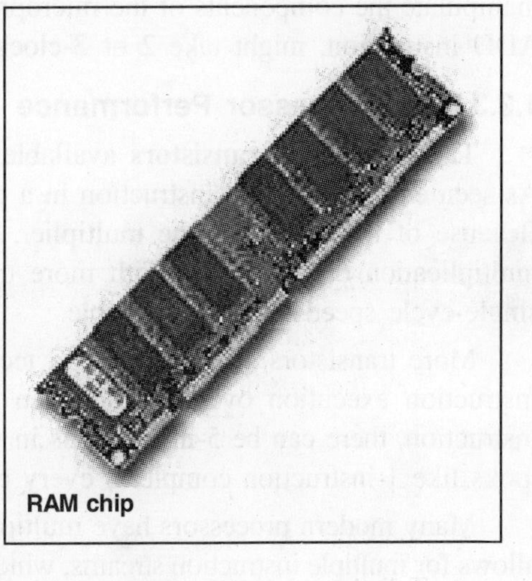

RAM chip

Fig.1.3.1 Typical ROM and RAM chips

1.3.2 Microprocessor Instructions

Even the incredibly simple microprocessor shown in the previous example will have a fairly large set of instructions that it can perform. The collection of instructions is implemented as bit patterns, each one of which has a different meaning when loaded into the instruction register. Humans are not particularly good at remembering bit patterns, so sets of short words are defined to represent the different bit patterns. This collection of words is called the **assembly language** of the processor. An **assembler** can translate the words into their bit patterns very easily, and then the output of the assembler is placed in memory for the microprocessor to execute.

How Instructions Work

The instruction decoder needs to turn each of the op codes into a set of signals that drive the different components inside the microprocessor. Let us take the ADD instruction as an example and look at what it needs to do:

1. During the 1st clock cycle, we need to actually load the instruction. Therefore the instruction decoder needs to:
   - Activate the tri-state buffer for the program counter
   - Activate the RD line
   - Activate the data-in tri-state buffer
   - Latch the instruction into the instruction register
2. During the 2nd clock cycle, the add instruction is decoded. It needs to do very little:
   - Set the operation of the ALU to addition
   - Latch the output of the ALU into the C register
3. During the 3rd clock cycle, the program counter is incremented (in theory this could be overlapped into the 2nd clock cycle).

Every instruction can be broken down as a set of sequenced operations like these that manipulate the components of the microprocessor in the proper order. Some instructions, like this ADD instruction, might take 2 or 3-clock cycles. Others might take 5 or 6-clock cycles.

1.3.3 Microprocessor Performance

The number of **transistors** available has a huge effect on the performance of a processor. As seen earlier, a typical instruction in a processor like an 8088 took 15-clock cycles to execute. Because of the design of the multiplier, it took approximately 80-cycles just to do one 16-bit multiplication on the 8088. With more transistors, much more powerful multipliers capable of single-cycle speeds become possible.

More transistors also allow for a technology called **pipelining.** In a pipelined architecture, instruction execution overlaps. So even though it might take 5-clock cycles to execute each instruction, there can be 5-instructions in various stages of execution simultaneously. That way it looks like 1-instruction completes every clock cycle.

Many modern processors have multiple instruction decoders, each with its own pipeline. This allows for multiple instruction streams, which means that more than 1-instruction can complete during each clock cycle. This technique can be quite complex to implement, so it takes lots of transistors.

1.4 CONVENTIONAL/UNCONVENTIONAL ARCHITECTURE

The conventional von Neumann computer architecture continued for over 30-years. Its main features were

- Single computing unit incorporating processor, memory, and communication units.
- Linear memory of a fixed size
- Central control
- Machine language

The conventional von Neumann architecture reached the limit for achieving higher processing speed because of their sequential fetching of instructions and data through a common memory interface. The advanced microprocessors had to choose a completely different philosophy of architecture design to increase the processing power: parallelism of some form to avoid bottleneck of von Neumann architecture.

A different method of classifying the computer architecture is ***Flynn's*** taxonomy based on how the computer relates its instructions to the data being processed. Figure 1.4 explains this.

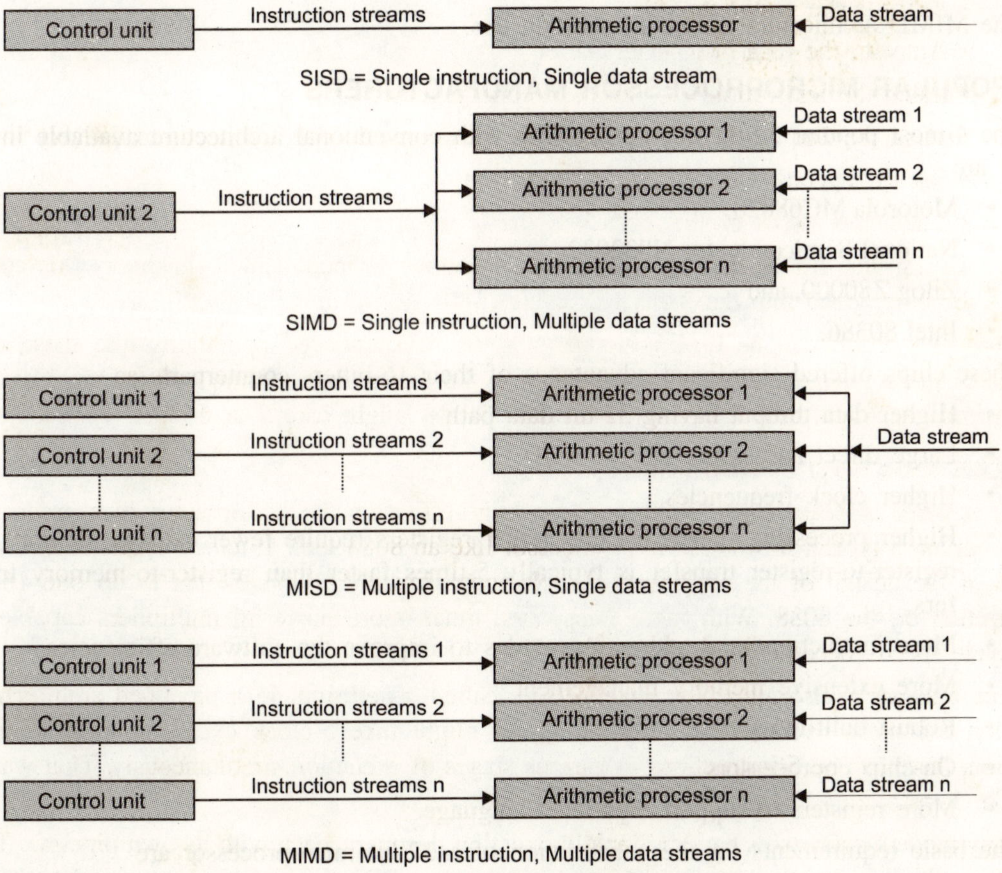

Fig. 1.4 SISD, SIMD, and MIMD Architecture

SISD

This stands for single instruction and single data streams. This is based on the conventional von Neumann architecture. Here, there is only one stream of instruction and each arithmetic instruction initiates one operation. Techniques like pipelining and cache memory may be used to speed up the performance.

SIMD

This stands for single instruction and multiple data streams. This architecture has single instruction stream. But this single instruction is vectored type that initiates many operations. Each element of a vector is regarded as a member of a separate data stream resulting into multiple data streams. For an instance, one instruction can operate on array of data.

MIMD

The word MIMD stands for multiple instruction and multiple data stream. The multiple instruction streams necessarily requires the presence of multiple processors and therefore several data streams are also required. The simple example of such configuration is the multiprocessor and array of microprocessors.

The MIMD architecture is not in common use.

1.5 POPULAR MICROPROCESSOR MANUFACTURERS

The 4-most popular 32-bit microprocessors with conventional architecture available in the market are

- Motorola MC68020,
- National Semiconductor NS32032,
- Zilog Z80000, and
- Intel 80386.

These chips offered significant advantages of their 16-bitters counterparts as

- Higher data thruput having 32-bit data path
- Large direct addressing (2^{32} = 4GB)
- Higher clock frequencies
- Higher processing speeds because larger registers require fewer calls to memory and register-to-register transfer is typically 5-times faster than register-to-memory transfers.
- More instructions and addressing modes to improve the software efficiencies,
- More extensive memory management
- Robust built-in cache
- On-chip coprocessors
- More registers to support high-level language.

The basic requirements for the construction of a 32-bit microprocessor are

- ALU,
- Efficient memory management unit (MMU),

- Floating-point coprocessor,
- Interrupt controller,
- Accurate timing control unit.

Some manufacturers have included several of these feature on the microprocessor chips whereas others have included provisions for interfacing individual chips for such elaborate functions. Each manufacturer approaches the design in different way that is suitable from their point of view. After a new processor is introduced, it is modified, diced, and scaled down or scaled up for long time and then sold under different brand names.

Pentium 4 : Prescott

Prescott is the working title for the next generation of the Pentium 4-architecture. Prescott is manufactured with a line-width of 90nm compared to the 1st generation (Willamette) that was manufactured width of 180nm. The previous generation (Northwood) had a line width of 130nm. At every new generation Intel has been able to fit twice as many transistors as the previous generation. The processor also becomes cheaper to manufacture by decreasing the circuit surface. Usually the manufacturer chooses to do both. The 1st generation had a L2 cache of 256KB while today's generation has a cache of 512KB. Prescott will have a 1MB cache and an 800MHz front side bus (FSB). In conjunction, this will improve the processor performance significantly. Fewer transistors also allow higher clock frequencies and thus even better performance. The 1st model of Prescott is expected to have a clock frequency of 3.5GHz. An integrated security function will also prevent intruders from reading or writing to the hard drive.

Designed for multimedia

Prescott is complemented with 13-additional processor instructions unavailable in Pentium-4. According to Intel these instructions are of significant use for the future use of PCs. This might indicate that the instructions are intended to improve video performance.

Rumors have previously claimed that the new instructions would be 64-bit instructions in order to compete with the upcoming 32/64-bit processor "Claw hammer" from AMD. Instructions in 64-bits makes it possible to address larger quantities of data than what today's 32-bit processor can handle. The "Hammer" from AMD is expected to be much faster than today's processor and compete with Intel's 64-bit Itanium-processors while being much cheaper. This extra 64-bit instruction only takes up a few percent of the circuit size, thus making manufacturing cheap for Intel.

64-bit processors

The 64-bit processors have been with us since 1992, and in 21st century they have started to become mainstream. Both Intel and AMD have introduced 64-bit chips, and the Mac G5 supports a 64-bit processor. The 64-bit processors have 64-bit ALUs, 64-bit registers, 64-bit buses and so on.

One reason why the world needs 64-bit processors is because of their **enlarged address spaces.** The 32-bit chips are often constrained to a maximum of $2^{32} = 2^{30} \times 2^2$ = 4GB of RAM access. That sounds like a lot, given that most home computers currently use only 256MB to 512MB of RAM. However, a 4GB limit can be a severe problem for **server** machines and machines **running large databases.** Even home machine will start bumping up against the 2GB or 4GB limit pretty soon if current trends continue. A 64-bit chip has none of these constraints because a 64-bit RAM address space is essentially infinite for the foreseeable future. The $2^{64} = 2^{60} \times 2^{4}$ bytes of RAM is something on the order of a quadrillion gigabytes of RAM.

With a 64-bit wide address bus and high-speed data buses on the motherboard, 64-bit machines also offer faster I/O (Input/Output) speeds to things like hard disk drives and video cards. These features can greatly increase system performance.

Servers can definitely benefit from 64-bits, but what about normal users? Beyond the RAM solution, it is not clear that a 64-bit chip offers 'normal users' any real, tangible benefits at the moment. They can process data (very complex data features lots of real numbers) faster. People doing video editing and people doing photographic editing on very large images benefit from this kind of computing power. High-end games will also benefit, once they are re-coded to take advantage of 64-bit features. But the average user who is reading e-mail, browsing the Web and editing Word documents is not really using the processor in that way. In addition, operating system like Windows XP has not yet been upgraded to handle 64-bit CPUs. Because of the lack of tangible benefits, it will be 2010 or so before we see 64-bit machines on every desktop.

A processor for 4GHz and more

Prescott is expected to use clock frequencies of 4GHz and more. Internal units to handle integers, registers, address generator and more will be added. So is it a 4GHz Pentium 4 or an 8GHz Pentium 5?

Will the CPU become more expensive?

Will Prescott become more expensive than today's Pentium 4? The answer is probably no since the size of the circuit used is less than that for Pentium 4. If AMD will manage to compete successfully with Intel also next year the Prescott will not become more expensive than today's processor while offering much better performance.

Pentium: Dual Core

Intel tests on its first dual-core desktop processors show that it would deliver some real benefit when we run software on them that is designed to take advantage of the two cores, or when we are performing multiple tasks simultaneously → running a virus check while surfing the Web, for instance.

As the name implies, dual-core processors incorporate two physical processors and two L2 memory caches into one piece of silicon, functioning, in theory, like two separate processors. Intel's first dual-core chip, the 3.2 GHz Pentium Extreme Edition 840 (which carries 1MB of L2 cache per core), goes one step further by including Intel's Hyper-Threading technology in each core, which theoretically brings us a "virtual second processor per core."

Multiply your multiple experience

Combined with a digital media adaptor and home network, an Intel Pentium D processor-based PC allows two people to share the content from one PC – in the same room, or even from different places in the house. For example, one person can check e-mail while the other uses a remote control to access digital photos stored on the same PC and view them on the TV from the comfort of their living room.

2

16-BIT μPs (8086 TO 80286)

2.1 INTRODUCTION

The power of a microprocessor depends on many factors. Important among these factors are *word size, cost, physical size, speed, instruction set,* and *addressable memory range* etc. The speed of the microprocessor is defined as the number of instructions executed per second. It is expressed in million instruction per second (MIPS). The speed of those microprocessors involving floating-point instructions, are expressed in million floating-point operations performed per second (MFLOP). It is also called as thruput. Its powerful instruction set facilitates the programmer/ user to write the programs efficiently under different conditions. Apart from speed and powerful instruction set, one of the other factors that decide the capability of the microprocessor is the number of pins. During 70s, the trend was to develop microprocessors with 40-pin packages only such as Intel's 8085, Zilog Z-80, and Digital Equipments (LSI-11) microprocessors. With the passage of time, more number of function (operations) were implemented on the microprocessor chips and hence 8-bit microprocessors 8085, Z-80, LSI-11 etc. were handicapped with *40-pins, number of instructions*, and *addressable memory range* to provide such facilities. As the 40-pin package limited either the addressable memory ranges or dictated to multiplex many functions. Fortunately, rapid advancement in semiconductor technology came as rescue to microprocessor designer to provide a platform to overcome these bottlenecks in the form of 16-bitters having much *higher thruput, larger memory addressing space, faster data handling capability,* and *better instruction set*. The obvious option was to break the barrier of 40-pin package. National Semiconductor (NS16000) and Zilog (Z-8001) came out with 48-pin packages, Motorola produced 64-pin package (MC68000) and Intel came out with 68-pin package (80286) as the 16-bitters. Thus, the special features of the 16-bitters can be summarized as

- Increased execution speed,
- Increased addressable memory range,
- Powerful instruction set,
- Support multiprocessing,
- Support Auxiliary processor for numeric operations, and
- Compatible to high level languages.

The 8086, Intel's first 16-bit microprocessor, was developed using HMOS (high-density short channel MOS) technology containing 29,000 transistors housed in 40-pin DIP package. It can directly address 1MB memory space. It uses segmented memory. The most interesting feature of the 8086 is that it prefetches up to 6-instructions byte from the memory and *queues* them in order to speed up the instruction execution. A variant of the 8086, the 8088, has 8-lines rather than 16-lines external data bus. The 8088, launched in 1979, was incorporated in the IBM PCs launched in 1982. This worked as **smash hit** and brought the Intel into the rank of the Fortune 500 and the Fortune magazine named the Intel **'The Business triumphs of seventies'**. The later microprocessor that remained in the scene of the PCs was the 80286.

The 8086 operates with single +5V power supply. It does not have on-chip clock generator. The 8284 clock generator/ driver chip generates the clock for the 8086. Its instruction execution time varies from 2-clock cycles to 30-clock cycles. The 8086 and 8088 are similar in organization and identical in instructions. The 8088 became very popular only because it found its birth in PCs. For software designer, the 8086 and 8088 are identical except for the reason that the 8088 executes slower than the 8086 for the same clock speed.

The 8086 has 20-bit address lines. The $AD_{15} \Leftrightarrow AD_0$ are 16-multiplexed address and data buses and 4-multiplixed address and status lines $A_{19}/S_6 \Leftrightarrow A_{16}/S_3$. The 4-bit most significant address lines $A_{19} \Leftrightarrow A_{16}$ and the 16-bit least significant address lines $AD_{15} \Leftrightarrow AD_0$ make it the 20-bit addresses. A 16-bit word is sent or received on $AD_{15} \Leftrightarrow AD_0$. The least significant byte of the word is transmitted on $AD_7 \Leftrightarrow AD_0$ and the most significant byte on the $AD_{15} \Leftrightarrow AD_8$. Either $AD_7 \Leftrightarrow AD_0$ or $AD_{15} \Leftrightarrow AD_8$ can be used to transfer only 1-byte and not the complete word (2-bytes) depending on whether the byte corresponds to even or odd memory address respectively. A very simplified representation of the 8086 is shown in Fig. 2.1 (*a*). It is evident now that the maximum address is $= 2^{20} - 1 \cong 2^{20} = 1048575$. In other words, the 8086 can address nearly 1million bytes (1MB) directly, which is much larger than the addressing space of the 8-bit microprocessors i.e., $2^{16} = 64KB$. Any 2-consecutive bytes in the memory form a word of the 8086. The address of the least significant byte (numerically lower address) is the **word address.** As an example, the word located at 00002H consists of 2-byte (55H and A5H) located at 00002H and 00003H in Fig. 2.1 (*b*).

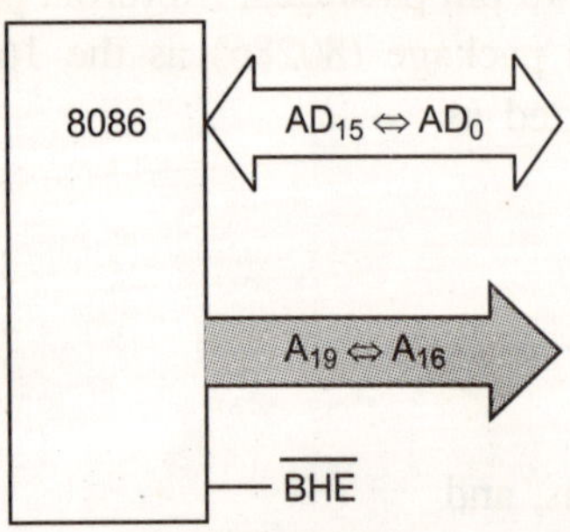

Fig. 2.1 (*a*) 8086 simplified model

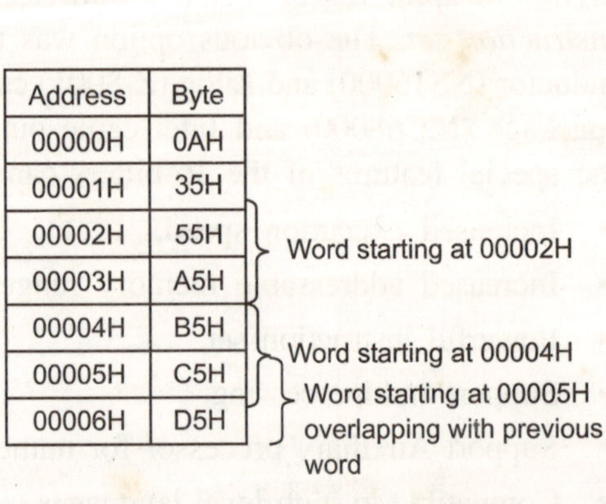

Fig. 2.1 (*b*) Words in 8086 memory

A word can also begin at odd address. The word located at address 00005H in Fig. 2.1 (*b*) consists of 2-bytes located at 00005H and 00006H. The contents of this word are D5C5H. The word located at 00004H obviously overlaps with word located at address 00005H. Here least significant byte is B5H is of the word starting at lower address 00004H and the most significant byte C5H is located at higher address 00005H respectively.

Most of the instructions of the 8086 operate both on **byte** and **word operands**. For an example, the contents of only one location 00002H can be moved to the accumulator AL (55H) (L.O. byte of the accumulator). Similarly, the complete word (2B) starting at 00002H can be moved to AX = (AH(A5H) + AL (55H) = 2-bytes in the accumulator).

2.2 PINOUTS OF 8086/88

Pinouts of the 8086 and 8088 are shown in Figs. 2.2 (*a*) and (*b*). All pins are TTL compatible. The 8086 operates in 2 -modes : *minimum mode and maximum mode*. In minimum mode of operation only one 8086 microprocessor is used to generate all control signals such as HOLD, HLDA, $\overline{RD}$, $\overline{WR}$, M/$\overline{IO}$, DT/$\overline{R}$, $\overline{DEN}$, ALE, and $\overline{INTA}$. In maximum mode the 8086 uses another processor called bus controller 8288 to generate advanced control signals that are not available on the pins of the 8086 directly. The 8288 generates advanced control signals such as $\overline{IORC}$, $\overline{IOWTC}$, $\overline{MRDC}$ and singnals DT/$\overline{R}$, $\overline{DEN}$, ALE using status signals $\overline{S_2}$, $\overline{S_1}$ and $\overline{S_0}$ from the 8086. Pin- 33 of the 8086 is labeled as MN/$\overline{MX}$ to select one of the 2 modes. If this pin- 33 is connected to HIGH, the 8086 functions in the minimum mode.

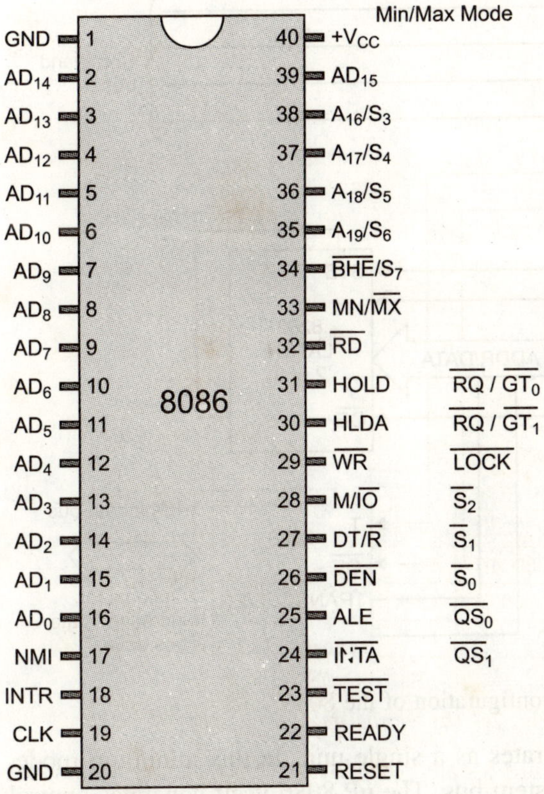

Fig. 2.2 (*a*) Pinouts of 8086

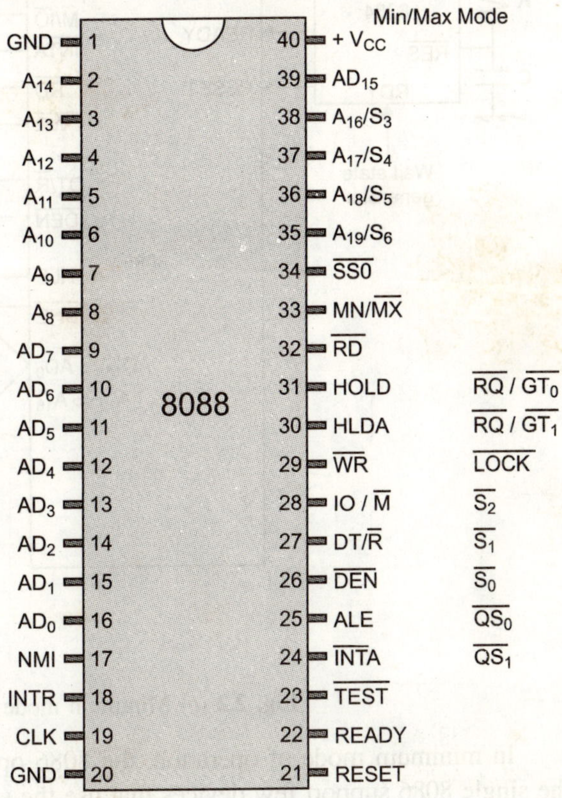

Fig. 2.2 (*b*) Pinouts of 8088

The control pins from 24 to 31 of 8086 do other functions also such as

$HOLD \Rightarrow \overline{RQ}/\overline{GT_0} \Rightarrow 31$, $HLDA \Rightarrow \overline{RQ}/\overline{GT_1} \Rightarrow 30$, $WR \Rightarrow \overline{LOCK} \Rightarrow 29$, $M/\overline{IO} \Rightarrow \overline{S_2} \Rightarrow 28$, $DT/R \Rightarrow \overline{S_1} \Rightarrow 27$, $DEN \Rightarrow \overline{S_0} \Rightarrow 26$, $ALE \Rightarrow QS_0 = 25$, $\overline{INTA} \Rightarrow QS_1 \Rightarrow 24$

Minimum/Maximum Mode (MN/$\overline{\text{MX}}$)

When pin-33 labeled as MN/$\overline{\text{MX}}$ is tied to logical-1 (+V_{CC}), it is referred to as minimum mode of operation of the 8086. In minimum mode, the 8086 provides 8085 style bus control signals directly as in Fig. 2.2(*c*). In maximum mode, encoded status lines replace the basic bus control signals as in Fig. 2.2(*d*). These status lines are then decoded by the 8288-system bus controller to generate bus control signals. Maximum mode must be used if the 8289 multibus arbiter or the 8087 coprocessor is to be used, since they connect to the encoded status lines. The maximum mode has the advantage of providing better-defined bus timing. The minimum mode is used primarily for the applications where minimum chip count is important.

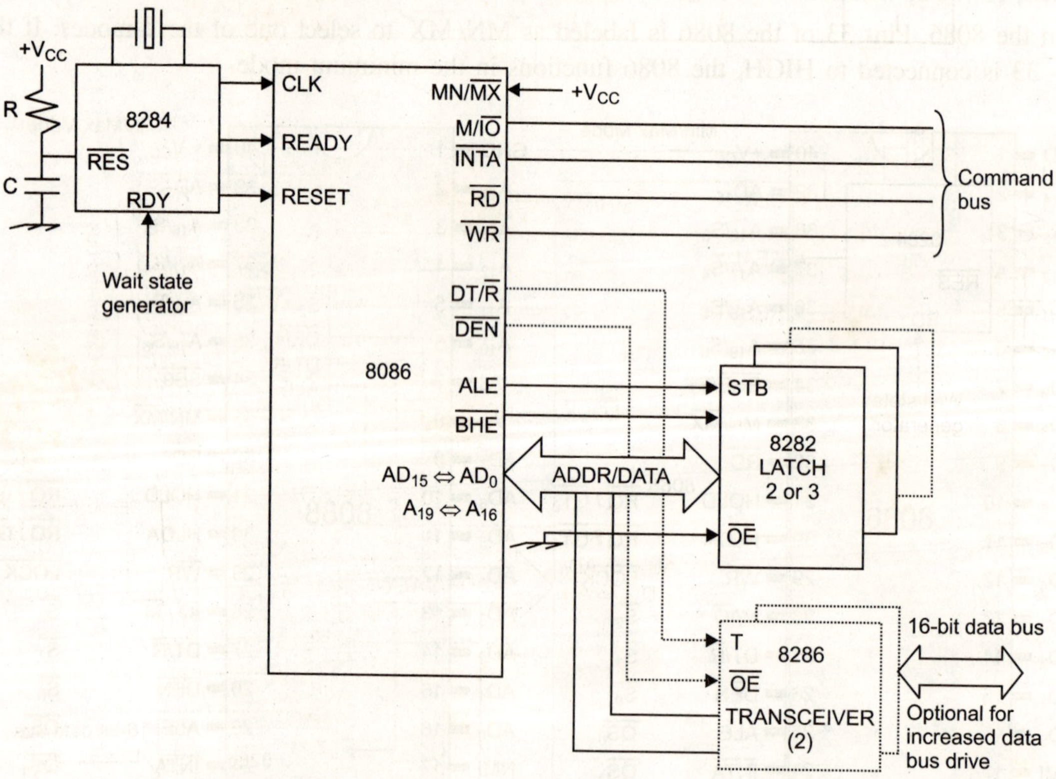

Fig. 2.2 (*c*) Minimum mode configuration of the 8086

In minimum mode of operation, the 8086 operates as a single unit. In this minimum mode, the single 8086 support few devices that use the system bus. The μP 8086 itself generates control signals such as $\overline{\text{INTA}}$, ALE, $\overline{\text{DEN}}$, DT/$\overline{\text{R}}$, M/$\overline{\text{IO}}$, $\overline{\text{WR}}$, HLDA, and HOLD at pins-24 through 31 with MN/$\overline{MX}$ pin tied to high logic. Pin-35 through 38 are used as address lines. The configuration of minimum mode of operation of the 8086 is shown in Fig. 2.2(*c*).

Table 2.2(*a*) Status information in maximum mode of 8086

$\overline{S_2}$	$\overline{S_1}$	$\overline{S_0}$	*μP cycle*	*8288 command*
0	0	0	Interrupt acknowledge	$\overline{INTA}$
0	0	1	Read I/O port	$\overline{IORC}$
0	1	0	Write I/O port	$\overline{IOWC}$, $\overline{AIOWC}$
0	1	1	Halt	None
1	0	0	Instruction fetch	$\overline{MRDC}$
1	0	1	Read memory	$\overline{MRDC}$
1	1	0	Write memory $\overline{MWDC}$	$\overline{MWTC}$, $\overline{AMWTC}$
1	1	1	Inactive	None

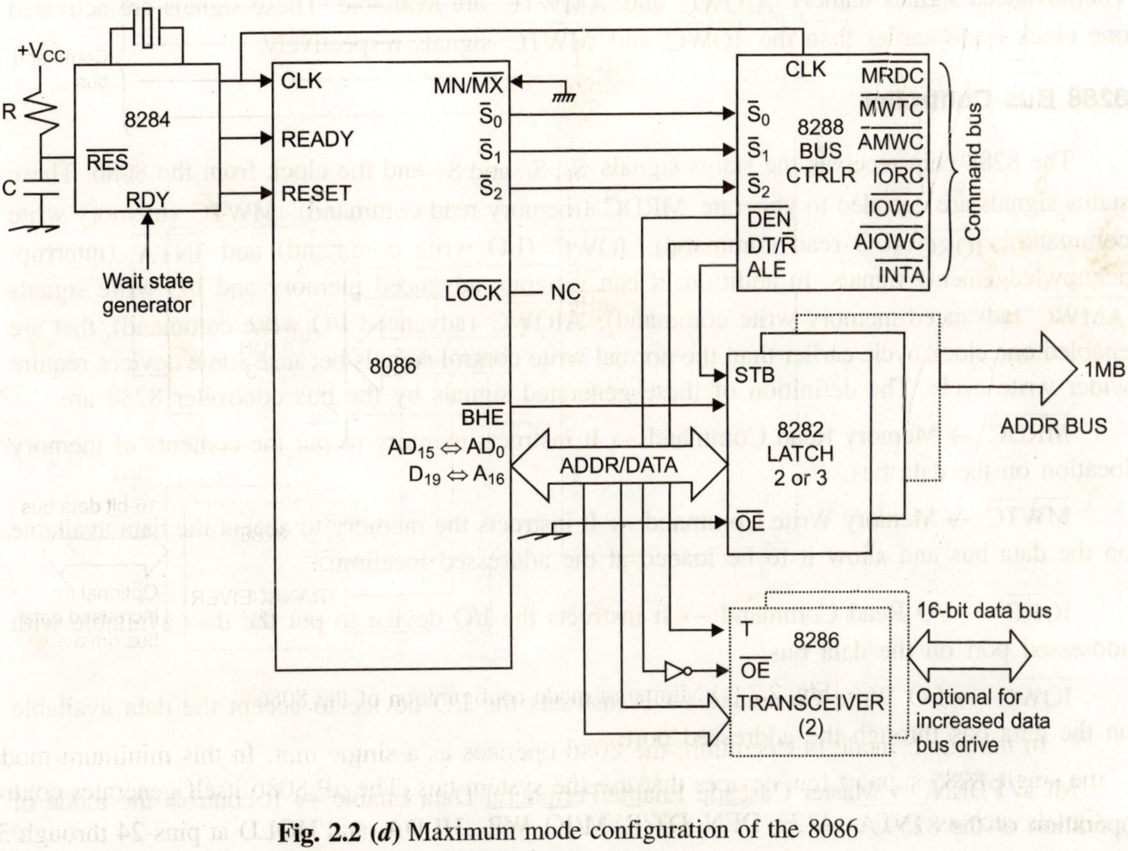

Fig. 2.2 (*d*) Maximum mode configuration of the 8086

The pin-33 is tied to **ground** in maximum mode of operation of the 8086. Why a bus controller is used in the maximum mode?

If we examine the pins of the 8086/8088 in its maximum mode we find that the signals required for controlling the memory, I/O, interrupts, interfaces etc. ($\overline{WR}$, IO/$\overline{M}$, DT/$\overline{R}$, $\overline{DEN}$, ALE, and INTA) are no longer directly available. Instead, three status signals ($\overline{S_2}$, $\overline{S_1}$, $\overline{S_0}$) are available. These status signals control the bus controller 8288 that decodes them to generate equivalent signals required for above controls as given in Table 2.2(*a*). For an instance, $\overline{S_2}$ $\overline{S_1}$ $\overline{S_0}$ = 001 on the 8086/8088 means I/O read cycle to be performed. On receipt of this combination of the status signals, the 8288 generates $\overline{IORC}$ on one of its pins to switch it to logic-0 level. Similarly, $\overline{S_2}$ $\overline{S_1}$ $\overline{S_0}$ = 111 means no bus activity to be performed by the 8086/8088 and accordingly the 8288 also does not generate any command signals.

Maximum Mode ($\overline{MX}$)

The $\overline{IORC}$, and $\overline{IOWC}$ are I/O read control and I/O write control respectively. The $\overline{MRDC}$ and $\overline{MWTC}$ are memory read control and memory write controls. All these control signals instruct the memory to accept or send data from or to the bus for both of these write command signals. The advanced signals namely $\overline{AIOWC}$ and $\overline{AMWTC}$ are available. These signals are activated one clock cycle earlier than the $\overline{IOWC}$ and $\overline{MWTC}$ signals respectively.

8288 Bus Controller

The 8288 chip receives the status signals $\overline{S_2}$, $\overline{S_1}$, and $\overline{S_0}$ and the clock from the 8086. These status signals are decoded to generate $\overline{MRDC}$ (memory read command), $\overline{MWTC}$ (memory write command), $\overline{IORC}$ (I/O read command), $\overline{IOWC}$ (I/O write command) and $\overline{INTA}$ (interrupt acknowledgement) signals. In addition, it can generate advanced memory and I/O write signals $\overline{AMWC}$ (advanced memory write command), $\overline{AIOWC}$ (advanced I/O write command), that are enabled one clock cycle earlier than the normal write control signals because some devices require wider write cycle. The definition of these generated signals by the bus controller 8288 are :

$\overline{MRDC}$ → Memory Read Command → It instructs memory to put the contents of memory location on the data bus.

$\overline{MWTC}$ → Memory Write Command → It instructs the memory to accept the data available on the data bus and allow it to be loaded at the addressed location.

$\overline{IORC}$ → I/O Read Command → It instructs the I/O device to put the data available with addressed port on the data bus.

$\overline{IOWC}$ → I/O Write Command → It instructs the I/O device to accept the data available on the data bus through the addressed port.

MCE/$\overline{PDEN}$ → Master Cascade Enable/Peripheral Data Enable → It controls the mode of operation of the 8259.

$\overline{AEN}$, IOB, and CEN → These signals help the µP system with single processor in the system. $\overline{AEN}$ and IOB are grounded and CEN is pulled high.

$\overline{AIOWC}/\overline{AMWC}$ → Advance I/O Write Command/Advance Memory Write Command → These signals are similar to $\overline{IOWC}$ and $\overline{MWTC}$ except they are activated one clock pulse earlier. The 8288 in T_2 will set $\overline{DEN}$ = 1 to enable the transceiver. For an input to the 8288, it will activate $\overline{MRDC}$ or $\overline{IORC}$. These signals are activated until T_4. For an output the $\overline{AMWC}$ or $\overline{AIOWC}$ is activated from T_2 to T_3 and $\overline{MWTC}$ or $\overline{IOWC}$ is activated from T_3 to T_4.

The status bits $\overline{S_2}$, $\overline{S_1}$, and $\overline{S_0}$ remain active till T_3 and becomes inactive during T_3 to T_4. If ready pin is not activated before T_3, wait state will be inserted between T_3 and T_4.

The $\overline{S_2}$, $\overline{S_1}$, and $\overline{S_0}$ are set at the beginning of the bus cycle. Once detecting the change on inactive state of $\overline{S_2}$, $\overline{S_1}$, $\overline{S_0}$ = 1, the 8288 bus controller will output pulse on its ALE and apply a required signal on the DT/$\overline{R}$ pin during T_1.

The 8288 generates bus control signals $\overline{DEN}$, DT/$\overline{R}$, ALE, MCE/$\overline{PDEN}$. The functions of the first 3-signals are the same as those in the minimum mode. The signal MCE/$\overline{PDEN}$ has 2-functions depending on the mode in which 8288 is operating. The 8288 can either operate in I/O bus mode or system bus mode. When CEN (command enable) and IOB (I/O bus) input pins are wired high, the 8288 operates in I/O bus mode. In this mode, the signal PDEN functions in the same way as $\overline{DEN}$ but it is active only during I/O instruction. This facility enables 8288 to control 2-sets of buses : ***system bus*** and ***I/O bus*** separately.

The three status signals $\overline{S_2}$, $\overline{S_1}$, and $\overline{S_0}$ are decoded to generate relevant bus control signals such as $\overline{WR}$, M/$\overline{IO}$, DT/$\overline{R}$, $\overline{DEN}$, ALE, and $\overline{INTA}$. Pin QS_0 (ALE) in maximum mode indicates status of the prefetch queue along with QS_1 ($\overline{INTA}$) as

QS_1	*QS_0*	*Functions*
0	0	no operation
0	1	the first byte of the instruction being executed
1	0	the queue is flushed
1	1	subsequent byte is taken from the queue

The signals QS_1 and QS_0 remain valid during clock period that follows any queue operation. In other words, the code informs the external circuitry about the type of the information removed from the queue during the previous clock cycle.

Pin $\overline{RQ}/\overline{GT_0}$ (HOLD) functions as request/ grant line in the maximum mode. This is the bi-directional line that can be used to share local bus between 2 processors via a handshake sequence. The 8086 will be forced to relinquish bus (i.e., address, data bus and status lines are tri-stated) by inputting a low pulse at this pin. The 8086 acknowledges the request by outputting a low to the requestor and relinquishes control of the bus. But EU continues to execute instructions until an instruction requires bus access or queue is flushed. When the requesting processor relinquishes the bus it does so by another low pulse on this line. A similar line is $\overline{RQ}/\overline{GT_1}$ (HLDA) pin. Its function is identical to the $\overline{RQ}/\overline{GT_0}$ in the maximum mode but has the lower priority. Pin $\overline{LOCK}$ ($\overline{WR}$) functions as an output pin in both maximum and minimum modes. In maximum mode, the $\overline{LOCK}$ outputs low to prevent the 8086 from losing system bus control while executing an instruction. The BIU activates $\overline{LOCK}$ signal when EU executes one byte $\overline{LOCK}$ prefixed instructions. The $\overline{LOCK}$ can be used to coordinate access to a shared memory location in a multiprocessing system by preventing a processor from modifying a value stored in a memory location while other processor is reading it.

The maximum mode, the 8086 supports implementation of a large system with more than one processor. The 8086 in the maximum mode provides necessary signals required for coordinating operations of multiprocessors communicating and cooperating each other.

Address and Data Bus ($AD_{15} - AD_0$)

We see from pinouts of the 8086 and 8088 in Figs.2.2 (a) and (b) that the 8088 has multiplexed 8-bit address and data bus $AD_7 \Leftrightarrow AD_0$ from pins 9 to 2 similar to the 8085. The other high order address bits are available on $A_{19} \Leftrightarrow A_{16}$. The 8086 has a 16-bit multiplexed address and data bus $AD_{15} \Leftrightarrow AD_0$ from pins 16 to 2 and 39. Its 4-upper address bits are also multiplexed with status signals labeled A_{16}/S_3, A_{17}/S_4, A_{18}/S_5, and A_{19}/S_6 from pins 38 to 35. During the first clock cycle (T_1) of a bus cycle (read or write), the total 20-bit address is available on the address lines along with high logic of ALE. In all other clocks (T_2, T_3, T_w, and T_4) of the first machine cycle, the $AD_{15} \Leftrightarrow AD_0$ function as the 16-data bus for memory and I/O operations.

Address/Status Signals (A_{19}/S_6, A_{18}/S_5, A_{17}/S_4, and A_{16}/S_3)

These pins are also multiplexed address and status bus. They are high active along with ALE = 1 and work as high order address lines (A_{19}, A_{18}, A_{17}, and A_{16}) during T_1 clock of the first machine cycle. The status information is available during T_4, T_w, T_3, and T_2 clocks on these lines. The status of S_4 and S_3 are used to decode different segments as described in Table 2.2(*b*).

Table 2.2(*b*) Decoding of S_3 and S_4

S_4	S_3	*Function*	Condition of 8088 with low reset Flags 0000H
0	0	Extra segment is accessed	IP FFFFH
0	1	Stack segment is accessed	CS 0000H
1	0	Code segment is accessed	DS 0000H
1	1	Data segment is accessed	SS 0000H
			ES 0000H
			S_6 always = 0,
			S_7 always = 1
S_5 = 1F (Interrupt flag status at the beginning of each cycle, IF = 0, S_5 = 0 and IF = 1, S_5 = 1			

BHE/S_7 → This pin is low active. The $\overline{BHE}/S_7$ pin is used as bus high enable signals during 1st clock cycle of instruction execution. The 8086 outputs a LOW on this pin during read, write, and interrupt acknowledge cycle in which data are to be transferred in a high order byte of $AD_{15} \Leftrightarrow AD_0$ of the data bus. The $\overline{BHE}$ along with AD_0 is used to select the memory banks.

$\overline{WR}$ → The write bar pin is low active and functions as an output pin in both maximum and minimum modes. This indicates that the μP is ready to deliver data for memory or I/O devices.

$\overline{LOCK}$ → The $\overline{LOCK}$ pin outputs a low in maximum mode to prevent the 8086 from losing system bus control while executing an instruction. The BIU activates $\overline{LOCK}$ signal when EU executes one byte LOCK prefixed instructions. The $\overline{LOCK}$ can be used to coordinate access to a shared memory location in a multiprocessing system by preventing a processor from modifying a value stored in a memory location while other processor is reading it.

$\overline{RD}$ → The read bar pin is also low active. This indicates that the μP is ready to read data from memory or I/O devices. During read operation, one other control signal, DEN ***(data enable)***, is also supplied. It enables external device to supply data to the μP.

$M/\overline{IO}$ → A high on this pin means the processor is accessing the memory and a low on this pin indicates that the processor accessing the I/O devices.

HOLD → Whenever an external device wants to take control over the system bus, it signals this fact to the μP by asserting a high logic (1) on this pin. The 8086/8088 will be forced to relinquish the system bus (i.e., address, data bus and status lines are tri-stated) by inputting a low pulse at this pin.

$HLDA/\overline{RQ}/\overline{GT}_0$ → The 8086 acknowledges the HOLD request by outputting a high (1) on the HLDA pin to the requestor and relinquishes control of the bus. But EU continues to execute instructions until an instruction requires bus access or queue is flushed. When the requesting processor relinquishes the bus it does so by another low pulse on this line. The function of $\overline{RQ}/\overline{GT}_0$ is to indicate that the processor is in the maximum mode but has the lower priority.

READY : If a logic-0 is placed on this pin by a slow memory device or an I/O device, the μP enters into wait state and remains idle; When a logic-1 is placed on this pin, it does not effect the normal processing of the μP.

CLOCK : The clock input provides basic input timing for processor operation and bus control activity. The range of frequency of for different 8086 versions is 5MHz to 10MHz.

INTR : It is an input to the 8086/8088. It is used to indicate by an external device to signal that it needs to be serviced. This input is checked during the final clock period each instruction acquisition cycle. A logic-1 on this pin represents an active interrupt request. When the μP recognizes this request, it indicates this fact to the external circuits by outputting a logic (0) at the $\overline{INTA}$ pin.

$\overline{TEST}$: The $\overline{TEST}$ input is also related to external interrupt. For example, execution of WAIT instruction causes the 8086/8088 to check the logic level at the $\overline{TEST}$ input. If $\overline{TEST}$ = 1, μP suspends operation and goes into ***idle state***. The μP does not execute instruction; rather it repeatedly checks the logic level of the $\overline{TEST}$ input waiting for the transition back to logic-0.

When $\overline{TEST} = 0$, execution resumes with the next instruction in the program. This pin is most often connected to the numeric coprocessor.

NMI : The non-maskable interrupt input cannot be masked by software. It is similar to the INTR except the NMI does not check to confirm whether IF bit is at logic-1. If NMI is activated, interrupt input uses interrupt vector-2 .

RESET : The reset input is used to provide hardware reset for the μP. This input forces the μP to terminate the current instruction and starts execution from FFFF0. Switching RESET to logic-0 initializes the internal registers of the μP.

In the maximum mode, the 8086 supports implementation of a large system with more than one processor. The 8086 in the maximum mode provides necessary signals required for coordinating operations of multiprocessors communicating and cooperating each other.

2.3 ARCHITECTURE OF 8086/88

The functional block diagram of the 8086 and 8088 are shown in Figs. 2.3 (*a*) and (*b*). The architecture of the 8086/8088 can be divided in 2-parts

- BIU **(Bus Interface Unit),**
- EU **(Execution Unit).**

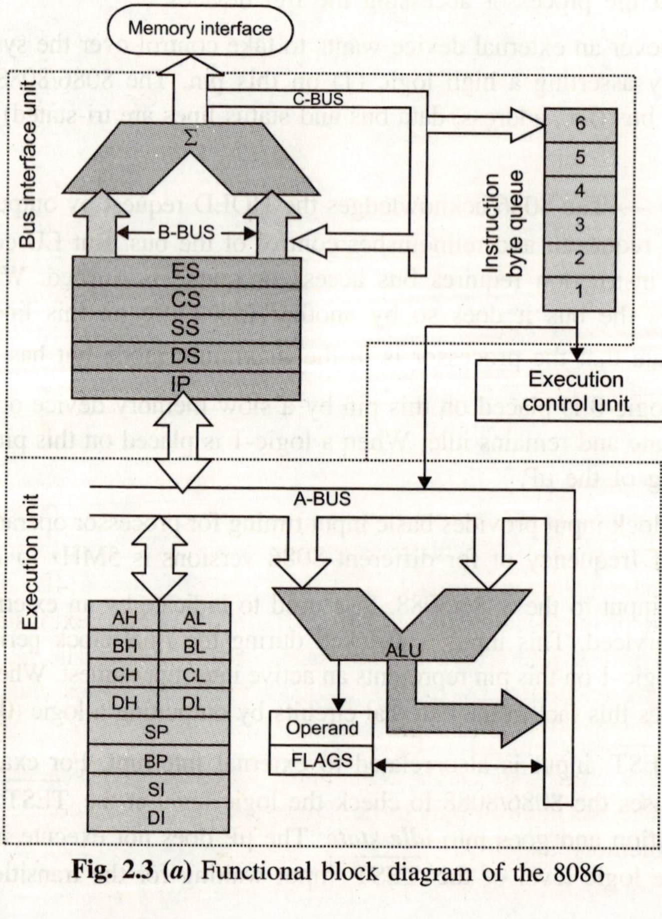

Fig. 2.3 (*a*) Functional block diagram of the 8086

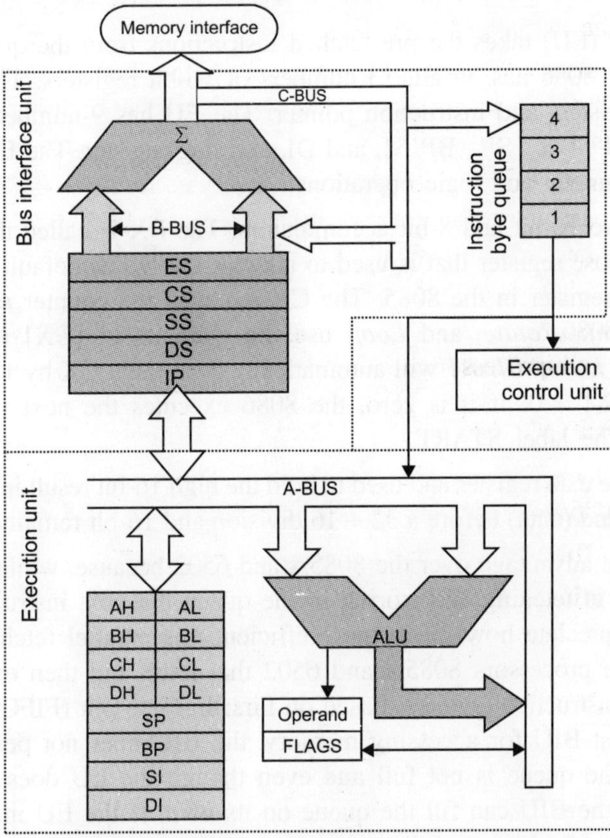

Fig. 2.3 (*b*) Functional block diagram of the 8088

These units (BIU & EU) function independently. Since, these units function independently, they can be assumed as 2 separate parallel processors.

BIU

The function of BIU is that it *fetches instructions, reads data from memory or I/O ports and writes data to memory or I/O ports and buffers them in the queue*. While the execution unit decodes or executes an instruction that does not involve uses of the bus, the BIU fetches up to 6 -instruction byte for the preceding instruction. The BIU stores these prefetched instruction byte in FIFO register called the ***queue***. The queue is of 2-word (4B) in case of the 8088 or 3-word (6B) in case of the 8086 locations in the BIU. Fetching the next instruction while executing the current instruction is referred to as the ***instruction pipeline***. The BIU fills this space with instructions awaiting execution by the EU. The 8088 fetches bytes and provides the EU with words. The 8086 fetches words and provides EU with words.

The BIU has a *dedicated adder*. The main function of this adder is to produce 20-bit physical address. The bus control logic of the BIU generates all bus control signals such as READ and WRITE for memory and I/O.

EU

The execution unit (EU) takes the pre-fetched instructions from the queue of BIU, decodes and executes them. The 8086 has, in all, 13 numbers of 16-bit registers. The BIU has CS , DS, SS.and ES segment registers and instruction pointer. The EU has 9-numbers of 16-bit registers. These are AX, BX , CX , DX , SP , BP, SI, and DI, and flag register. The EU has a 16 -bit ALU for performing the arithmetic and logic operations.

The AX and AL are 16-bit and 8-bit accumulators. The BX is called the base register. This is the only general-purpose register that is used to address the DS as default in the 8086 memory just like H - L pair of register in the 8085. The CX is called the counter register because some instructions such as *Shift, Rotate*, and *Loop* use the contents of [CX] as a counter. For an example, the instruction *LOOP START* will automatically decrement CX by 1 without affecting the flags and checks if [CX] = 0. If it is zero, the 8086 executes the next instruction, otherwise execution branches to the label START.

The DX is called the data register and used to hold the high 16-bit result in 16×16 multiplication or the high 16-bit dividend (data) before a 32 ÷ 16 division and 16-bit remainder after the division.

This system has the advantage over the 8085A and 6502 because, while the EU is executing an instruction, the BIU is fetching and storing in the queue the next instructions. This is called ***pipelining***. One can appreciate how much more efficient this parallel fetching and executing is helpful compared to the processors 8085A and 6502 that fetch and then execute instruction by instruction. The BIU's instruction queue is based on First-In-First-Out (FIFO). If the queue is full and EU does not request BIU for accessing memory, the BIU does not perform any bus cycle. On the other hand, if the queue is not full and even though the EU does not request BIU for accessing the memory, the BIU can fill the queue on its own. If the EU interrupts the BIU, the BIU 1st completes the prefetching and then attains to the service of the EU. If JUMP or CALL SUBROUTINE is called, the BIU resets the queue and begins refilling after passing the new instruction to the EU.

It is time to look at those components of the μP that are most visible to us as the assembly language programmer. The most visible component of the μP is the register set. Like our hypothetical processors, the 80x86 chips have a set of on-board registers. The register set for each processor in the 80x86 family is a superset of those in the proceeding μPs. The best place to start is with the register set for the 8088, 8086, 80188, 80186, and 80286 since these 5 -processors have the same set of registers. In the discussion that follows, the term 8086 will imply any of these 4μPs. Intel's designers have classified the registers on the 8086 into 5-categories

- **general purpose registers** (AX, BX , CX , DX)
- **index registers** (SI, DI)
- **pointer registers** (BP, SP)
- **segment registers** (CS, DS, ES, SS), and
- **special purpose registers (flag register, IP)**

The general-purpose registers are those which may appear as operands of the *arithmetic, logical*, and *related instructions*. Although these registers are *general purpose*, every one has its own special purpose. Intel uses the term general purpose loosely. The 8086 uses the segment registers to access blocks of memory called, surprisingly enough segments. The final classes of 8086 registers are the miscellaneous registers.

General-purpose registers of 8086

8086 μP has 8-general purpose registers. Each register has its own name as AX, BX, CX, DX, SI, DI, BP, and SP.

AX—the accumulator register (divided into **AH/AL**)

1. Generates shortest machine code
2. Operates on Arithmetic and logic functions
3. One number must be in AL or AX
4. Used for Multiplication and Division
5. Used for Input and Output transfers

Although we can do most **arithmetic** and **logical** operations in *other registers,* it is often more efficient to use the **AX** register for such computations. The important characteristic of the ALU of the 8086 is its implementation of 16-bit fixed-point multiplication and division.

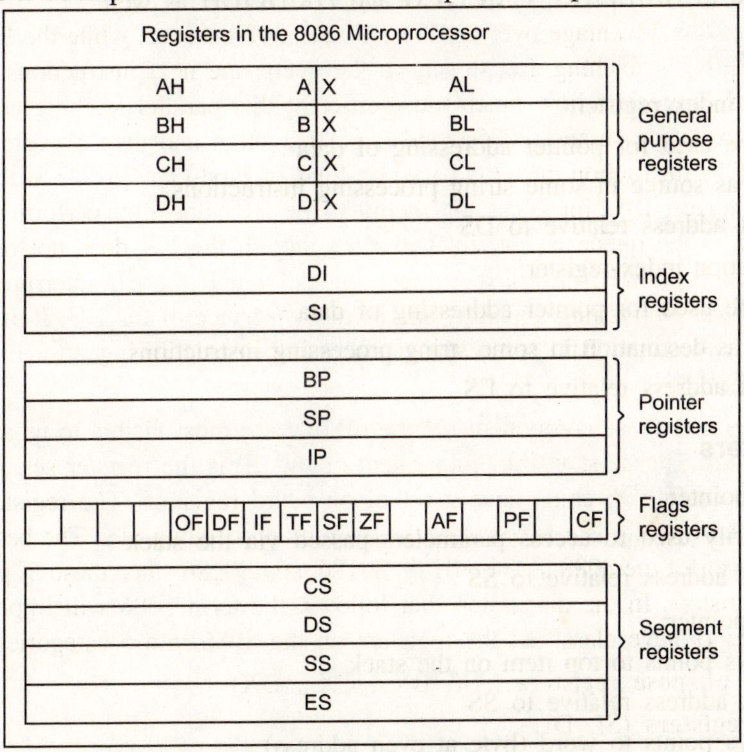

Fig. 2.3 (*c*) Registers inside the 8086 microprocessor

BX—the base address register (divided into **BH/BL**).

The **BX** register (Base) has some special purposes as well. It is commonly used to hold indirect addresses.

CX—the count register (divided into **CH/CL**):

1. Used for Iterative code segments using the LOOP instruction

2. Used for repetitive operations on strings with the REP command
3. Used for count (in CL) of bits to shift and rotate

The **CX** register (Count), as its name implies, counts things. We often use it to count off the number of iterations in a loop or specify the number of characters in a string.

DX—the data register (divided into **DH/DL**):

1. DX: AX concatenated into 32-bit register for some MUL and DIV operations
2. Specifying ports in some IN and OUT operations

The **DX** register (Data) has 2-special purposes: it holds the **overflow** from certain **arithmetic operations,** and it holds I/O addresses when accessing data on the 80x86 I/O bus.

Note that the 8-bit registers do not form an independent register set. Modifying AL will change the value of AX; so will modifying AH. The value of AL exactly corresponds to bits-0 through-7 of AX. The value of AH corresponds to bits 8-through-15 of AX. Therefore, any modification to AL or AH will modify the value of AX. Likewise, modifying AX will change *both* AL and AH. However, changing AH will not affect the value of AL, and vice versa. This statement applies to BX/BL/BH, CX/CL/CH and DX/DL/DH as well.

Index register

SI—source index register:

1. Can be used for pointer addressing of data
2. Used as source in some string processing instructions
3. Offset address relative to DS

DI—destination index register:

1. Can be used for pointer addressing of data
2. Used as destination in some string processing instructions
3. Offset address relative to ES

Pointer registers

BP—base pointer:

1. Primarily used to access parameters passed via the stack
2. Offset address relative to SS

SP—stack pointer:

1. Always points to top item on the stack
2. Offset address relative to SS
3. Always points to word (byte at even address)
4. An empty stack will had SP = FFFEH

Special Purpose Registers

IP—the instruction pointer:

1. Always points to next instruction to be executed
2. Offset address relative to CS

IP register always works together with **CS** segment register and it points to currently executing instruction.

Segment Registers

The 8086 has 4-special *segment registers:* CS, DS, ES, and SS as depicted in Fig.2.3 (*d*). These stand for Code *Segment, Data Segment, Extra Segment,* and *Stack Segment,* respectively. These registers are all 16-bits wide. They deal with selecting blocks (segments) of main memory. A segment register (i.e., CS) points at the beginning of a segment in memory.

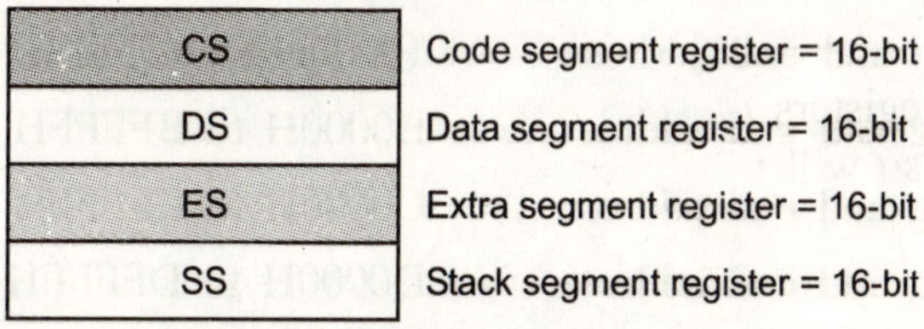

Fig. 2.3 (*d*) Segment registers

CS—points at the segment containing the current program.

DS—generally points at segment where variables are defined.

ES—extra segment register, it is upto a coder to define its usage.

SS—points at the segment containing the stack.

Although it is possible to store any data in the segment registers, this is never a good idea. The segment registers have a very special purpose - pointing at accessible blocks of memory. Segment registers work together with general-purpose register to access any memory value. *For example* if we would like to access memory at the physical address **12345H** (hexadecimal), we could set the **DS = 1230H** and **SI = 0045H.** This way we can access much more memory than with a single register, which is limited to 16-bit values. The CPU makes a calculation of the physical address by multiplying the segment register by 10H and adding the general-purpose register to it (1230H*10H+0045H = 12345H).

The address formed with 2 registers is called an **effective address.** By default

- **BX, SI** and **DI** registers work with **DS** segment register;
- **BP** and **SP** work with **SS** segment register.

Other general-purpose registers cannot form an effective address. Also, although **BX** can form an effective address, **BH** and **BL** cannot.

The 8086 can address a memory up to 1MB (2^{20}). Since, each segment register is 16-bit wide, the entire 1MB can be divided into 16-segments. The segment address can start at the bottom with the reference in the code segment as CS = 0000H. The lowest offset could be 00H. Hence, effective address = CS×10H+00H = 00000H. The highest location of this segment of the memory = 0FFFFH. The starting and end address of each segment are given as:

1st segment starting and end address = 00000H to 0FFFFH

2nd segment starting and end address = 10000H to 1FFFFH

3rd segment starting and end address = 20000H to 2FFFFH

4th segment starting and end address = 30000H to 3FFFFH
5th segment starting and end address = 40000H to 4FFFFH
6th segment starting and end address = 50000H to 5FFFFH
7th segment starting and end address = 60000H to 6FFFFH
8th segment starting and end address = 70000H to 7FFFFH
9th segment starting and end address = 80000H to 8FFFFH
10th segment starting and end address = 90000H to 9FFFFH
11th segment starting and end address = 10000H to AFFFFH
12th segment starting and end address = B0000H to BFFFFH
13th segment starting and end address = C0000H to CFFFFH
14th segment starting and end address = D0000H to DFFFFH
15th segment starting and end address = E0000H to EFFFFH
16th segment starting and end address = F0000H to FFFFFH

Flag Registers

Flag Registers determine the current state of the processor. They are modified automatically by μP after mathematical operations. This allows to determine the type of the result and to determine conditions to transfer control to other parts of the program. Generally we cannot access these registers directly. The flags register of 8086 is shown in Fig. 2.3 (*e*).

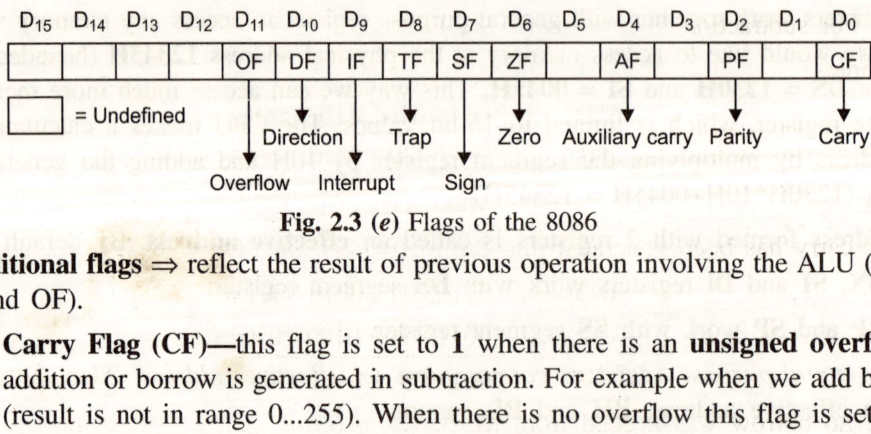

Fig. 2.3 (*e*) Flags of the 8086

Conditional flags ⇒ reflect the result of previous operation involving the ALU (SF, ZF, PF, CF, AF and OF).

1. **Carry Flag (CF)**—this flag is set to **1** when there is an **unsigned overflow** due to addition or borrow is generated in subtraction. For example when we add bytes **255+1** (result is not in range 0...255). When there is no overflow this flag is set to **0**.
2. **Parity Flag (PF)**—this flag is set to **1** when there is even number of one bit in result, and to **0** when there is odd number of one bit.
3. **Auxiliary Flag (AF)**—set to **1** when there is an **unsigned overflow** for low nibble (4 bits).
4. **Zero Flag (ZF)**—set to **1** when result is **zero.** For non-zero result this flag is set to **0.**
5. **Sign Flag (SF)**—set to **1** when result is **negative.** When result is **positive** it is set to **0.** (This flag takes the value of the most significant bit.)

6. **Overflow Flag (OF)**—set to **1** when there is a **signed overflow.** For example, when you add bytes **100+50** (result is not in range -128... 127).

Control flags => control the execution of special functions (DF, IF, and TF).

7. **Trap Flag (TF)**—used for on-chip debugging.
8. **Interrupt enable Flag (IF)**—when this flag is set to **1** CPU reacts to interrupts from external devices.
9. **Direction Flag (DF)**—this flag is used by some instructions to process data chains, when this flag is set to **0**-the processing is done forward, when this flag is set to **1** the processing is done backward.

The flag register of EU holds the status of the ALU operation. Instruction pointer (IP) is a 16-bit register and used similar to the PC in 8085.

OF (Overflow Flag)

This flag is set if the operation results in an overflow of addition or subtraction of 2-signed bit numbers. In the sign shown below, 7-bits are used for data and the MSB is used for sign of the data. Left side of the data shown below represents positive sign of the data whereas the side right represents negative value of data.

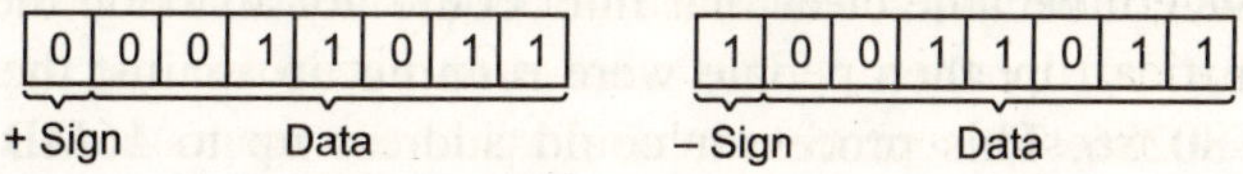

- For addition, the overflow flag (OF) is set if there is carry into MSB (sign bit).
- For subtraction, it is set when MSB needs a borrow.

Example

65D1		0110 0101 1101 0001
+2359	+	0010 0011 0101 1001
		1000 1001 0010 1010

Here, carry has been generated into the MSB (sign bit), OF = 1.

6721		0110 0111 0010 0001
–354A	–	0011 0101 0100 1010
		0011 0001 0101 1111

Here, no borrow was needed from MSB. OF = 0.

Note. The sign, zero, auxiliary carry, parity, and carry flags function like the same flags in 8085A. They even occupy the same bit positions in the low byte of the flag register. Only the overflow, direction, interrupts enable/disable and trap flags are new.

2.4 MEMORY SEGMENTATION

Among other things, segmentation provides a powerful memory management mechanism.

- It allows programmers to partition their programs into modules that operate independent of one another.

- Segments provide a way to easily implement object-oriented programs.
- Segments allow 2-processes to easily share data.

All in all, segmentation is a really **neat feature.** On the other hand, if we ask 10-programmers what they think of segmentation, at least 9-of the 10-will claim it is terrible. Why such a response? Well, it turns out that segmentation provides one other ***nifty feature.*** It allows us to extend the addressability of a processor. In the case of the 8086, segmentation let Intel's designers extend the maximum addressable memory from 64KB to 1MB. Gee, that sounds good. Why is everyone complaining? Well, a little history lesson is in order to understand what went wrong. In 1976, when Intel began designing the 8086 processor, memory was very expensive. Personal computers (at the time), typically had 4KB of memory. Even when IBM introduced the PC 5-years later, 64KB was still quite a bit of memory, 1MB.was a tremendous amount. Intel's designers felt that 64KB memory would remain a large amount throughout the lifetime of the 8086. The only mistake they made was completely underestimating the lifetime of the 8086. They thought it would last about 5-years, like their earlier 8080 processor. They had plans for lots of other processors at the time, and 8086 was not a suffix on the names of any of those. Surely 1MB would be more than enough to last until they came out with something better.

Unfortunately, Intel did not count on the IBM PC and the massive amount of software to appear for it. By 1983, it became clear that Intel could not abandon the 80×86 architecture. They were stuck with it, but by then people were running up against the 1MB limit of 8086. So Intel gave us the 80286. This processor could address up to 16MB of memory. Surely, more than enough. The only problem was that all that wonderful software written for the IBM PC was written in such a way that it could not take advantage of any memory beyond 1MB.

It turns out that the maximum amount of addressable memory is not everyone's main complaint. The real problem is that the 8086 was a 16-bit processor, with 16-bit registers and 16-bit addresses. This limited the processor to addressing $2^{16} = 2^{10} \times 2^{6} = 64$KB chunks of memory. Intel's clever use of segmentation extended this to 1MB, but addressing more than 64KB at one time takes some effort. Addressing more than 256KB at one time takes a lot of effort.

Despite what we might have heard, segmentation is not bad. In fact, it is a really great memory management scheme. What is bad is Intel's 1976 implementation of segmentation still in use today? We cannot blame Intel for this → they fixed the problem in the 80's with the release of the 80386. The real culprit is MSDOS that forced programmers to continue to use 1976 style segmentation. Fortunately, newer operating systems such as LINUX, UNIX, Windows 9x, Windows NT, Windows 2000, and OS/2 do not suffer from the same problems as MSDOS. Furthermore, users finally seem to be more willing to switch to these newer operating systems, so programmers can take advantage of the new features of the 80×86 family.

With the history lesson aside, it is probably a good idea to figure out what segmentation is all about. Consider the current view of memory. It looks like a linear array of bytes. A single index (address) selects some particular byte from that array. Let us call this type of addressing linear or flat addressing. Segmented addressing uses 2-components to specify a memory location

- a segment value and
- an offset within that segment.

Ideally, the segment and offset values are independent of one another. The best way to describe segmented addressing is with a 2-dimensional array shown in Fig. 2.4(*a*).

A full-segmented address contains a segment component and an offset component. This text will write segmented addresses as segment : offset. On the 8086 through the 80286, these 2-values are 16-bit constants. On the 80386 and later, the offset can be a 16-bit constant or a 32-bit constant.

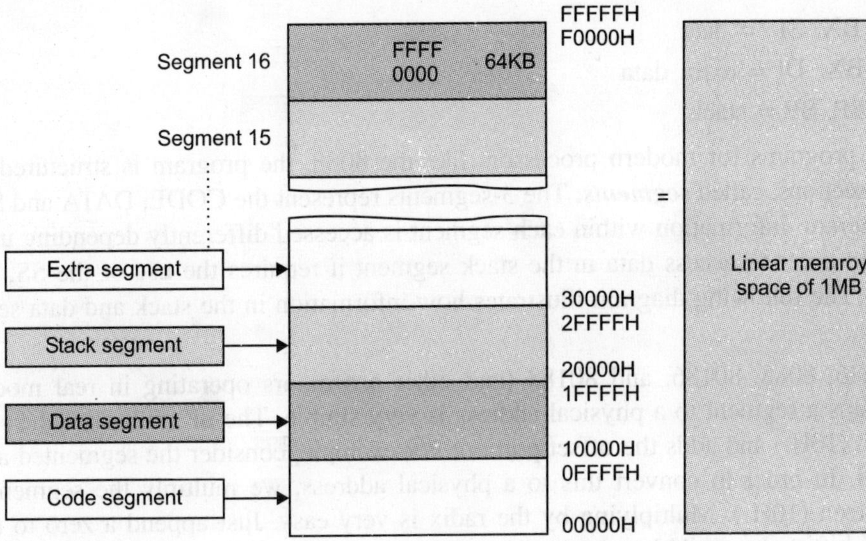

Fig. 2.4 (*a*) Segmented memory

Now we may be wondering, 'Why make this process more complex?' Linear addresses seem to work fine, why bother with this 2-dimensional addressing scheme? Well, let us consider the way we typically write a program. If we were to write, say, a SIN(X) routine and we needed some temporary variables, we probably would not use global variables. Instead, we would use local variables inside the SIN(X) function. In a broad sense, this is one of the features that segmentation offers → the ability to attach blocks of variables (a segment) to a particular piece of code. We could, for example, have a segment containing local variables for SIN, a segment for SQRT, a segment for DRAWWindow, etc. Since the variables for SIN appear in the segment for SIN, it is less likely our SIN routine will affect the variables belonging to the SQRT routine. Indeed, on the 80286 and later operating in protected mode, the μP can prevent one routine from accidentally modifying the variables in a different segment.

The size of the offset limits the maximum size of a segment. On the 8086 with 16-bit offsets, a segment may be no longer than 64KB; it could be smaller (and most segments are), but never larger. The 80386 and later processors allow 32-bit offsets with segments as large as 4GB.

The segment portion is 16-bits on all 80x86 processors. This lets a single program have up to 65,536 different segments in the program. Most programs have less than 16 segments (or thereabouts) so this is not a practical limitation.

Of course, despite the fact that the 80x86 family uses segmented addressing, the actual (physical) memory connected to the μP is still a linear array of bytes. There is a function that converts the segment value to a physical memory address. The processor then adds the offset to this physical address to obtain the actual address of the data in memory as depicted in Fig. 2.4(*b*). This text will refer to addresses in our programs as segmented addresses or logical addresses. The actual linear address that appears on the address bus is the physical address.

Processors like the 8088 have separate registers that deal with each section of a program. The valid combination of registers for addressing different segments is

- CS and IP = instructions
- DS, BX, SI = data
- ES, BX, DI = extra data
- SS, SP, BP = stack

In writing programs for modern processors like the 8088, the program is structured with a minimum of 3-sections, called ***segments.*** The 3-segments represent the CODE, DATA and STACK areas of the program. Information within each segment is accessed differently depending upon the segment type. In order to access data in the stack segment it requires the use of the SS, SP and or BP registers. The following diagram illustrates how information in the stack and data segments is accessed.

On the 8086, 8088, 80186, and 80188 (and other processors operating in real mode), the function that maps a segment to a physical address is very simple. The μP multiplies the segment value by sixteen (10H) and adds the offset portion. For example, consider the segmented address: 1000H : 1F00H. In order to convert this to a physical address, we multiply the segment value (1000H) by sixteen (10H). Multiplying by the radix is very easy. Just append a zero to the end of the number (shifting by a nibble). Appending a zero to 1000H produces 10000H. Now we add 1F00H to this to obtain (10000H + 1F00) 11F00H. So 11F00H is the physical address that corresponds to the segmented address 1000H : 1F00H. The addressing of data in stack segment **(SS : BP)** and address of data in data segment **(DS : BX)** is shown in Fig. 2.4(*b*) **Warning :** A very common mistake people make when performing this computation is to forget they are working in hexadecimal, not decimal. It is surprising to see how many people add 9 + 1 and get 10H rather than the correct answer 0AH.

Intel, when designing the 80286 and later processors, did not extend the addressing by adding more bits to the segment registers. Instead, they changed the function the μP uses to convert a logical address to a physical address. If we write code that depends on the 'multiply by 16D and add in the offset' function, our program will only work on an 80x86 processor operating in real mode, and we will be limited to 1MB of memory.

In the 80286 and later processors, Intel introduced protected mode segments. Among other changes, Intel completely revamped the algorithm for mapping segments to the linear address space. Rather than using a function (such as multiplying the segment value by 10H), the protected mode processors use a look up table to compute the physical address. In protected mode, the 80286 and later processors use the segment value as the index into an array. The contents of the selected array element provide (among other things) the starting address for the segment. The μP adds this value to the offset to obtain the physical address.

Note that our applications cannot directly modify the segment descriptor table (the lookup table). The protected mode operating system (UNIX, LINUX, Windows, OS/2, etc.) handles that operation.

The best programs never assume that a segment is located at a particular spot in memory. We should leave it up to the operating system to place our programs into memory and not generate any segment addresses on our own.

Unlike the 8085, total memory space of the 8086 of IMB (2^{20}) is segmented into 16 logical segments. Each segment register is 16-bit wide and hence its addressing capability is = 2^{16} = 65,536D = FFFFH = 64KB. A segment can begin at any address that is divisible by 16. The *code segment register* is used for addressing the memory location of the executable program. The *data segment register* is used to point the locations where data are stored. The *extra segment register* points the locations that form another locations of data. The stack *segment register* points where stack data is stored. The program can access to code and data in the segment by changing the segment register contents to point to the desired segment. The proper combination of segment and offset address is called *physical address*.

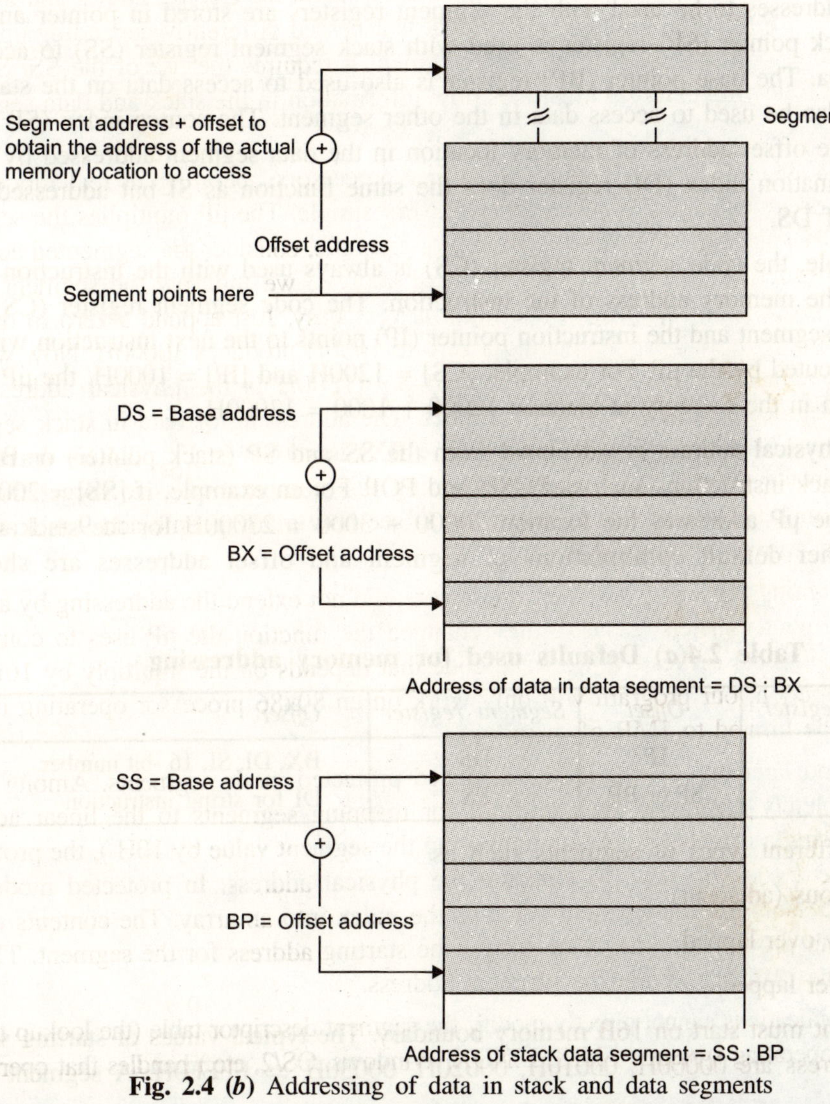

Fig. 2.4 (*b*) Addressing of data in stack and data segments

In a typical Von Neumann Architecture (VNA) machine, the μP connects to memory via the bus. The 80x86 selects some particular memory element using a binary number on the address bus. Different 80x86 μPs have different address busses that control the maximum number of elements in the memory array. However, regardless of the number of address lines on the bus, most computer systems do not have one byte of memory for each addressable location. For example, 80386 processors have 32-address lines allowing up to $2^{32} = 2^2 \times 2^{10} \times 2^{10} \times 2^{10} = 4GB$ of memory. Very few 80386 systems actually have 4GB. Usually, we will find one to 256MB in an 80x86 based system.

The 1st megabyte of memory, from address 000000H to 0FFFFFH is special on the 80×86. This corresponds to the entire address space of the 8088, 8086, 80186, and 80188 microprocessors. Most DOS programs limit their program and data addresses to locations in this range. Addresses limited to this range are named real addresses after the 80 x 86 real mode.

The *offset* addresses to be used with the segment registers are stored in pointer and index registers. The stack pointer (SP) register is used with stack segment register (SS) to access the stack memory area. The base pointer (BP) register is also used to access data on the stack. The BP register can also be used to access data in the other segment. The source index (SI) register is used to hold the offset address of memory location in the data segment addressed by the DS register. The destination index (DI) register does the same function as SI but addressed by ES register instead of DS.

As an example, the code segment register (CS) is always used with the instruction pointer (IP) to generate the memory address of the instruction. The code segment register (CS) points to the start of the segment and the instruction pointer (IP) points to the next instruction within the code segment executed by the μP. For example, [CS] = 1200H and [IP] = 1000H, the μP fetches its next instruction in the memory at location 12000 + 1000 = 13000H.

The 20-bit physical address is calculated from the SS and SP (stack pointer) or BP (base pointer) for the stack instructions such as PUSH and POP. For an example, if [SS] = 2000H and [BP] = 3000H, the μP addresses the location 20000 + 3000 = 23000H for the stack segment memory. The other default combinations of segment and offset addresses are shown in Table 2.4(*a*).

Table 2.4(*a*) Defaults used for memory addressing

Segment register	*Offset*	*Segment register*	*Offset*
CS	IP	DS	BX, DI, SI, 16 -bit number.
SS	SP or BP	ES	DI for string instruction

There are different types of segments such as

- continuous (adjacent)
- partially over lapped
- fully over lapped

Every segment must start on 16B memory boundary. The typical values of starting segment on a physical address are 00000H, 00010H, 00020H, 00030H ⇔ FFFF0H. A segment can be pointed by more than one segment register. As an example, DS and ES can point the same segment in the memory if string located in that segment is used as a source segment in one string instruction and a destination segment in an another string instruction. Let us note that the ES for a string instruction must point a destination segment.

Overlapping segmentation

The utilization of segment register essentially divides the memory space into overlapping segments, each segment of 64KB = 2^{16} = 65,536 i.e., 0 ⇔ 65,535 locations beginning at an address that is divisible by 16. Henceforth, the contents of a segment register will be referred as the segment address and the segment address multiplied by 16 is the beginning segment address. An illustration of overlapping segment is shown in Fig. 2.4(*c*).

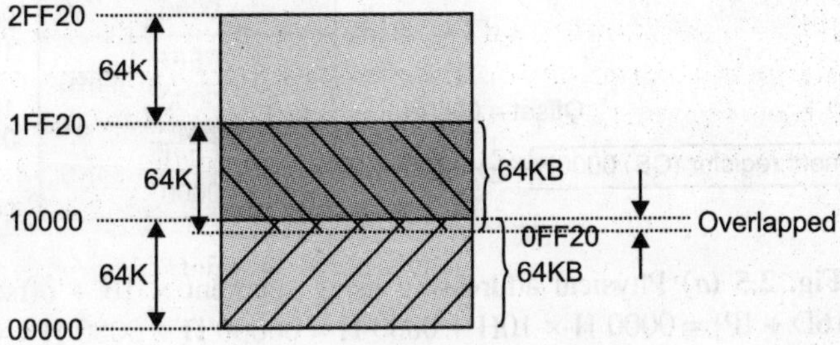

Fig. 2.4 (*c*) Overlapping segmentation

2.5 PHYSICAL ADDRESS

There is 2-types of memory address associated with each program due to the presence of segmentation. These are called *logical address* and *physical address*. The logical address ranges from 0000H to FFFFH (65,536D) that corresponds to 64K locations within one segment. For small programs, the programmers need to worry about these addresses. The physical address corresponds to the actual address that the 8086 outputs on its $A_{19} \Leftarrow A_{16}$, $AD_{15} \Leftrightarrow AD_0$ address lines. Combining the logical address as an offset to the base segment address forms this address.

Example

What memory address will be accessed, if IP = 0080H and CS = 0000H ? Each segment register is internally appended with OH on its right most end to form the 20-bit memory address allowing it to access the start of the segment at almost any location within the 1MB of memory. The 20-bit address of this instruction will be formed as indicated in Fig. 2.5(*a*).

It is multiplied by 16D because content of the segment has to be shifted 4-bits (2^4) left. Hence, provisions have been made to automatically append the least significant nibble (OH) on the right most side of the segment contents.

The direct addressing mode specifies a 16-bit number and hence called effective address of 16-bits. But the address on the address bus, called the physical address must be of 20-bits. How, then, a 20-bit address be accommodated? The BIU computes the 20-bit physical address using 16-bit contents of any one of the segment registers (CS , DS, SS, & ES) and IP. The contents of any one of the segment register is shifted 4-bits left and then the 16-bit contents of the IP is added to it to get the physical address as explained in Fig. 2.5(*b*).

The contents of the segment register including the 4-least significant 0-bit is the base address for a segment of addresses. Changing the contents of the segment register allows the base of a segment to exist anywhere in the physical memory.

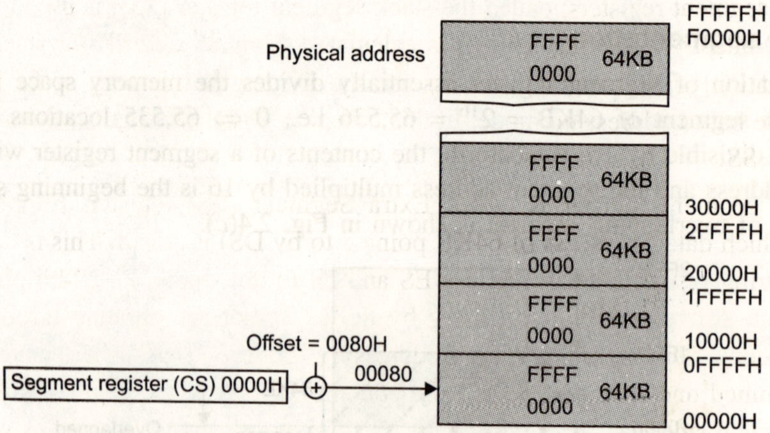

Fig. 2.5 ***(a)*** Physical addressing using segment × 16 + offset
[CS × 16D + IP] = 0000 H × 10H + 0080 H = 00000 H + 0080 H = 00080 H

One of the 4-segment registers, called the code segment register (CS), is used to establish the current base for the segment of containing instructions. In this case, the logical address is contained in the instruction pointer (IP). The IP is another name for the PC used in other processors. In other words, the contents of CS are multiplied by 16D for computing the 20-bit physical address. This means that all instructions of a program are relative to the contents of CS register multiplied by 16D and then offset is added provided by the 16-bit contents of IP.

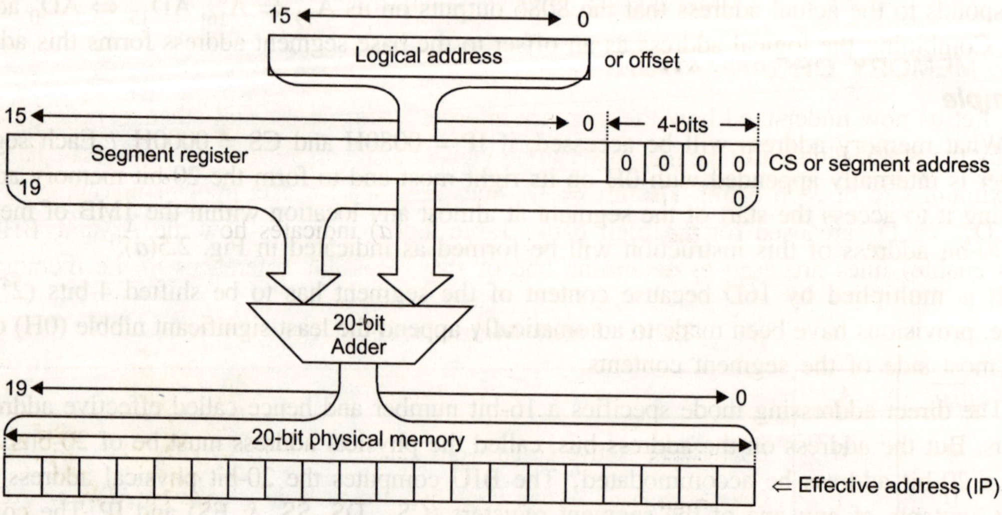

Fig. 2.5 ***(b)*** Formation of physical address

The DS is an another segment register used to point the current data segment. Operands for most operations are fetched from this segment. The 16-bit contents of Source Index (SI) or Destination Index (DI) are used as offset for computing the 20-bit physical address.

The 3rd segment registers; called the stack segment register (SS); is used to point the current stack. The 20-bit physical stack address is calculated from SS and SP for stack instruction such as PUSH and POP. The programmer can use the BP register instead of SP for accessing the stack using the based addressing mode. In this case, the 20-bit physical stack address is calculated from BP and SS.

The 4th segment register is called Extra Segment (ES) register. ES points to the extra segment in which data (in excess of 64KB pointed to by DS) is stored. This is dedicated for string operations. String instructions always use ES and DI to determine the 20-bit physical address for the destination. All references to memory by the processor are done so through the use of the segment registers. Thus, no matter what addressing mode is used, an effective (logical) address is first determined and then added with the contents of one of the segment registers to produce a 20-bit physical address. There is no restriction on the contents of the segment registers. Therefore, at any time, 4-segments of memory may be defined. They do not need to occupy continuous memory locations and they may overlap each other.

Example

Obtain the 20-bit physical address generated by the BIU if the contents of CS = 124BH and IP = 341CH.

$$\begin{array}{rl} 341CH & \Rightarrow \text{EA i.e., (IP)} \\ +\ 124B0H & \Rightarrow \text{CS shifted 4-bit left} \\ \hline 158CCH & \Rightarrow \text{Physical address} \end{array}$$

2.6 MEMORY ORGANIZATION

Let us now understand how the memory interprets the address and other related signals sent out by the 8086. Physically the memory can be viewed to be organized as 2-banks of 512KB each (maximum size of each bank). The $D_7 \Leftrightarrow D_0$ lines are used for data transfer involving low bank and $D_{15} \Leftrightarrow D_8$ are used for the high bank. Table 2.6(*a*) indicates how the A_0 and $\overline{BHE}$ (bus high enable) lines are used to determine one of the 3 possible references to the memory.

Table 2.6(*a*) Memory references

$\overline{BHE}$	A_0	*Processing*
0	0	Both Banks Active 16-bit word transfer on $AD_{15} \Leftrightarrow AD_0$
0	1	Only High bank Active (One byte from/to odd address on $AD_{15} \Leftrightarrow AD_8$)
1	0	Only Low bank Active (One byte from/to even address on $AD_7 \Leftrightarrow AD_0$)
1	1	No Bank Active

Figure 2.6(*a*) indicates how the address space is physically implemented on a 16-bit data bus by dividing the total address space into 2-banks, each one up to 512KB . One of the banks (low bank) is selected by $A_0 = 0$ and $\overline{BHE} = 1$ connected to $D_7 \Leftrightarrow D_0$. The other bank (high bank) is selected by $A_0 = 1$ and $\overline{BHE} = 0$ and is connected to $D_{15} \Leftrightarrow D_8$.

Case-1 : Reading an odd addressed byte

Let us consider the execution of MOV instruction for transferring data from memory to one of the registers of the 8086 (MOV DH, [BX]). The 20-bit address calculated by BX and DS is 20005H. As the binary representation of 20005H contains a 1 in the least significant bit, $A_0 = 1$. Also as it is stated that only one byte has to be transferred from the memory, $\overline{BHE} = 0$. Thus only high bank (Odd Bank) is selected and only one byte of data can be transferred from odd memory. The contents of the $AD_{15} \Leftrightarrow AD_8$ = 2AH as in Fig. 2.6(*b*).

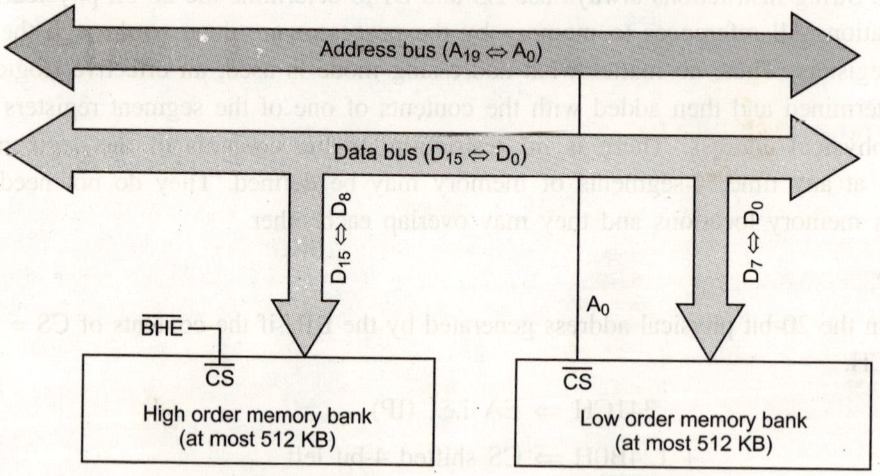

Fig. 2.6 (*a*) Memory organization of 8086

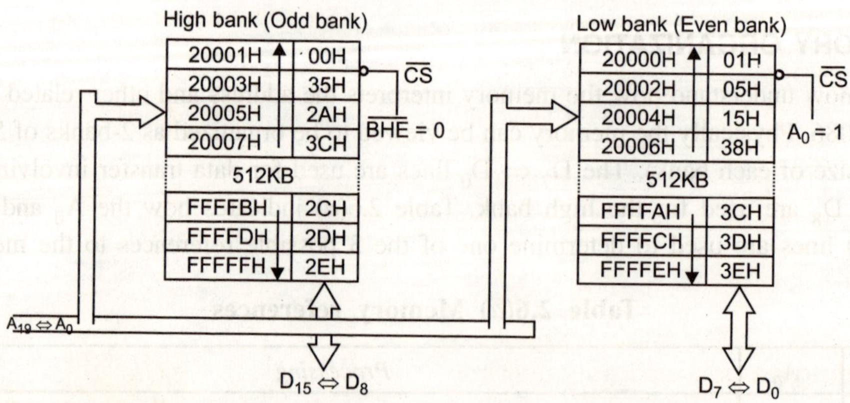

Fig. 2.6 (*b*) Odd addressed byte transfer

Case-2 : Reading even addressed byte

In case-1, if the 20-bit address calculation comes out to be even i.e., 20004H, for this $A_0 = 0$, $\overline{BHE} = 1$ and only one byte is to be transferred from the memory. It is clear that only even bank (low Bank) is selected and only one byte located at 20004H will be transferred. The contents of the data transferred on $D_7 \Leftrightarrow D_0$ will be 15H as in Fig. 2.6(*c*).

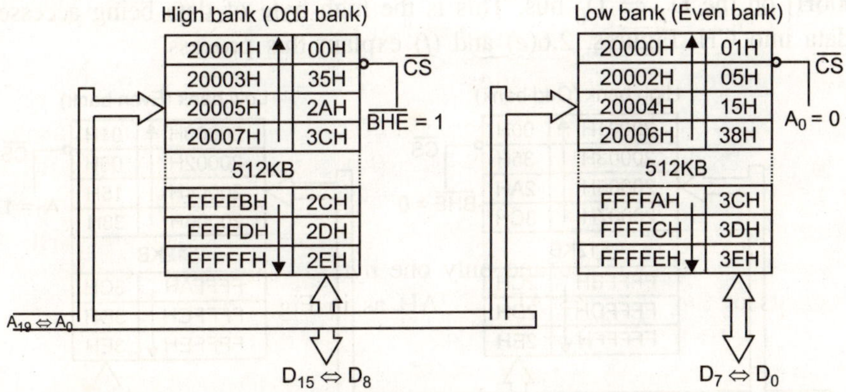

Fig. 2.6 (*c*) Even addressed byte transfer

Case-3 : Writing even addressed word

For 16-bit word transfer, let us suppose that the 8086 executes the instruction MOV [BX], CX. Also let us assume that [BX] = 0004H , and [DS] = 2000H. The 20-bit physical address for the word is formed by shifting [DS] 4-bit left and then adding [BX] to it to form the 20-bit = 20004H. The 8086 outputs $A_0 = 0$, $\overline{BHE} = 0$ to enable both banks simultaneously. The 8086 copies [CL] to $D_7 \Leftrightarrow D_0$ and [CH] to the $D_{15} \Leftrightarrow D_8$ lines with $\overline{WR} = 0$ and $M/\overline{IO} = 1$. Thus, the enabled memory banks obtain the 16-bit data and copies the [CL] at location 20004H and [CH] at location 20005H as in Fig. 2.6(*d*).

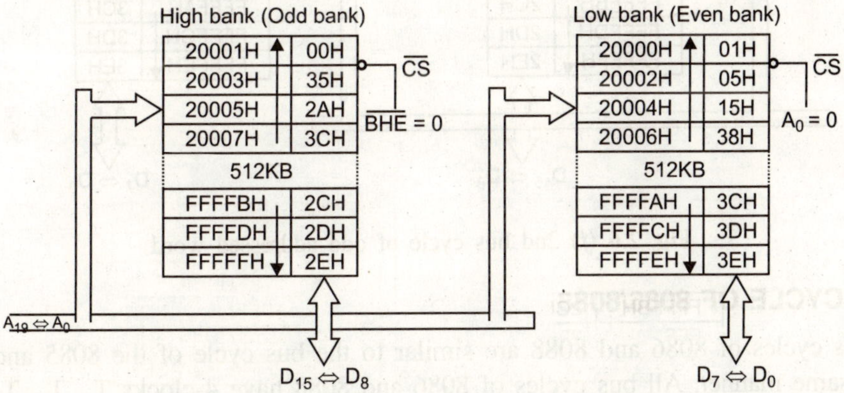

Fig. 2.6 (*d*) Even addressed word transfer

Case-4 : Reading an odd addressed word

Let us suppose that 20-bit physical address calculated in case-3 is 20005H. The 8086 does this transfers in 2-bus cycle. In the 1st cycle $A_0 = 1$ and $\overline{BHE} = 0$ enables the odd bank (high bank) and disables the low bank (even bank). The 8086 outputs $\overline{RD} = 0$ and $M/\overline{IO} = 1$. In this cycle, the odd memory bank places [20005H] on $D_{15} \Leftrightarrow D_8$ bus. We must note that this is the low byte of the data accessed. The 8086 reads the data into CL. In the 2nd bus cycle, the 8086 outputs $A_0 = 0$ and $\overline{BHE} = 1$, enabling the low bank (even bank) and disabling the high bank (odd bank). The 8086 also outputs $\overline{RD} = 0$ and $M/\overline{IO} = 1$. The selected even addressed memory bank places [20006H] on the $D_7 \Leftrightarrow D_0$ bus. This is the high byte of data being accessed. The 8086 reads that data into CH. Figures. 2.6(*e*) and (*f*) explain this process.

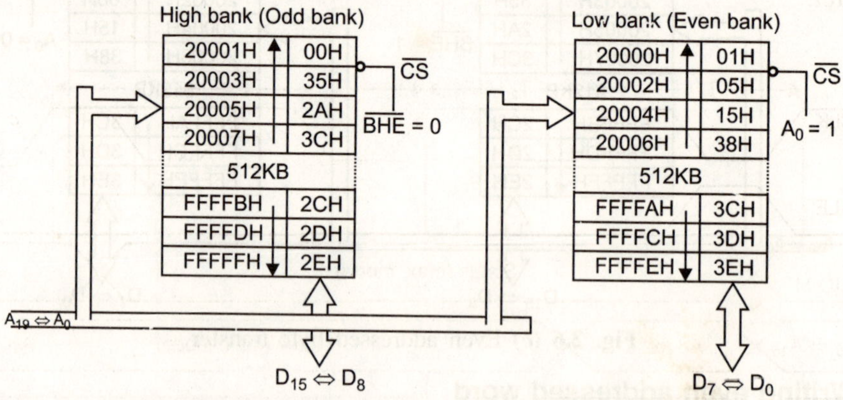

Fig. 2.6 (*e*) 1st bus cycle of odd addressed word

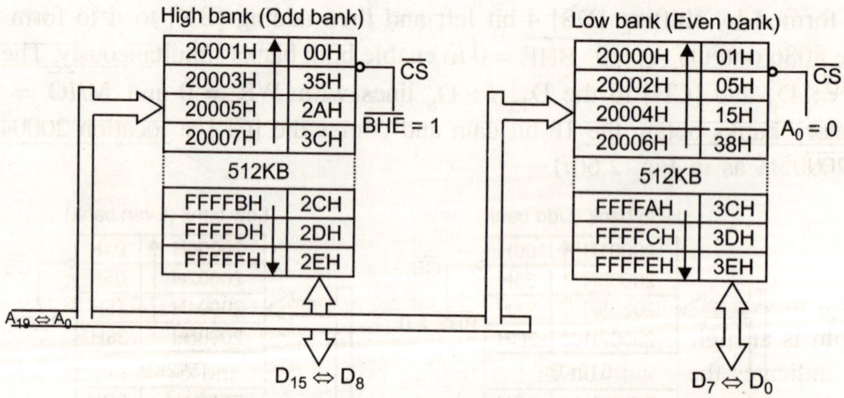

Fig. 2.6 (*f*) 2nd bus cycle of odd addressed word

2.7 BUS CYCLE OF 8086/8088

The bus cycles of 8086 and 8088 are similar to the bus cycle of the 8085 and interpreted also in the same manner. All bus cycles of 8086 and 8088 have 4-clocks T_1, T_2, T_3, and T_4 as indicated in Figs. 2.7 (*a*) and (*b*). If, at all there is a WAIT state, it occurs between T_3 and T_4. Each WAIT is one time period long. Each time period represents the duration of each clock cycle. In the 8088, each clock cycle is from 200ns to 500ns long depending on the frequency of the crystal being used.

Read cycle

The read bus cycle of the 8088 is shown in Fig. 2.7(*a*). The positions of different signals are explained as

ALE (Address Latch Enable)

A HIGH (enable) is asserted on ALE at the beginning of the T_1 and returns to inactive state towards the end of the T_1. The trailing edge of ALE is used to latch the multiplexed address and data bus $AD_7 \Leftrightarrow AD_0$ and $A_{19} \Leftrightarrow A_{16}$ as only $AD_7 \Leftrightarrow AD_0$ are used for external data transfer.

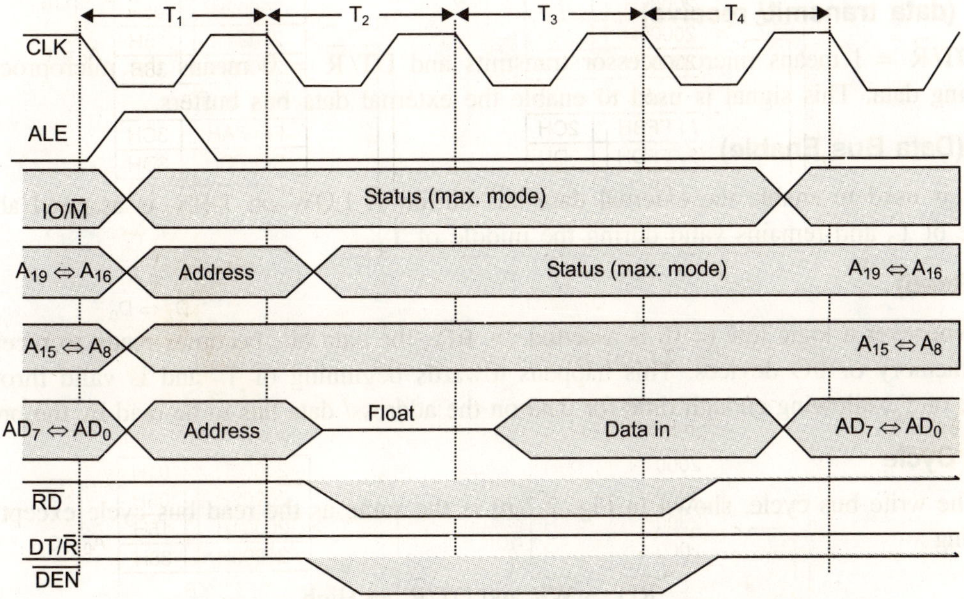

Fig. 2.7 (*a*) Read bus cycle of 8088

The $A_{19} \Leftrightarrow A_{16}$ provides address and $S_6 \Leftrightarrow S_3$ provides status signals. During hold acknowledge, this pin is at high impedance state. S_6 always remains at logic-0 and S_7 always remains at logic-1. S_5 indicates the condition of the IF flag bits, and S_4 and S_3 indicates which segment is accessed during the current bus cycle. These 2-status bits could be used to address 4-separarte 1MB memory bank as per the conditions depicted in Table 2.7(*a*).

Address is asserted in the beginning of T_1 and remains valid until the beginning of T_2. In the maximum mode, system contains status information asserted on the bus during T_2 and in minimum mode no valid data is present during T_2, T_3, and T_4.

IO/M (8088) & M/IO (8086) memory or input/output

This pin indicates that the microprocessor address bus contains the address of either memory or I/O devices. During hold acknowledge, this pin is at high impedance-state.

$A_{15} \Leftrightarrow A_8$ Address bus of 8088

Upper half memory address is valid on $A_{15} \Leftrightarrow A_8$ lines throughout the bus cycle. During hold acknowledge, this pin is at high Impedance State.

$AD_7 \Leftrightarrow AD_0$ Address/data bus of 8088

Whenever ALE is high, these pins contain the address of the lower order memory address or I/O port number and data whenever ALE = 0. During hold acknowledge, this pin is at high impedance state. Thereafter for approximately one clock cycle the lines are in the high impedance-state. This delay exists to allow the bus time to adjust the change in the direction of data. The data is then asserted during T_3 and is valid during T_4.

$DT/\overline{R}$ (data transmit/ receive)

$DT/\overline{R} = 1$ means microprocessor transmits and $DT/\overline{R} = 0$ means the microprocessor is receiving data. This signal is used to enable the external data bus buffers.

$\overline{DEN}$ (Data Bus Enable)

It is used to enable the external data bus buffer. A LOW on $\overline{DEN}$ is asserted about the middle of T_2 and remains valid during the middle of T_4.

$\overline{RD}$ (Read)

Whenever a logic low (= 0) is asserted on $\overline{RD}$, the data bus becomes ready to receive data from memory or I/O devices. This happens towards beginning of T_2 and is valid through the middle of T_4, allowing enough time for data on the address/ data bus to be read by the processor.

Write Cycle

The write bus cycle, shown in Fig. 2.7(*b*) is the same as the read bus cycle except for the following

$$\overline{RD} \Rightarrow \overline{WR} \text{ and } DT/\overline{R} \Rightarrow \text{High.}$$

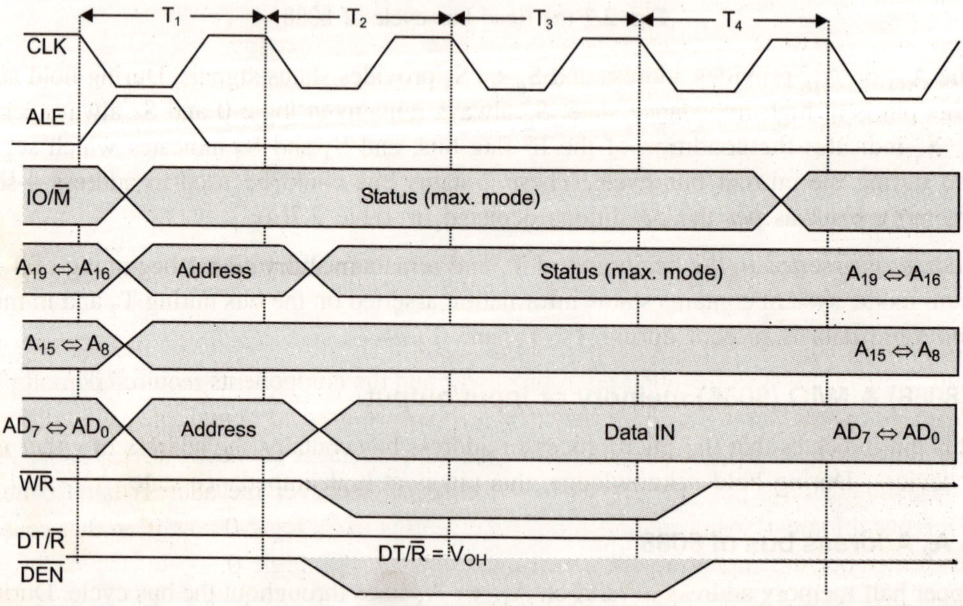

Fig 2.7 (*b*) Write bus cycle of 8088

$\overline{DEN}$ ⇒ is asserted early in T_2 (since data is present) and is valid through the middle of T_2.

The bus cycle of the 8086 is the same as that of the 8088 except for the following exceptions.

The 2-signals $\overline{BHE}$ and AD_0 developed for 8086 are not required for 8088. The $\overline{BHE}$ together with AD_0 (LSB of address/data bus) is used to select the proper byte(s) of address according to Table 2.7(*a*).

The 8088 signal IO/$\overline{M}$ is replaced by signal M/$\overline{IO}$ in 8086.

The address/data bus in 8086 is 16-bit i.e., $AD_{15} \Leftrightarrow AD_0$ in place of 8-bit address/data $AD_7 \Leftrightarrow AD_0$ in 8088.

Table 2.7(*a*) Selection of proper byte of address

$\overline{BHE}$	A_0	*Action*
0	0	Whole word
0	1	Upper byte from/to odd address
1	0	Lower byte from/to even address
1	1	None.

2.8 DEMULTIPLEXING OF BUSES

The address/data bus on the 8086/ 8088 is multiplexed (shared) to reduce its number of pins. Unfortunately, the hardware designer had to extract or demultiplex information from these multiplexed pins. Why not leave the buses multiplexed? Memory and I/O require that the address remains valid and stable throughout a read or write cycle.

If the buses are multiplexed, the address changes at the memory and I/O that cause them to read or write data in the wrong locations. All computer systems have 3 buses

- an address bus to address either memory and I/O through I/O port number,
- a data bus to transfers data between the microprocessor and memory and I/O,
- a control bus that provides control signals to the memory and I/O.

These buses must be present in order to interface to memory and I/O .

Demultiplexing 8088

Figure 2.8(*a*) illustrates the 8088 microprocessor and the components required demultiplexing its buses. In this case, two 74LS373 transparent latches are used to demultiplex the address/data bus connections $AD_7 \Leftrightarrow AD_0$ and the multiplexed address/status connections $A_{19}/S_6 \Leftrightarrow A_{16}/S_3$. These transparent latches pass the inputs to the outputs whenever the address latch enable pin (ALE) becomes a logic-1. After a short time, ALE returns to its logic-0 condition that causes the latches to remember the inputs at the time of the change to a logic-0.

In this case, $A_7 \Leftrightarrow A_0$ are remembered in the bottom latch and $A_{19} \Leftrightarrow A_{16}$ in the top latch. This yields a separate address bus with connections $A_{19} \Leftrightarrow A_0$. These address connections allow the 8088 to address 1MB memory space. The fact that the data bus is separate allows it to be connected to any 8-bit peripheral device or memory component.

Demultiplexing 8086 buses

Like the 8088, the 8086 system requires separate address, data, and control buses. It differs primarily in the number of multiplexed pins. In 8088 only $AD_7 \Leftrightarrow AD_0$ and $A_{19}/S_6 \Leftrightarrow A_{16}/S_3$ are multiplexed. In the 8086, the multiplexed pins include $AD_{15} \Leftrightarrow AD_0$, $A_{19}/S_6 \Leftrightarrow A_{16}/S_3$ and BHE/S_7. All of these signals must be demultiplexed. Figure 2.8(*b*) illustrates a demultiplexed 8086 with all 3 buses: address $A_{19} \Leftrightarrow A_0$ and $\overline{BHE}$, data ($D_{15} \Leftrightarrow D_0$), and control ($M/\overline{IO}$, $\overline{RD}$, and $\overline{BHE}$). This circuit is almost identical to the one drawn in Fig. 2.8(*a*), except that an additional 74LS373 latch has been added to demultiplex the address/data bus $AD_{15} \Leftrightarrow AD_8$ and a $\overline{BHE}/S_7$ inputs has been added to the top 74LS373 to select the high-order memory bank in the 16-bit memory bank in the 16-bit memory system of the 8086. Here, the memory and I/O system see the 8086 as a device with a 20-bit address bus ($A_{19} \Leftrightarrow A_0$), a 16-bit data bus ($D_{15} \Leftrightarrow D_0$) and a 3 line control bus ($M/\overline{IO}$, $\overline{RD}$, and $\overline{WR}$).

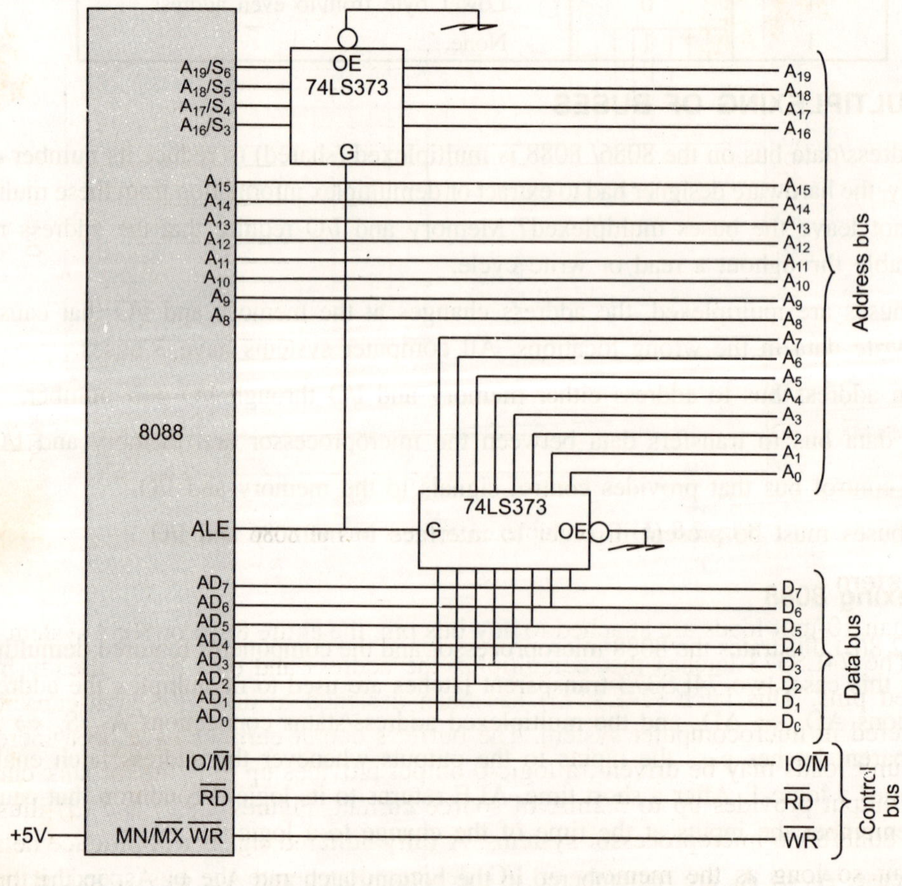

Fig. 2.8 (*a*) Demultiplexed address bus of 8088

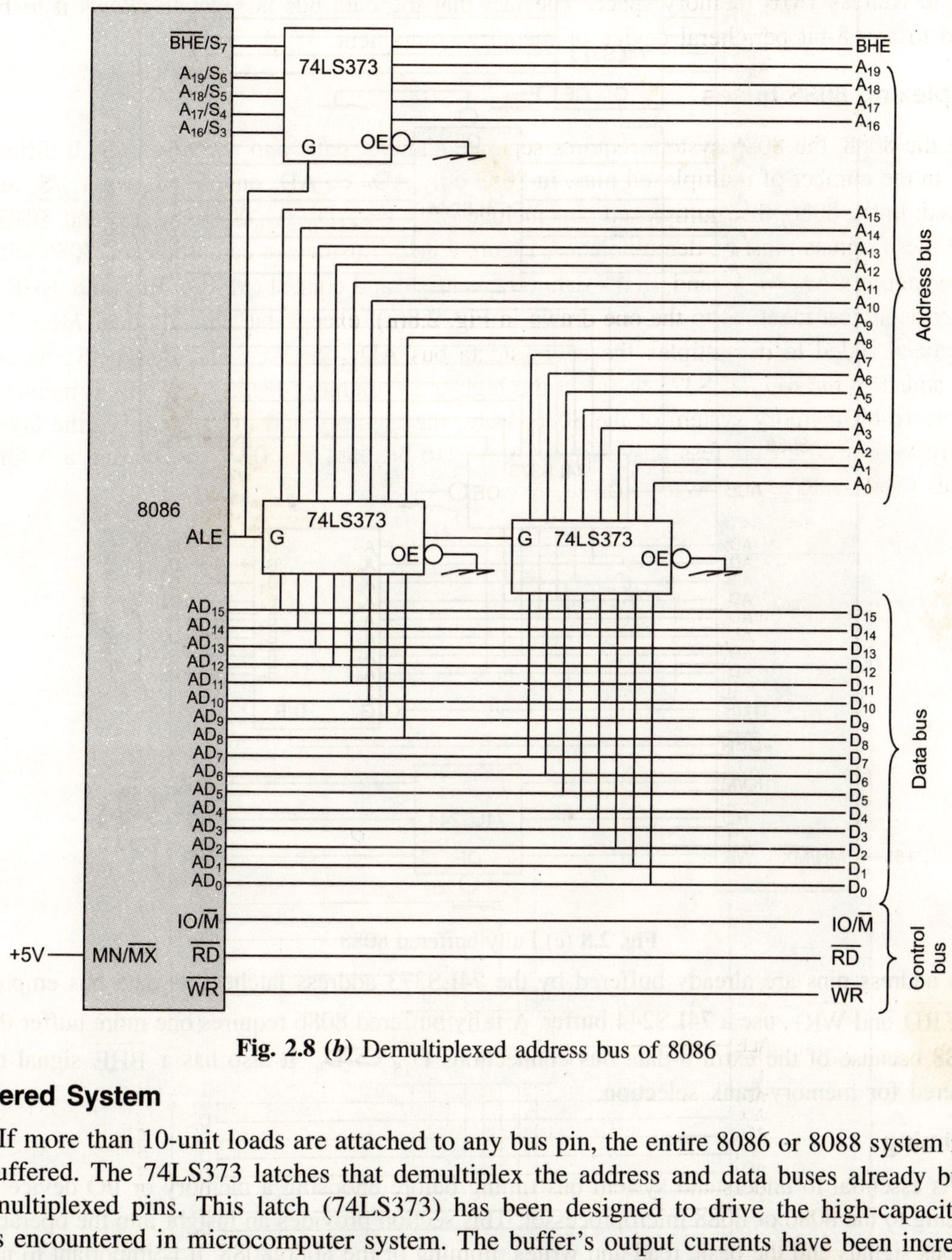

Fig. 2.8 (*b*) Demultiplexed address bus of 8086

Buffered System

If more than 10-unit loads are attached to any bus pin, the entire 8086 or 8088 system must be buffered. The 74LS373 latches that demultiplex the address and data buses already buffer the multiplexed pins. This latch (74LS373) has been designed to drive the high-capacitance buses encountered in microcomputer system. The buffer's output currents have been increased so more TTL unit loads may be driven : a logic-0 output provides up to 32mA of sink current, and a logic-1 output provides up to 5.2mA of source current. Figures 2.8(*c*) and (*d*) illustrate fully buffered 8088/8086 microprocessor systems. A fully buffered signal will produce delay to the system. But so long as the memory or I/O devices do operate the bus speed, it hardly matters.

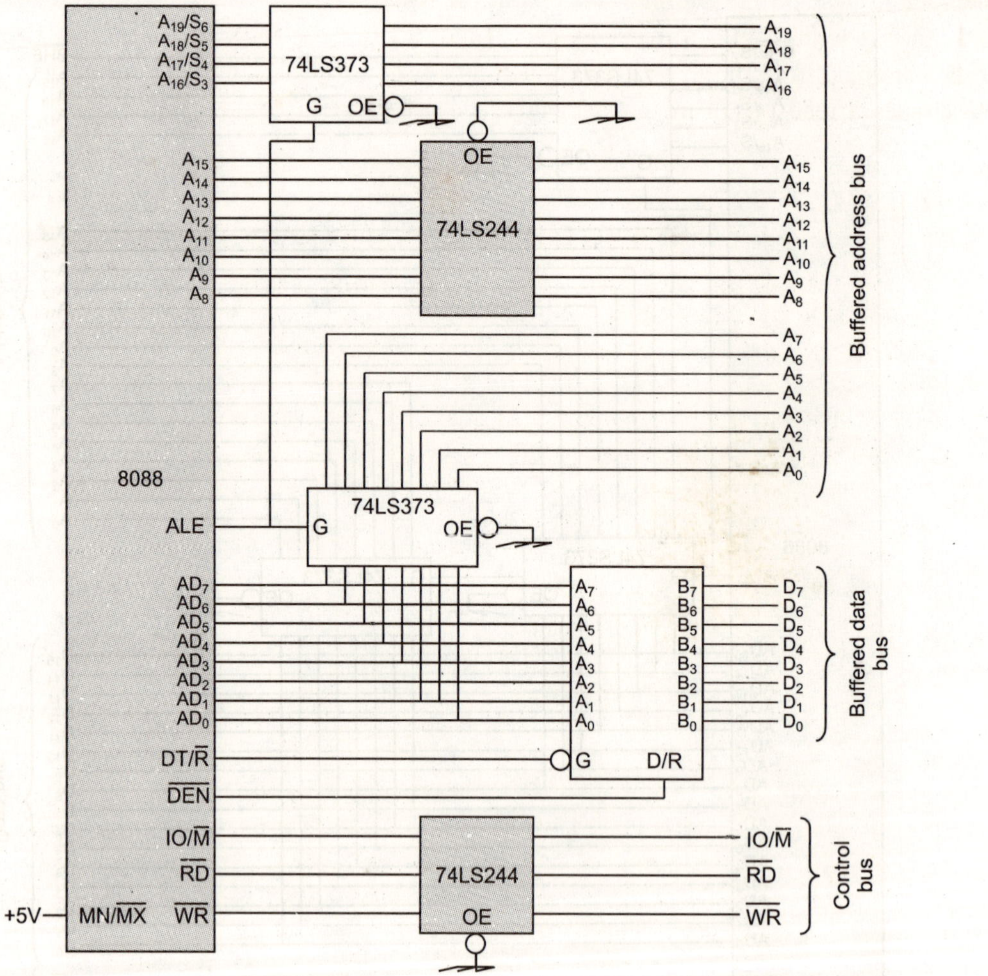

Fig. 2.8 (*c*) Fully buffered 8088

Its address pins are already buffered by the 74LS373 address latches; its data bus employs $(\overline{M}/IO, \overline{RD}, \text{and } \overline{WR})$, use a 74LS244 buffer. A fully buffered 8086 requires one more buffer than the 8088 because of the extra 8-data bus connections $D_{15} \Leftrightarrow D_8$. It also has a $\overline{BHE}$ signal that is buffered for memory-bank selection.

Bus Timing

It is essential to understand system bus timing before choosing a memory or I/O device for interfacing to the 8086 or 8088 microprocessor. This section provides an insight into the operation of the bus signals and the basic read and writes timming of the 8086/8088. It is important to note that we discuss only the times that affect memory and I/O interfacing in this section.

Basic bus operation

The address and data of the 8086 and 8088 function in exactly the same manner as those of any other microprocessor. If data are written to the memory, the microprocessor outputs the memory address on the address bus, output the data to be written into memory on the data bus, and issues a write ($\overline{WR}$) to memory and IO/$\overline{M}$ = 0 for the 8088 and $\overline{IO}$/M = 1 for the 8086. If data are read from the memory, the microprocessor outputs the memory address on the address bus, issues a read ($\overline{RD}$) memory signals, and accepts the data via the data bus.

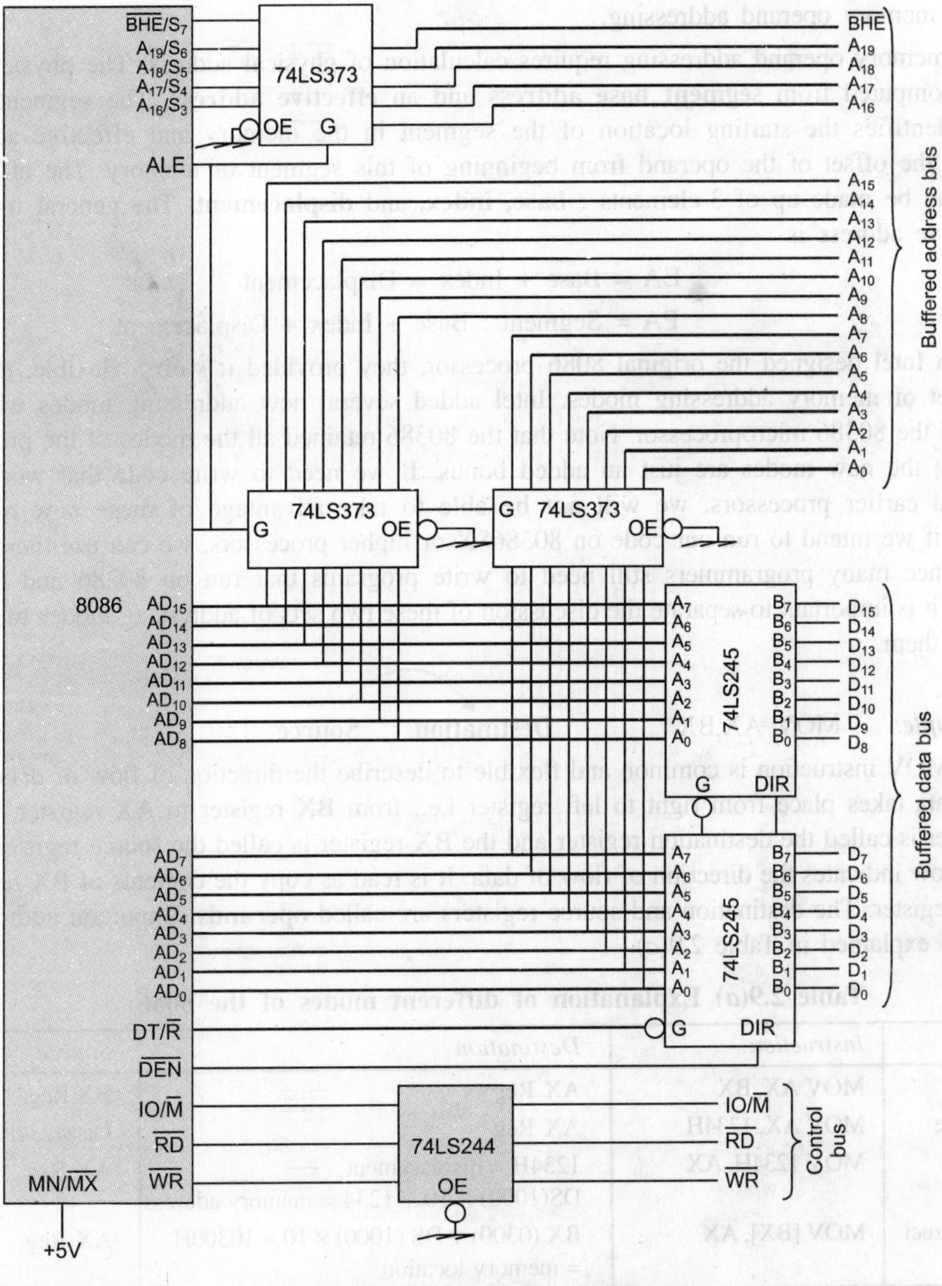

Fig. 2.8 (*d*) Fully buffered 8086

2.9 ADDRESSING MODES

The addressing mode is a method of specifying an operand. The addressing mode is categorized under 3-types :

- register operand addressing,
- immediate operand addressing, and
- memory operand addressing.

The memory operand addressing requires calculation of physical address. The physical address is computed from **segment base address** and an **effective address.** The segment base address identifies the starting location of the segment in the memory and effective address represents the offset of the operand from beginning of this segment of memory. The effective address can be made up of 3-elements : **base, index,** and **displacement.** The general formula for effective address is

$$EA = Base + Index + Displacement$$

$$PA = Segment : Base + Index + Displacement$$

When Intel designed the original 8086 processor, they provided it with a flexible, though limited, set of memory addressing modes. Intel added several new addressing modes when it introduced the 80386 microprocessor. Note that the 80386 retained all the modes of the previous processors; the new modes are just an added bonus. If we need to write code that works on 80286 and earlier processors, we will not be able to take advantage of these new modes. However, if we intend to run our code on 80386SX or higher processors, we can use these new modes. Since many programmers still need to write programs that run on 80286 and earlier machines, it is important to separate the discussion of these two sets of addressing modes to avoid confusing them.

Example: MOV AX,BX **Destination** **Source**

The MOV instruction is common and flexible to describe the direction of flow of data. The flow of data takes place from right to left register i.e., from BX register to AX register. Thus, AX register is called the destination register and the BX register is called the source register. The curved arrow indicates the direction of flow of data. It is read as copy the contents of BX register into AX register. The destination and source registers are called **operands.** Important addressing modes are explained in Table 2.9(*a*).

Table 2.9(*a*) Explanation of different modes of the 8086

Type	*Instruction*	*Destination*	*Source*
Register	MOV AX, BX	AX Reg	BX Reg.
Immediate	MOV AX, 1234H	AX Reg	Data1234H
Direct	MOV 1234H, AX	1234H = displacement DS(1000) × 10 + 1234 = memory address	AX Reg.
Reg. Indirect	MOV [BX], AX	BX (0300) + DS (1000) × 10 = 10300H = memory location	AX Reg.

(*Contd...*)

Base+Index	MOV [BX + SI], AX	BX (0300) + SI (0200) + DS (1000) × 10 = 10500H = memory location	AX Reg.
Reg. Relative	MOV [BX + 04], AX	BX(0300) + 0004 + DS(1000) × 1 0 = 10304H = memory location	AX Reg.
Base Relative + Index	MOV[BX + 04 + SI], AX	BX(0300) + 0004 + SI(0200) + DS(1000) × 10 = 10504H = memory location	AX Reg.

Immediate addressing

In this mode of addressing, immediately data as an operand is available after the opcode. The operand looks like a part of the instruction.

Example : MOV AX, 1234H

In this example, the data 1234H is moved to the AX register.

Register addressing

Most 8086 instructions can operate on the 8086's general-purpose register set. By specifying the name of the register as an operand to the instruction, we may access the contents of that register. Consider the 8086 MOV (move) instruction

MOV Destination, source

This instruction copies the data from the source operand to the destination operand. The 8 and 16-bit registers are certainly valid operands for this instruction. The only restriction is that both operands must be of the same size. Now, let us look at some actual 8086 MOV instructions

```
MOV AX,BX        ; Copies the value of BX into AX
MOV DL,AL        ; Copies the value of AL into DL
MOV SI,DX        ; Copies the value of DX into SI
MOV SP,BP        ; Copies the value of BP into SP
MOV DH,CL        ; Copies the value of CL into DH
MOV AX,AX        ; Yes, this is legal!
```

Remember that the registers are the best place to keep often-used variables. As we will see a little later, instructions using the registers are shorter and faster than those that access memory. Throughout this article we will see the abbreviated operands reg. and r/m (register/ memory) used wherever we may use one of the 8086's general-purpose registers.

In addition to the general-purpose registers, many 8086 instructions (including the MOV instruction) allow us to specify one of the segment registers as an operand. There are 2-restrictions on the use of the segment registers with the MOV instruction. First of all, we may not specify CS as the destination operand, 2nd, **only one of the operands can be a segment register.** We **cannot** move data from **one segment register to another with a single** MOV instruction. In order to copy the value of CS to DS , we would have to use some sequence like

MOV AX, CS, MOV DS, CS → is an illegal instruction.

MOV DS, AX

We should never use the segment registers as data registers to hold arbitrary values. They should only contain segment addresses. Throughout this text we will see the abbreviated operand SREG used wherever segment register operands are allowed (or required).

Memory addressing

The 8086 provides 17-different ways to access memory. This may seem like quite a bit at 1st, but fortunately most of the addressing modes are simple variants of one another so they are very easy to learn. Should be learn them? We should! The key to good assembly language programming is the proper use of memory addressing modes. In fact the memory addressing could be various combinations of the following :

- direct addressing (displacement addressing),
- register indirect addressing,
- based addressing,
- indexed addressing, and
- based indexed addressing.

Variations on these 5-forms provide the 17-different addressing modes on the 8086. See, from 17 down to 5. It is not so bad after all. In order to reference memory, the 8086 calculates the physical address (PA) of the memory location where the operand lies. The **physical address** is computed from the **segment base address** and **effective address.** The segment base address identifies the starting address of the segment in the memory and effective address represents the **offset** of the operand from beginning of this segment of memory. The effective address can be specified in many ways. One way is to encode effective address of the operand directly in the instruction. This is the simplest type of memory addressing and called **direct addressing** mode. The effective address can be made from 3-components: Base, Index, and Displacement. The effective address and physical address computation is expressed as in Fig.2.9(*a*).

EA = Base + Index + Displacement, and

PA = Segment address: EA

$$\text{PA} = \begin{Bmatrix} \text{CS} \\ \text{DS} \\ \text{SS} \\ \text{ES} \end{Bmatrix} : \begin{Bmatrix} \text{BX} \\ \text{BP} \end{Bmatrix} + \begin{Bmatrix} \text{SI} \\ \text{DI} \end{Bmatrix} + \begin{Bmatrix} \text{8-bit displacement} \\ \text{16-bit displacement} \end{Bmatrix}$$

Fig. 2.9 (***a***) Physical and effective address computation of memory operand

Direct addressing/Displacement addressing

The direct addressing mode is similar to the immediate addressing mode with the exception that the opcode follow an effective address instead of data. This is the most common addressing mode and the one that is easiest to understand. The displacement-only addressing mode consists of a 8-bit or 16-bit constant that specifies the offset of the actual address of the target location. The MOV AX,[1234] copies 2-byte of data from memory location whose offset value is 1234H to the AX-register. The lower byte is at the location with the offset 1234H and the higher byte will be at the location 1235H. The other instruction MOV AL, DS:[8088H] loads the AL register with the content of memory location 8088H. Likewise, the instruction MOV DS:[1234H], DL stores the value in the DL-register to memory location with the offset 1234H. The general expression for computation of direct memory operand is expressed as in Fig. 2.9(*b*). The EA is

$$EA = 1234H, \quad EA = 8088H,$$

$$PA = \begin{Bmatrix} CS \\ DS \\ SS \\ ES \end{Bmatrix} : \{\text{Direct address}\} = \begin{Bmatrix} CS \\ DS \\ SS \\ ES \end{Bmatrix} : \{1234\} \text{ or } \begin{Bmatrix} CS \\ DS \\ SS \\ ES \end{Bmatrix} : \{8088\}$$

By default, all displacement values provide offsets into the data segment. If we want to provide an offset into a different segment, we must use a segment override prefix before our address. For example, to access location 1234H in the extra segment (ES), we would use an instruction of the form MOV AX,ES:[1234H]. Likewise, to access this location in the code segment we would use the instruction MOV AX,CS: [1234H]. The DS : prefix in the previous examples is not a segment override. The μP uses the data segment register by default. These specific examples require DS : because of MASM's syntactical limitations.

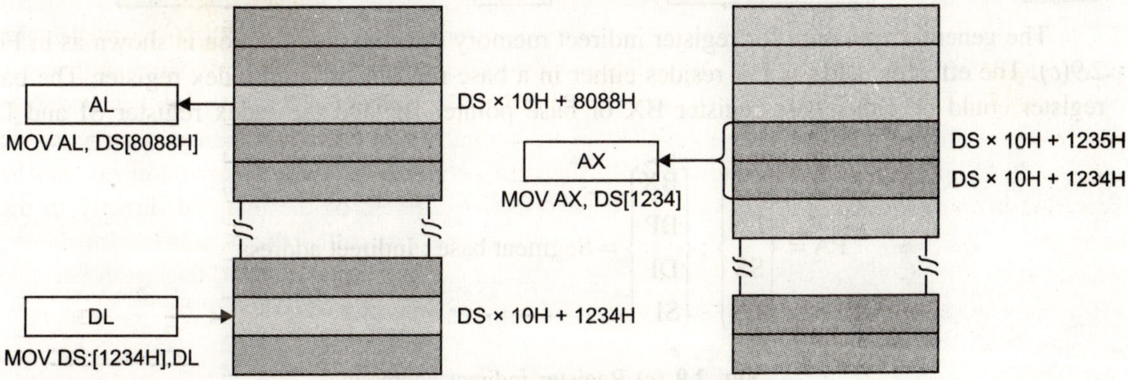

Fig. 2.9 (*b*) Direct memory addressing

Register indirect addressing

The 80x86 μPs allow us to access memory indirectly through a register using the register indirect addressing modes. There are 4-forms of this addressing mode on the 8086, best demonstrated by the following instructions

MOV AL, [BX]

MOV AL, [BP]

MOV AL, [SI]

MOV AL, [DI]

The EA address resides in either a base register or in an index register. The base register is BX or base pointer register is BP and the index register is SI or DI.

As with the 80x86 [BX] addressing mode, these 4-addressing modes reference the byte at the offset found in the BX , BP, SI, or DI registers respectively. The **[BX], [SI],** and **[DI]** modes use the DS segment by default. The **[BP]** addressing mode uses the stack segment **(SS)** by default.

We can use the segment override prefix symbols if we wish to access data in different segments. The following instructions demonstrate the use of these overrides

MOV AL, CS:[BX]

MOV AL, DS:[BP]

MOV AL, SS:[SI]

MOV AL, ES:[DI]

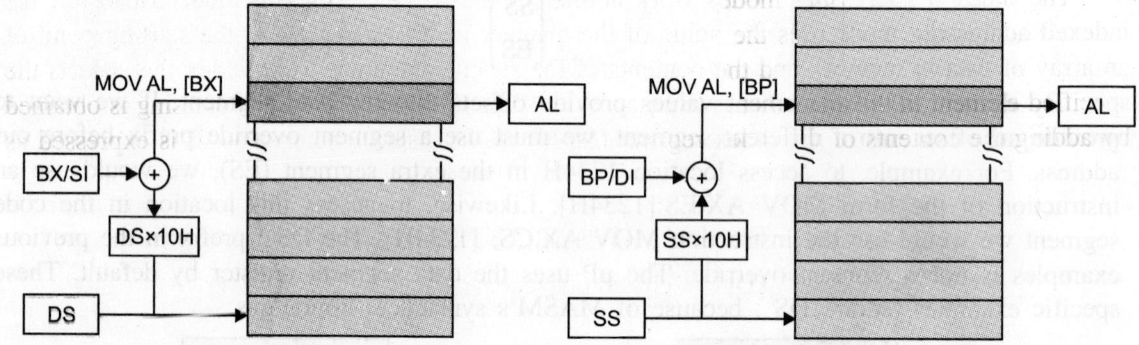

The general expression for register indirect memory operand computation is shown as in Fig. 2.9(*c*). The effective address EA resides either in a base register or in an index register. The base register could be either base register BX or base pointer BP and the index register SI and DI.

$$PA = \begin{Bmatrix} CS \\ DS \\ SS \\ ES \end{Bmatrix} : \begin{Bmatrix} BX \\ BP \\ DI \\ SI \end{Bmatrix} = \text{Segment base : Indirect address}$$

Fig. 2.9 (*c*) Register indirect addressing

Intel refers to [BX] and [BP] as base addressing modes and BX and BP as base registers (in fact, BP stands for base pointer). Intel refers to the [SI] and [DI] addressing modes as indexed addressing modes (SI stands for source index, DI stands for destination index). However, these addressing modes are functionally equivalent. This text will call these forms of register indirect modes to be consistent.

Note: The [SI] and [DI] addressing modes work exactly the same way by just substituting SI and DI for BX above.

Base Addressing mode

The EA in based addressing mode is obtained by adding the direct or indirect displacement to the contents of either base register BX or base pointer BP. As depicted below

$$EA = \begin{bmatrix} BX \\ BP \end{bmatrix} + \begin{bmatrix} \text{8-bit displacement} \\ \text{16-bit displacement} \end{bmatrix}$$

$$PA = \begin{bmatrix} CS \\ DS \\ SS \\ ES \end{bmatrix} : \begin{bmatrix} BX \\ BP \end{bmatrix} + \begin{bmatrix} \text{8-bit displacement} \\ \text{16-bit displacement} \end{bmatrix}$$

Indexed addressing

The indexed addressing modes work similar to the based addressing mode. However, the indexed addressing mode uses the value of the displacement as a pointer to the starting point of an array of data in memory and the contents of the specified register as an index that selects the specified element in the array that is to be accessed. The EA in the indexed addressing is obtained by adding the contents of the index register (SI/DI) to the displacement value. It is expressed as

$$PA = \begin{Bmatrix} CS \\ DS \\ SS \\ ES \end{Bmatrix} : \begin{Bmatrix} SI \\ DI \end{Bmatrix} + \begin{Bmatrix} \text{8-bit displacement} \\ \text{16-bit displacement} \end{Bmatrix}, \begin{matrix} \text{MOV AL, dis[SI]} \\ \text{MOV AL, dis[DI]} \end{matrix}$$

The offsets generated by these addressing modes are the sum of the constant and the specified register. The addressing modes involving BX, SI, and DI all use the data segment, the disp[BP] addressing mode uses the stack segment by default. As with the register indirect addressing modes, we can use the segment override prefixes to specify a different segment:

```
MOV AL,SS : dis[BX]
MOV AL,ES : dis[BP]
MOV AL,CS : dis[SI]
MOV AL,SS : dis[DI]
```

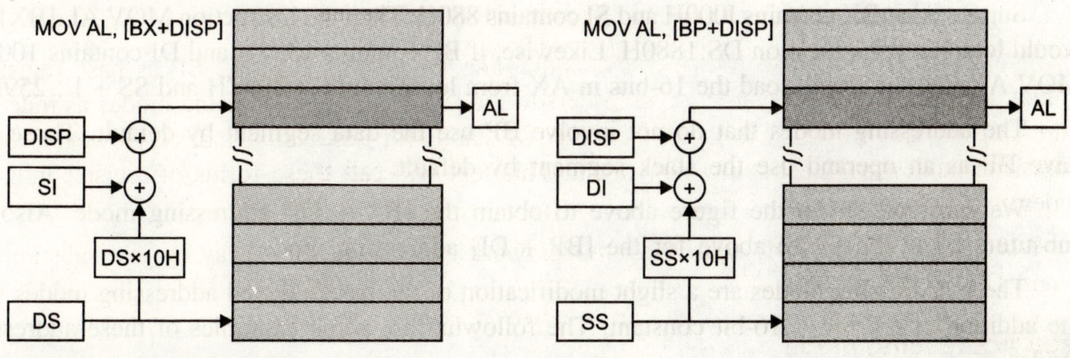

We may substitute BX or BP in the figure above to obtain the [BX + disp] and [BP + disp] addressing modes.

Note that Intel still refers to these addressing modes as based addressing and indexed addressing. Intel's literature does not differentiate between these modes with or without the constant. If we look at how the hardware works, this is a reasonable definition. From the programmer's point of view, however, these addressing modes are useful for entirely different things. That is why this text uses different terms to describe them. Unfortunately, there is very little consensus on the use of these terms in the 80x86 world.

Based indexed addressing

The based indexed addressing modes are simply combinations of the register indirect addressing modes. These addressing modes form the offset by adding together a base register (BX or BP) and an index register (SI or DI). The combination of the base addressing and the indexed addressing results into the based indexed addressing mode. The EA and the PA in this mode are

EA = [BX] + [SI] + 1234H, and

$$PA = \begin{Bmatrix} CS \\ DS \\ SS \\ ES \end{Bmatrix} : \begin{Bmatrix} BX \\ BP \end{Bmatrix} + \begin{Bmatrix} SI \\ DI \end{Bmatrix}$$

MOV AL, [BX][SI]
MOV AL, [BX][DI]
MOV AL, [BP][SI]
MOV AL, [BP][DI]

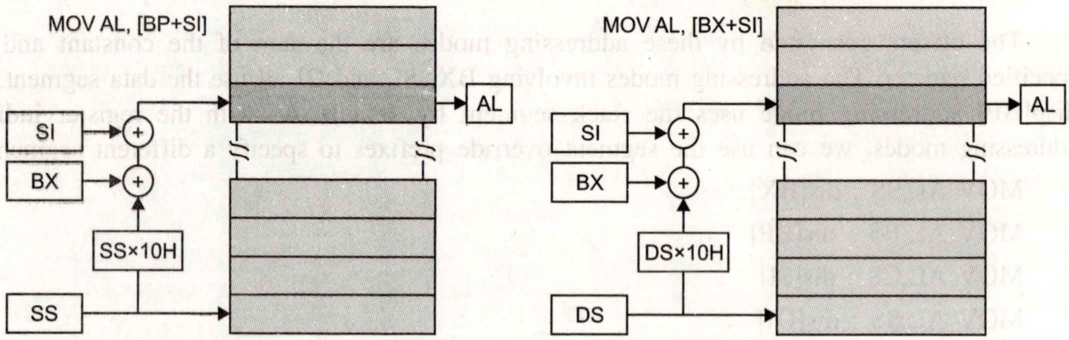

Suppose that BX contains I000H and SI contains 880H. Then the instruction MOV AL,[BX][SI] would load AL from location DS:1880H. Likewise, if BP contains 1598H and DI contains 1004H, MOV AX,[BP + DI] will load the 16-bits in AX from locations SS : 259CH and SS + 1 : 259DH.

The addressing modes that do not involve BP use the data segment by default. Those that have BP as an operand use the stack segment by default.

We substitute DI in the figure above to obtain the [BX + DI] addressing mode. Also we substitute DI in the figure above for the [BP + DI] addressing mode.

These addressing modes are a slight modification of the base/indexed addressing modes with the addition of a 8-bit or 16-bit constant. The following are some examples of these addressing modes

$$PA = \begin{Bmatrix} CS \\ DS \\ SS \\ ES \end{Bmatrix} : \begin{Bmatrix} BX \\ BP \end{Bmatrix} + \begin{Bmatrix} SI \\ DI \end{Bmatrix} + \begin{Bmatrix} \text{8-bit displacement} \\ \text{16-bit displacement} \end{Bmatrix}$$

MOV AL, [BX + SI + disp]
MOV AL, [BX + DI + disp]
MOV AL, [BP + SI + disp]
MOV AL, [BP + DI + disp]

We may substitute DI in the figure below to produce the [BP + DI + disp] addressing mode.

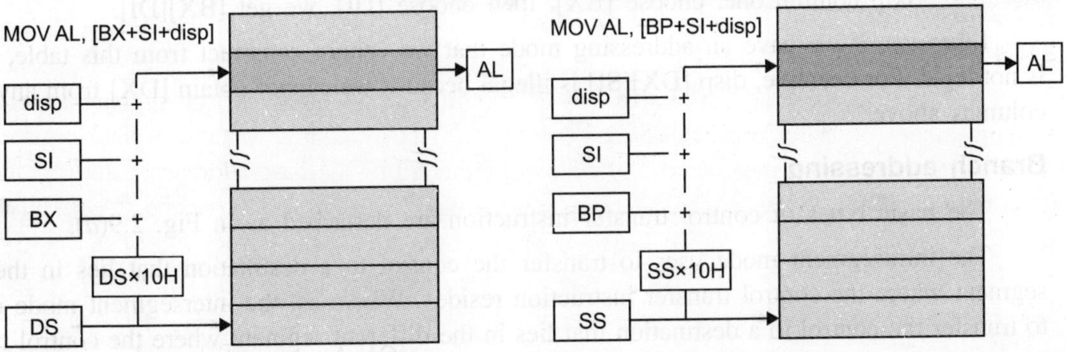

Suppose BP contains 1000H, BX contains 2000H, SI contains 120H, and DI contains 5H. Then MOV AL,10H[BX + SI] loads all from address DS:2130; MOV CH,125H[BP + DI] loads CH from location SS:112AH; and MOV BX,CS:2[BX][DI] loads BX from location CS:2007H.

MASM syntax for 8086

Microsoft's assembler uses several different variations to denote indexed, based/indexed, and displacement plus based/indexed addressing modes. We will see all of these forms used interchangeably throughout this text. The following list some of the possible combinations that are legal for the various 80×86-addressing modes

disp[BX], [BX][disp], [BX + disp], [disp][BX], and [disp + BX], [BX][SI], [BX + SI], [SI][BX] and [SI + BX], disp[BX][SI], disp[BX + SI], [disp + BX + SI], [disp + BX][SI], disp[SI][BX], [disp + SI][BX], [disp + SI + BX], [SI + disp + BX], [BX + disp + SI] etc.

MASM treats the '[ ]' symbols just like the '+' operator. This operator is commutative, just like the '+' operator. Of course, this discussion applies to all the 8086 addressing modes, not just those involving BX and SI. We may substitute any legal registers in the addressing modes above.

Shortcut to the 8086 memory addressing modes

There are a total of 17-different legal memory addressing modes on the 8086 :disp, [BX], [BP], [SI], [DI], disp[BX], disp[BP], disp[SI], disp[DI], [BX][SI], [BX][DI], [BP][SI], [BP][DI], disp[BX]|SI], disp[BX][DI], disp[BP][SI], and disp[BP][DI]. We could memorize all these forms so that we know which are valid (and, by omission, which forms are invalid). However, there is an easier way besides memorizing these 17 forms. Consider the chart:

$$\text{disp} \begin{bmatrix} \text{BX} \\ \\ \text{BP} \end{bmatrix} \begin{matrix} [\text{SI}] \\ [\text{DI}] \end{matrix}$$

If we choose zero or one items from each of the columns and wind up with at least one item, we have got a valid 8086 memory addressing mode. Some examples

- Choose disp from column one, nothing from column 2, [DI] from column 3, we get disp[DI].
- Choose disp , [BX], and [DI], we get disp[BX][DI].

- Skip column one & two, choose [SI], we get [SI].
- Skip column one, choose [BX], then choose [DI], we get [BX][DI].

Likewise, if we have an addressing mode that we cannot construct from this table, then it is not legal. For example, disp [DX][SI] is illegal because we cannot obtain [DX] from any of the columns above.

Branch addressing

The basic types of control transfer instruction are described as in Fig. 2.9(*d*).

The intrasegment mode uses to transfer the control to a destination that lies in the same segment where the control transfer instruction resides. Where as, the intersegment mode is used to transfer the control to a destination that lies in the different segment where the control transfer instruction lies.

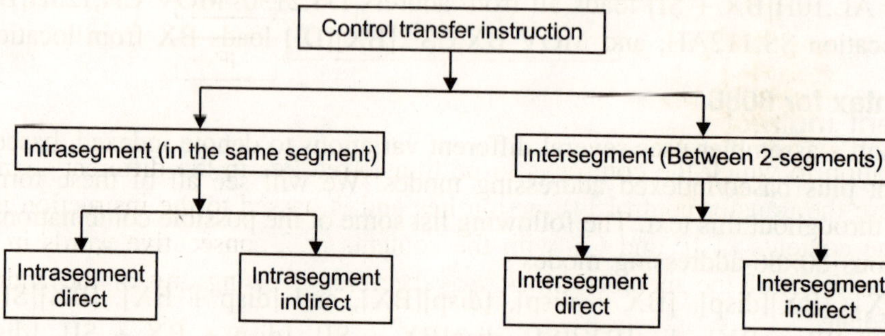

Fig. 2.9 (*d*) Control transfer instructions

Intrasegment direct

In the intrasegment direct mode, the destination location to which the control is transferred lies in the same segment where the control transfer instruction lies and appears directly in the instruction as an immediate displacement value. The displacement is relative to the contents of the IP. The effective address too which the control is transferred is given as

8- or 16-bit displacement + contents of the IP.

In case of the jump, if the signed displacement is of 8-bits ($-128 < d < +127$), it is called short jump. Where as if the signed displacement is of 16-bit ($-32768 < d < +32767$), it is called long jump.

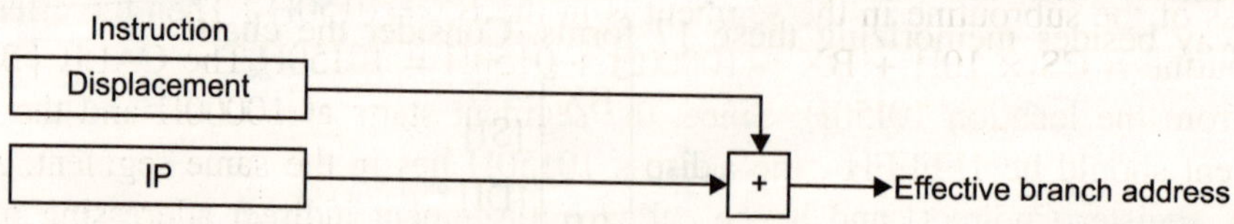

Intrasegment indirect

In this mode, the control to be transferred lies in the same segment where the control instruction lie and it is passed indirectly to the instruction. It uses unconditional branch instructions. The effective branch address is that of register or memory location that is accessed using any of the above data related addressing modes except the immediate mode. The contents of the IP is replaced by the effective branch address.

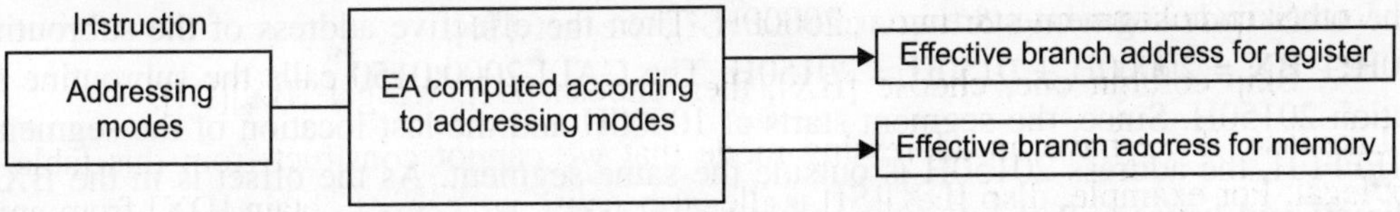

Intersegment direct

The location to which the control is to be transferred lies in the different segment than in which the control transfer instruction lies, is called intersegment. This addressing mode provides facility of branching from one segment to the other segment. The CS and IP specifies the destination address directly in the instruction. It replaces the contents of IP with the part of the instruction and the contents of CS with another part of the instruction.

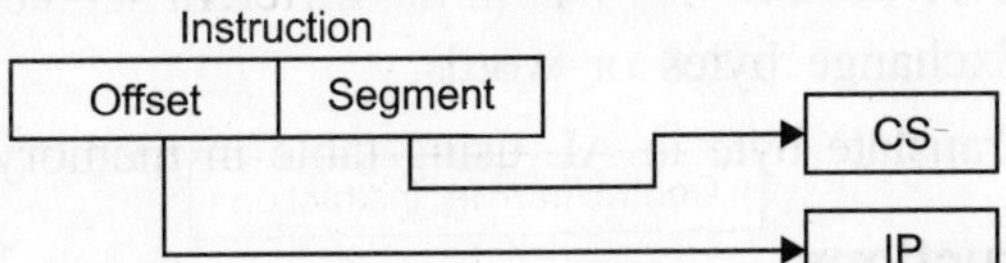

Intersegment indirect

The location to which the control is to be transferred lies in the different segment than the segment where the transfer control instruction lies and is passed to the instruction indirectly i.e., it replaces the contents of IP and CS with the contents of 2-consecutive words in the memory. The starting address of the memory block may be referred using any of the addressing mode except immediate.

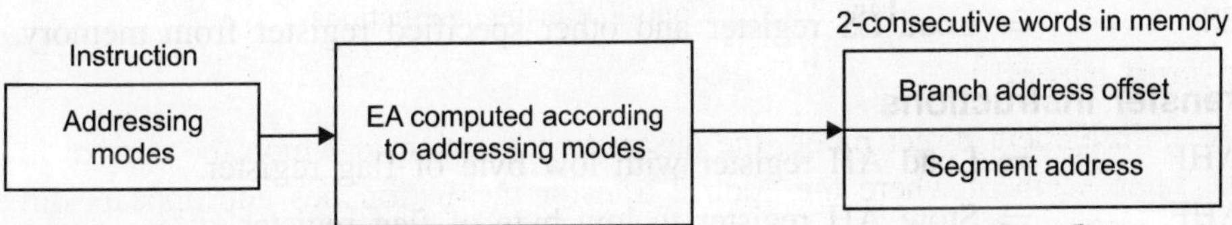

Example

How a subroutine in the same segment where the main program resides can be called? Explain it.

Let us assume that the code segment with CS = 1000H contains the main program. If the offset address of the subroutine in the segment is in the BX = 0150H . Then the effective address of the subroutine = CS × 10H + BX = 10000H + 0150H = 10150H.The CALL {BX} calls the subroutine from the location 10I50H. Since, the segment starts at 10000H and the last location of the segment should be 1FFFFH , the address 10150H lies in the same segment. As the offset is in the BX register (indirect) and hence called intrasegment indirect addressing mode.

Example

How a subroutine in the different segment where the main program resides can be called? Explain it.

Let us assume that the code segment with CS = 1000H contains the main program. If the offset address of the subroutine in the segment is in the BX = 0150H and the subroutine resides in the other code segment starting at 20000H. Then the effective address of the subroutine = CS × 10H + BX = 20000H + 0150H = 20150H. The CALL2000:0150 calls the subroutine from the location 20150H. Since, the segment starts at 10000H and the last location of the segment should be 1FFFFH, the address 20150H is outside the same segment. As the offset is in the BX register (indirect) and hence called intersegment indirect addressing mode.

Data transfer instructions

General Purpose Byte/Word Transfer Instructions

MOV ⇒ Copy byte or word from source to destination.

PUSH ⇒ Copy specified word into stack.

POP ⇒ Copy word from top of stack to specified.

XCHG ⇒ Exchange bytes or words.

XLAT ⇒ Translate byte in AL using table in memory.

I/O port transfer instructions

IN ⇒ Copy a byte or word from specified port to accumulator.

OUT ⇒ Copy a byte or word from accumulator to specified port.

Special address transfer instructions

LEA ⇒ Load effective address of operand into specified register.

LDS ⇒ Load DS register and other specified register from memory.

LES ⇒ Load ES register and other specified register from memory.

Flag transfer instructions

LAHF ⇒ Load AH register with low byte of flag register.

SAHF ⇒ Store AH register to low byte of flag register.

PUSHF ⇒ Copy flag register to top of stack.

POPF ⇒ Copy word at top of stack to flag register.

Arithmetic instructions

Addition

ADD ⇒ Add specified byte to byte or word to word.

ADC ⇒ Add byte + byte + carry flag or word + word + carry.

INC ⇒ Increment byte or word by unity(1).

AAA ⇒ ASCII adjust after addition.

DAA ⇒ Decimal (BCD) adjust after addition.

Subtraction

SUB ⇒ Subtract byte from byte or word from word.

SBB ⇒ Subtract byte or word and carry flag from byte or word.

DEC ⇒ Decrement byte or word.

NEG ⇒ Negate byte or word (inverts each bit +1).

CMP ⇒ Compare two specified bytes or words.

AAS ⇒ ASCII adjust after subtraction.

DAS ⇒ Decimal (BCD) adjust after subtraction.

Multiplications

MUL ⇒ Multiply byte by byte or word by word unsigned.

IMUL ⇒ Multiply signed byte by byte or signed word by word.

AAM ⇒ ASCII adjust after multiply.

Divisions

DIV ⇒ Divide unsigned word by byte, or unsigned double word by word.

IDIV ⇒ Divide signed word by byte or signed double word by word.

AAD ⇒ ASCII adjust before division.

CBW ⇒ Fill upper byte of word with copies of sign bit of lower byte.

CWD ⇒ Fill upper word of double word with sign bit of lower word.

Bit Manipulation instructions

Logical instructions

NOT ⇒ Invert each bit of a byte or word.

AND ⇒ AND each bit in byte or word with corresponding bit in another byte or word.

OR ⇒ OR each bit in a byte or word with corresponding bit in another byte or word.

TEST ⇒ AND operands to update flags, but don't change operands.

Shift instructions

SHL/SAL ⇒ Shift bit of word or byte left, put 0s in LSBs

SHR ⇒ Shift bits of word or byte right, put 0s in MSBs.

SAR ⇒ Shift bits of word or byte right, copy old MSB into new MSB.

Rotate instructions

ROL ⇒ Rotate bits of byte or word left, MSB to LSB and to CF.

ROR ⇒ Rotate bits of byte or word right, LSB to MSB and to CF.

RCR ⇒ Rotate bits of byte or word right, LSB to CF and CF to MSB.

String instructions

REP ⇒ Repeat following instruction until CX = 0.

REPE/REPZ ⇒ Repeat instruction until CX = 0 or ZF ≠ 1.

REPNE/REPNZ ⇒ Repeat until CX = 0 or ZF = 1.

MOVS/MOVSB/MOVSW ⇒ Move byte or word from one string to another.

COMPS/COMPSB/COMPSW ⇒ Compare two string bytes or words.

SCAS/SCASB/SCASW ⇒ Scan string. Compare a string byte with byte in AL or a string with word in AX.

LODS/LODSB/LODSW ⇒ Load string byte into AL or string word into AX.

STOS/STOSB/STOSW ⇒ Store byte from AL or word from AX into string.

A series of consecutive data in the memory is called **string.** This may be arranged in the form of byte or word. Two parameters are required for referring to a string :

- Starting or end address, and
- Length of the string

The length of the string (number of bytes) is loaded into the count register CX. The contents of instruction pointer and the count register are modified after each iteration till the required condition for processing is further satisfied. The incrementing or decrementing of pointer depends on the contents of the direction flag (DF). For byte operation, the index registers (SI or DI) are updated by 1-byte whereas for word operation, it is updated by 2-bytes. The counter decrements its content by unity for both byte and word.

REP : **Repeat Instruction Prefix** → The REP is used as the prefix to instruction. The instruction to which REP is prefixed, repeats its execution until CX = 0. (after each iteration CX decrements by unity). When CX = 0, the execution proceeds to the next instruction in the sequence. The 2-more options of REPE or REPZ and REPNE or REPNZ are used as prefix to the instructions.

MOVSB/MOVSW : **Move String of Byte or Word** → A string of byte residing in memory locations can be moved to another section of the memory location using this instruction. The source and destination address of the string are stored in SI and DI registers. The source address is computed as [DS]× 10 + [SI] while the destination address is computed as [ES] × 10 + [DI]. Thus, the combination of DS : SI and ES : DI moves a string of byte from location [DS] × 10 + [SI] to the location [ES] × 10 + [DI]. The REP prefix is used with MOVS instruction to repeat it by value given in the count register CX. No flags are affected after this instruction. If DS = 0, the index register increments and if DS = 1, the index register decrements the value after each iteration.

CMPS: **Compare String of Byte** → The CMPS instruction compares 2-string of data bytes stored at 2-different location of the memory. The length of the string must be loaded in the CX register. When both strings to be compared become equal, zero flag is set and CX = 0.

SCAS : **Scan String Byte or String Word** → This instruction scans a string of byte or word for an operand byte or word specified in the AX or AL. The ES : DI points the string. The length of the string is stored in the CX. The DF controls incrementing or decrementing the sequence of scanning. When the length of string is matched, CX = 0, zero flag is set and execution stops. The REPNZ prefix is used for SCAS instruction.

LODS: **Load String of Byte or Word** → This instruction loads the AX/AL from the memory location pointed by the combination of DS . SI register pair. The SI is modified after each iteration depending upon the condition of the DF. No other flags are affected.

STOS: **Store String Byte or Word** → This instruction stores the AX/AL register to the memory pointed by the combination of ES : DI. No flags are affected.

Program execution transfer instructions

Unconditional transfer instructions

CALL	⇒ Call a subprogram, save return address on stack.
RET	⇒ Return from subprogram to main program
JMP	⇒ Jump to specified address to get next instructions.

Conditional transfer instructions

JA/JNBE	⇒ Jump if above/jump if not below or equal.
JAE/JNB	⇒ Jump if above or equal/jump if not below.
JB/JNAE	⇒ Jump if below/jump if not above or equal.
JBE/JNA	⇒ Jump if below or equal/jump if not above.
JC	⇒ Jump if carry flag (CF) = 1.
JE/JZ	⇒ Jump if equal/jump if zero flag (ZF) = 1.
JG/JNLE	⇒ Jump if greater/jump if neither less than nor equal.
JGE/JNL	⇒ Jump if greater than or equal/jump if not less than.
JL/JNGE	⇒ Jump if less than/jump if not greater than nor equal.
JLE/JNG	⇒ Jump if less than or equal/jump if not greater than.
JNC	⇒ Jump if no carry i.e., CF = 0.
JNE/JNZ	⇒ Jump if not equal/jump if not zero (zero flag = 0)
JNO	⇒ Jump if overflow (jump if overflow flag = 0)
JNP/JPO	⇒ Jump if not parity/jump if parity odd (PF = 0).
JNS	⇒ Jump if no sign (jump if sign flag = 0).
JO	⇒ Jump if overflow flag = 1.
JP/JPE	⇒ Jump if parity/jump if parity even (PF = 1).
JS	⇒ Jump if sign flag = 1.

Iteration control instruction

These instructions can be used to execute a series of instruction some number of times. Here mnemonics separated by a "/" represent the same instruction. Use the one that best fits the specific application.

LOOP	⇒ Loop through a sequence of instructions until CX = 0.
LOOPE/LOOPZ	⇒ Loop a sequence of instruction while zero flag = 1 and CX ≠ 0.
LOOPNE/LOOPNZ	⇒ Loop a sequence of instructions while zero flag = 0 and CX ≠ 0.
JCXZ	⇒ Jump to specified address if CX = 0.

If you are not tired of instruction, continue skimming through the rest of the list.

INT ⇒ Interrupt program execution call service procedure.

INTO ⇒ Interrupt program execution if overflow flag = 1.

IRET ⇒ Return from interrupt service procedure to main program.

Processor control instructions

Flag set/clear instructions

STC ⇒ Set carry flag (CF) = 1.

CLC ⇒ Clear carry flag (CF) to 0.

CMC ⇒ Complement the state of CF.

STD ⇒ Set direction flag (DF) to 1 (decrement string pointer)

CLD ⇒ Clear direction flag (DF) to 0.

STI ⇒ Set interrupt enable flag to 1 (enable INTR input).

CLI ⇒ Clear interrupt enable flag to 0 (disable INTR input).

External hardware synchronization instruction

HLT ⇒ Halt (do nothing) until interrupt or reset.

WAIT ⇒ Wait until signal on the $\overline{\text{TEST}}$ pin is low.

ESC ⇒ Escape to external coprocessor such as 8087 or 8089.

LOCK ⇒ An instruction prefix. Prevents another processor from taking bus control while adjacent instruction executes.

NOP ⇒ No action except fetch and decode.

Immediate Mode

Examples

```
MOV destination, source
MOV Reg,data
MOV AX, 0000H          ; Copy 0000H into AX
MOV BH,FFH             ; Copy FFH into BH
MOVDX,FFFFH            ; Copy FFFFH into DX
MOV BP, Base address   ; Copy value of Base Address into BP
```

The later microprocessor that remained in the scene of the PCs was the 80286.

Direct mode

Register Operands

- MOV reg., reg.
- MOV mem., reg.
- MOV mem., data

Examples

MOV DX,[2000H] ; Copy contents of 2000H to DL and 2001H to DH .

MOV [2000H],AH ; Copy contents of AH into 2000H location.

MOV BP,[temp] ; Copy contents of memory pointed by temp into BP.

Indirect mode

Register Indirect Addressing

MOV [mem reg. pointer], reg.

MOV reg., [mem reg. pointer]

Examples

MOV CL, [BX] ; Copy memory contents pointed by BX into CL.

We can use the segment override prefix symbols if we wish to access data in different segments. The following instructions demonstrate the use of these overrides

MOV AL,CS:[BX]

MOVAL,DS:[BP]

MOV AL,SS:[SI]

MOV AL,ES:[DI]

Intel refers to [BX] and [BP] as base addressing modes and BX and BP as base registers (in fact, BP stands for base pointer). Intel refers to the [SI] and [DI] addressing modes as indexed addressing modes (SI stands for source index, DI stands for destination index). However, these addressing modes are functionally equivalent. This text will call these forms of register indirect modes to be consistent. DX; Copy contents of DX into memory pointed by SI and SI + 1 i.e.,

$$DL \Rightarrow [SI], DH \Rightarrow [SI + 1]$$

Direct (Displacement) and index mode

Adding a constant to it and the contents of the index register, which can be BX or BP, can form the displacement. Thus,

MOVCL,[BX + SI–2] ; Copy contents of memory pointed by BX + SI – 2 into CL.

MOV[SI + indent], DX ; Copy contents of DX into memory pointed by SI + indent where
; indent is defined constant.

MOV[BP + 2],CX ; Copy CX into memory pointed by BP + 2.

MOV AX,[BP + SI] ; Copy word in memory pointed by BP + SI i.e., [BP + SI]
⇒ AL and ; [BP + SI + 1] ⇒ AH.

2.10 CLOCK GENERATOR (8284A)

It is well known to us that the 8086 does not have on chip clock generator and an external clock generator in IC form or discrete component form should be connected to it. The 8284A, a clock generator, is an ancillary component to the 8086/ 8088/8089. Without IC form of the clock generator 8284A, many additional circuits would have been required to generate the clock (CLK) for an 8086/ 8088-based system. The 8284A provides additional basic functions or signals such as : clock generation, RESET Synchronization, READY Synchronization, and a TTL level peripheral clock signal. Figure 2.10(*a*) illustrates the pinout of the 8284A-clock generator. Figure 2.10(*b*) illustrates connection of reset and clock signals to 8086 and clock generator 8284. It is clear from Fig. 2.10(*b*) that a 15MHz crystal frequency is converted to 5MHz clock by the clock generator 8284A.

Pin functions of 8284A

The 8284A is a 18 pin integrated circuit designed specifically for use with the 8086/ 8088 microprocessor. The definitions and its function are

Address enable

Pins are provided to qualify the bus ready signals, RDY1 and RDY2 , respectively. The use of these 2 pins, which are used to cause wait states, along with the RDY1 and RDY2 inputs. Wait states are generated by the READY pin of the 8086/8088 microprocessor, which is controlled by these two inputs.

RDY1 and RDY2 : Bus Ready inputs are provided in conjunction with are $\overline{\text{AEN1}}$ and $\overline{\text{AEN2}}$ pins to cause wait states in an 8086/8088-based system.

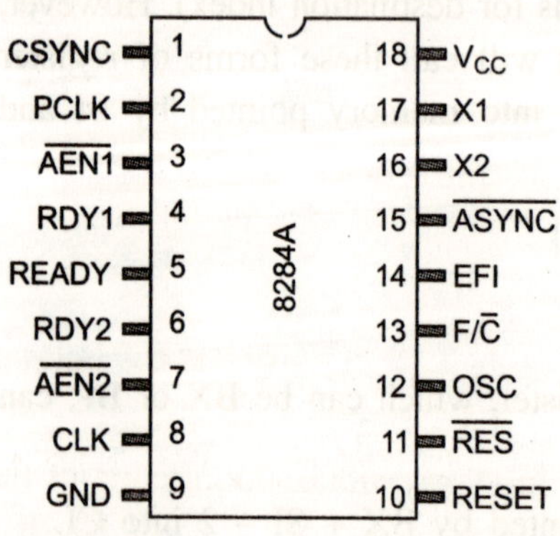

Fig. 2.10 (*a*) Pinouts of the 8284A

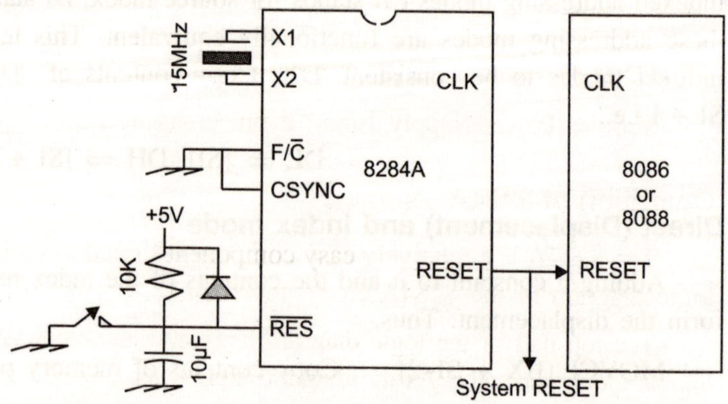

Fig. 2.10 (*b*) CLK generator 8284 for 8086/88

$\overline{\text{ASYNC}}$ ⇒ Ready Synchronization Select : an input used to select either one or two stages of synchronization for the RDY1 and RDY2 inputs.

READY : This output pin connects to the 8086 READY input. This signal is synchronization with the RDYl and RDY2 inputs.

X1 and X2 ⇒ Crystal Inputs : pins connect to an external crystal used as the timing source for the clock generator and all its functions.

$F/\overline{C}$ ⇒ Frequency/Crystal : selects the clocking source for the 8284A. If this pin is held high, an external clock is provided to the EFI inputs pin, and if it is held low, the internal crystal oscillator provides the timing signal.

CLK ⇒ Clock : an output pin that provides the CLK input signal to the 8086/8088 microprocessor and other components in the system. The CLK pin has an output signal that is l/3rd of the crystal of EFI input frequency and has a 33% duty cycle, which is required by the 8086/ 8088.

EFI ⇒ External Frequency Input : an input used when the $F/\overline{C}$ pin is pulled high. EFI supplies the timing whenever the $F/\overline{C}$ pin is high.

PCLK ⇒ Peripheral Clock : a signal that is one sixth of the crystal or EFI input frequency and has a 50% duty cycle. The PCLK output provides a clock signal to the peripheral equipment in the system.

OSC ⇒ Oscillator Output : a TTL level signal that is at the same frequencies of the generators in some multiple processor system.

$\overline{\textbf{RES}}$ ⇒ Reset Input : the active-low reset input to the 8284A. The $\overline{RES}$ pin is often connected to a RC network that provides power- on resetting.

RESET ⇒ Reset Output: the signal connected to the 8086/ 8088 RESETs input pin.

CSYNC ⇒ Clock Synchronization : a pin used whenever the EFI input provides synchronization in system with multiple processors. If the internal crystal oscillator is used, this pin must be grounded.

GND ⇒ Ground : a pin must be grounded.

V_{CC} ⇒ Power Supply Input: a pin connected to +5V with a tolerance of ±10%.

Operation of 8284A

The 8284A is a relatively easy component to understand. Figure 2.10(*c*) illustrates the internal logic diagram of the 8284A-clock generator.

The top half of the logic diagram represents the clock and reset synchronization section of the 8284A clock generator. As the diagram shows, the crystal oscillator has 2 inputs : X1 and X2. If a crystal is attached to X1 and X2 , the oscillator generates a square-wave signal at the same frequency as the crystal. The square-wave signal is fed to a AND gate and also an inverting buffer that provides the OSC output signal. The OSC signal is sometimes used as an EFI input to other 8284A.

A close inspection of the AND gate reveals that when $F/\overline{C}$ is a logic-0, the oscillator output is steered through to divide by 3 counter. If $F/\overline{C}$ is a logic-1, then EFI is steered through to the counter.

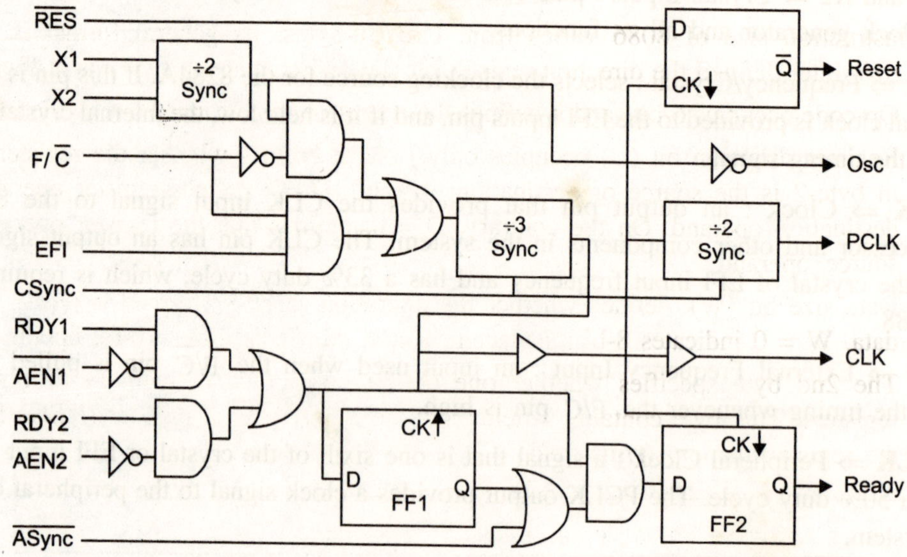

Fig. 2.10 (*c*) Block diagram of 8284 clock generator

The output of the divide by 3 counter generates the timing for ready synchronization, a signal for another counter (divide by 2), and the CLK. signal to the 8086/ 8088 microprocessor. The CLK signal is also buffered before it leaves the clock generator. Notice that the output of the first counter feeds the second. These two cascaded counters provide the divide by 6 output at PCLK, the peripheral clock output. Figure 2.10 (*b*) shows how an 8284A is a connected to the 8086/8088. Notice that F/C and CSYNC are grounded to select the crystal oscillator, and that a 15MHz crystal provides the normal 5MHz clock signal to the 8086/ 8088 as well as a 2.5MHz peripheral clock signal.

Reset section

The reset section of the 8284A is very simple. It consists of an Schmitt trigger buffered single D type flip-flop (DFF) circuit. The DFF ensures that the timing requirements of the 8086/ 8088 RESET input are met. This circuit applies the RESET signal to the microprocessor on the negative edge (l-to-0 transition) the clocks. Therefore, this circuit meets the timing requirements of the 8086/ 8088.

Notice in Figure 2.10 (*c*) that a RC circuit provides a logic-0 to $\overline{\text{RES}}$ input pin when power is first applied to the system. After a short time, RES input becomes a logic-1 because the capacitor charges towards ±5V through the register. A push button switch allows the microprocessor to be reset by the operator. Correct reset timing requires the Figure 2.10 (*c*). The clock generator (8284A) and the 8086 or 8088 microprocessor illustrating the connection for the clock and reset signals. A 15MHz crystal provides the 5MHz clock for the microprocessor. RESET input to become a logic-1 no later than four clock after system power is applied and to be held high for atleast 50μs.

2.11 INSTRUCTION FORMAT OF 8086

The instruction size of 8086 varies from 1 to 6 -bytes. Its general format is given in Figure 2.11(*a*). Intel defined the direction register bit (D) and data size bit (W) as in Fig. 2.11(*a*).

- Op-code occupies 6-bits and it defines the operation to be carried out by the instruction,
- Register direction bit (D) occupies only 1-bit. It defines whether the register operand in byte-2 is the source or destination operand. D = 1 defines register operand is the destination operand. On the contrary, D = 0, defines that the register operand is the source operand.
- Data size bit (W), defines whether the operation to be performed is on 8-, or 16-bit data. W = 0 indicates 8-bit displacement and W = 1 specifies 16-bit operation.
- The 2nd byte specifies whether one of the operand is in memory or both are in registers. This byte contains 3-fields : Mode (**MOD**) field, register (**REG**) field, and the register/memory (**r/m**) field.

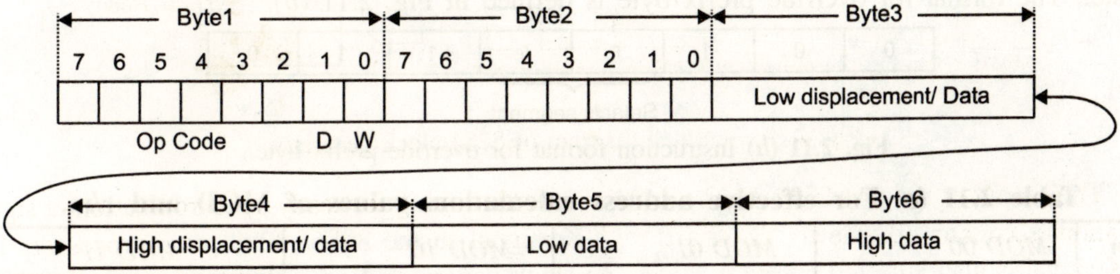

Fig. 2.11 (*a*) 8086 instruction format

The MOD is interpreted as

00 Memory addressing without displacement

01 Memory addressing with 8-bit displacement

10 Memory addressing with 16-bit displacement

11 Register addressing with W = 0 for 8-bit and W = 1 for 16-bit data.

The REG field occupies 3 bits. It defines the register for the 1st operand that is specified as the source or destination by the D bit (in byte 1). Definition of REG with W is explained in Table 2.11(*a*). The segment selection is given in Table 2.11(*b*).

Table 2.11 (a) Definition of registers with 'W'

REG	W = 0	W = 1
000	AL	AX
001	CL	CX
010	DL	DX
011	BL	BX
100	AH	SP
101	CH	BP
110	DH	SI
111	BH	DI

Table 2.11 (b) Segment selection

rr values	Segment
00	ES
01	CS
10	SS
11	DS

- The r/m field also occupies 3-bits. The **r/m** field along with MOD field defines the 2nd operand in Table 2.11 (*c*).
- Encoding of the r/m field in Table 2.11(c) depends on how the mode field is set. If MOD = 11 (register-to-register mode), then r/m identifies the 2nd register operand. If MOD selects memory mode, then r/m field specifies how the effective address of the memory operand is to be calculated.
- Bytes-3 to 6 of any instruction are optional field that normally contain the displacement values of a memory operand and/ or the actual value of an immediate constant operand.

Segment Override

One segment is used as default segment for every memory addressing as described in Table-2.11(d). In some cases, other than the default segment register is used by selecting a segment override prefix byte. The override prefix byte immediately proceeds the op-code byte of the instruction. In such situation the processor uses the prefix byte to select the particular segment register. The format for override prefix byte is defined in Fig. 2.11 (*b*).

0	0	1	r	r	1	1	0

Selects segment (r r)

Fig. 2.11 (***b***) Instruction format for override prefix byte

Table 2.11 (*c*) For effective address calculation, values of MOD and r/m.

r/m	*MOD 00*	*MOD 01*	*MOD 10*	*MOD 11*	
				W = 0	*W = 1*
000	[BX]+[SI]	[BX]+[SI]+D8	[BX]+[SI]+D16	AL	AX
001	[BX]+[DI]	[BX]+[DI]+ D8	[BX]+[DI]+D16	CL	CX
010	[BP]+[SI]	[BP]+[SI]+ D8	[BP]+[SI]+D16	DL	DX
011	[BP]+[DI]	[BP]+[DI]+ D8	[BP]+[DI]+D16	BL	BX
100	[SI]	[SI]+D8	[SI]+D16	AH	SP
101	[DI]	[DI]+D8	[DI]+D16	CH	BP
110	Direct Addressing	[BP]+D8	[BP]+D16	DH	SI
111	[BX]	[BX]+D8	[BX]+D16	BH	DI

Table 2.11(*d*) Segment register for various memory addressing modes.

r/m	*MOD 00*	*MOD 01*	*MOD 10*	*Segment Register used*
000	[BX]+[SI]	[BX]+[SI]+D8	[BX]+[SI]+D16	DS
001	[BX]+[DI]	[BX]+[DI]+D8	[BX]+[DI]+D16	DS
010	[BP]+[SI]	[BP]+[SI]+D8	[BP]+[SI]+D16	SS
011	[BP]+[DI]	[BP]+[DI]+D8	[BP]+[DI]+D16	SS
100	[SI]	[SI]+D8	[SI]+D16	DS
101	[DI]	[DI]+D8	[DI]+D16	DS
110	D16 Direct Addressing [DS]	[BP]+D8 Stack pointer register [SS]	[BP]+D16 Stack segment register [SS]	DS or SS as in MOD column
111	[BX]	[BX]+D8	[BX]+D16	DS

Data transfer instruction

MOV

Copy byte or word from specified source to specified destination. The destination can be a register or memory and source can be a register or memory or immediate operand (data). Both source and destination can not be memory locations.

MOV reg,8-bit/ 16-bit data

The immediately available 8-bit or 16-bit data is moved to any register. The object code could be 2 to 3-bytes. The encoding format is

011W rrr data byte low data byte high if W = 1

MOV AL,35H

8-bit data movement, W = 0, AL = 000, 8-bit data = 35

MOV AL,35H, Object code	= 10110000 35	= B0,35
MOV CL,35H, Object code	= 1011 0001 35	= B1,35
MOV DL,35H, Object code	= 1011 0010 35	= B2,35
MOV BL,35H, Object code	= 1011 0011 35	= B3,35
MOV AH,35H, Object code	= 1011 0100 35	= B4,35
MOV CH,35H, Object code	= 1011 0101 35	= B5,35
MOV DH,35H, Object code	= 1011 0110 35	= B6,35
MOV BH,35H, Object code	= 1011 0111 35	= B7,35
MOV AX,2535H, Object code	= 1011 1000 2535	= B8,35,25
MOV CX,2535H, Object code	= 1011 1001 2535	= B9,35,25
MOV DX,2535H, Object code	= 1011 1010 2535	= BA,35,25
MOV BX,2535H, Object code	= 1011 1011 2535	= BB,35,25
MOV SI,2535H, Object code	= 1011 1110 2535	= BE,35,25

Example

0100	B0,35	MOV AL,35
0102	F4	HLT

Result is 35H in AL register

0100	B8,35,25	MOV AX,2535
0103	F4	HLT

Result in AX i.e., (AL) = 35H and (AH) = 25H.

MOV mem/reg, data

This instruction is not normally used for MOV reg,data because this instruction is separately available. Its object code could be 3 to 6-bytes. The encoding format is

1100 011 W mod reg r/m data high data if W = 1

This is direct memory addressing. Hence, mod = 00, reg = 000 and r/m = 110. Substituting these encoding format results object code as

Object code = 1100 011 W 00 000 110 8-bit/16-bit data

(*a*) MOV [0200], 35. Here, W = 0

Object code = 1100 0110 00 000 110, 35 = C6,06,00,02,35

MOV [0200], 2535, (W = 1)

Object code = 1100 0111 00 000 110 2535 = C7,06,00,02,35,25

(*b*) MOV [BX], 16-bit data, (W = 1)

Object code = 1100 0111 00 000 111, 16-bit data = C7,07, 16-bit data

MOV [BX], 8-bit data (W = 0)

Object code = 1100 0110 00 000 111, 8-bit data = C6,07, 8-bit data

(*c*) MOV [BX]+D8, 16-bit data (W = 1)

Object code = 1100 0111 01 000 111, D8, 16-bit data = C7,47,D8, 16-bit data

MOV [BX]+D8, 8-bit data (W = 0)

Object code = 1100 0110 01 000 111, D8, 8-bit data = C6,47,D8, 8-bit data

(*d*) MOV [BX]+D16, 16-bit data (W = 1)

Object code = 1100 0111 1000 0111, D16, 16-bit data = C7,87,D16, 16-bit data

MOV [BX]+D16, 8-bit data (W = 0)

Object code = 1100 0110 1000 0111, D16, 8-bit data = C6,87,D16, 8-bit data

(*e*) MOV [BX+SI], 16-bit data (W = 1)

Object code = 1100 0111 0000 0000, 16-bit data = C7,00, 16-bit data

MOV [BX+SI], 8-bit data (W = 0)

Object code = 1100 0110 0000 0000, 8-bit data = C6,00, 8-bit data

(*f*) MOV [BX+DI], 16-bit data (W = 1)

Object code = 1100 0111 0000 0001, 16-bit data = C7,01, 16-bit data

MOV [BX+DI], 8-bit data

Object code = 1100 0110 0000 0001, 8-bit data = C6,01, 8-bit data (*g*) MOV [BP+SI], 16-bit data

Object code = 1100 0111 0000 0010, 16-bit data = C7,02, 16-bit data

MOV [BP+SI], 8-bit data

Object code = 1100 0110 0000 0010, 8-bit data = C6,02, 8-bit data

(*h*) MOV [BP+DI], 16-bit data

Object code = 1100 0111 0000 0011, 16-bit data = C7,03, 16-bit data

MOV [BP+DI], 8-bit data

Object code = 1100 0110 0000 0011, 8-bit data = C6,03, 8-bit data

(*i*) MOV [SI], 16-bit data

Object code = 1100 0111 0000 0100, 16-bit data = C7,04, 16-bit data

MOV [SI], 8-bit data

Object code = 1100 0110 0000 0100, 8-bit data = C6,04, 8-bit data

(*j*) MOV [DI], 16-bit data

Object code = 1100 0111 0000 0101, 16-bit data = C7,05, 16-bit data

MOV [DI], 8-bit data

Object code = 1100 0110 0000 0101, 8-bit data = C6,05, 8-bit data

(*k*) MOV [BX+SI]+D16, 16-bit data

Object code = 1100 0111 1000 0000, D16, 16-bit data = C7,80,D16, 16-bit data

MOV [BX+SI]+D16, 8-bit data

Object code = 1100 0110 1000 0000, D16, 8-bit data = C6,80,D16, 8-bit data

(*l*) MOV [BX+DI]+D16, 16-bit data

Object code = 1100 0111 1000 0001, D16, 16-bit data = C7,81,D16, 16-bit data

MOV [BX+DI]+D16, 8-bit data

Object code = 1100 0110 1000 0001, D16, 8-bit data = C6,81,D16, 8-bit data

(*m*) MOV [SI]+D16, 16-bit data

Object code = 1100 0111 1000 0100, D16, 16-bit data = C7,84,D16, 16-bit data

MOV [SI]+D16, 8-bit data

Object code = 1100 0110 1000 0100, D16, 8-bit data = C6,84,D16, 8-bit data

MOV [SI]+D8, 16-bit data

Object code = 1100 0111 0100 0100, D8, 16-bit data = C7,44,D8, 16-bit data

MOV [SI]+D8, 8-bit data

Object code = 1100 0110 0100 0100, D8, 8-bit data = C6,44,D18, 8-bit data (*n*) MOV [DI]+D16, 16-bit data

Object code = 1100 0111 1000 0101, D16, 16-bit data = C7,85,D16, 16-bit data

MOV [DI]+D16, 8-bit data

Object code = 1100 0110 1000 0101, D16, 8-bit data = C6,85,D16, 8-bit data

MOV [DI]+D8, 16-bit data

Object code = 1100 0111 0100 0101, D8, 16-bit data = C7,45,D8, 16-bit data

MOV [DI]+D8, 8-bit data

Object code = 1100 0110 0100 0101, D8, 8-bit data = C6,45,D8, 8-bit data

MOV Acc,mem

Move 8-bit or 16-bit data from memory to AL or AX. The object code is of 3-bytes. Its encoding format is

1010 000W low addr high addr

MOV AX,[0106], Object code = 1010 0001, 06, 01 = A1, 06, 01

MOV mem/reg1,meme/reg2

Move data from register to register

Memory to register

Register to memory

It refers to segment registers. The object code is 2 to 4-bytes. Its encoding format is

1000 10 dW mod reg r/m

MOV AX,CX

16-bit operation, W = 1, CX source = reg = 001, d = 0, AX decides mod = 11, r/m = 000

MOV AX,CX

16-bit operation, W = 1, CX source reg, d = 1, AX = reg. = 000, CX decides mod = 11, r/m = 001

	d = 0	d = 1
MOV AX,CX , Object code =	1000 1001 11 001 000 = 89, C8	1000 1011 11 000 001 = 8B, Cl
MOV AX,DX, Object code =	1000 1001 11 010 000 = 89, D0	1000 1011 11 000 010 = 8B, C2
MOV AX,BX, Object code =	1000 1001 11 011 000 = 89, D8	1000 1011 11 000 011 = 8B, C3
MOV CX,AX , Object code =	1000 1001 11 000 001 = 89, C1	1000 1011 11 001 000 = 8B, C8
MOV CX,DX, Object code =	1000 1001 11 010 001 = 89, D1	1000 1011 11 001 010 = 8B, CA
	d = 0	d = 1

MOV CX,BX, Object code = 1000 1001 11 011 001 = 89, D9

MOV DX,AX, Object code = 1000 1001 11 000 010 = 89, C2

MOV DX,CX, Object code = 1000 1001 11 001 010 = 89, CA

MOV DX,BX, Object code = 1000 1001 11 011 010 = 89, DA

MOV BX,AX, Object code = 1000 1001 11 000 011 = 89, C3

MOV BX,CX, Object code = 1000 1001 11 001 011 = 89, CB

MOV BX,DX, Object code = 1000 1001 11 010 011 = 89, D3

MOV AL,CL, Object code = 1000 1000 11 001 000 = 88, C8

MOV AL,DL, Object code = 1000 1000 11 010 000 = 88, D0

MOV AL,BL, Object code = 1000 1000 11 011 000 = 88, D8

MOV CL,AL, Object code = 1000 1000 11 000 001 = 88, C1

MOV CL,DL, Object code = 1000 1000 11 010 001 = 88, D1

MOV CL,BL, Object code = 1000 1000 11 011 001 = 88, D9

MOV DL,AL, Object code = 1000 1000 11 000 010 = 88, C2

MOV DL,CL, Object code = 1000 1000 11 001 010 = 88, CA

MOV DL,BL, Object code = 1000 1000 11 011 010 = 88, DA

MOV BL,AL, Object code = 1000 1000 11 000 011 = 88, C3

MOV BL,CL, Object code = 1000 1000 11 001 011 = 88, CB

MOV BL,DL, Object code = 1000 1000 11 010 011 = 88, D3

1000 1011 11 001 011 = 8B, CB

1000 1011 11 010 000 = 8B, D0

1000 1011 11 010 001 = 8B, D1

1000 1011 11 010 011 = 8B, D3

1000 1011 11 011 000 = 8B, D8

1000 1011 11 011 001 = 8B, D9

1000 1011 11 011 010 = 8B, DA

1000 1010 11 000 001 = 8A, C1

1000 1010 11 000 010 = 8A, C2

1000 1010 11 000 011 = 8A, C3

1000 1010 11 001 000 = 8A, C8

1000 1010 11 001 010 = 8A, CA

1000 1010 11 001 011 = 8A, CB

1000 1010 11 010000 = 8A, D0

1000 1010 11 010 001 = 8A, D1

1000 1010 11 010 011 = 8A, D3

1000 1010 11 011000 = 8A, D8

1000 1010 11 011001 = 8A, D9

1000 1010 11 011010 = 8A, DA

Example

```
0100    BB,33,22    MOV BX,2233H
0103    8B,C3       MOV AX,BX
0105    F4          HLT
```

Result in AX = 2233H

MOV AX,[BX]

16-bit operation, W = 1, AX destination reg. =000, d = 1, mod and r/m field is as per the 1st operand [BX]. Mod = 00, and r/m = 111.

MOV AX,[BX], Object code = 1000 1011 00 000 111 = 8B, 07

MOV CX,[BX], Object code = 1000 1011 00 001 111 = 8B, 0F

MOV DX,[BX], Object code	= 1000 1011 00 010 111 = 8B, 17
MOV AX,[0105], Object code	= 1000 1011 00 000 110 = 8B, 06
MOV CX,[0105], Object code	= 1000 1011 00 001 110 = 8B, 0E
MOV DX,[0105], Object code	= 1000 1011 00 010 110 = 8B, 16
MOV BX,[0105], Object code	= 1000 1011 00 011 110 = 8B, 1E
MOV [BX],CX, Object code	= 1000 1001 00 001 111 = 89, 0F
MOV [BX],AX, Object code	= 1000 1001 00 000 111 = 89, 07
MOV [BX],DX, Object code	= 1000 1001 00 010 111 = 89, 17
MOV [0108],CX, Object code	= 1000 1001 00 001 110 = 89, 0E
MOV [0108],AX, Object code	= 1000 1001 00 000 110 = 89, 06
MOV [0108],DX, Object code	= 1000 1001 00 010 110 = 89, 16
MOV [0108],BX, Object code	= 1000 1001 00 011 110 = 89, 1E

Example

```
0100    BB,08,02             MOV BX,0208
0103    C7,07,55,44          MOV[BX],4455H
0107    F4                   HLT
```

Result in 0208 = 55H and 0209 = 44H

```
0100    C7,06,03,02,55,44    MOV [0203],4455H
0106    F4                   HLT
```

Result in 0203 = 55H and 0204 = 44H

```
0100    C6,06,03,02,55       MOV [0203],55H
0105    F4                   HLT
```

Result in 0203 = 55H

```
0100    BB,03,02             MOV BX,0203
0103    C6,07,55             MOV[BX],55H
0106    F4                   HLT
```

Result in 0203 = 55H

```
0100    BB,03,02             MOV BX,0203
0103    C6,47,08,55          MOV[BX+08],55H
0107    F4                   HLT
```

Result in 020B = 55H

0100	BB,03,02	MOV BX,0203
0103	C7,47,08,55,44	MOV[BX+08],4455H
0108	F4	HLT

Result in 020B = 55H and 020C = 44H

0100	BB,03,02	MOV BX,0203
0103	C6,87,0A,02,55	MOV[BX+020A],55H
0108	F4	HLT

Result in 040D = 55H

MOV Seg reg,mem/reg

Contents of memory or register are transferred to segment register. The object code is 2 to 4-byte long. Its encoding format is

1000 1100 mod 0 rr r/m

MOV SS,DX, Object code = 1000 1100, 11 010 010 = 8C, D2

MOV SS,[BX], Object code = 1000 1100, 00 010 111 = 8C, 17

MOV mem/reg,Seg reg

Contents of segment register is transferred to memory or register. The object code is 2 to 4-byte long. Its encoding format is

1000 1110 mod 0 rr r/m

MOV [BX],DS, Object code = 1000 1110, 00 011 111 = 8C, 1F

PUSH

Copy the specified word at the top of the stack. The specified word can be a general purpose register or a segment register or a memory location. The SP is automatically decremented by 2.

PUSH reg

The 16-bit immediate data pushed at the top of stack. This instruction is used to save data of all registers. This is single byte instruction. The encoding format is

010 10 rrr

PUSH AX, Object code = 010 10 000 = 50

PUSH CX, Object code = 010 10 001 = 51

PUSH DX, Object code = 010 10 010 = 52

PUSH BX, Object code = 010 10 011 = 53

PUSH mem/reg

This is 2 to 4-byte instruction. The encoding format is

1111 1111 mod 110 r/m

PUSH [BX]+D8

The content of memory location+8-bit displacement is pushed to the top of the stack. [BX+D8] decides the mod = 01, r/m = 111

Object code = 1111 1111, 0111 0111 = FF, 77

PUSH 0108

Direct memory decides mod = 00, r/m = 110

Object code = 1111 1111, 00 110110 = FF, 36

PUSH segment reg

The content of segment register is pushed on the top of the stack. The encoding format is

000 rr 110

PUSH CS, Object code = 000 01 110 = 0E

PUSHF

The content of Flag register is pushed on the top of the stack. The encoding format is

1001 1100

PUSHF, Object code = 1001 1100 = 9C

POP

It is inverse of PUSH. Copy the specified word from the top of the stack to the specified destination. The destination can be a general purpose register or a segment register (except CS) or a memory location. The SP is automatically incremented by 2.

POP mem/reg

The content of flags pushed on the stack. The encoding format is

1000 1111 mod 000 r/m

POP [BX]+D8

The content of the top of the stack with 8-bit displacement is retrieved.

Object code = 1000 1111 01 000 111 = 8F, 47

POP reg

The 16-bit data from top of the stack is moved to any 16-bit reg.

010 11 rrr

POP AX, Object code = 010 11 000 = 58

POP Seg reg

The 16-bit data from top of the stack is moved to any 16-bit segment register.

000 rr 111

POP CS, Object code = 000 01111 = 0F

POPF

The 16-bit data from top of the stack is moved to flags.

1001 1101

POPF, Object code = 1001 1101 = 9D

EXCHG reg/mem

Exchange the byte or word with the contents of the accumulator with either contents of register or memory location except segment register. Its encoding format is

1000 011 W mod reg r/m

XCHG DX,AX

Object code = 1000 0111, 11 000 010 = 87,C2

XCHG AL,[BX]

Object code = 1000 0110, 00 000 111 = 86,07

XLAT

This instruction replaces a byte in AL register with a byte from the top of the lookup table. It adds the 8 bit data of AL in the (BX) register. The BX register contains the starting offset address of the lookup table. After execution, the data from corresponding location of the lookup table is loaded into the accumulator. Its encoding format is

1101 0111

Object code = 1101 0111 = D7

IN Acc, DX

The IN instruction copies a byte or a word from the specified port to the accumulator. Input the data from port whose 16 bit address is in DX register to the accumulator. It is one byte instruction. Its encoding format is

1110 110 W

IN AX, DX, Object code = 1110 1101 = ED

IN AL,DX, Object code = 1110 1100 = EC

IN Acc.PORT

Input the data to accumulator from port whose address is 8-bit only. It is one byte instruction. Its encoding format is

1110 010 W 8-bit port addr

IN AX,04H, Object Code = 1110 0101, 04 = E5,04

After execution, (AL) = 08H and (AH) = 00H (Port 04H has data 0008H).

OUT DX, Acc

Output contents of the accumulator to the port whose 16 bit address is in DX register. It is one byte instruction. Its encoding format is

1110 111 W

OUT DX,AC , Object code = 1110 1111 = EF

OUT PORT, Acc

Output contents of the accumulator to the port whose address is only 8 bit. It is 2B instruction. Its encoding format is

1110 011 W 8-bit port addr

OUT 08, AX, Object code = 1110 0111, 08 = E7.08

After execution (AL) = (08H) and (AH) = (09H) (Port 08H has data 0908H).

LEA

Loads the effective address (offset of the physical address) into the specified register of 16-bit. In other words contents of memory location with 16-bit offset are loaded in a 16-bit register addressed by the memory operand The object code is 2B to 4B long. The encoding format is

1000 1101 mod reg r/m

LEA SI,[BX +0203]

Object Code = 8D, 10110111, 0203 = 8D, B7,03,02

Example 1

LEA BX, BPS → means the label BPS has an offset value of 16-bit. This 16-bit offset value will be loaded into the BX register.

Example 2

LEA SI, [BX + 0203] → this instruction loads the offset [BX + 0203H] into the SI register.

LDS/LES (reg, mem)

These instruction loads DS/ES register along with another specified register with DS/ES to hold most significant 16-bits and the specified register to hold the least significant 16-bits from the memory. In other words, in one go these instructions load 32-bits of data to the combination of registers DS/ES along with the specified 16-bit register in the instruction.

Contents of memory location loaded in a specified register and subsequent locations are loaded in DS simultaneously. The encoding format is

C5 mod reg r/m

LDS AX,[BX], Object code = C5, 00 000 111 = C5, 07

Example

0100	BB,01,02	MOV BX, 0201H
0103	C5,07	LDS AX,[BX]
0105	F4	HLT

Memory locations	Data	Memory locations	Data
0201	11H	0203	33H
0202	22H	0204	44H

After execution of these instructions, the combination (DS) : (AX) = 4433H : 2211H

Result (AX) = 2211H, (DS)= 4433H

The register such as DI and ES are loaded with 32-bit pointer from the memory. The object code is 2B to 4B long. The encoding format is

1100 0100 mod reg r/m LES DI,[BX]

Object code = C4, 00 111111 = C4, 3F

LAHF

It loads the lower 8-bits of the flags into the AH. Sometimes it is used to know the status of the conditional operations. Its encoding format is

1001 1111

Object code = 9F

SAHF

It stores the contents of AH to the low byte of the flags. Its encoding format is

1001 1110

Object code = 9E

Arithmetic Instructions

The 8086/8088 perform four types of operations of numbers

- Unsigned binary
- Signed binary
- Unsigned packed BCD
- Unsigned unpacked BCD

ADD

Add byte with byte or word with word. The source operand may be an immediate data, a register or a memory location. Flags affected are AF, CF, OF, PF, SF, ZF.

ADD Acc, data

Add immediately available 8, 16-bit data to accumulator. This instruction is 2B to 3B long and takes 4clock cycles. The encoding format is

0000 010 W data high data if W = 1

ADD AL,25

8-bit operation, W = 0

Object code = 0000 0100,8-bit data = 04,8-bit data = 04,25

ADDAX,2535, W = 1 , Object code = 0000 0101,16-bit data = 05,16-bit data = 05,35,25

ADD mem/reg, data

Add immediately available 8, 16-bit data to memory/Register. This instruction is 2B to 4B long and takes four clock cycles. The encoding format is

100000 SW mod 000 r/m data data if S:W = 01

ADD mem 1/reg 1, mem2/reg2

Add 8, 16 bit content of memory/register to memory/register. The encoding format is

0000 00 dW mod reg r/m

ADD AX,CX

16-bit operation, W = 1, CX = source, d = 0, AX decides mod = 11, r/m = 000, CX = reg = 001

ADD AX,CX, Object code = 0000 0001,11001000 = 01, C8

ADD AX,DX, Object code = 0000 0001,1101 0000 = 01, D0

ADD AX,BX, Object code = 0000 0001,1101 1000 = 01, D8

Example

0100	B8,44,A3	MOV AX,A344H
0103	BB,55,A4	MOV, BX,A455H
0106	01,D8	ADD AX,BX
0107	F4	HLT

Result in AX = 4799H. Here, carry has taken place.

ADD reg,[mem]

ADD CX,[0105]

16-bit operation, W = 1, CX = destination = reg.= 001, d =1, memory decides mod = 00, r/m = 110,

ADD CX,[0105], Object code = 0000 0011, 0000 1110 = 03, 0E

ADD AX,[0105], Object code = 0000 0011, 0000 0110 = 03, 06

ADD DX[0105], Object code = 0000 0011, 0001 0110 = 03,16

ADD CX,[BX], Object code = 0000 0011, 0000 1111 = 03, 0F

ADD CX,[BX]+D8, Object code = 0000 0011, 0100 1111, D8 = 03, 4F, D8

ADD CX,[BX]+D16, Object code = 0000 0011,1000 1111, D6 = 03, 8F, D16

ADD AX,[BX]+D8, Object code = 0000 0011,0100 0111, D8 = 03, 47, D8

ADD AX,[BX]+D16, Object code = 0000 0011,1000 0111, D16 = 03, 87, D16

ADD DX,[BX]+D8, Object code = 0000 0011,0101 0111, D8 = 03, 57, D8

ADD DX,[BX]+D16, Object code = 0000 0011,1001 0111, D16 = 03, 97, D16

ADD [mem],CX

ADD [0108],CX

16-bit operation, W = 1, CX = source reg. = 001, d = 0, memory decides mod = 00, r/m = 110.

ADD [0108], CX, Object code = 0000 0001, 0000 1110 = 01, 0E

ADD [0108], AX, Object code = 0000 0001, 0000 0110 = 01, 06

ADD [0108], BX, Object code = 0000 0001, 0001 1110 = 01, 1E

ADD [0108], DX, Object code = 0000 0001, 0001 0110 = 01, 16

ADD[BX],CX

16-bit operation, W = 1, CX = source reg = 001, d = 0, [BX] decides mod = 00, r/m = 111.

ADD[BX],CX, Object code = 0000 0001,0000 1111 = 01, 0F

ADD[BX+D8],CX, Object code = 0000 0001,0100 1111 = 01, 4F

ADD[BX+D16],CX, Object code = 0000 0001,1000 1111 = 01, 8F

Example

```
C:\WINDOWS>DEBUG
-a : Addition without carry
0F6C:0100    MOV AX,A233
0F6C:0103    MOV BX,A455
0F6C:0106    ADD AX,BX
0F6C:0108    HLT
0F6C:0109
-GCS:0108
AX = 4688  BX = A455  CX = 0000 DX = 0000  SP = FFEE BP = 0000  SI=0000  DI=0000
DS = 0F6C  ES = 0F6C SS = 0F6C CS = 0F6C IP = 0108 OV UP EI PL NZ NA PE CY
0F6C:0108 F4          HLT
```

```
C:\WINDOWS>DEBUG
-A : Addition without carry
0F6C:0100    MOV BX,2233
0F6C:0103    MOV DX,3344
0F6C:0106    MOV CX,A344
0F6C:0109    MOV AX,A455
0F6C:010C    ADD CX,AX
0F6C:010E    ADD DX,BX
0F6C:0110    HLT
0F6C:0111
-GCS:0110
AX = A455 BX = 2233  CX = 4799  DX = 5577  SP = FFEE  BP = 0000  SI=0000  DI=0000
DS = 0F6C ES = 0F6C  SS = 0F6C CS = 0F6C IP = 0110 NV UP EI PL NZ NA PE NC
0F6C:0110 F4              HLT
```

ADC

Byte + byte + carry flag or word + word + carry flag.

ADC Acc, Data

Adds immedi ately available 8-, or 16-bit data with carry to accumulator AL or AX. It is 3-byte long instruction. Its encoding format is

0001 010 W data high data if W = 1

ADC AX,2535H

16-bit operation, W = 1, object code = 0001 0101, 3525 = 15,35,25.

ADC AL,35H

8-bit operation, W = 0, object code = 0001 0100, 35 = 14,35

ADC mem/reg,data

Adds immediately available 8-, or 16-bit data with carry to register or memory contents. It is 3 to 6-byte long instruction. Its encoding format is

1000 00 sW mod 010 r / m data high data if W = 1 S = 0

Here, W = 1 for16-bit operation, S = 0, for 8-bit operation S = 1,

ADC [0108], 2535H

16-bit operation, W = 1, S = 0, memory decides mod = 00, r/m = 110

ADC [0108],2535 Object code = 1000 0 001, 00 010 110,35,25 = 81,16,08,01,35,25

ADC [BX],2535H, Object code = 1000 0 001, 00 010 111,35,25 = 81,17,35,25

ADC [BX]+D8,2535H, Object code = 1000 0 001, 01 010 111, D8,35,25 = 81,57,D8,35,25

ADC [BX]+D16,2535H, Object code = 1000 0 001, 10 010 111,35,25 = 81,97,D16,35,25

ADC mem1/reg1,mem1/reg2

Adds with carry register to register or memory to register or register to memory. It is 2 byte long instruction. Its encoding format is

0001 00 dW mem reg r / m

ADC [DI],AX

16-bit operation, W = 1, AX = source, d = 0, [DI] decides mod = 00, r/m = 101, reg. = AX = 000

ADC [DI],AX Object code = 0001 0 001, 00 000 101 = 11,05

ADC CX,[BX], Object code = 0001 0 011, 00 001 111 = 13,0F

ADC [BX],CH, Object code = 0001 0 000, 00 101 111 = 10,2F

```
C:\WINDOWS>DEBUG
-A : Addition with carry
0F6C:0100      MOV BX,2233
0F6C:0103      MOV DX,4455
0F6C:0106      MOV CX,A344
0F6C:0109      MOV AX,A455
0F6C:010C      ADC CX,AX
0F6C:010E      ADC DX,BX
0F6C:0110      HLT
0F6C:0111
-GCS:0110
AX = A455 BX = 2233 CX = 4799 DX = 6689 SP = FFEE BP = 0000 SI=0000 DI=0000
DS = 0F6C ES = 0F6C SS = 0F6C CS = 0F6C IP = 0110 NV UP EI PL NZ NA PO NC
0F6C:0110 F4              HLT
```

SUB Acc, data

Subtracts immediately available 8- or 16-bit data from AL or AX. The encoding format is

0010 110 W data high data if W = 1

SUB AX,2535H

Object code = 0010 1101,35,25 = 2D,35,25

SUB mem/reg,data

Subtracts immediately available 8- or 16-bit data from contents of register or memory. The encoding format is

1000 00 SW mod 101 r/m data high data if W = 1, S = 0

SUB [BX], 2535H

16-bit operation, W = 1, S = 0, [BX] decides mod = 00, r/m = 111

Object code = 1000 0001,00 101111,35,25 = 81,2F,35,25

SUB [0108], 2535H

16-bit operation, W = 1, S = 0, memory decides mod = 00, r/m = 110

Object code = 1000 0001,00 101110,35,25 = 81,2E,35,25

SUB mem/reg1,mem/reg2

Subtracts register from register, register from memory, memory from register. The encoding format is

0010 10 dW mod reg r / m

SUB [BX],AX

16-bit operation, W = 1, AX = source, d = 0, [BX] decides mod = 00, r/m = 111

Object code = 0010 1001, 00 00 0 111 = 29,07

SUB [0108],AX, Object code = 0010 1001, 00 00 0 110 = 29,06

SUB AX,[BX], Object code = 0010 1011, 00 00 0 111 = 2B,07

SUB AX,[0108], Object code = 0010 1011, 00 00 0 110 = 2B,06

SBB Acc, data

Subtracts immediately available 8- or 16-bit data from AL or AX with borrow. The encoding format is

0001 110 W data high data if W = 1

SBB AX,2535H

16-bit operation, W = 1,

Object code = 0001 1101, 35,25 = 1D,35,25

SBB mem/reg,data

Subtracts immediately available 8- or 16-bit data from contents of memory or register with borrow. The encoding format is

1000 00 sW mod 011 r / m data high data if W = 1, S = 0

SBB [BX],AX

16-bit operation, W = 1, s = 0, [BX] decides mod = 00, r/m = 111

Object code = 1000 0001, 00 01 1111 = 81,1F

SBB [0108],AX, Object code = 1000 0001, 00 01 1110 = 81,1E

SUB AX,[BX], Object code = 1000 0001, 0001 1111 = 81,1F

SBB AX,[0108], Object code = 1000 0001, 00 01 1110 = 81,1E

CMP

Compares two specified bytes or words. Both source and destination can be an immediate data, a register or a memory location, but both operands cannot be memory location. After subtraction of source from destination, both the source and destination remain unaltered, only flags are affected.

CMP Acc, data

Compares 8-bit or 16-bit immediately available data from AL or AX. It is performed by subtracting the immediate data from the accumulator and use the result by setting the flags. The contents of the accumulator is not affected. The encoding format is

0011 110 W data high data if W = 1

CMP AX, 1234, Object code = 0011 1101,34,12 = 3D,34,12

CMP AL, 12, Object code = 0011 1100,12 = 3C,12

CMP mem/reg1, data

Compares 8-bit or 16-bit immediately available data from memory or register. The encoding format is

1000 00 SW mod 111 r/m data high data if W = 1, S = 0

CMP [0200], 2244H, Object code = 1000 0011, 00111110, 44, 22 = 83, 3E, 44, 22

CMP mem/reg1, mem/reg2

Compares 8-bit or 16-bit data in mem/reg2 with data in mem/reg1. The encoding format is

0011 10 SW mod reg r/m

CMP [BX], BX, Object code = 0011 1001, 00 011 111 = 39, 1F

INC

Increment the specified byte or word by unity. The operands can be a register or memory location. The flags affected are AF, SF, ZF, PF, and OF. The carry flag is not affected. Hence, even if FFH or FFFFH are incremented, the result becomes zero without any carry.

INC reg

Increment the contents of the register by 1. It is one byte instruction. The encoding format is 0100 0 rrr

INC AX, Object code = 0100 0000 = 40

INC CX, Object code = 0100 0001 = 41

INC DX, Object code = 0100 0010 = 42

INC BX, Object code = 0100 0011 = 43

INC mem/reg

Increment the contents of the memory location or register by 1. It is 2 to 4-byte instruction. The encoding format is

1111 111W mod 000 r/m

INC [BX], Object code = 1111 1111, 00 000 111 = FF,07

DEC

Decrement the specified byte or word by unity. The operands can be a register or memory location. The rules applicable for INC are also applicable for DEC.

DEC reg

Decrement the contents of the register by 1. It is one byte instruction. The encoding format is

0100 1 rrr

DEC AX, Object code = 0100 1000 = 48

DEC CX, Object code = 0100 1001 = 49

DEC DX, Object code = 0100 1010 = 4A

DEC BX, Object code = 0100 1011 = 4B

DEC mem/reg

Decrement the contents of the memory location or register by 1. It is 2 to 4 byte instruction. The encoding format is

1111 111W mod 001 r/m

DEC [BX], Object code = 1111 1111, 00 001 111 = FF,0F

MUL mem/reg

Multiplies contents of any memory location or register to AL or AX. Both multiplicand and multiplier contains unsigned numbers. For 16-bit multiplication, the result is available in [DX]: [AX]. DX contains higher 16-bits and AX contains lower 16-bit. In either case of 8-bit or 16-bit multiplication, if higher order result is 00 or FF or 0000 or FFFF, the overflow flag (OF) and the carry flag (CF) are set to 0 otherwise they are set to 1. Other flags are undefined. The encoding format is

1111 011 W mod 100 r / m

Here 1st operand is always AL or AX.

MUL CL

8-bit operation, W = 0, CL decides mod = 11, r/m = 001,

MUL CL, Object code	= 1111 0110, 1110 0001	= F6, E1
MUL DL, Object code	= 1111 0110, 1110 0010	= F6, E2
MUL BL, Object code	= 1111 0110, 1110 0011	= F6, E3
MUL CX, Object code	= 1111 0111, 1110 0001	= F7, E1
MUL DX, Object code	= 1111 0111, 1110 0010	= F7, E2
MUL BX, Object code	= 1111 0111, 1110 0011	= F7, E3
MUL [BX]16-bit, Object code	= 1111 0111, 0010 0111	= F7, 27
MUL [BX]8-bit, Object code	= 1111 0110, 0010 0111	= F6, 27
MUL [BX]+D8, 16-bit, Object code	= 1111 0111, 0110 0111, D8	= F7, 67, D8
MUL [BX]+D16, 16-bit, Object code	= 1111 0111, 1010 0111, D16	= F7, A7, D16

Example

Multiply 2535H with 3535H.

```
0100    B9,35,25    MOV CX,2535H
0103    B8,35,35    MOV AX,3535H
0106    F7,E1       MUL CX
0108    F4          HLT
```

Result [DX]:[AX] = 07BBH:ACF9H

Multiply 35H with 25H.

```
0100    B1,25       MOV CX,25H
0102    B0,35       MOV AX,35H
0104    F6,E1       MUL CX
0106    F4          HLT
```

Result in [AX] = 07A9H

IMUL mem/reg

Multiplies AL or AX by content of memory location or register. Either or both operands can be signed numbers. For 8-bit operation, the higher order 8-bit result resides in AH and lower 8-bit result resides in AL. For 16-bit operation, higher 16-bit result resides in DX and lower 16-bit result resides in AX. If the magnitude of the product does not require all the bits of the destination, the unused bits will be filled with copies of the sign bit. For 8-bit signed multiplication, if (AH) = 00 or FF, or in 16-bit signed multiplication, (DX) = 0000 or FFFF, the OF and CF flags are set to 0 otherwise it remains as 1. Other flags are undefined. The encoding format is

1111 011 W mod 101 r / m

IMUL CL

8-bit operation, W = 0, CL decides mod = 11, r/m = 001.

IMUL CL, Object code	= 1111 0110, 11101001	= F6, E9
IMUL DL, Object code	= 1111 0110, 11101010	= F6, EA
IMUL BL, Object code	= 1111 0110, 11101011	= F6, EB
IMUL CX, Object code	= 1111 0111, 11101001	= F7, E9
IMUL DX, Object code	= 1111 0111, 11101010	= F7, EA
IMUL BX, Object code	= 1111 0111, 11101011	= F7, EB
IMUL [BX]8-bit operation, Object code	= 1111 0110, 00101111	= F6,2F
IMUL [BX]16-bit operation, Object code	= 1111 0111, 00101111	= F7,2F

Example

Multiply -75H with 56H.

0100	B1,56	MOV CL,56H
0102	B0,75	MOV AL,75H
0104	F6,D8	NEG AL
0106	F6,E9	IMUL CL
0108	F7,D8	NEG AX
010A	F4	HLT

Result in AX = –274EH

Multiply –2535H with –3535H.

0100	B9,35,35	MOV CX,3535H
0103	F7,D9	NEG CX
0105	B8,35,25	MOV AX,2535H
0108	F7,D8	NEG AX
010A	F7,E9	IMUL CX
010C	F4	HLT

Result in [DX]:[AX] = 07BBH:ACF9H

Multiplying a signed byte by a signed word

```
MOV CX, MULTIPLIER
MOV AX, MULTIPLICAND
CBW
IMULCX.
```

DIV mem/reg

Divides the contents of AX or DX : AX by the contents of memory location or register. Here both dividend and divisor are unsigned numbers. For 8-bit divisor, the quotient resides in AL and remainder in AH. For 16-bit divisor the quotient resides in AX and remainder resides in DX. The encoding format is

1111 011W mod 110 r/m

Here, W = 0 refers to 8-bit divisor and W = 1 refers to 16-bit divisor

DIV CL	1111 0110, 11110 001	= F6, F1
DIV DL	1111 0110, 11110 010	= F6, F2
DIV BL	1111 0110, 11110 011	= F6, F3
DIV CX	1111 0111, 11110 001	= F7, F1
DIV DX	1111 0111, 11110 010	= F7, F2
DIV BX	1111 0111, 11110 011	= F7, F3
DIV [BX]8-bit	1111 0110, 00 110 111	= F6,37
DIV [BX]16-bit	1111 0111, 00 110 111	= F7,37

Example

Divide 3535H by 55H

0100	B9,55	MOV CX,55H
0102	B8,35,35	MOV AX,3535H
0106	F7,F1	DIV CX
0108	F4	HLT

Result (AL) = Quotient = A0H, (AH) = Remainder = 15H

Divide 2535 2636H by 2535H.

0100	B8,36,26	MOV AX,2636H
0103	BA,35,25	MOV DX,2535H
0106	B9,35,25	MOV CX,2535H
0109	F7,F1	DIV CX
010B	F4	HLT

IDIV MEM/REG

Divides the contents of AX or DX : AX by the contents of memory location or register. This instruction allows signed 16-bit number to be divided by an 8-bit signed number or a 32-bit signed number to be divided by a signed 16-bit number. For 8-bit divisor, the quotient resides in AL and remainder in AH. For 16-bit divisor the quotient resides in AX and remainder resides in DX. Remainder has the same sign as dividend. The encoding format is

1111 011W mod 111 r/m

IDIV CL	1111 0110, 11111001	= F6, F9
IDIV DL	1111 0110, 11111010	= F6, FA
IDIV BL	1111 0110, 11111011	= F6, FB
IDIV CX	1111 0111, 11111001	= F7, F9
IDIV DX	1111 0111, 11111010	= F7, FA
IDIV BX	1111 0111, 11111011	= F7, FB
IDIV [BX]8-bit	1111 0110, 00111111	= F6,3F
IDIV [BX]16-bit	1111 0111, 00111111	= F7,3F

Example

Divide -26F7H by 55H.

Since dividend is negative, both quotient and remainder will be negative numbers.

```
0100    B8,F7,26    MOV AX,26F7H
0103    F7,D8       NEG AX
0105    B1,55       MOV CL,55H
0107    F6,F9       IDIV CL
0109    F6,D8       NEG AL
010B    F6,DC       NEG AH
010D    F4          HLT
```

Result = (AL) = Quotient = –75H, (AH) = Remainder = –1EH.

Divide 26F7H by –55H.

Since, divisor is negative, only quotient will be negative number.

```
0100    B8,F7,26    MOV AX,26F7H
0103    B1,55       MOV CL,55H
0105    F6,D9       NEG CL
0107    F6,F9       IDIV CL
0109    F6,D8       NEG AL
010D    F4          HLT
```

Result = (AL) = Quotient = –75H, (AH) = Remainder = 1EH.

NEG mem/reg

This instruction performs 2's complement of the operand. It is 2 to 4-byte long instruction. The encoding format is

1111 011 W mod 011 r / m

NEG AL

8-bit operation, W = 0, AL decides mod = 11, r/m = 000

NEG AL = 1111 0110, 11 011 000 = F6, D8

NEG CL = 1111 0110, 11 011 001 = F6, D9

NEG DL = 1111 0110, 11 011 010 = F6, DA

NEG BL = 1111 0110, 11 011 011 = F6, DB

NEG AX = 1111 0111, 11 011 000 = F7, D8

NEG CX = 1111 0111, 11 011 001 = F7, D9

NEG DX = 1111 0111, 11 011 010 = F7, DA

NEG BX = 1111 0111, 11 011 011 = F7, DB

Example

```
0100    B8,39,57    MOV AX,5739H
0103    F7,D8       NEG AX
0105    F4          HLT
```

Result in AX = A8C7H

BCD

Binary coded decimal (BCD) information is stored in either packed or unpacked forms in the memory. Packed BCD occurs when two BCD digits are stored per memory byte. In unpacked BCD only one BCD digit is stored per byte. The valid binary codes for BCD are 0000 (0_{10}) to 1001 (9_{10}). A few packed and unpacked BCD numbers are listed in Table-2.11(e).

Table 2.11(e) Packed and unpacked BCD numbers

	Packed BCD	*Unpacked BCD*
12	00010010	00000001 00000010
1, 22	00000001 00100010	00000001 00000010 00000010
12, 22	00010010 00100010	00000001 00000010 00000010 00000010

The BCD data always appear in packed form and are stored as two BCD digits per byte.

DAA

Decimal adjust after addition (DAA) instruction follows BCD addition (ADD or ADC). The adjust instruction works on AL only after BCD addition and subtraction. If the content of LSN is more than 9, 6 is added to it. Similarly, if MSN is more than 9, 60 is added to it. The objective code = 27H.

```
C:\WINDOWS>DEBUG
-A :
0F6C:0100    MOV DX,1234
0F6C:0103    MOV BX,5078
0F6C:0106    MOV  AL,BL
0F6C:0108    ADD  AL,DL
0F6C:010A    DAA
0F6C:010B    MOV  CL,AL
0F6C:010D    MOV  AL,BH
0F6C:010F    ADC  AL,DH
0F6C:0111    DAA
0F6C:0112    MOV  CH,AL
0F6C:0114    HLT
0F6C:0115
-GCS:0114
AX = 0063 BX = 5078 CX = 6312 DX = 1234 SP = FFEE BP = 0000 SI = 0000 DI = 0000
DS = 0F6C ES = 0F6C SS = 0F6C CS = 0F6C IP = 0114 NV UP EI PL NZ NA PE NC
0F6C:0114 F4          HLT
```

DAS

Decimal adjust after subtraction (DAS) instruction follows BCD subtraction (SUB or SBB). The adjust instruction works on AL only after BCD subtraction. If the content of LSN is more than 9, 6 is subtracted to it. Similarly, if MSN is more than 9, 60 is subtracted to it. The objective code = 2FH. (updates AF, CF, PF, ZF)

```
C:\WINDOWS>DEBUG
-A
0F6C:0100    MOV DX,1234
0F6C:0103    MOV BX,5078
0F6C:0106    MOV  AL,BL
0F6C:0108    SUB  AL,DL
0F6C:010A    DAS
```

```
0F6C:010B    MOV  CL,AL
0F6C:010D    MOV  AL,BH
0F6C:010F    SBB  AL,DH
0F6C:0111    DAS
0F6C:0112    MOV  CH,AL
0F6C:0114    HLT
0F6C:0115
-GCS:0114
AX = 0038 BX = 5078 CX = 3844 DX = 1234 SP = FFEE BP = 0000 SI = 0000 DI = 0000
DS = 0F6C ES = 0F6C SS = 0F6C CS = 0F6C IP = 0114 NV UP EI PL NZ AC PO NC
0F6C:0114 F4              HLT
```

ASCII Arithmetic

The ASCII numbers ranges in values from 30H to 39H for the numbers 0 to 9.

AAA

ASCII Adjust after addition. This instruction function as per following steps

If the LSN of (AL) is more than 9, 6 is added to it and a 1 is added to AH.

The MSN = 0 (always).

The object code = 37H.

Example

ADD ASCII 33H and 37H.

```
C:\WINDOWS>DEBUG
-A
0F6C:0100    MOV AX,0037H
0F6C:0103    ADD AX,0033H
0F6C:0106    AAA
0F6C:0107    OR AX,3030H
0F6C:010A    HLT
0F6C:010B
-GCS:010A
AX = 3130 BX = 0000 CX = 0000 DX = 0000 SP = FFEE BP = 0000 SI = 0000 DI = 0000
DS = 0F6C ES = 0F6C SS = 0F6C CS = 0F6C IP = 010A NV UP EI PL NZ AC PE NC
0F6C:010A F4              HLT
```

AAM

ASCII Adjust after multiplication (AAM) instruction follows the multiplication instruction after multiplying two 1-digit unpacked BCD numbers. The multiplication on ASCII numbers should not be performed unless the MSN of each number is cleared. For instance, if we want to multiply 5H into 5H, we should write as 05H into 05H. The objective code = D4H, 0AH.

Example

Multiply unpacked 05H to unpacked 05H.

```
C:\WINDOWS>DEBUG
-A
0F6C:0100     MOV AL,05
0F6C:0102     MOV CL,05
0F6C:0104     MUL CL
0F6C:0106     AAM
0F6C:0108     HLT
0F6C:0109
-GCS:0108
AX = 0205 BX = 0000 CX = 0005 DX = 0000 SP = FFEE BP = 0000 SI = 0000 DI = 0000
DS = 0F6C ES = 0F6C SS = 0F6C CS = 0F6C IP = 0108 NV UP EI PL NZ NA PE NC
0F6C:0108 F4             HLT
```

AAD

ASCII Adjust before division (AAD) instruction appears before division. The AAD instruction requires a two digit unpacked BCD number (not ASCII) before executing. After adjusting the AX with AAD, it is divided by an unpacked BCD number to generate single digit result in AL with any remainder in AH. The objective code = D5H, 0AH.

Example

Divide ASCII 72H by 9H.

```
C:\WINDOWS>DEBUG
-A
0F6C:0100     MOV BL,9H
0F6C:0102     MOV AX,0702H
```

```
0F6C:0105     AAD
0F6C:0107     DIV BL
0F6C:0109     HLT
0F6C:010A
-GCS:0109
AX = 0008 BX = 0009 CX = 0000 DX = 0000 SP = FFEE BP = 0000 SI = 0000 DI = 0000
DS = 0F6C ES = 0F6C SS = 0F6C CS = 0F6C IP = 0109 NV UP EI PL NZ NA PE NC
0F6C:0109 F4          HLT
```

AAS

ASCII Adjust after subtraction (AAS) instruction adjust the AX after ASCII subtraction. It functions as follows:

If the LSN of (AL) is more than 9, 6 is subtracted from it and a 1 is subtracted from AH.

The MSN = 0 (always) of AL.

The objective code = 3FH.

```
C:\WINDOWS>DEBUG
-A / ASCII7 – ASCII9 = 37H – 39H
0F6C:0100     MOV AX,0037
0F6C:0103     SUB AL,39
0F6C:0105     AAS
0F6C:0106     OR AL,30
0F6C:0108     HLT
0F6C:0109
-GCS:0108
AX = FF38 BX = 0000 CX = 0000 DX = 0000 SP = FFEE BP = 0000 SI = 0000 DI = 0000
DS = 0F6C ES = 0F6C SS = 0F6C CS = 0F6C IP = 0108  NV UP EI PL NZ AC PO NC
0F6C:0108 F4          HLT
C:\WINDOWS>DEBUG
-A
0F6C:0100     MOV AX,0037
0F6C:0103     SUB AX,39
```

```
0F6C:0106     HLT
0F6C:0107
-GCS:0106
AX = FFFE BX = 0000 CX = 0000 DX = 0000 SP = FFEE BP = 0000 SI = 0000 DI = 0000
DS = 0F6C ES = 0F6C SS = 0F6C CS = 0F6C IP = 0106 NV UP EI NG NZ AC PO CY
0F6C:0106 F4          HLT
C:\WINDOWS>DEBUG
```

LODS/LODSB/LODSW

A byte or a word is transferred from memory addressed by SI and SI is automatically incremented. The encoding format is

1010 110 W

LODSB = 1010 1100 = AC

LODSW = 1010 1101 = AD

STOS/STOSB/STOSW

A byte or a word is transferred from AL or AX to memory addressed by DI and DI is automatically incremented. The encoding format is

1010 101 W

STOSB = 1010 1010 = AA

STOSW = 1010 1011 = AB

REP/REPE/REPZ/REPNE/REPNZ

This instruction means repeat, repeat while equal, repeat while zero, repeat till not equal, repeat while not zero. The encoding format is

1111 001 Z

Z = 1, repeat until CX = 0, This refers to REPZ/REPE instructions

Z = 0, repeat until CX ≠ 0, This refers to REPNE/REPNZ instruction.

REPNZ = 1111 0010 = F2

MOVS/MOVSB/MOVSW

A byte or a word is transferred from memory location addressed by SI to the location addressed by DI. SI and DI are automatically incremented/decremented as per DF. MOVS instruction moves data from memory to memory using REP prefix. The encoding format is

1010 0100 W

MOVSB = 1010 0100 = A4

MOVSW = 1010 0101 = A5

Example

Move 5-bytes from memory starting at 0300 to memory starting at 0400.

0100	FC	CLD
0101	BE,00,30	MOV SI,0300
0104	BF,00,04	MOV DI,0400
0107	8B,05,00	CX,0005
010A	F2	REPNZ
010B	A4	MOVSB
010C	F4	HLT

Move 3-words of 16-bit from memory starting at 0300 to memory starting at 0400.

0100	FC	CLD
0101	BE,00,30	MOV SI,0300
0104	BF,00,04	MOV DI,0400
0107	8B,03,00	CX,0003
010A	F2	REPNZ
010B	A5	MOVSW
010C	F4	HLT

Here it is essential to clear the direction flag by CLD.

Bit Manipulation Instructions

There are broadly three groups of bit manipulation instructions of 8086. These are

- Logical instructions,
- Shift instructions,
- Rotate instructions.

The logical instructions of 8086 include AND, OR, X-OR, NOT, and TEST.

NOT mem/reg

This instruction performs 1's complement of the operand. It is 2 to 4-byte long instruction. The encoding format is

1111 011 W mod 010 r/m

NOT AL

8-bit operation, w = 0, AL decides mod = 11, r/m = 000

NOT AL	= 1111 0110, 11 010 000 = F6, D0
NOT CL	= 1111 0110, 11 010 001 = F6, D1
NOT DL	= 1111 0110, 11 010 010 = F6, D2
NOT BL	= 1111 0110, 11 010 011 = F6, D3
NOT AX	= 1111 0111, 11 010 000 = F7, D0
NOT CX	= 1111 0111, 11 010 001 = F7, D1
NOT DX	= 1111 0111, 11 010 010 = F7, D2
NOT BX	= 1111 0111, 11 010 011 = F7, D3
NOT [BX]8-bit operation	= 1111 0110, 00 010 111 = F6,17
NOT [BX]16-bit operation	= 1111 0111, 00 010 111 = F7,17

Example

0100	B8,39,57	MOV AX,5739H
0103	F7,D0	NOT AX
0105	F4	HLT

Result in AX = A8C6H

AND Acc,data

ANDs each bit of a byte or of a word of immediately available data with AL or AX. The instruction is 2 to 3-byte long. The encoding format is

0010 010 W data high data if W = 1

AND AX,1234H, Object code = 0010 0101, 34,12 = 25, 34,12

AND AL,12H, Object code = 0010 0100, 12 = 24,12

AND mem/reg,data

ANDs each bit of a byte or of a word of immediately available data with the contents of memory location or register. The instruction is 3 to 6-byte long. The encoding format is

1000 000 W mod 100 r / m data high data if W = 1

AND BX,1234H, Object code = 1000 0001,11 100 011, 34,12 = 81, E3,34,12

AND [BX],1234H, Object code = 1000 0001, 00 100 111,34,12 = 81, 27,34,12

AND mem/reg1,mem/reg2

ANDs each bit of a byte or of a word of the contents of memory location or register with the contents of another memory location or register. The instruction is 2 to 4-byte long. The encoding format is

0010 00dW mod reg r/m

AND AX,[BX], Object code = 0010 0011, 00 000 111 = 23, 07

AND AL,BL, Object code = 0010 0010, 11 000 011 = 22, C3

AND CX,DX, Object code = 0010 0011, 11 001 010 = 23, CA

AND AX,[DI], Object code = 0010 0011, 00 000 101 = 23, 05

OR Acc,data

ORs each bit of a byte or of a word of immediately available data with the contents of AL or AX. The encoding format is

0000 110 W data high data if W = 1

OR AX,1234H, Object code = 0000 1101,34,12 = 0D,34,12

OR AL,12H, Object code = 0000 1100,12 = 0C,12

OR mem/reg,data

ORs each bit of a byte or of a word of immediately available data with the contents of memory or register. The encoding format is

1000 000 W mod 001 r/m data data if W = 1

OR BX,1234H, Object code = 1000 0001, 11 001 011,34,12 = 81, CB,34,12

OR [BX],1234H, Object code = 1000 0001, 00 001 111,34,12 = 81,0F,34,12

OR mem/reg1,mem.reg2

ORs each bit of a byte or of a word of the contents of memory location or register with the contents of memory or register. The encoding format is

0000 10 dW mod reg r/m

OR DL,AL, Object code = 0000 1010, 11 000 010 = 0A,C2

OR DX,[BX], Object code = 0000 1011, 00 010 111 = 0B,17

XOR Acc,data

XORs each bit of a byte or of a word of immediately available data with the contents of AL or AX. The encoding format is

0011 010 W data high data if W = 1

XOR AX,1234H, Object code = 0011 0101, 34,12 = 35,34,12

XOR AL,12H, Object code = 0011 0100, 34,12 = 34,34,12

XOR mem/reg,data

XORs each bit of a byte or a word of immediately available data with the contents of memory or register. The encoding format is

1000 000 W mod 110 r/m data high data if W = 1

XOR BX,1234H, Object code = 1000 0001, 11 110 011,34,12 = 81, F3,34,12

XOR [BX],1234H, Object code = 1000 0001, 00 110 111,34,12 = 81,37,34,12

XOR mem/reg1,mem.reg2

XORs each bit of a byte or of a word of the contents of memory location or register with the contents of memory or register. The encoding format is

0011 00 dW mod reg r / m

XOR CH,DL, Object code = 0011 0000, 11 010 101 = 30, D5

XOR DX,[BX], Object code = 0011 0011, 00 010 111 = 33,17

TEST

The TEST instruction performs AND operation with the difference that the AND operation changes the contents of the destination operand, while the TEST operation does not. It only affects the flag register that indicates the result of the TEST operation. The TEST operation is similar to the CMP instruction, but CMP instruction test the entire byte or word while the TEST instruction normally tests a single bit.

TEST Acc,data

This stands to AND the immediate data with the AL or AX and do not store the ANDed data in the accumulator. The encoding format is

1010 100 W data high data if W = 1

Overflow and carry flag reset to 0.

TEST mem/reg,data

This stands to AND contents of memory or register with the immediate data without affecting the contents of the operands. The encoding format is

1111 001W mod 000 r / m data high data if W = 1

TEST reg,mem/reg

This stands to AND contents of a register with contents of memory or register without affecting the contents of the operands. The encoding format is

1000 010 W mod 010 r/m

JUMPs and RETURNs

Unconditional jumps are also of 3-types : ***short jump, near jump** and **far jump***. The short jump is of 2-byte and allows jumps or branching to memory location within 256-bytes i.e., –128 byte to +127 byte from the memory location following the jump. The 3-byte *near jump* allows a jump within ±32KB.

The 5-byte *far jump* allows a jump to any memory location within the memory system. The short and near jump are often called intrasegment jump and far jump is called intersegment jump.

The short jump is also given the name relative jump as it can be moved any where in the current code segment. This is because the jump address is not stored with the opcode. Instead an 8-bit displacement follows the opcode. The short jump, thus, is a 2-byte instruction that contains an opcode followed by an 8-bit displacement. Figure 2.11 (*c*) shows the structure of the short jumps as

Opcode disp 8

Short Jump : JMP dis-8

This is a 2-byte intrasegment instruction. It adds the signed 8-bit displacement to IP. The encoding format is

1110 1011 disp 8

Object code = EB, disp 8

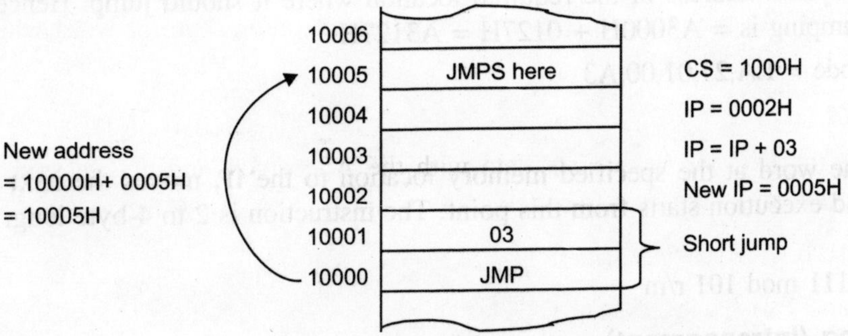

Fig. 2.11 (*c*) Short jumps structure

NEAR JUMP

The near jump is just like the short jump but the distance is farther than that of the short jump as indicated in Fig. 2.11 (*d*). It passes control to an instruction in the current code segment located within ±32K byte from the near jump instruction. The near jump is a 3-byte instruction that contains an opcode followed by a signed 16-bit displacement. Its opcode follows 16-bit displacement. The encoding format is

1110 1001 low disp high disp

Object code = E9, disp 16

Jumped address = 10003 H + 00003H = 10006H

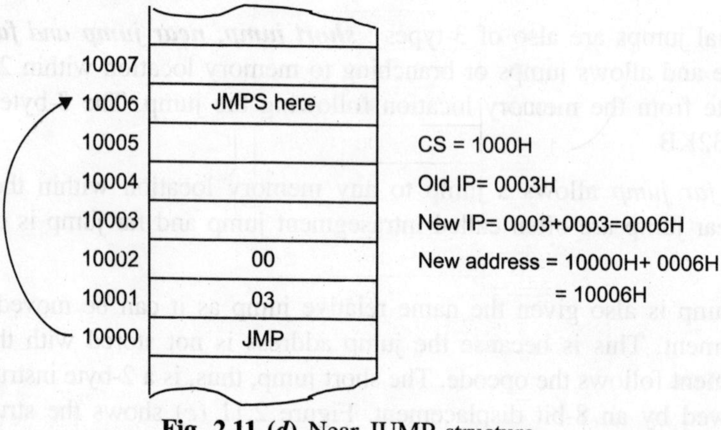

Fig. 2.11 (*d*) Near JUMP structure

Far jump

The far jump obtains a new segment and offset address to accomplish this jump as shown in Figure 2.11 (*e*). The instruction is of 5-bytes. The 1st byte is the opcode. The 2nd and 3rd byte contain the new offset address to be stored in IP. The 4th and the 5th bytes contain the new address to be stored in the CS register. The encoding format is

1110 1010 IP low IP high CS low CS high

Above statements are illustrated by Fig. 2.11 (*e*). The IP address is 0127H and (CS) = A300H. The content of CS is shifted by 4-bits left to make it A3000H and then IP is added to it to get the physical address of the required location where it should jump. Hence, the required location for jumping is = A3000H + 0127H = A3127H.

Object code = EA,27,01,00,A3

JMP mem

Moves the word at the specified memory location to the IP, moves the next word into the CS register and execution starts from this point. The instruction is 2 to 4-byte long. The encoding format is

1111 1111 mod 101 r/m

JMP mem/reg (Intrasegment)

Jumps to the location specified by the content of the register or memory. It is 2 to 4-byte long instruction. The encoding format is

1111 1111 mod 100 r/m

JMP AX, Object code = 1111 1111, 11100 000 = FF, E0

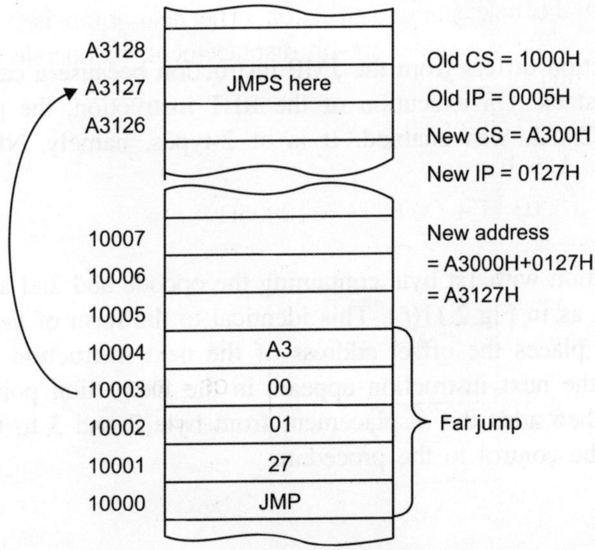

Fig. 2.11 (*e*) Far JUMP structure

Conditional jumps

Conditional jumps are always short jumps. This limits the range from –128 to +127 bytes from the location following the condition of jump. The conditional jumps take the help of sign (S), zero (Z), carry (C), parity (P), and overflow (O) bits. Conditional jumps are explained in Table 2.11(*f*).

Table 2.11(*f*) Explanation of conditional jumps

S.No.	*Mnemonic*	*Comments*	*Flags Tested*	*Opcode*
1	JZ/JE	Jump if zero/ Jump if equal	Z = 1	74
2	JNZ/JNE	Jump if no zero/ Jump if not equal	Z = 0	75
3	JC	Jump if carry	C = 1	72
4	JNC	Jump if no carry	C = 0	73
5	JS	Jump if sign or negative	S = 1	78
6	JNS	Jump if no negative	S = 0	79
7	JP/JPE	Jump if parity even/ Parity equal	P = 1	7A
8	JPO/JNP	Jump if parity odd/ not even	P = 0	7B
9	JO	Jump if overflow	O = 1	70
10	JNO	Jump if no overflow	O = 0	71
11	JA	Jump if above (more)	C = 0 and Z = 0	77
12	JAE	Jump if above or equal	C = 0	73
13	JB	Jump if below	C = 1	72
14	JBE	Jump if below or equal	C = 1 and Z = 1	76
15	JG	Jump if greater than	Z = 0 or S (XOR) O = 0	7F
16	JGE	Jump if greater than or equal	S (XOR) O = 0	7D
17	JL	Jump if less than	S ≠ 0	7C
18	JLE	Jump if less than or equal	Z = 0 or S ≠ 0	7E

CALL

The CALL instruction differs from the JMP instruction because a call instruction saves the return address on the stack. On execution of the RET instruction, the program returns to the place wherefrom the CALL was enabled. It is of 2-types, namely, NEAR CALL and FAR CALL.

NEAR CALL

It is 3-byte instruction with 1st byte containing the opcode and 2nd and 3rd byte containing the 16-bit displacement as in Fig 2.11(*f*). This identical to the form of *near jump*. When NEAR CALL executes, it 1st places the offset address of the next instruction on the stack (1003H). The offset address of the next instruction appears in the instruction pointer (IP). After saving this return address, it then adds the displacement from byte 2 and 3 to the IP (0FFFH+1003H = 2002H) to transfer the control to the procedure.

FAR CALL

The far CALL is like far jump. It is 5B instruction. The opcode is followed IP address and the contents of the CS. Bytes 2 and 3 contains new IP address (1100H) and byte 4 and 5 (0004H) contain the new contents of CS. The new address will 10000H + 0004H = 10004H. The far call instruction places the contents of both IP and CS on the stack before jumping to address indicated by byte 4-5 of the instruction. IP = 1000H and CS = 0005H pushed on the stack in Fig. 2.11 (*h*). Its encoding format is

10011010 offset low offset high seg low seg high

CALL dis-16

This instruction pushes the address of the next instruction on the stack. Adds the next 2-program memory bytes, assuming them as 16-bit displacement, to the IP ignoring the carry. The program execution starts from the new memory location available in the IP. The encoding format is

11101000 disp low disp high

CALL mem/reg

Stores the address of the instruction following the CALL on the top of the stack. Stores the contents of the specified register or memory location to the PC and continues execution from this point. The object code is 2B to 4B long. The encoding format is

1111 1111 mod 010 r/m

RET (Near Return)

It POPs a word (16-bit) from the top of the stack into the IP (near return) and places it in IP or 32-bit number (far return) and places it in IP and CS. The execution starts where from it left the main program. Its object code is C3 for intrasegment and CB for intersegment. The stack has 0003H = IP. CS = 1000H. The new address = 10000H + 0003H = 10003H in Fig. 2.11 (*g*).

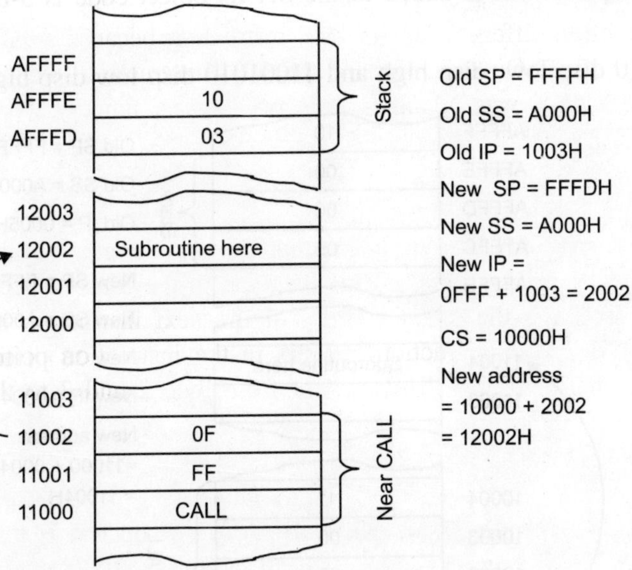

Fig. 2.11 (*f*) Near CALL stores the old IP on the stack

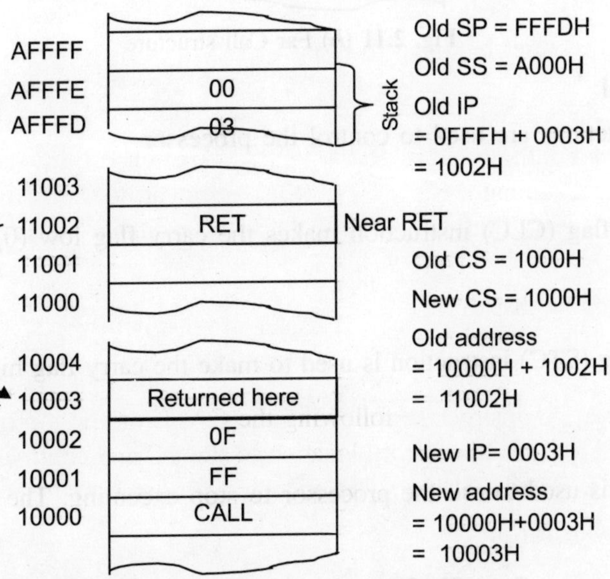

Fig. 2.11 (*g*) Near Return stores the old IP on the stack

RET disp-16 (Far Return)

Return from intrasegment subroutine and add 16-bit displacement to the stack pointer as in Fig. 2.11 (*i*). The μP POPs a word from the top of the stack into the IP and increment the stack by 2. The 16-bit dispalcement is added to the SP. Its object code is 3-byte long. The encoding format is

1100 0010 disp low disp high and 11001010 disp low disp high for interseg.

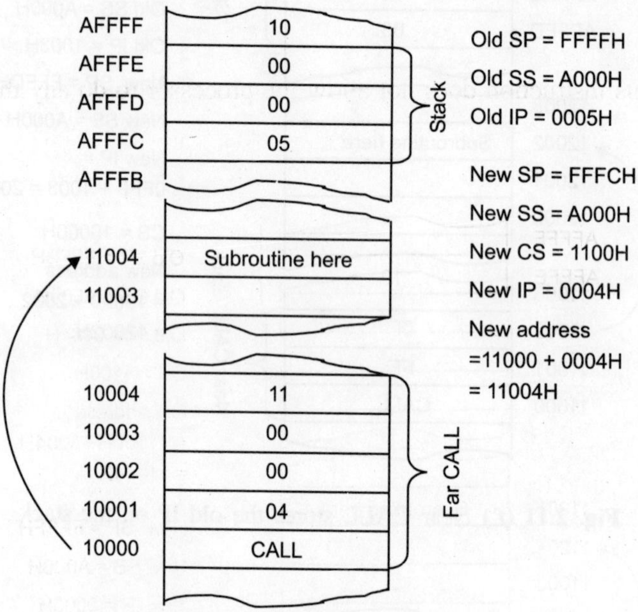

Fig. 2.11 (*h*) Far Call structure

Processor Control

Following instructions are used to control the processor

CLC

Clear the carry flag (CLC) instruction makes the carry flag low (0). The object code is = F8H.

STC

Set the carry flag (STC) instruction is used to make the carry flag high (1). The object code is = F9H.

HLT

Halt instruction is used to ask the processor to stop executing. The object code is = F4H.

CMC

Complement the carry flag (CMC) instruction is used to complement the state of the carry flag. The object code is = F5H.

STD

Set the direction flag to 1. This instruction is used to set the SI or DI is automatically decremented by unity. If STD = 0, SI or DI will automatically be incremented. The object code is = FDH.

LOCK

This instruction is used to avoid any other processor to take over the bus while adjacent instruction is being executed. The object code is = F0H .

NOP

No operation. This instruction does not allow the processor to do any thing. The object code is = 90H .

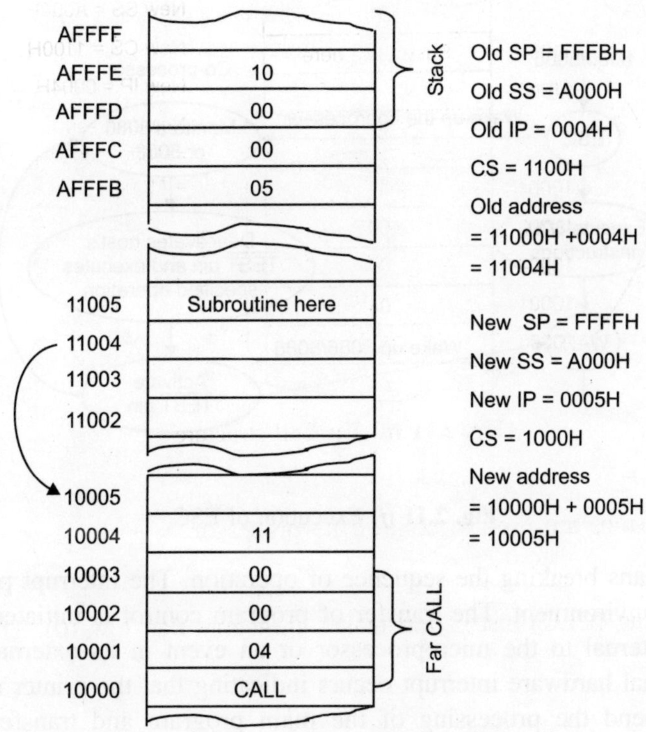

Fig. 2.11 (*i*) Far RET structure

$\overline{\text{TEST}}$: This input is examined by an WAIT instruction. If the $\overline{\text{TEST}}$ input goes low, execution will continue, else allow the processor remain in the idle state. This input is synchronized internally during each clock cycle on leading edge of the clock.

ESC

This instruction can be spelled as *escape*. On execution of the ESC instruction the µP hands over the bus to external processor such as a coprocessor or any peripheral device. Both CPU and coprocessor execute their instructions from the same program that written in a super set of the 8086/8088 instruction set. An instruction to be executed by the coprocessor is indicated when an escape (ESC) instruction appears in the program sequence. Only the host CPU can fetch instruction, but the coprocessor also receives all instructions and monitors the instruction sequencing of the host. The ESC instruction contains 2-operands. The 1st of the 2-operands is an external operation code, indicating the action to be taken by the coprocessor. The 2nd operand specifies a memory location. The 8086/8088 will fetch a word from this location for the coprocessor and may pass the coprocessor an address for storing the result. If the 2nd operand is a register, the register address is treated as part of the external op code and the CPU does not do any thing. The above explanation is illustrated by Fig. 2.11 (*j*). Thus, once the coprocessor finds that the op code is for itself, it sends a $\overline{\text{TEST}}$ signal to the host processor to force it to go to the idle state.

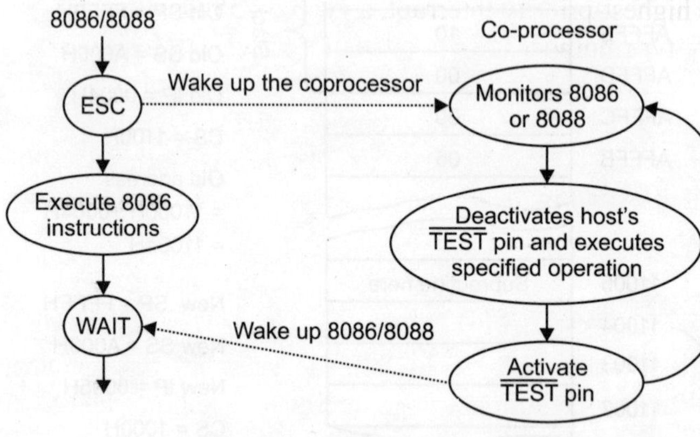

Fig. 2.11 (*j*) Execution of ESC

The interrupt means breaking the sequence of operation. The interrupt provides a facility to change the program environment. The transfer of program control is initiated by the occurrence of either an event internal to the microprocessor or an event in its external hardware. For an example, if the external hardware interrupt occurs indicating that the printer requires service, the uP is forced to suspend the processing of the main program and transfers the control to a subroutine that performs the service required by the output device (printer). The section of the program where the control is transferred is called **interrupt service routine.** In the case of this example of printer, the routine is called the **printer driver** that is a piece of software which when executed drives the printer output interface.

The 8086 or 8088 are capable of implementing any combination of 256 interrupts. The whole interrupts are divided into 5 groups as in Fig. 2.11(*k*) such as

- external hardware interrupts,
- non-maskable interrupt,
- software interrupts,
- internal interrupts, and
- reset

The user defines the functions of external hardware, software, and nonmaskable interrupts. For an example, the hardware interrupts are often assigned to devices such as the keyboard, printer, and timers. On the other hand, the function of the internal interrupts and reset are not user defined. They perform dedicated system functions.

Hardware, software, and internal interrupts are serviced on a **priority** basis. The Priority is achieved in 2-ways. 1st, the interrupt-processing sequence implemented in the 8088/8086 tests for the occurrence of the various groups based on the hierarchy shown in Fig. 2.11 (*k*). We see from Fig. 2.11 (*k*) that internal interrupts are highest-priority group, and the external hardware interrupts are the lowest-priority group.

2nd each of the interrupts is given a different priority level by assigning it a type number. Type- 0 identifies the highest-priority interrupt, and type-255 identifies the lowest-priority interrupt. A few of the interrupt type number are not available for use with software or hardware interrupts. This is because they are reserved for special interrupt function of the 8088/8086, such as internal interrupts. For instance, within the internal interrupt group, the interrupt known as **divide error** is assigned to type number 0. Therefore, it has the highest priority of the internal interrupts. Another interrupt, called **overflow**, is assigned the type number-4. Overflow is the lowest-priority internal interrupt.

Interrupt Vector Table

An **address pointer table** is used to link the interrupt type number to the locations of their service routine in the program-storage memory. Figure 2.11(*k*) shows the map of the pointer table in the memory of the 8088 or 8086. This table contains 256 **address pointers** (vectors) that are identified as vector-0 through 255. These address pointers identify the starting location of their service routines in the program memory. The contents of this table may either be held as firmware in EPROM or loaded into RAM as part of the system initialization routine. Figure 2.11(*k*) depicts that the pointer is located at the low-address end of the memory address space. It starts at the address 00000H and ends at 003FEH. This represents 1KB of memory.

Each of the 256-pointers requires 2-words (4-byte) of memory space and is always stored at an even-address boundary. The higher of the 2-word vector is called the **base address**. It identifies the memory segment in which the service routine resides. This is the reason, it is loaded in the code segment (CS) register within the μP. The lower-addressed word of the vector is the **offset** of the 1st instruction of the service routine from the beginning of the code segment defined by the base address loaded into the CS. This offset is loaded into the instruction pointer (IP) register. For example, the offset and the base address for the type number 255 , IP_{255} and CS_{255}, are stored at word addresses 003FCH and 003FEH respectively. When loaded into the μP, it points to the instruction at CS_{255} : IP_{255}.

More careful look into the pointer table reveals that 1st 32 (vector 0 to F → 16 and 10 to 1F → 16) pointers either have dedicated functions or are reserved. For instance, pointer-0, 1, 3, and 4 are used by the 8086's and 8088's internal interrupts : **divide error, single step, break-point,** and **overflow.** Pointer-2 is used to identify the starting address location of the nonmaskable interrupt's service routine. The pointers from 5 through 31 represent served portion of the pointer table and should not be used. The remainder 224 pointers in the address range 00080H through 003FFH are available to the user for storage of software or hardware interrupt vectors. These pointers correspond to type numbers 32 through 255.

Example : What is the location in the memory for CS_{50} and IP_{50}.

As each vector requires 4-consecutive bytes of memory, its address can be found by multiplying by 4. Since, CS_{50} and IP_{50} represent the type-50 interrupt pointer,

Address = 50 × 4 = 200D = C8H

Hence, IP_{50} address is 000C8H and CS_{50} address is 000CAH

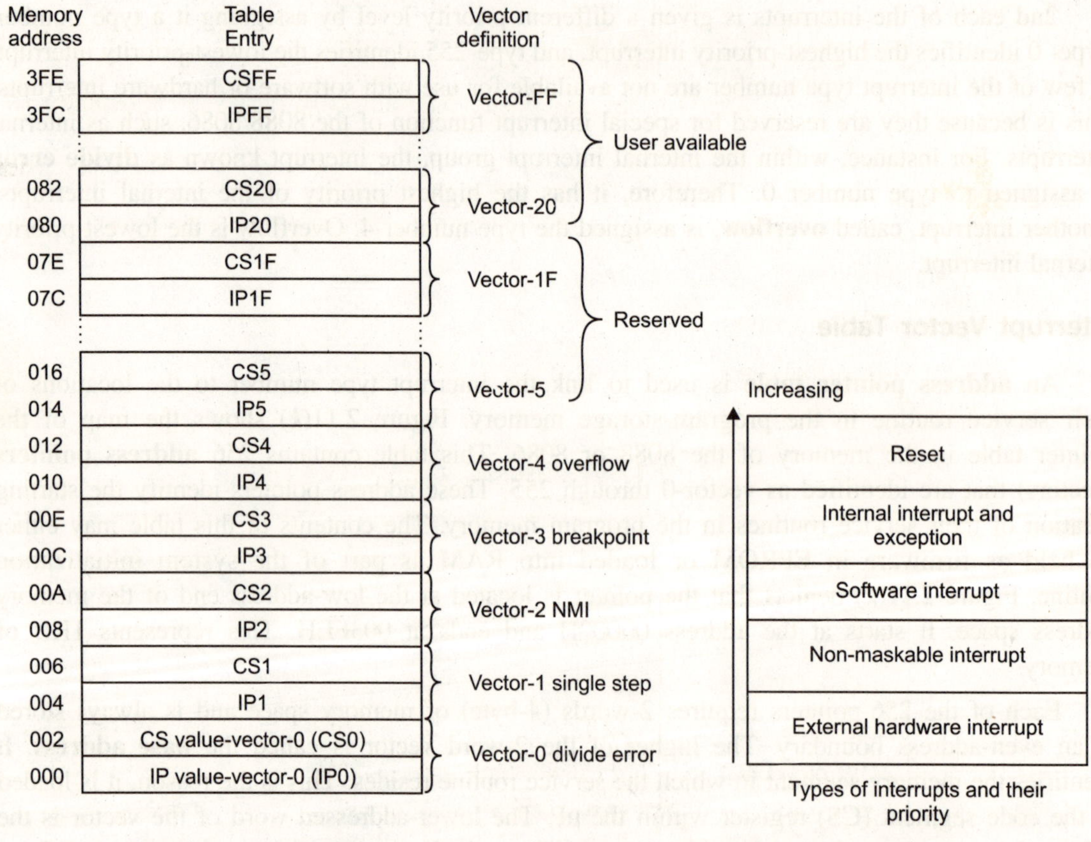

Fig. 2.11 (*k*) Interrupt vector table of 8086/8088 and priority of interrupt types

2.12 ASSEMBLER DIRECTIVES

The basic function of an assembler is to convert the assembly language instructions of the source program to the corresponding machine instructions. However, a practical assembly program do not only contain assembly language statements, rather contains statements that are not converted to the equivalent machine instruction. The statements in the assembly language program that do not have equivalent machine instruction are called **assembler directives.**

An assembler directive is simply a message to the assembler that tells it how to go about translating instructions into machine code. Assembly directives are not translated into machine code because they are not instructions. Typical assembler directives tell the assembler where to put a program in memory, what space to allocate to variables, allow you to preset memory with data, define symbols, generate fixed tables and data, mark the end of the program, and determine the form of the program listing. Pseudo-operations also appear in the fixed assembler mnemonic table; they do not have binary equivalents but do have corresponding assembler subroutines that produce the prescribed actions. When code is assembled, a **location counter** in the assembler keeps track of where the code and data is to go in memory.

The use of directives gives the programmer some control over the operation of the assembler, increasing flexibility in writing the programs. The assembler directives are written in the operation code field of any assembly language program for certain direction to the assembler. These directives are also known as **pseudo-operations**, as they appear in the operation code field. The pseudo-operations are not translated into equivalent binary instruction codes. The assembler directives are instructions meant for the assembler program only. It is not assembled as an instruction.

There are number of assembler directive, based on the type of operation that it specifies. Table 2.12 (*a*) presents a partial list of MASM directives. The categories of directives are

- Data directives,
- Conditional directives,
- Macro directive, and
- Listing directives.

Table 2.12(*a*) Assembler directives

Type	*Directives*			
Data	DB or BYTE	NED	EXTRN	PROC
	DD or DWORD	ENDP	GROUP	PUBLIC
	DQ or QWORD	ENDS	INCLUDE	RADIX
	DT or TBYTE	EQU	LABEL	RECORD
	DW or WORD	= (Equal sign)	NAME	SEGMENT
	ASSUME	EVEN	ORG	STRUC
Conditional	ELSE	IFB	IFNB	IF2
	ENDIF	IFDIFIFE	IFNDEF	
	IF	IFIDN	IF1	
Macro	ENDM	IRP	LOCAL	PURGE
	EXITM	IRPC	MACRO	REPT
Listing	.CREF	.LIST	.SALL	.XALL
	LALL	%OUT	.SFCOND	.XCREF
	.LFCOND	PAGE	TITLE	.XLIS

Data directives

The primary purpose of the data directive is to define the values for *constants*, *variables*, and *labels*. These directives can also be used to assign a size to a variable and reserving storage location in the memory. Table 2.12(*b*) lists the most commonly used data directives.

Table 2.12(*b*) Commonly used data directives

Directives	*Meaning*	*Functions*
EQU	Equate	Assigns a permanent value to a symbol
=	Equal to	Set or redefine the value of a symbol
DB or BYTE	Define byte	Define byte size variable or location
DW or WORD	Define word	Define word (2-byte) size variable or location
DD or DWORD	Define double word	Define double word (4-byte) size variable or location

EQU

The important 2-data directives are 'equal' and 'equal to'. They are written in the form of statements as EQU and '=' respectively. Both of these assembler directives simply equate a symbolic name to a numeric value. In other words the EQU allows an 8- or 16-bit value to be assigned to a label. This label can then be used throughout the program, making it more readable. In addition, if it is necessary to define a new value, only the EQU statement needs to be changed and the program is reassembled. For an instance,

```
BP  EQU  0700H
```

assigns the value 0700H to the name BP. The value in the operand (0700H) can be assigned using arithmetic or logic operations.

Example

```
UP  EQU      BP + 05H
```

In this statement UP is assigned the value BP + 05H.

These 2-operations can be performed using the '=' directive as

```
BP           EQU      0700H
UP  =        BP + 05H
```

The BP and UP can be referenced anywhere in the program once values are assigned to them. The difference between EQU and '=' directives lies in the fact that the value assigned to the symbol EQU cannot be changed whereas '=' is used to define the symbol, its value can be changed anywhere in the program.

Example

```
BP          EQU     0000H
UP EQU      BP + 05H
:   :       :
UP EQU    BP + 01H (illegal)
```

Here, BP is equal to 0000H and UP equal to 0005H. The UP cannot be changed in the last statement using EQU as 0001H .

Example

```
BP          =       0000H
UP  =       BP + 05H
:   :       :
UP  =       BP + 01H (legal)
```

In the last statement of this example, UP is assigned a new value equal to BP + 01H = 0001H. The assembler can process this.

The other important data directives are defined by byte DB, defined by word DW, defined by double word DD. Sometimes DD is also called long word DL. The purpose of these directives is to define the size of variables as being byte, word, or double word in length, allocates space for them, and assign them initial values. If, however, the initial values are not known, the DB, DW, and DD statements simply allocate a byte or a word or double word of the memory to the variable name.

Example

DB directive is

```
BP DB       07H
```

Here BP is defined as byte size and assigned the value 07H. Here, it is important to note that the value assigned to a DB, DW , or DD statement must not be larger than the maximum value that can be stored in the specified size location.

Example

The maximum decimal value of a byte is 255D for an unsigned number and –128D or +127D for a signed number. An example of DB statement is

```
BP DB       ?
```

In this case, a byte of memory is allocated to the variable BP, but no value is assigned to it. The use of question mark (?) as the operand means that an initial value is not to be assigned.

If we want to initialize a block of memory location with the same value, the assembler provides a way using the byte, word, or double word directive to repeat a value.

Example

TABLE-A DB 10DUP(?), 5 DUP (5)

This statement asks the assembler to allocate a memory space of 15-bytes for TABLE-A. The 1st 10-bytes of this block is left uninitialized (vacant) and the last 5-bytes are all filled with the value 5. The DUP (duplicate) statement tells the assembler to duplicate number of times the value enclosed in the parentheses. If each element of the table is to be initialized (filled) with different value, the DB command can be written as

TABLE-A DB 0, 1, 2, 3, 4, 5, 6, 7, 8, 9

TABLE-A is now set assigning 10-byte of memory location by 10-decimal number 0, 1, 2. 3, 4, 5, 6, 7, 8, 9.

Segment control directives

The memory of 8086/ 8088 based microcomputer has 3-types of segment partitions such as

- Code segment,
- Data segment, and
- Stack segment.

The code segment contains the machine code instructions, the data segment contains data, and stack segment is used as the temporary storage place. Table 2.12(*c*) gives few segment directive attributes.

Table 2.12(*c*) Segment directives

Directive	*Functions*
SEGMENT	Defines the beginning of a segment and specifies its kind at what type of memory it is be stored and how it is to be positioned w.r.t. the other similar segment in the memory.
ENDS	Specifies end of the segment
ASSUME	Specifies the segment address register for a particular segment.

The 'SEGMENT' directive identifies the beginning of the segment where as the 'ENDS' directive specifies end of the segment.

Example

```
SEGA        SEGMENT        PARA PUBLIC 'CODE'
            MOVAX,BX
  :
SEGA        ENDS
```

Here, SEGA is the name given to a segment. The SEGMENT directive is followed by the operand PARA PUBLIC 'CODE'. The PARA (paragraph) defines that this segment be aligned in memory on a 16-byte address boundary. This part of the operand is called the *align-type* attribute.

The other types of attributes are given in Tables 2.12(*d*) and (*e*). The PUBLIC that follows the PARA in the operand of the SEGMENT directive defines *combine-type* attribute.

The last part of the operand in the SEGMENT directive is the 'CODE' that specifies the *class-type* of attribute.

Table 2.12(*d*) Align-type attributes

Attributes	*Functions*
PARA	Segment begins on a 16-byte address boundary in memory (4 LSBS of the address are equal to 0)
BYTE	Segment can begin anywhere in the memory
Word	Segment begins on word (2-byte) address boundary (LSBs = 0)
PAGE	Segment begins on a 256-byte address boundary (8 LSBs = 0)

Table 2.12(*e*) Combine-type attributes

PUBLIC	Concatenates segment with the same name
COMMON	Overlaps from beginning segments with the same name
AT	Locates the segment at the 16-bit paragraph number evaluated from the expression
STACK	The segment is a part of the run-time stack segment
Memory	Locates the segment at an address above all other segments

At the end of the group of statements of for the code segment ENDS directive is essential. The 'ASSUME' directive assigns the segment registers that holds the base address of the program segments.

Example

```
ASSUME      CS:SEGA, DS : SEGB, SS : SEGC
```

Here, CS holds the base address of segment SEGA, DS holds the base address of segment SEGB, and SS holds the base address of segment SEGC. The 'ASSUME' directive is written at the beginning of a code segment just after the SEGMENT directive.

Example

```
SEGA        SEGMENT   PARA PUBLIC 'CODE'
ASSUME      CS : SEGA
MOV         AX.BX
:
SEGA        ENDS
```

The data definition directives create storage at the time of assembly and can initialize a variable to a string value. The data definition is summarized as

Directives	*Description*	*Number of bytes*
DB	Define byte	1
DW	Define word	2
DD	Define double word	4
DQ	Define quad word	8
DT	Define ten bytes	10

Origin (ORG)

Sets the current origin to a new value. This is used to set the program or register address during assembly. For example, ORG 0100H tells the assembler to assemble all subsequent code starting at address 0100H.

The legal instructions now will be written from the starting address 0100H.

Here, the logical question arises as to what will happen if we do not subscribe to the directive ORG prior to writing the legal instructions. In such situations, the assembler assumes on it's starting address as 0000H. For any starting address other than 0000H, we must write the ORG directive.

The ORG directive tells the assembler to begin assembling the mnemonics at the specified in the operand field. There can be any number of ORG commands within a particular program. In the following program 2000H is used to specify the origin of the assembling of the program.

```
20000    ORG 20000H      ; Starts at location 2000H
20000    MOV CX,25H      ; (CX) = 25H
20002    MOV AX,35H      ; (AX) = 35H
20004    MUL CX          ; Multiply contents of CX to contents of AX
20006    HLT
```

END

The END is also an optional assembler directive. It is the last statement in an assembly language program. All statements following the END will be ignored by the assembler.

Normally we place all definitions at the start of the program so that we can find them easily and we can use them as a part of the documentation. We can divide the definition into different groups such as input/output/units, variable names, fixed memory, address, and parameters.

All programs require an ORG and an END directive. The ORG statement presets the assemblers location counter. Anything that appears in column 1 is a user-defined label that refers to that line (unless it is a comment). Once the starting address is selected, the assembler exactly knows where to place the object code byte. The starting address is chosen by the ORG (origin directive). This directive is placed in the opcode field. Consider the following example (this is a textbook example that is chosen to illustrate concepts rather than do something useful). It is a program with only assembler directives and, therefore, do es nothing.

```
          ORG 0400                 Start of the program (there is no program here)
          ORG 0100                 Start of the data area
Time      DC B6                    Define variable 'Time' and give it a value 6
Time1     DC B9                    Define variable 'Time 1' and give it a value 9
Day       DC W3                    Define 'Day' and store 3 in that location
Week      DS B7                    Define 'Week' as an array of 7 bytes
Friday    EQU 5                    Equate the name 'Friday' to the number 5
Sunday    EQU 7                    Equate the name 'Sunday' to the number 7
Test      DC B Friday + Sunday     Define 'Test' and give it the value??
Test1     CD W Time + Friday       Define 'Test 1 and give it the value???
          END 0400                 End of program and entry point
```

Note that you can use an expression as an operand in the 68000 assembler (e.g., Friday + Sunday). The assembler evaluates the expression and the result used. For example, MOVE,B # Friday-Sunday, D0 D0 is identical to MOVE.B #-2, D0.

Let us assemble the program and see what code is generated (you should be able to do this by hand). The purpose of this example is to get you to think about the values represented by?? and ???.

```
1.  00000400                ORG 0400           ; Start of the program (there is none!)
2.
3.  00001000                ORG      1000      ; Start of the data area
4.  00001000 06             Time :   DC.B 6    ; Define variable Time and give it a value 6
5.  000001001 09            Time 1   DC.B 9    ; Define variable Timel and give it a value 9
6.  00001002 0003           Day :    DC.W 3    ; Define Day and store 3 in that location
7.  00001004 00000007       WEEK :   DC.B 7    ; Define Week as an array of 7 bytes
8.  00000005                Friday : EQU 5     ; Equate the name "Friday" to the number 5
9.  00000007                Sunday : EQU 7     ; Equate the name "Sunday" to the number 7
10. 00001100B 0C            TEST :   DC.B      ; Define Test and give it the value ??
                                     Friday + Sunday
11. 0000100C 1005           TEST1 :  DC.W Time ; Define Test1 and give it the value ???
                                     + Friday
12.
13. 00000400                END 0400           ; End of program and entry point
```

Let us examine the above program. The ORG 1000H sets the location counters to 1000H. All data and instructions are loaded sequentially after this point until the next ORG. 'TIME' is a user-defined label that indicates the address of the assembly directive DC.B 6 . This address is, of course, 1000H and the value 6 is loaded into address 1000H. Note that the value is a byte value and the next free address is 1001H.

The value of TIME 1 is 1001H (i.e., the next free location). The number 9 is loaded into that location by DC.B 9. The next free address is 1002H, the numerical value of DAY. At location 'DAY' the word 3 is stored. That is, the 16-bit value 0000000000000011H is put in memory location 1002H The next free location is 1004H (Why? Because we have stored a word that is 2B). The location we have called 'WEEK' is 1004H. The directive DC.B 7 simply moves the location counter on by 7B so that the next free address is 100BH. The purpose of this directive is to reserve 7B starting at address WEEK for future use. Assembler directives FRIDAY EQU 5 and SUNDAY EQU 7 do not do anything at all until you write 'FRIDAY' or 'SUNDAY'.

Now consider the line TEST DC.B FRIDAY + SUNDAY. The value of TEST (i.e., the address of the current memory location) is 100BH. A byte value is loaded at this address. The value is FRIDAY + SUNDAY that is 5 + 7 = 12 (i.e., 0CH). Finally, what is TESTI and what is loaded there? The next free address after TEST is 100CH. The word loaded into that address is TIME + FRIDAY. We know FRIDAY is 5. What is TIME ? It is not 6! TIME is a label that defines a memory location. The value of time is 1000H. Therefore, the value of TIME is 1000 + 5 = 1005. Memory location 100C contains the value 1005. The next free location is 100EH. The common pseudo-operations given in Table 2.12(*f*).

Table 2.12(*f*) Common assembler directives

Directives	*Effect*
ORG	Defines starting point in memory
EQU	Equates a data to a label
DB	Define byte(s) (8-bits)
DW	Defines word(s) (16-bits)
DD	Defines double word(s) (32-bits)
DQ	Defines quad word(s) (64-bits)
DT	Defines ten byte(s) (80-bits)
END	End of the program
ENDP	End of a procedure
ENDS	End of a segment
FAR	Specifies a far address as in JMP FAR
NEAR	Specifies a near address as in JMP NEAR
PROC	Defines beginning of a procedure

PTR	Indicates a memory pointer
STACK	Indicates that a segment is the stack segment
.286	Selects the 80286 instruction set
.286P	Selects the protected mode 80286 instruction set
.386	Selects the 80386 instruction set
.386P	Selects the protected mode 80386 instruction set
.486	Selects the 80486 instruction set
.486P	Selects the protected mode 80486 instruction set
.287	Selects the 80287 numeric coprocessor
.387	Selects the 80387 numeric coprocessor
PAGE	Skips to next page of listing
RES (reserve)	Reserves memory space for variable data
LIST	Provides listing of assembly language program
NAME	Assigns a name to the module when programs consisting of several modules

A program may have several origins as indicated below

```
ORG     RESET
   •
   •  Start of program
   •
ORG     INT1
   •
   •  Service routine for Interrupt 1
   •
ORG     Main
   •  Main program
   •
ORG     VARS
   •
   •  RAM data storage
```

Memory model directives

MODEL : defines the memory model

TINY = all data and code in one 64KB segment

SMALL = 64KB data and 64KB code

LARGE = 1MB data and 1MB code

DOSSEG

Orders segments in a DOS fashion, using paragraph alignment

CODE : defines the code segment

DATA : defines the data segment using name _ DATA

STACK size : defines the stack segment to be of size bytes

Modular program directive

Large program are broken into small segments called *modules*. Each module performs a specific function and has its own code and data segments. During execution of one module, the other module can be called or data residing in the other module can be accessed. A portion of the program that can be called for execution from the other module is called *procedure*. The beginning and end of a procedure must be marked with directive statement. The PROC (procedure) directive marks the beginning of the procedure and the *end of procedure* (ENDP) marks the end. The procedures are of 2-types

- Near procedure, and
- Far procedure.

Whenever a procedure is called, the return address has to be saved on the stack. After completing the code in the called procedure, the address is used to return execution back to the point of initiation.

In a near procedure call, the code-offset address (contents of IP) is saved on the stack. Hence, the near procedure is called from the same code segment. On the other hand, far procedure the contents of both code segment (CS) and the code offset (IP) are saved on the stack. For this reason, far procedure can be called from any code segment. Depending upon the type of procedure, the return (RET) instruction at the end of the procedure restores the IP or both CS and IP from the stack.

The procedure can be called from other modules provided its name is made *public* by using PUBLIC directive. This is done before entering the procedure.

Example

```
PUBLIC          SUB_NEAR
SUB_NEAR        PROC
                :
                RET
SUB_NEAR        ENDP
```

In this example PROC does not have either NEAR or FAR attribute in its operand field. For no presence of any operand, the NEAR is assumed as the default attribute. Thus, this procedure can be called only from other modules in the same code segment.

The FAR attribute can be used to call from module in the other segment as

```
PUBLIC  SUB_FAR
        SUB_FAR    PROC     FAR
                   :
                   RET
        SUB_FAR    ENDP
```

Procedure (PROC) is also called subroutine. The PROC and ENDP (end of procedure) indicate the start and end of the procedure (subroutine). Both directives require a label to indicate the name of the subroutine. The PROC directive must be followed with NEAR or FAR. A NEAR directive is one that resides in the same code segment as the program. A FAR PROC may reside at any location in memory system. We often tell NEAR PROC as local, and FAR PROC as *global*. The global term is used to define a PROC that can be used by any program, while the local defines a PROC that is used by the current program.

Example

```
; Procedure that adds BX, CX, and DX with the sum in the AX
0000              ADDEM  PROC FAR
0000    03,09     ADD  BX,CX
0002    03,0A     ADDBX.DX
0004    8B,C3     ADDAX,BX
0006    CB        RET
0007              ADDEM  ENDP
```

In this simple program, though the PROC is short, it does illustrate how to use a PROC and ENDP.

A special case of the procedure used in MASM 6.0. The procedure directive specifies and automatically save any registers used within the procedure. The USES statement specifies what registers are used by the procedure so the assernbler can automatically save before the procedure begins and restores before the procedure ends with the RET instruction.

Example

The ADD PROC USES AX,BX,CX statement automatically pushes AX,BX and CX on the stack before the procedure begins and pops them from the stack before RET instruction executes at the end of the procedure.

```
; procedure that includes the USES directive to save BX.CX.DX on the stack
0000              ADDS PROC NEAR USES BX,CX,DX
0003    03, D8    ADDBX,AX
```

```
0005    03, CB    ADDCX,BX
0007    03.D1     ADD DX.CX
0009    8B,C2     MOV AX,DX
000B    RET
000F    ADDS      ENDP
```

Group

The group directive is used to tell the assembler to group the logical segments named after the directive into one logical group segments. The assembler sends a message to the linker and/or locator telling to link the segments such that segments are physically in the same 64KB segment.

Example

```
SMALL SYSTEM GROUP  CODE DATA STACK-SEG
ASSUME CS:SMALL-SYSTEM, DS:SMALL-SYSTEM, SS:SMALL-SYSTEM
```

LABEL

The LABEL directive is used to give a name to the current content of the location counter. At the start of the assembly process, the assembler initializes a location counter to keep the track of the memory locations assigned to the program. The content of the location counter is updated as the assembly of program execution proceeds. Whenever the assembler comes across the LABEL directive, it assigns the declared label with the current contents of the location counter. The type of LABEL may be NEAR, FAR, BYTE, or WORD label etc.

The FAR jump can utilize the LABEL directive. The FAR jump cannot take directly to a label with a colon. The label CONTINUE if

```
CONTINUE    LABEL FAR
```

The LABEL directive is used to refer the data segment with data type, byte or word as

```
DATA        SEGMENT
DATAS       DB 20H  DUP (?)
DATA-LAST   LABEL BYTE FAR
DATA ENDS
```

After reserving 20H location for DATAS, the next location will be assigned a label DATA-LAST with its type as byte and FAR .

EXTRN (External) and PUBLIC (Public)

The EXTRN directive tells the assembler that the names, procedure, and label declared after this directive have already been declared in some other module of the assembly program. However, the names, procedure, and labels declared in the other modules must be declared public using PUBLIC directive.

Module-1	*Module-2*
PUBLIC SUB	EXTRN SUB : FAR
CSEG 1 SEGMENT	CSEG 2 SEGMENT
⋮	⋮
SUB PORC FAR	CALL SUB
MOV AX, BX	⋮
⋮	CSEG 2 ENDS
SUB RET	The external call is made from module-2 to the procedure SUB that resides in module-1. Hence, procedure SUB defined as PUBLIC in module-1.
ENDP	
⋮	
CSEG 1 ENDS	

Note that 2-modules have different code segments, CSEGI and CSEG2.

PTR (Pointer)

The pointer is used to declare the type of label, variable or memory operand. Either BYTE or WORD prefixes the operator PTR.

Example

MOV AL, BYTE PTR [SI] → Move the contents of memory addressed by SI (8-bit) to AL.

MOV AX, WORD PTR[2000H] → Moves 16-bit contents of memory 2000H to AX i.e., (2000H) → AL and (2001H) → AH.

The PTR operator is also used to specify the type of jumps.

Example

JMP WORD PTR [BX] → NEAR jump

JMP DWORD PTR [BX] → FAR jump

GLOBAL

The labels, variable, constants, or procedure declared GLOBAL may be used by other modules of the program. Once a variable is declared GLOBAL , it can be used by any module of a particular program.

Example

ROUTINE PROC GLOBAL

Here, ROUTINE is declared GLOBAL.

LOCAL

The labels, variable, constants, or procedure declared LOCAL in a module are used that module only. Some other module at a later time may declare a particular data type LOCAL that previously was declared by another module or modules. Thus, the same label may serve different purpose for different modules of a program. A number of variables can be declared local by a single statement as

LOCAL a, b, DATA, ARRAY, ROUTINE.

Parameter passing

A program can enter into a subroutine by labeling the 1st instruction and ending the program with RET instruction. The main program must call the subroutine and initialize the stack pointer to ensure that return address is stored properly.

However, a more general subroutine is easier to use and is not restricted to particular memory location. The main program can place data and address in registers before calling the subroutine, and the subroutine can place the result in the registers. The data and address required by the subroutine is called **parameters** and the process of making these items available to the subroutine is called **passing parameters**. The more complex subroutine may require too many parameters or the registers available. Such subroutine uses an assigned area in memory or the stack or parameters and results.

Parameter passing refers to the exchange of data between modules. There are many ways this information can be exchanged.

Global data

The data is stored in memory accessible to all modules. The disadvantage of this technique is that any module that makes debugging harder may modify the data.

Consider the following simple program that adds 2-numbers together, storing the result. All data has been declared as common.

```
TITLE       Common Data
MODEL       Large
STACK       200H
DATA
NUM1        dW22
NUM2        dW32
RESULT      dW0
CODE
addnum      PROC FAR
            MOV AX,[NUM1]
            MOV BX,[NUM2]
            ADD AX, BX
            MOV [RESULT], AX
            RET
addnum      ENDP
START:      MOV AX, @ DATA
```

```
        MOV DS,AX
        CALL addnum          ; add NUMI and NUM2
        MOV AX,4C00H
        INT21H
        END START
```

Register variables

This technique involves passing and returning values using processor registers. Routines must ensure that they do not corrupt any registers other than those that have been specified. The programmer 1st determines which registers will be used and which can be altered (contents destroyed).

Consider the following implementation of the previous addition program to use register variables.

```
TITLE       Common Data
MODEL       Large
STACK       200H
DATA
NUMI        dW22
NUM2        dW32
RESULT      dW0
CODE
addnum      PROC FAR        ; accepts num1 in AX.NUM2 in BX , returns result in DX
            PUSH AX
            ADD AX,BX
            MOV DX,AX
            POP AX
            RET
addnum      ENDP
START :     MOV AX, @ DATA
            MOV DS.AX
            MOV AX,[NUM1]
            MOV BX,[NUM2]
            CALL addnum     ;add NUMI and NUM2
            MOV[result],DX
            MOV AX,4C00H
            INT 21H
            END START
```

The advantage is that only the calling module alters the data, whilst the module addnum only works on copies of the data. In this way, it is easier to track which modules affect the data variables. Disadvantage → limited no. of registers limits the no. of parameter that can pass.

Stack variables

Parameters may also be passed using the stack. This involves pushing the values onto the stack before the module is called. This may also involve pushing space onto the stack for a return result.

The module then accesses the parameters on the stack using the appropriate addressing mode.

Upon return to the calling module, the stack space is deallocated using appropriate POP or stack pointer adjustment instructions.

There are two ways in which data may be referenced using the stack.

Call by value

This refers the placing of copies of the data value on the stack. Only the copy is worked with, the original remains unmodified.

Call by reference

This refers to the passing of the address of the variable using the stack. This address is used to access the data, thus the original data is used.

Call by value is normally used for simple data types, whilst call by reference is used for data types like arrays and records, because of the amount of memory space they occupy (and stack space is normally limited).

Consider the following program for an MC6802 processor that uses Call by Value to add two variables together.

```
        CPU      6802
        HOF      MOT
        ORG      100H
        Num1:    DFB 10
        Num 2:   DFB 20
        Result:  DFB 0
Start:  PSHA                ; make room for result on stack
        LDAA     Num1
        LDAB     Num2
        PSHA                ; place copy Num1 on stack
        PSHB                ; place copy of Num 2 on stack
        JSR      Addup
        PULB                ; remove copy of Num 2
        PULA                ; remove copy of Num1
        PULA                ; get result from Addup
        STAA     Result
Exit:   BRA      Exit
```

```
Addup: TSX              ; transfer SP into IX register
       PSHA             ; save registers
       PSHB
       LDAA 02,X        ; get Num2
       LDAB 03,X        ; get Num1
       ABA              ; add Num1 and Num2
       STAA 04,X        ; store on stack for return
       PULB             ; recover original register values
       PULA
       RTS
       END Start
```

Parameter passing for 8088

Accessing the stack frame inside a module

Lets look at how a module handles the stack frame. Because each module will use the BP register to access any parameters, its first chore is to save the contents of BP.

```
PUSH BP
```

It then transfers the address of SP into BP; BP now points to the top of the stack.

```
MOV BP,SP
```

thus the 1st 2-instructions in a module will be the combination,

```
PUSH BP
MOV BP,SP
```

Allocation of local storage inside a module

Local variables are allocated on the stack using a

```
SUB SP, N
```

instruction. This decrements the stack pointer by the number of bytes specified by N. For example, a module might want to use temporary storage space for an integer **i**, which equates to the machine code instruction

```
SUB SP, 2
```

Pictorially, the stack frame looks like,

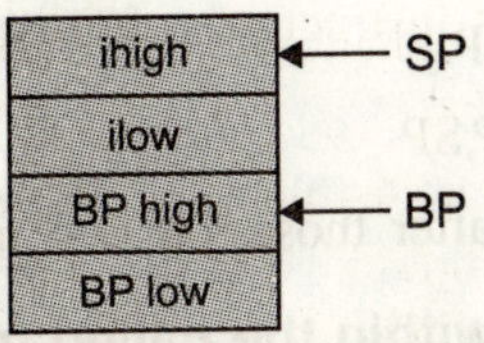

The local variable 'i' can be accessed using **SS : BP - 2**, so the statement,

i = 24;

is equivalent to

```
MOV [BP - 2], 18H
```

Note that 24-decimal is 18H.

Deallocation of local variables when the module terminates

When the module terminates, it must deallocate the space it allocated for the variable 'i' on the stack. Referring to the above diagram, it can be seen that BP still holds the top of the stack as it was when the module was 1st entered. BP has been used for two purposes,

- to access parameters relative to it
- to remember where SP was upon entry to the module

The deallocation of any local variables (in our case the variable 'i') will occur with the following code sequence,

```
MOV SP,BP      ; this recovers SP, deallocating 'i'
POP BP         ; SP now is the same as on entry to module
```

Passing of parameters to a module

Consider the following module call in a high level langauge.

```
ADD_two (10, 20);
```

The language pushes parameters (the values 10 and 20) right to left, thus the sequence of statements that implement this are,

```
PUSH AX        ; assume AX contains 2nd parameter i.e., integer value 20
PUSH CX        ; assume CX contains 1st parameter i.e., integer value 10
CALL add_two
```

The stack frame now looks like,

Stack	
Return	← SP
address	
00H	← 1st parameter. Integer 10D
0AH	
00H	← 2nd parameter. Integer 20D
14H	

Remembering that the first two statements of module add_two() are,

```
add_two:    PUSH BP
            MOV BP,SP
```

The stack frame now looks like (after those 1st two instructions inside add_two)

Accessing of passed parameters within the called module

It should be clear that the passed parameters to module add_two() are accessed relative to BP, with the 1st parameter residing at [BP+4], and the 2nd parameter residing at [BP+6].

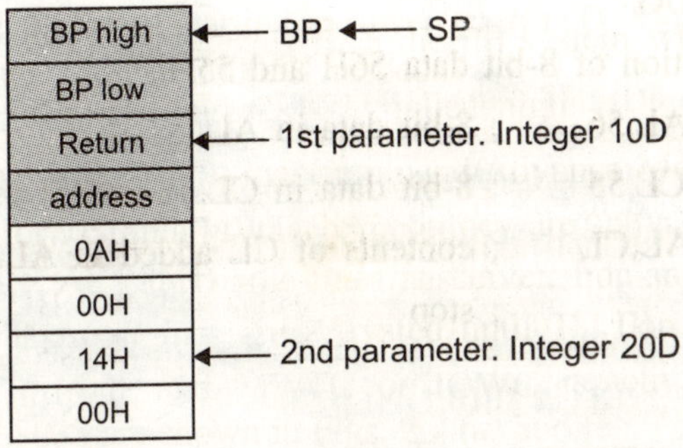

Programs

Programs of any microprocessor can be written and tested on any PC provided the PC on which the programs are to be tested contains the same microprocessor or of the higher version.

The debug is a program that allows programmers to program, debug, and execute assembly language programs. It allows the program to examine the contents of any registers of the microprocessor before and after the execution of any program. It is also used to run the program under single instruction mode.

Once you write debug on the c:\ Windows prompt as c:\ Windows\debug, the control is transferred to the debug prompt indicated by a hyphen '-'. Now, you have to write 'A' indicating assembling of the program. The microprocessor of the PC generates an address with the combination of the CS : OFFSET. This is called default address and it is like 1115 : 0100. If you want to write the program from a particular location (0110), write 0110 after the letter A as 1115 : 0100 A 0110.

The default address will change to 1115 : 0110. Start writing the program from this address. At the end of the program, press the Enter key, it will exit the 'A' command and starts displaying the debug command '-'. The GCS along with the last address of the program, will execute the program and contents of all registers will be displayed.

If you want to examine the contents of any register before execution of the program, even then you have to write at least one instruction and then assemble the program from the next address using 'gcs' control as explained below.

```
C:\WINDOWS>DEBUG
-a
1115:0100 MOV AX,0120
1115:0103
-gcs:0103
AX = 0120 BX = 0000 CX = 0000 DX = 0000 SP = FFEE BP = 0000 SI = 0000 DI = 0000
DS = 1115 ES = 1115 SS = 1115 CS = 1115 IP = 0103  OV UP EI NG NZ NA PO NC
1115:0103      D4BA          AAM BA
```

The contents of any registers of a microprocessor after execution of any program is automatically displayed as explained in subsequent examples.

```
C : \WINDOWS>DEBUG
1-A : Program for Addition of 8-bit data 56H and 55H.
0F6C : 0100    MOV AL,56      ; 8-bit data in AL.
0F6C : 0102    MOV CL,55      ; 8-bit data in CL.
0F6C : 0104    ADD AL,CL      ; contents of CL added to AL.
0F6C : 0106    HLT            ; stop
0F6C : 0107
-GCS : 0106
AX = 00AB BX = 0000 CX = 0055 DX = 0000 SP = FFEE BP = 0000 SI = 0000 DI = 0000
DS = 0F6C ES = 0F6C SS = 0F6C CS = 0F6C IP = 0106 OV UP EI NG NZ NA PO NC
0F6C : 0106 F4      HLT
Result in AL = ABH
C : \WINDOWS>DEBUG
2-A : Program for Addition of 16-bit data 2525H and 1526H.
0F6C : 0107    MOV AX,2525    ; 16-bit data in AX.
0F6C : 010A    MOV CX,1526    ; 16-bit data in CX.
0F6C : 010D    ADD AX,CX      ; contents of CX added to AX.
0F6C : 010F    HLT            ; stop
0F6C : 0110
-GCS : 010F
AX = 3A4B BX = 0000 CX = 1526 DX = 0000 SP = FFEE BP = 0000 SI = 0000 DI = 0000
DS = 0F6C ES = 0F6C SS = 0F6C CS = 0F6C IP = 010F NV UP EI PL NZ NA PE NC
0F6C : 010F F4      HLT
Result in AX = 3A4BH
C : \WINDOWS>DEBUG
3-A : Program for Addition of 16-bit data 1298H and 193EH.
0F6C : 0100    MOV BX,0300    ; BX contains the offset address.
0F6C : 0103    MOV CX,1298    ; 16-bit data in CX.
0F6C : 0106    ADD CX,[BX]    ; Addition in CX.
0F6C : 0108    HLT            ; stop.
0F6C : 0109
-GCS : 0108
AX = 0000 BX = 0300 CX = 2BD6 DX = 0000 SP = FFEE BP = 0000 SI = 0000 DI = 0000
DS = 0F6C ES = 0F6C SS = 0F6C CS = 0F6C IP = 0108 NV UP EI PL NZ AC PO NC
0F6C : 0108 F4      HLT
```

```
-ECS : 0300
0F6C : 0300 3E.   19.
Result in CX = 1298 + 193E = 2BD6H
C : \WINDOWS>DEBUG
```

4-A : Program for Addition of 16-bit data 081EH at [BX+05] and 1235H.

```
0F6C : 0109     MOV BX,0300         ; offset address in BX.
0F6C : 010C     MOV AX,1235         ; 16-bit data in AX.
0F6C : 010F     ADD AX,[BX+05]      ; 16-bit data from offset+displacement added in AX.
0F6C : 0112     HLT                 ; stop.
0F6C : 0113
-GCS : 0112
AX = 1A53 BX = 0300 CX = 2BD6 DX = 0000 SP = FFEE BP = 0000 SI = 0000 DI = 0000
DS = 0F6C ES = 0F6C SS = 0F6C CS = 0F6C IP = 0112 NV UP EI PL NZ AC PE NC
0F6C : 0112 F4       HLT
-ECS : 0305
0F6C : 0305 1E.   08.   D3.
Result in AX = 081E+1235 = 1A53H
C : \WINDOWS>DEBUG
```

5-A : Program for Addition and store the result.

```
0F6C : 0100     MOV CX,1289     ; 16-bit data in CX.
0F6C : 0103     ADD [0300],CX   ; (CX) added and stored at 0300 and 0301.
0F6C : 0107     HLT             ; stop
0F6C : 0108
-ECS : 0300
0F6C : 0300 3E.   19.   D8.
-GCS : 0107
AX = 0000 BX = 0000 CX = 1289 DX = 0000 SP = FFEE BP = 0000 SI = 0000 DI = 0000
DS = 0F6C ES = 0F6C SS = 0F6C CS = 0F6C IP = 0107  NV UP EI PL NZ AC PO NC
0F6C : 0107 F4       HLT
-ECS : 0300
0F6C : 0300 C7.   2B.   D8.
Result in 0300 = 2BC7H
C : \WINDOWS>DEBUG
```

6-A : Program for Addition store the result.

```
0F6C : 0100    MOV BX,0300    ; 16-bit offset address in BX.
0F6C : 0103    MOV CX,1289    ; 16-bit data in CX.
0F6C : 0106    ADD[BX],CX     ; (CX) added and stored at 0300 and 0301.
0F6C : 0108    HLT            ; stop
0F6C : 0109
-ECS : 0300
0F6C : 0300 3E.   19.   D8.
-GCS : 0108
AX = 0000 BX = 0300 CX = 1289 DX = 0000 SP = FFEE BP = 0000 SI = 0000 DI = 0000
DS = 0F6C ES = 0F6C SS = 0F6C CS = 0F6C IP = 0108  NV UP EI PL NZ AC PO NC
0F6C : 0108 F4      HLT
-ECS : 0300
0F6C : 0300 C7.   2B.   D8.
```

Result in 0300 = 2BC7H

C : \WINDOWS>DEBUG

7-A : Program for Multiplication of A8H and 25H.

```
0F6C : 0100    MOV AL,A8      ; 8-bit multiplicand in AL.
0F6C : 0102    MOV CL,25      ; 8-bit multiplier in CL.
0F6C : 0104    MUL CL         ; (CL) multiplied in AL.
0F6C : 0106    HLT            ; stop
0F6C : 0107
-GCS : 0106
AX = 1848 BX = 0000 CX = 0025 DX = 0000 SP = FFEE BP = 0000 SI = 0000 DI = 0000
DS = 0F6C ES = 0F6C SS = 0F6C CS = 0F6C IP = 0106  OV UP EI PL NZ NA PO CY
0F6C : 0106 F4      HLT
```

Result in AX = 1848H

C : \WINDOWS>DEBUG

8-A : Program for Multiplication of 2535H and 3535H.

```
0F6C : 0107    MOV AX,2535    ; 16-bit multiplicand in AX.
0F6C : 010A    MOV CX,3535    ; 16-bit multiplier in CX.
0F6C : 010D    MUL CX         ; (CX) multiplied in DX : AX.
0F6C : 010F    HLT            ; stop.
0F6C : 0110
```

```
-GCS : 010F
AX = ACF9 BX = 0000 CX = 3535 DX = 07BB SP = FFEE BP = 0000 SI = 0000 DI = 0000
DS = 0F6C ES = 0F6C SS = 0F6C CS = 0F6C IP = 010F OV UP EI PL NZ NA PO CY
0F6C : 010F F4    HLT
Result in DX : AX = 07BBH : ACF9H
C : \WINDOWS>DEBUG
9-A : Program for Multiplication of 2535H and 6595H.
0F6C : 0100   MOV AX,2535       ; 16-bit multiplicand in AX.
0F6C : 0103   MOV BX,0300       ; offset address in BX.
0F6C : 0106   MUL AX,[BX+05]    ; contents of offset + displacement multiplied in DX
                                  and AX.
0F6C : 0108   HLT               ; stop
0F6C : 0109
-ECS : 0300
0F6C : 0300 3E.11   19.22   D8.
-GCS : 0108
AX = 5CF9 BX = 0300 CX = 0000 DX = 0568 SP = FFEE BP = 0000 SI = 0000 DI = 0000
DS = 0F6C ES = 0F6C SS = 0F6C CS = 0F6C IP = 0108  OV UP EI PL NZ NA PO CY
0F6C : 0108 F4    HLT
Result in DX : AX = 0568H : 5CF9H
C : \WINDOWS>DEBUG
10-A : Program for Multiplication of -75H and 56H.
0F6C : 0100   MOV AL,75         ; 8-bit signed multiplicand in AL.
0F6C : 0102   MOV CL,56         ; 8-bit unsigned multiplier in CL.
0F6C : 0104   NEG AL            ; 2's complement of unsigned multiplicand.
0F6C : 0106   IMUL CL           ; (CL) multiplied in AX.
0F6C : 0108   NEG AX            ; 2's complement of AX.
0F6C : 010A   HLT               ; stop
0F6C : 010B
-GCS : 010A
AX = 274E BX = 0000 CX = 0056 DX = 0000 SP = FFEE BP = 0000 SI = 0000 DI = 0000
DS = 0F6C ES = 0F6C SS = 0F6C CS = 0F6C IP = 010A NV UP EI PL NZ AC PE CY
0F6C : 010A F4    HLT
Result in AX = 274EH
C : \WINDOWS>DEBUG
```

11-A : Program for Multiplication of -2535H and -3535H.

```
0F6C : 0100   MOV AX,2535   ; 16-bit unsigned multiplicand in AX.
0F6C : 0103   NEG AX        ; 2's complement of 16-bit unsigned multiplicand.
0F6C : 0105   MOV CX,3535   ; 16-bit unsigned multiplier in CX.
0F6C : 0108   NEG CX        ; 2's complement of (CX).
0F6C : 010A   IMUL CX       ; 2's complement of (CX) multiplied in DX and AX.
0F6C : 010C   HLT           ; stop.
0F6C : 010D
-GCS : 010C
AX = ACF9 BX = 0000 CX = CACB DX = 07BB SP = FFEE BP = 0000 SI = 0000 DI = 0000
DS = 0F6C ES = 0F6C SS = 0F6C CS = 0F6C IP = 010C OV UP EI NG NZ AC PO CY
0F6C : 010C F4      HLT
```

Result in DX : AX = 07BBH : ACF9H

C : \WINDOWS>DEBUG

12-A : Program for Multiplication of -2535H and 3535H.

```
0F6C : 0100   MOV AX,2535   ; 16-bit unsigned multiplicand in AX.
0F6C : 0103   NEG AX        ; 2's complement of (AX).
0F6C : 0105   MOV CX,3535   ; 16-bit unsigned multiplier in CX.
0F6C : 0108   IMUL CX       ; (CX) multiplied in DX : AX.
0F6C : 010A   NEG AX        ; 2's complement of AX.
0F6C : 010C   NEG DX        ; 2's complement of DX.
0F6C : 010E   HLT           ; stop
0F6C : 010F
-GCS : 010E
AX = ACF9 BX = 0000 CX = 3535 DX = 07BC SP = FFEE BP = 0000 SI = 0000 DI = 0000
DS = 0F6C ES = 0F6C SS = 0F6C CS = 0F6C IP = 010E NV UP EI PL NZ AC PO CY
0F6C : 010E F4      HLT
```

Result in DX : AX = 07BCH : ACF9H

C : \WINDOWS>DEBUG

13-A : Program for dividing 3535H by 55H.

```
0F6C : 0100   MOV AX,3535   ; 16-bit unsigned dividend in AX.
0F6C : 0103   MOV CL,55     ; 8-bit divisor in CL.
0F6C : 0105   DIV CL        ; (AX) divided by (CL).
0F6C : 0107   HLT           ; stop
```

```
0F6C : 0108
-GCS : 0107
AX = 15A0 BX = 0000 CX = 0055 DX = 0000 SP = FFEE BP = 0000 SI = 0000 DI = 0000
DS = 0F6C ES = 0F6C SS = 0F6C CS = 0F6C IP = 0107 NV UP EI PL NZ NA PO NC
0F6C : 0107 F4        HLT
```

Result in AX i.e., (AL) = Quotient = A0H and (AH) = Remainder = 15H

C : \WINDOWS>DEBUG

14-A : Program for dividing 25352636H by 3535H.

```
0F6C : 0108    MOV AX,2636   ; LS 16-bit dividend in AX.
0F6C : 010B    MOV DX,2535   ; MS 16-bit data in DX.
0F6C : 010E    MOV CX,3535   ; 16-bit divisor in CX.
0F6C : 0111    DIV CX        ; (DX : AX) divided by (CX).
0F6C : 0113    HLT           ; stop
0F6C : 0114
-GCS : 0113
AX = B305 BX = 0000 CX = 3535 DX = 0D2D SP = FFEE BP = 0000 SI = 0000 DI = 0000
DS = 0F6C ES = 0F6C SS = 0F6C CS = 0F6C IP = 0113  NV UP EI PL NZ NA PO NC
0F6C : 0113 F4        HLT
```

Result in AX i.e., (AX) = Quotient = B305H and Remainder = 0D2DH

C : \WINDOWS>DEBUG

15-A : Program for dividing -26F7H by 55H.

```
0F6C : 0100    MOV AX,26F7   ; 16-bit unsigned dividend in AX.
0F6C : 0103    MOV CL,55     ; 8-bit divisor in CL.
0F6C : 0105    NEG AX        ; 2's complement of dividend.
0F6C : 0107    IDIV CL       ; (AX) divided by (CL).
0F6C : 0109    NEG AL        ; 2's complement of AL.
0F6C : 010B    NEG AH        ; 2's complement of AH.
0F6C : 010D    HLT           ; stop.
0F6C : 010E
-GCS : 010D
AX = 1E75 BX = 0000 CX = 0055 DX = 0000 SP = FFEE BP = 0000 SI = 0000 DI = 0000
DS = 0F6C ES = 0F6C SS = 0F6C CS = 0F6C IP = 010D NV UP EI PL NZ AC PE CY
0F6C : 010D F4        HLT
```

Result in AX i.e., (AL) = Quotient = –75H, (AH) = Remainder = –1EH

```
C : \WINDOWS>DEBUG
16-A : Program for dividing 26F7H by -55H.
0F6C : 0100     MOV AX,26F7   ; 16-bit unsigned dividend in AX.
0F6C : 0103     MOV CL,55     ; 8-bit divisor in CL.
0F6C : 0105     NEG CL        ; 2's complement of (CL).
0F6C : 0107     IDIV CL       ; (AX) divided by 2's complement of (CL).
0F6C : 0109     NEG AL        ; 2's complement of AL.
0F6C : 010B     HLT           ; stop.
0F6C : 010C
-GCS : 010B
AX = 1E75 BX = 0000 CX = 00AB DX = 0000 SP = FFEE BP = 0000 SI = 0000 DI = 0000
DS = 0F6C ES = 0F6C SS = 0F6C CS = 0F6C IP = 010B NV UP EI PL NZ AC PO CY
0F6C : 010B F4        HLT
Result in AX i.e., (AL) = -75H = Quotient and (AH) = 1E = Remainder
C : \WINDOWS>DEBUG
17-A : Program for block move of byte using loop instruction.
0F6C : 0100     CLD           ; clear direction flag.
0F6C : 0101     MOV SI,0300   ; source address in SI.
0F6C : 0104     MOV DI,0400   ; destination address in DI.
0F6C : 0107     MOV CX,0005   ; count value loaded in CX.
0F6C : 010A     MOVSB         ; 8-bit data copied from memory addressed by SI to
                                addressed by DI.
0F6C : 010B     LOOPNZ 010A   ; loop unless count value CX = 0.
0F6C : 010D     HLT           ; stop
0F6C : 010E
-ECS : 0300
0F6C : 0300 3E.00  19.11  D8.22  1A.33  8E.44  1E.55  08.66 D3.77
0F6C : 0308 75.
-GCS : 010D
AX = 0000 BX = 0000 CX = 0000 DX = 0000 SP = FFEE BP = 0000 SI = 0305 DI = 0405
DS = 0F6C ES = 0F6C SS = 0F6C CS = 0F6C IP = 010D NV UP EI PL NZ NA PO NC
0F6C : 010D F4        HLT
-ECS : 0400
0F6C : 0400 00.   11.   22.   33.   44.   01.   C7.
```

```
C : \WINDOWS>DEBUG
18-A : Program for block move of byte using loop instruction
0F6C : 010E    CLD             ; clear direction flag.
0F6C : 010F    MOV SI,0300     ; source address in SI.
0F6C : 0112    MOV DI,0400     ; destination address in DI.
0F6C : 0115    MOV CX,0003     ; count value loaded in CX
0F6C : 0118    MOVSW           ; 16-bit data moved from memory addressed by SI to
                                 addressed by DI.
0F6C : 0119    LOOPNZ 0118     ; loop unless count value CX = 0.
0F6C : 011B    HLT             ; stop
0F6C : 011C
-ECS : 0300
0F6C : 0300 00.   11.   22.   33.   44.   55.   66.   77.
-GCS : 011B
AX = 0000 BX = 0000 CX = 0000 DX = 0000 SP = FFEE BP = 0000 SI = 0306 DI = 0406
DS = 0F6C ES = 0F6C SS = 0F6C CS = 0F6C IP = 011B NV UP EI PL NZ NA PO NC
0F6C : 011B F4      HLT
-ECS : 0400
0F6C : 0400 00.   11.   22.   33.   44.   55.   C7.   06.
C : \WINDOWS>DEBUG
19-A : Program For Block Move of Bytes using Repeat Instruction
0F6C : 011C    CLD             ; clear direction flag
0F6C : 011D    MOV SI,0300     ; source address in SI.
0F6C : 0120    MOV DI,0400     ; destination address in DI.
0F6C : 0123    MOV CX,0005     ; count value for number of bytes.
0F6C : 0126    REPNZ           ; repeat data unless CX = 0.
0F6C : 0127    MOVSB           ; move data bytes addressed by SI to DI.
0F6C : 0128    HLT             ; stop
0F6C : 0129
-ECS : 0300
0F6C : 0300 00.   11.   22.   33.   44.   55.   66.   77.
0F6C : 0308 75.
-GCS : 0128
AX = 0000 BX = 0000 CX = 0000 DX = 0000 SP = FFEE BP = 0000 SI = 0305 DI = 0405
DS = 0F6C ES = 0F6C SS = 0F6C CS = 0F6C IP = 0128 NV UP EI PL NZ NA PO NC
0F6C : 0128 F4      HLT
```

```
-ECS : 0400
0F6C : 0400 00.   11.   22.   33.   44.   55.   C7.   06.
C : \WINDOWS>DEBUG
20-A : Program for block move of 16-bit word using repeat instruction
0F6C : 0129    CLD            ; clear direction flag
0F6C : 012A    MOV SI,0300    ; source address in SI.
0F6C : 012D    MOV DI,0400    ; destination address in DI.
0F6C : 0130    MOV CX,0003    ; count value for number of bytes.
0F6C : 0133    REPNZ          ; repeat data unless CX = 0.
0F6C : 0134    MOVSW          ; move data of 16-bit addressed by SI to DI.
0F6C : 0135    HLT            ; stop
0F6C : 0136
-ECS : 0300
0F6C : 0300 00.   11.   22.   33.   44.   55.   66.   77.
0F6C : 0308 75.
-GCS : 0135
AX = 0000 BX = 0000 CX = 0000 DX = 0000 SP = FFEE BP = 0000 SI = 0306 DI = 0406
DS = 0F6C ES = 0F6C SS = 0F6C CS = 0F6C IP = 0135 NV UP EI PL NZ NA PO NC
0F6C : 0135 F4      HLT
-ECS : 0400
0F6C : 0400 00.   11.   22.   33.   44.   55.   C7.   06.
C : \WINDOWS> DEBUG
21-A : Program to find out 1's complement of a string of bytes.
0F6C : 0100    MOV SI,0300
0F6C : 0103    MOV DI,0400
0F6C : 0106    MOV CX,0010
0F6C : 0109    LODSB
0F6C : 010A    NOT AL
0F6C : 010C    STOSB
0F6C : 010D    LOOPNZ 0109
0F6C : 010F    HLT
0F6C : 0110
-ecs : 0300
```

```
0F6C : 0300 01.   02.   03.   04.   05.   06.   07.   08.
0F6C : 0308 09.   0A.   0B.   0C.   0D.   0E.   0F.   10.
-gcs : 010f
AX = 00EF BX = 0000 CX = 0000 DX = 0000 SP = FFEE BP = 0000 SI = 0310 DI = 0410
DS = 0F6C ES = 0F6C SS = 0F6C CS = 0F6C IP = 010F NV UP EI PL NZ NA PO NC
0F6C : 010F F4        HLT
-ecs : 0400
0F6C : 0400 FE.   FD.   FC.   FB.   FA.   F9.   F8.   F7.
0F6C : 0408 F6.   F5.   F4.   F3.   F2.   F1.   F0.   EF.
C : \WINDOWS>DEBUG
```

22-A : Program to find out 2's complement of a string of bytes.

```
0F6C : 0100     MOV SI,0300     ; source address.
0F6C : 0103     MOV DI,0400     ; destination address
0F6C : 0106     MOV CX,0010     ; count value of bytes.
0F6C : 0109     LODSB           ; load AL with data addressed by SI and SI = SI+1.
0F6C : 010A     NEG AL          ; 2's complement of (AL).
0F6C : 010C     STOSB           ; store (AL) addressed by DI and DI = DI+1
0F6C : 010D     LOOPNZ 0109     ; loop unless CX = 0.
0F6C : 010F     HLT             ; stop.
-ECS : 0300
0F6C : 0300 3E.01  19.02  D8.03  1A.04  8E.05  1E.06  08.07
0F6C : 0308 75.09  40.0A  B8.0B  01.0C  42.0D  B9.0E  FF.0F  FF.10
-GCS : 010F
AX = 00F0 BX = 0000 CX = 0000 DX = 0000 SP = FFEE BP = 0000 SI = 0310 DI = 0410
DS = 0F6C ES = 0F6C SS = 0F6C CS = 0F6C IP = 010F NV UP EI NG NZ NA PE CY
0F6C : 010F F4        HLT
-ECS : 0400
0F6C : 0400 FF.   FE.   FD.   FC.   FB.   FA.   F9.   F8.
0F6C : 0408 F7.   F6.   F5.   F4.   F3.   F2.   F1.   F0.
C : \WINDOWS>DEBUG
```

23-A : Program to convert binary to gray code

```
1091 : 0100     MOV BX,0300     ; offset address in BX.
1091 : 0103     MOV SI,0150     ; source address in SI.
1091 : 0106     LODSB           ; load AL addressed by SI and SI = SI+1.
```

```
1091 : 0107    XLAT           ; the contents addressed by (BX)+(AL) is moved to AL.
1091 : 0108    MOV [SI],AL    ; (AL) moved to memory addressed by SI.
1091 : 010A    HLT            ; stop.
1091 : 010B
-ecs : 0300
1091 : 0300 D3.00  A2.01  26.03  D3.02  A3.06  3A.07  D3.05 A2.04
1091 : 0308 25.0c  D3.0d  A2.0f  63.0e  DB.0a  A2.0b  64.09 DB.08
-ecs : 0150
1091 : 0150 07.
-gcs : 010A
AX = 0004 BX = 0300 CX = 0000 DX = 0000 SP= FFEE BP = 0000 SI= 0151 DI= 0000
DS = 1091 ES = 1091 SS = 1091 CS = 1091 IP = 010A NV UP EI PL NZ NA PO NC
1091 : 010A F4        HLT
-ecs : 0150
1091 : 0150 07.    04.
C : \WINDOWS>DEBUG
```

24-A : Program to add multi bytes (16-bit word).

```
0F6C : 0100    CLC            ; clear carry
0F6C : 0101    MOV CX,0004    ; 16-bit word count in CX.
0F6C : 0104    MOV SI,0300    ; source address in SI.
0F6C : 0107    MOV DI,0308    ; destination address in DI.
0F6C : 010A    MOV AX,[SI]    ; (AX) filled with data addressed by SI.
0F6C : 010C    ADC [DI],AX    ; (AX) added to 16-bit data addressed by DI and DI+1.
0F6C : 010E    INC SI         ; increment SI
0F6C : 010F    INC SI         ; increment SI.
0F6C : 0110    INC DI         ; decrement DI.
0F6C : 0111    INC DI         ; decrement DI.
0F6C : 0112    DEC CX         ; decrement CX.
0F6C : 0113    JNZ 010A       ; jump till CX ≠ 0.
0F6C : 0115    HLT            ; stop.
0F6C : 0116
0F6C : 0116
-ECS : 0300
0F6C : 0300 3E.01  19.02  D8.03  1A.04  8E.05  1E.06  08.07
```

```
0F6C : 0308 75.0A  40.0B  B8.0C  01.0D  42.0E  B9.0F  FF.10  FF.11
-GCS : 0115
AX = 0807 BX = 0000 CX= 0000 DX = 0000 SP = FFEE BP = 0000 SI = 0308 DI = 0310
DS = 0F6C ES = 0F6C SS = 0F6C CS = 0F6C IP = 0115 NV UP EI PL ZR NA PE NC
0F6C : 0115 F4      HLT
-ECS : 0308
0F6C : 0308 0B.   0D.   0F.   11.   13.   15.   17.   19.
0F6C : 0310 8B.
```

In this program, the 2nd stream of data bytes are washed out and result bytes are stored in those location

```
C : \WINDOWS>DEBUG
25-A : Program to add multi bytes (16-bit word).
0F6C : 0100     CLC             ; clear carry
0F6C : 0101     MOV SI,0300     ; source address in SI.
0F6C : 0104     MOV CX,[SI]     ; (SI) and (SI+1) transferred to CX i.e., count value.
0F6C : 0106     MOV DI,0500     ; destination address.
0F6C : 0109     INC SI          ; increment SI.
0F6C : 010A     INC SI          ; increment SI.
0F6C : 010B     MOV AX,[SI]     ; (SI) and (SI+1) transferred to AX.
0F6C : 010D     ADC AX,[SI+50]  ; (SI+50) is added to AX.
0F6C : 0110     STOSW           ; (AX) is stored at DI and DI = DI+1.
0F6C : 0111     LOOPNZ 0109     ; loop till CX ≠ 0.
0F6C : 0113     HLT             ; stop.
0F6C : 0114
-ECS : 0300
0F6C : 0300 3E.08  19.00  D8.3C  1A.42  8E.9A  1E.86  08.35  D3.40
-ECS : 0352
0F6C : 0352 C3.15  50.6A  33.84  C9.3B  FC.92  AC.85
-GCS : 0113
AX = 454F BX = 0000 CX = 0000 DX = 0000 SP = FFEE BP = 0000 SI = 0310 DI = 0510
DS = 0F6C ES = 0F6C SS = 0F6C CS = 0F6C IP = 0113 NV UP EI PL NZ NA PO CY
0F6C : 0113 F4      HLT
-ECS : 0500
0F6C : 0500 51.   AC.   1E.   C2.   C7.   C5.
```

In this program none of the data streams are washed out.

C : \WINDOWS>DEBUG

26-A : Program for arranging a string of bytes in descending order.

```
1091 : 0100    MOV SI,0300    ; SI loaded with source address.
1091 : 0103    MOV BX,[SI]    ; BX loaded with contents of SI and SI+1.
1091 : 0105    DEC BX         ; BX decremented.
1091 : 0106    MOV CX,[SI]    ; CX loaded with contents of SI and SI+1.
1091 : 0108    DEC CX         ; CX decremented.
1091 : 0109    MOV SI,0302    ; SI loaded with data addressed by SI and SI+1.
1091 : 010C    MOV AL,[SI]    ; AL loaded from (SI).
1091 : 010E    INC SI         ; increment SI.
1091 : 010F    CMP AL,[SI]    ; compare (AL) with data addressed by SI.
1091 : 0111    JAE 0119       ; If (AL) is more or equal to (SI), jump to 0119.
1091 : 0113    XCHG AL,[SI]   ; Else exchange (AL) by data addressed by SI.
1091 : 0115    DEC SI         ; decrement SI.
1091 : 0116    MOV [SI],AL    ; move (AL) to memory addressed by SI.
1091 : 0118    INC SI         ; increment SI.
1091 : 0119    LOOP 010C      ; loop unless to 010C.
1091 : 011B    DEC BX         ; decrement BX.
1091 : 011C    MOV SI,0300    ; SI loaded from 0300 and 0301.
1091 : 011F    JNZ 0106       ; jump if CX ≠ 0 to 0105.
1091 : 0121    HLT            ; stop
1091 : 0122
-ECS : 0300
0F6C : 0300 3E.06  19.00  D8.20  1A.25  8E.28  1E.15  08.07  D3.05
-GCS : 0121
AX = 0007 BX = 0000 CX = 0000 DX = 0000 SP = FFEE BP = 0000 SI = 0300 DI = 0000
DS = 0F6C ES = 0F6C SS = 0F6C CS = 0F6C IP = 0121 NV UP EI PL ZR NA PE NC
0F6C : 0121 F4      HLT
-ECS : 0300
0F6C : 0300 06.  00.  28.  25.  20.  15.  07.  05.
```

C : \WINDOWS>DEBUG

27-A : Program for arranging a string of bytes in ascending order.

```
0F6C : 0100    MOV SI,0300    ; SI loaded with source address.
```

```
0F6C : 0103    MOV BX,[SI]     ; BX loaded with contents of SI and SI+1.
0F6C : 0105    DEC BX          ; BX decremented.
0F6C : 0106    MOV CX,[SI]     ; CX loaded with contents of SI and SI+1.
0F6C : 0108    DEC CX          ; CX decremented.
0F6C : 0109    MOV SI,0302     ; SI loaded with data addressed by SI and SI+1.
0F6C : 010C    MOV AL,[SI]     ; AL loaded from (SI).
0F6C : 010E    INC SI          ; increment SI.
0F6C : 010F    CMP AL,[SI]     ; compare (AL) with data addressed by SI.
0F6C : 0111    JB 0119         ; If (AL) is less, jump to 0119.
0F6C : 0113    XCHG AL,[SI]    ; Else exchange (AL) by data addressed by SI.
0F6C : 0115    DEC SI          ; decrement SI.
0F6C : 0116    MOV [SI],AL     ; move (AL) to memory addressed by SI.
0F6C : 0118    INC SI          ; increment SI.
0F6C : 0119    LOOPNZ 010C     ; loop unless to 010C.
0F6C : 011B    DEC BX          ; decrement BX.
0F6C : 011C    MOV SI,0300     ; SI loaded from 0300 and 0301.
0F6C : 011F    JNZ 0106        ; jump if CX ≠ 0 to 0105.
0F6C : 0121    HLT             ; stop.
0F6C : 0122
-ECS : 0300
0F6C : 0300 3E.06  19.00  D8.20  1A.25  8E.28  1E.15  08.07  D3.05
-GCS : 0121
AX = 0025 BX = 0000 CX = 0000 DX = 0000 SP = FFEE BP = 0000 SI = 0300 DI = 0000
DS = 0F6C ES = 0F6C SS = 0F6C CS = 0F6C IP = 0121 NV UP EI PL ZR NA PE CY
0F6C : 0121 F4      HLT
-ECS : 0300
0F6C : 0300 06.  00.  05.  07.  15.  20.  25.  28.
0F6C : 0308 75.
C : \WINDOWS>DEBUG
```

28-A : Program to find out max value of the 16-bit word

```
0F6C : 0100    MOV SI,0300     ; source address in SI.
0F6C : 0103    MOV CX,0008     ; 16-bit word count value in CX.
0F6C : 0106    MOV AX,0000     ; AX initialized to 0000.
0F6C : 0109    CMP AX,[SI]     ; (AX) compared to (SI) and (SI+1).
```

```
0F6C : 010B   JAE 0111       ; If (AX) ≥ (SI) and (SI+1), jump to 0111.
0F6C : 010D   MOV AX,[SI]    ; move (SI) and (SI+1) to AX.
0F6C : 010F   INC SI         ; increment SI.
0F6C : 0110   INC SI         ; increment SI.
0F6C : 0111   LOOPNZ 0109    ; loop unless CX ≠ 0.
0F6C : 0113   MOV [SI],AX    ; move (AX) to memory addressed by SI and SI+1.
0F6C : 0115   HLT            ; stop
0F6C : 0116
-ECS : 0300
0F6C : 0300 3E.01  19.02  D8.03  1A.04  8E.05  1E.06  08.15  D3.07
0F6C : 0308 75.12  40.08  B8.09  01.0A  42.0B  B9.0C  FF.0D  FF.0E
-GCS : 0115
AX = 0E0D BX = 0000 CX = 0000 DX = 0000 SP = FFEE BP = 0000 SI = 0310 DI = 0000
DS = 0F6C ES = 0F6C SS = 0F6C CS = 0F6C IP = 0115  NV UP EI PL NZ AC PO CY
0F6C : 0115 F4      HLT
-ECS : 0310
0F6C : 0310 0D.   0E.   CD.   21.
C : \WINDOWS>DEBUG
```

29-A : Program to find out min value of the 16-bit word.

```
0F6C : 0100   MOV SI,0300    ; source address in SI.
0F6C : 0103   MOV CX,0008    ; 16-bit word count value in CX.
0F6C : 0106   MOV AX,FFFF    ; AX initialized with FFFF.
0F6C : 0109   CMP AX,[SI]    ; (AX) compared with (SI) and (SI+1).
0F6C : 010B   JB 010F        ; jump if (AX) is less than (SI) and (SI+1).
0F6C : 010D   MOV AX,[SI]    ; else move (SI) and (SI+1) in AX.
0F6C : 010F   INC SI         ; increment SI.
0F6C : 0110   INC SI         ; increment SI.
0F6C : 0111   LOOPNZ 0109    ; loop unless CX ≠ 0.
0F6C : 0113   MOV [SI],AX    ; move (AX) to memory pointed by SI and SI+1.
0F6C : 0115   HLT            ; stop
0F6C : 0116
-ECS : 0300
0F6C : 0300 3E.01  19.02  D8.03  1A.04  8E.05  1E.06  08.15  D3.07
0F6C : 0308 75.12  40.08  B8.09  01.0A  42.0B  B9.0C  FF.0D  FF.0E
```

```
-GCS : 0115
AX = 0201 BX = 0000 CX = 0000 DX = 0000 SP = FFEE BP = 0000 SI = 0310 DI = 0000
DS = 0F6C ES = 0F6C SS = 0F6C CS = 0F6C IP = 0115  NV UP EI PL NZ AC PO CY
0F6C : 0115 F4      HLT
-ECS : 0310
0F6C : 0310 01.   02.   CD.
C : \WINDOWS>DEBUG
```

30-A : Program to find out the maximum value of byte from a string of bytes

```
0F6C : 0100    MOV SI,0300     ; source address in SI.
0F6C : 0103    MOV CX,0010     ; count value of byte in CX.
0F6C : 0106    MOV AH,00       ; AH initialized with 00H.
0F6C : 0108    CMP AH,[SI]     ; (AH) compared with (SI).
0F6C : 010A    JAE 010E        ; jump if (AH) ≥ to (SI) to 010EH.
0F6C : 010C    MOV AH,[SI]     ; else, move (SI) to AH.
0F6C : 010E    INC SI          ; increment SI.
0F6C : 010F    LOOPNZ 0108     ; loop unless CX ≠ 0.
0F6C : 0111    MOV [SI],AH     ; (AH) transferred to memory pointed by SI.
0F6C : 0113    HLT             ; stop.
0F6C : 0114
-ECS : 0300
0F6C : 0300 3E.01  19.02  D8.03  1A.04  8E.05  1E.06  08.15  D3.07
0F6C : 0308 75.12  40.08  B8.09  01.0A  42.0B  B9.0C  FF.0D  FF.0E
-GCS : 0113
AX = 1500 BX = 0000 CX = 0000 DX = 0000 SP = FFEE BP = 0000 SI = 0310 DI = 0000
DS = 0F6C ES = 0F6C SS = 0F6C CS = 0F6C IP = 0113  NV UP EI PL NZ AC PO NC
0F6C : 0113 F4      HLT
0F6C : 0113 F4      HLT
-ECS : 0300
0F6C : 0300 01.   02.   03.   04.   05.   06.   15.   07.
0F6C : 0308 12.   08.   09.   0A.   0B.   0C.   0D.   0E.
0F6C : 0310 15.   D1.
C : \WINDOWS>DEBUG
```

31-A : Program to find out the minimum value from a string of bytes

```
0F6C : 0100    MOV SI,0300       ; source address in SI.
```

```
0F6C : 0103    MOV CX,0010     ; count value of byte in CX.
0F6C : 0106    MOV AH,FF       ; AH initialized with 00H.
0F6C : 0108    CMP AH,[SI]     ; (AH) compared with (SI).
0F6C : 010A    JB 010E         ; jump if (AH) less than (SI) to 010EH.
0F6C : 010C    MOV AH,[SI]     ; else, move (SI) to AH.
0F6C : 010E    INC SI          ; increment SI.
0F6C : 010F    LOOPNZ 0108     ; loop unless CX ≠ 0.
0F6C : 0111    MOV [SI],AH     ; (AH) transferred to memory pointed by SI.
0F6C : 0113    HLT             ; stop.
0F6C : 0114
-ECS : 0300
0F6C : 0300 3E.01  19.02  D8.03  1A.04  8E.05  1E.06  08.15  D3.07
0F6C : 0308 75.12  40.08  B8.09  01.0A  42.0B  B9.0C  FF.0D  FF.0E
-GCS : 0113
AX = 0100 BX = 0000 CX = 0000 DX = 0000 SP = FFEE BP = 0000 SI = 0310 DI = 0000
DS = 0F6C ES = 0F6C SS = 0F6C CS = 0F6C IP = 0113  NV UP EI PL NZ AC PO CY
0F6C : 0113 F4     HLT
-ECS : 0300
0F6C : 0300 01.  02.  03.  04.  05.  06.  15.  07.
0F6C : 0308 12.  08.  09.  0A.  0B.  0C.  0D.  0E.
0F6C : 0310 01.  D1.  CD.
C : \WINDOWS>DEBUG
```

32-A : Program to calculate number of bytes in a string of bytes

```
0F6C : 0100    MOV SI,0300     ; destination address in SI.
0F6C : 0103    MOV CX,FFFF     ; count in CX.
0F6C : 0106    MOV DI,0302     ; source address of data.
0F6C : 0109    MOV AL,03       ; identifier byte in AL.
0F6C : 010B    REPNE SCASB     ; Data byte addressed by DI compared with identifier till
                               ; matches with it decrementing CX and incrementing SI.
0F6C : 010D    NOT CX          ; 1's complement of CX
0F6C : 010F    MOV [SI],CX     ; (CX) moved to 0300.
0F6C : 0111    HLT             ; stop
0F6C : 0112
-ECS : 309
```

```
0F6C : 0309 02.03
-GCS : 0111
AX = 0003 BX = 0000 CX = 0008 DX = 0000 SP = FFEE BP = 0000 SI = 0300 DI = 030A
DS = 0F6C ES = 0F6C SS = 0F6C CS = 0F6C IP = 0111   NV UP EI PL ZR NA PE NC
0F6C : 0111 F4        HLT
-ecs : 0300
0F6C : 0300 08.   00.
```

The process is like storing an identifier at the last of the string of the byte. The program counts number of bytes till it matches with the identifier byte. Say, the identifier 03H is loaded at 0309 (at end of bytes 0308).

C : \WINDOWS> DEBUG

33-A : Program to multiply a single ASCII number to a string of ASCII.

```
0F6C : 0100    MOV SI,0300     ; starting address of string in SI.
0F6C : 0103    MOV DI,0308     ; destination address of result.
0F6C : 0106    MOV DL,34       ; DL loaded with 34H.
0F6C : 0108    MOV CX,0008     ; CX contains count value.
0F6C : 010B    AND DL,0F       ; (DL) logically ANDed with OFH i.e., MSN zeroed.
0F6C : 010E    MOV AL,[SI]     ; (AL) loaded with 1st ASCII number.
0F6C : 0110    INC SI          ; increment SI i.e., next location of ASCII number.
0F6C : 0111    AND AL,0F       ; MS nibble of AL zeroed.
0F6C : 0113    MUL DL          ; DL multiplied in AL.
0F6C : 0115    AAM             ; AH = AL/0AH = remainder
0F6C : 0117    ADD AL,[DI]     ; add data addressed by DI in AL.
0F6C : 0119    AAA             ; added with 00H which are in 1st DI location
0F6C : 011A    MOV [DI],AL     ; (AL) is unpacked decimal number moved to 0308.
0F6C : 011C    INC DI          ; increment DI
0F6C : 011D    MOV [DI],AH     ; (AH) = quotient in AAM operation moved to 0309.
0F6C : 011F    DEC CX          ; decrement CX.
0F6C : 0120    JNZ 010E        ; jump unless CX ≠ 0.
0F6C : 0122    HLT             ; stop.
0F6C : 0123
-ECS : 0300
0F6C : 0300 31.   32.   33.   34.   35.   36.   31.   32.
-ECS : 0308
0F6C : 0308 03.00
```

```
-GCS : 0122
AX = 0008 BX = 0000 CX = 0000 DX = 0004 SP = FFEE BP = 0000 SI = 0308 DI = 0310
DS = 0F6C ES = 0F6C SS = 0F6C CS = 0F6C IP = 0122  NV UP EI PL ZR NA PE NC
0F6C : 0122 F4        HLT
-ECS : 0308
0F6C : 0308 04.  08.  02.  07.  01.  06.  06.  08.

                               12345612
                               *4
                               48271668
```

This is obtained by multiply from left to right and dividing it by 10D with quotient added to the next left bit.

48271668 = 04 08 02 07 01 06 06 08

C : \WINDOWS>DEBUG

34-A : Program to divide a string of ASCII by a single ASCII.

```
0F6C : 0100   MOV DL,36       ; DL loaded with 36H.
0F6C : 0102   MOV SI,0300     ; starting address of ASCII string in SI.
0F6C : 0105   MOV DI,0308     ; destination address in DI i.e., result.
0F6C : 0108   MOV CX,0008     ; count value in CX.
0F6C : 010B   AND DL,0F       ; MSN of (DL) zeroed.
0F6C : 010E   XOR AH,AH       ; (AH) XORed = 00H
0F6C : 0110   LODSB           ; (AL) loaded with 1st ASCII byte.
0F6C : 0111   AND AL,0F       ; MSN of 1st ASCII number zeroed.
0F6C : 0113   AAD             ; AL = AH*0AH+AL, AH = 00H.
0F6C : 0115   DIV DL          ; divide AX by DL. AL = quotient, AH = remainder.
0F6C : 0117   STOSB           ; store AL pointed by DI and DI = DI+1.
0F6C : 0118   LOOPNZ 0110     ; loop unless CX = 0.
0F6C : 011A   HLT             ; stop.
0F6C : 011B
-ECS : 0300
0F6C : 0300 31.  32.  33.  34.  35.  36.  31.  32.
-GCS : 011A
AX = 0002 BX = 0000 CX = 0000 DX = 0006 SP = FFEE BP = 0000 SI = 0308 DI = 0310
DS = 0F6C ES = 0F6C SS = 0F6C CS = 0F6C IP = 011A NV UP EI PL NZ NA PE NC
0F6C : 011A F4        HLT
```

-ECS : 0308

0F6C : 0308 00. 02. 00. 05. 07. 06. 00. 02.

```
6)12345612(02057602    ××45      ××36
  12                    42        36
  ----                 ----      ----
  ××34                 ××36      ××12   02057602 = 00 02 00 05 07 06 00 02
    30                  36        12
    ----               ----      ----
    ××45                          ××
```

2.13 80186

The Intel 80186 processor is a high performance, highly integrated 16-bit microprocessor that is well suited for the embedded market segment. The 80186 is an evolution from the 8086. It is an embedded processor having the same instruction set of the 8086. It has on-chip timer system, interrupt controller, DMA controller, and clock generator. The development of the 80186 and its cousin 80188 was announced in 1983. Thus, 80186 and 80188 are processors with peripherals. The simplified architecture of the 80186 in the form of block diagram is illustrated in Fig. 2.13(*a*). The 8MHz 80186/80188 provides 2-times higher thruput than the standard 8086/8088.

The Intel 80186 processor has found its way into a broad spectrum of embedded designs such as modems, public and private PBX switching systems, cellular phones, POS and handheld terminals. The architecture can also be found in many real time environments such as robotics, factory automation, measurement control systems, sensors and test equipment, copiers, fax machines, printers and medical equipment. The Intel 80186 processor family, XL/EA/EB/EC, have been accepted in a broad range of applications, with over 10-thousand design wins and over 100 million units shipped. This legacy, coupled with upgrade paths with our Intel 386™ and Intel 486™ microprocessors ensures a path for your product's future design needs.

The 80C186/188XL processor family is pin-to-pin compatible with the previous 80186/188 processor family and adds an enhanced feature set. The high-performance CHMOS process allows the 80C186/188XL to run twice the clock rate of the HMOS 80186/188 while consuming less than one-fourth of the power.

The 80186 is completely object code compatible with the 8086. It has all registers of the 8086. It generates 20-bit physical address from a 16-bit segment register, and a 16-bit offset in the same way as the 8086. The $S_2 \Leftrightarrow S_0$ can be connected to the external bus controller chip such as 8288 for generating the maximum mode type signals. The 80186 can work at 2-different speeds.

- 8MHz(80186) ; and
- 6MHz(80186-6).

It can directly address 2^{20} = 1MB memory. It has added 7-new instructions and extends the PUSH, IMUL, and ROTATE/SHIFT instructions to include the immediate operands. Thus, in all, the 80186 has added 10-new instructions w.r.t. the 8086.

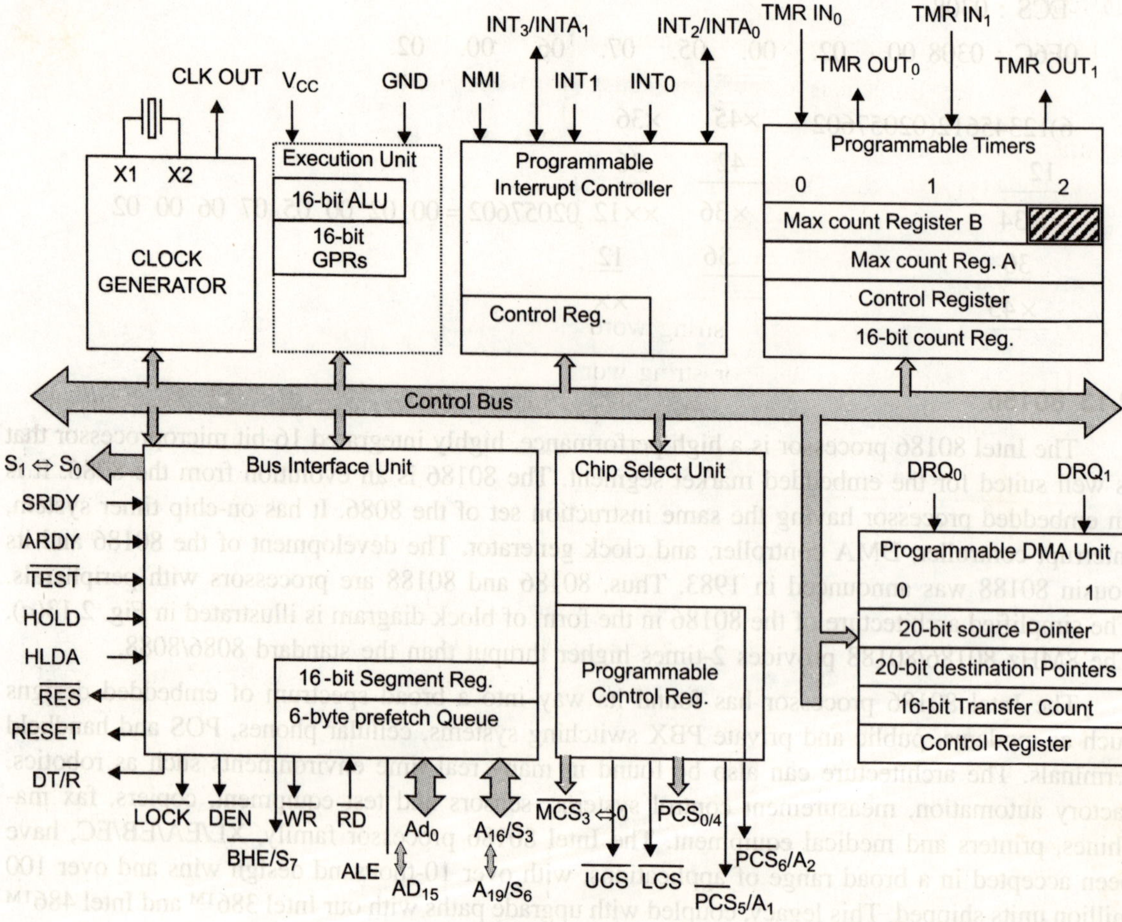

Fig. 2.13 (*a*) Block diagram of 80186

The architecture and instruction set of the 80188 and 80186 are identical except that the 80188 has only 8-bit data bus and the 808186 has 16-bit data bus. With this difference in mind, the 80186 will be used to discuss both the 80188 and 80186. Intel 80186 is the improved version of 8086, in the sense that it has faster instruction execution time, includes a few more instructions as compared to 8086. The 80186 has been strongly influenced by the need of modern software designer to exhibit *good, compact, fast and efficient codes* to be written.

Instructions

The 10-new instructions of the 80186 are divided into following groups

Data Transfer

PUSHA	⇒ Push all registers onto stack.
POPA	⇒ Pop all registers from stack.
PUSH immediate	⇒ Push immediate number onto stack.

Arithmetic

IMUL ⇒ Destination register, source, immediate data
⇒ Immediate data∗Source data into destination.

Logical

SHIFT/ROTATE ⇒ Destination, immediate data
⇒ Immediate data gives number of bit shift

String Instruction

INS ⇒ Input string byte or string word.
OUTS ⇒ Output string byte or string word.
Explanation of new instructions of the 80186

Name	*Mnemonics*	*Description*
Push all	PUSHA	It copies AX, CX, DX, BX, SP, BP, SI, and DI onto stack in that order. The SP value pushed is the contents of the SP register before AX is pushed.
Pop all	POPA	It Pops DI, SI, BP, SP, BX, DX, CX, and AX from the stack in that order. The popped SP value is discarded.
Input String	INS DST, PORT (INSB, INSW)	It copies a byte using INSB or a word using INSW from I/O port pointed by DX to memory location pointed by DI. After transfer, DI is updated using the same rules as other primitives.
Output String	OUTS PORT, SRC (OUTSB, OUTSW)	It copies a byte using OUTSB or word using OUTSW from memory location pointed by SI to I/O port pointed by DX. After transfer, SI is updated using same rules as other string primitives.
Detect Value out of range	BOUND REG, SRC	If (REG) ← (SRC) or (REG) → (SRC+2), then an interrupt of type-5 occurs.
Block entry	ENTER SIZE, level	1. Push BP. Save SP to a reg. TEMP. 2. If LEVEL = 0, go to step 8. 3. If LEVEL-1 = 0, go to step 7. 4. (BP) ⇐ (BP) – 2 5. Push to the stack pointed by BP. 6. LEVEL ⇐ LEVEL–1; go to step 3 7. Push TEMP. 8. (BP) ⇐ (TEMP.) 9 .(SP) ⇐ (SP) -SIZE
Block Exit	LEAVE	(SP) ⇐ (BP). Pop stack into BP
Note : No flags are affected.		

High Level Instructions

ENTER ⇒ Prepare stack for procedure entry.

BOUND ⇒ Detect values outside predefined range

Addressing modes

BOUND REG, SCR must point 2-consecutive words in the memory. For INS and OUTS, operands are actually ignored. For ENTER, SIZE is non-negative integer less than 65,536 and LEVEL is a non-negative integer less than 256.

Functions of instructions

IMUL ⇒ It is called signed multiply immediate instruction. It requires 3-operands. Its format is

IMUL REG,SRC,DATA

Its 1st operand must be REGISTER (REG) as the *destination*. The 2nd operand, (SRC) can be either memory or register of a word size as the *source*. The 3rd operand must be immediately available data *05H*. The IMUL multiplies the contents of source operand with immediately available data (05H) and lower 16-bits of the result is stored in the destination register, ignoring the upper 16-bit.

Example

IMUL BC,CX,05H.

Multiplies the contents of the CX register with the data 05H and stores the lower 16-bit result in the BC-register.

PUSH DATA

The immediate byte or word is pushed onto stack.

Example

PUSH 2032H; store 2032H on stack and decrement the SP by 2.

The 80186 included certain hardware improvement that facilitates to calculate the effective address by hardware instead of micro-codes. This resulted in much smaller execution time in clock cycles than that of the 8086 for most of the instructions.

Example

The IMUL instruction with a memory word operand and a memory to register ADD instruction will require 134 to 160 plus effective address (EA) and 16+EA clock cycles respectively for the 8086. The same instruction can be executed on the 80186 in 40 to 43 and 10-clock cycles respectively. Thus, it can be seen that the 80186 can exhibit same instructions 6 + EA clock faster as compared to that of the 8086.

ARCHITECTURE

The 80186 architecture consists of

- a clock generator,
- 2-independent DMA controller,
- a 16-bit timer unit consisting of three timers,

- an interrupt controller, and
- a chip selects unit into a single chip.

As 80186/80188 included an on chip clock generator. Just like the 8085, an external crystal is connected to it that divides the crystal frequency internally by 2. It means external crystal of 12 or 16MHz must be connected to generate 6 or 8MHz internal clock. The 80186/80188 can operate either at 6 or 8MHz internal clock. The requirement of crystal frequencies is

- 16MHz for 80186/80188
- 12MHz for 80186-6/80188-6 (an advance version)

The DMA unit of the 80186 has 2-DMA channels. Each DMA channel contains 20-bit source and destination pointers used to address the source and destination of the data transferred. Each of these 2-pointers can independently address either memory or I/O. After each DMA cycle, these pointers are incremented or decremented (remains constant for certain transfers). Each channel has a 16-bit count register that contains the number of DMA transfer to be performed. Data transfer can take place in the form of byte or 16-bit words. Each data transfer takes 2-bus cycle (a minimum of 8-clocks) : one cycle to read and the other to store data. Each channel has 16-bit control register. This register gives the following information

- whether 8, 16-bit data are to be transferred;
- whether DMA operation will be terminated with contents of count register = 0.
- whether source and destination pointer will be incremented/decremented or remains constant.
- whether interrupt will be generated or not after the last transfer.
- whether source or destination pointers will address memory or I/O.

The DMA channel can be programmed with different priorities. The DMA cycle will occur whenever the ST/ STOP bit in the control register is set to 1. Out of the 3-timer counters, two can be programmed to count the external events. The 3rd timer is not connected to any external pins. This timer is used to count the 80186 clock cycles. It can also be used to interrupt the 80186 after a programmed interval of time to provide a count pulse to DMA unit.

The mode control register allows the 80186 to operate in 2-modes. These modes are

- non-iRMX-86, and
- iRMX-86 modes.

The *non-iRMX* mode is also known as master mode. In this mode, the interrupt controller acts as the master interrupt controller for the system. In the *iRMX-86* mode, the interrupt controller operates as a slave to an external interrupt controller, such as the 8259, for the system. After reset, the 80186 starts operating automatically in the non-iRMX-86 (master) mode.

The internal interrupt controller of the 80186 allows the chip to receive interrupts from both external and internal sources. The 80186 accepts interrupts in the master mode only. The pin function of 80186 has been summarized in Table 2.13 (*a*).

Table 2.13(*a*) Pin functions of 80186

Symbol	*In/Out/ Z-state*	*Description*
V_{CC}		+5V (Power supply for ±10%)
V_{SS}		Ground
$\overline{RES}$	I	$\overline{RES}$ resets 80186/ 188. For proper reset, the $\overline{RES}$ must be held low for at least 50ms after power is applied. This pin connected to a RC circuit that generates a reset signal after power is applied. The reset location is identical to that of the 8086/ 88-FFFF0H
$AD_{15} \Leftrightarrow AD_0$	I/O-Z	Multiplexed address/ data bus. During T_1, the 80186 places $A_{15} \Leftrightarrow A_0$ on these pins, and during T_2, T_3, and T_4, the 80186 works as $AD_7 \Leftrightarrow AD_0$ and $A_{15} \Leftrightarrow A_8$ of the 8086.
A_{19}/S_6, A_{18}/S_5, A_{17}/S_4, A_{16}/S_3	O-Z	Multiplexed address and status pins that provide $A_{19} \Leftrightarrow A_{16}$ and Status $S_6 \Leftrightarrow S_3$. Unlike the 8086, $S_6 = 0$ indicates a processor cycle and a DMA cycle with $S_6 = 1$. The status S_5, S_4, and S_3 bits are always at logic-0.
$\overline{BHE}/S_7$	O-Z	When low, valid data are transferred through data bus. S_7 is always at logic-1 so no latch is required to demultiplex S_7.
$DT/\overline{R}$	O-Z	Data bus buffer direction control if attached to system like min of 8086.
$\overline{DEN}$	O-Z	Enables the external data bus buffers like min mode of 8086.
S_2, S_1, S_0	O-Z	Status bits to indicate the system with the type of bus transfer in effect. S_2 is used for $M/\overline{IO}$ ($IO/\overline{M}$ in 8088) signals and S_1 is used for $DT/\overline{R}$.
ALE/QS_0	O	Multiplexed pin. QS_0 is the queue status signal required for the 80187.
$\overline{WR}/QS_1$	O-Z	Used to write data in memory or I/O. The $\overline{WR}$ signal is active during T_3, TW, and T_4, and QSMD is selected when this I/O pin is grounded by the 80187.
$\overline{RD}/\overline{QSMD}$	O-I	Use to read data from memory or I/O. The $\overline{RD}$ signal is active during T_2, TW, and T_4, and QSMD is selected when the 80187 ground this I/O pin. $\overline{QS_0}$ and $\overline{QS_1}$ are available only during the queue status mode.
$\overline{LOCK}$	O-Z	Used to prevent other bus masters from accessing the system bus. If an instruction is prefixed with LOCK, the $\overline{LOCK}$ pin becomes low for the duration of locked instruction.
$\overline{TEST}$	I	$\overline{TEST}$ is normally connected to BUSY output of 80187. This pin is interrogated with the WAIT instruction.
NMI	I	Non-maskable interrupt. Always positive edge active.
HOLD, HLDA	I,O	Hold request and hold acknowledge are used by other bus masters for gaining control of the system bus.

(*Contd....*)

RESET	O	Indicates that the 80186 is being reset and can be used as a system reset signal.
X1,X2	I	External crystal connection terminals to generate CLK.
CLKOUT	O	Clock output provides a system clock whose frequency is one half of the frequency of the X2-X1 inputs.
INT_0,INT_1	I	Maskable interrupts requests.
$INT_2/INTA_0$ $INT_3/INTA_1$	I/O	Maskable interrupt requests or interrupt acknowledge signals for requests INT_0 and INT_1.
ARDY	I	Asynchronous ready is used by memory or a device to acknowledge CPU's read or write request. This signal is internally asynchronous. When this pin is tied to +5V, the 80186/ 188 functions normally and enters wait state when grounded.
SRDY	I	Synchronous ready. Can also be used as a ready signal but must be synchronized externally. If ARDY is used and SRDY is not, then SRDY should be grounded and vice versa.
TMR IN_0 TMR IN_1	I	These are used as external clocking source to timer-0 and 1.
$TMROUT_0$ TMR OUT_1	O	Outputs signals from timer-0 and 1. It can be programmed to output square wave or pulse.
DRQ_0,DRQ_1	I	DMA requests for 2-internal DMA channels.
$\overline{UCS}$	O	Selects the Upper portion of the memory map. This output is programmed to enable memory size of 1KB-256KB ending at location FFFFFH.
$\overline{LCS}$	O	This pin enables memory map starting at 00000H. This pin is programmed to select memory size of 1KB-256KB.
$MCS_3 \Leftrightarrow MCS_0$	O	Mid-range memory chip select.
$PCS_4 \Leftrightarrow PCS_0$	O	Peripheral chip selects signals.
PCS_5 / A_1 PCS_6 / A_2	O	Provides peripheral chip select signals or latched address signals A_1 and A_2.

The working of the 80186 is similar to the 8086 except in few cases. The major execution improvements in the working of the 80186 are

- 8086 ignores certain operation codes as 63H, 64H etc but 80186 accepts it and will present an interrupt of byte-6 (invalid instruction exception) which in case of 8086 are reserved interrupt pointers.
- In order to execute a rotate or shift instruction, the number of bits to shift is the count specified in the instruction modulo 32. This limits the number of shifts to 31. In case of 8086 it will need more than 1000 cycles to complete and it will cause a long delay if an interrupt of highest priority is requested.
- If 80186 is interrupted during the execution of a repeated string primitive with additional prefix, the return address will point to the 1st prefix of repeated primitive. Therefore, the interrupted string instruction can resume properly upon return from the interrupt.

- The LOCK signal in 80186 will not be activated until the locked instruction begins its operand reference buses cycles. But in case of 8086 (when used in maximum mode i.e. when $MN/\overline{MX}$ pin is held low) it can be initiated by a lock instruction prefix and is maintained until the end of the next instruction.
- The ESC [escape] instruction can be executed either in the same way or as on the 8086 or it can cause a byte-7 interrupt. This option is enabled by setting the most significant bit of the 80186 relocation register to 1 and is designed to facilitate emulation of floating point instructions. The interrupt's return address will point to the ESC instruction.
- The 80186 extends the range of negative numbers allowed as a quotient to include 8000H and 80H for the IDIV instruction.
- When a word is written into a location with offset = FFFFH, the 80186 writes the 2nd byte into the location with offset = 10000H instead of 0000. Similarly, if a PUSH with SP = 1 is executed, the 80186 writes the word into 2-consecutive physical locations.

These 7-instructions are responsible for improved hardware of the 80186 as compared to the 8086 in order to provide better execution process.

2.14 80286

In 1982 Intel announced the 80286, a high performance 16-bit microprocessor with on chip memory protection capabilities. The 80286 with 8MHz clock provides up to 6-times higher thruput than the 5MHz 8086. It has been primarily designed for multi-user/multi-tasking system. The 80286 is booted in real mode, and then it could be switched to protected mode by software command. However, it is not possible to switch the 80286 from protected mode back to real mode without resetting the microprocessor. The 80386 addressed this problem well and included the facilities to permit the switching from either mode to other via software commands. This announcement was welcome by the programmers. Figure 2.14 (*a*) shows functional block of the 80286. In contrast to the 8086, the 80286 has 24-address pins instead of 20. This allows the chip to access 2^{24} = 16,777,216 =16MB memory. The 80286 is upward compatible with 8086 and 8088. In other word, the software written for the 8086/ 8088 will run without any change on the 80286. The reverse is not true. Its address/ data pins are not multiplexed. Another major difference is the way in which the 80286 communicates with the coprocessor. It can be operated at 4 (80286-4). 6(80286-6), and 8MHz (80286) clock frequencies. The 80268 is housed in a quad type 68-pin surface mounted package with 24-bit address lines and 16-bit data lines.

Unlike the 8086 and 80186, the 80286 does not have on-chip clock generator circuitry. Hence, an external 82284 chip is required to generate the external clock. The 80286 has a single CLK pin for single-phase clock input. The external clock is divided by 2-internally to generate the internal clock. The 82284 also provides the 80286 $\overline{RESET}$ and READY signals.

The 80286 offers two modes of operations, namely, real mode for compatibility with existing 8086/8088 software base, and protected mode for enhanced system level features such as memory management, multitasking, and protection.

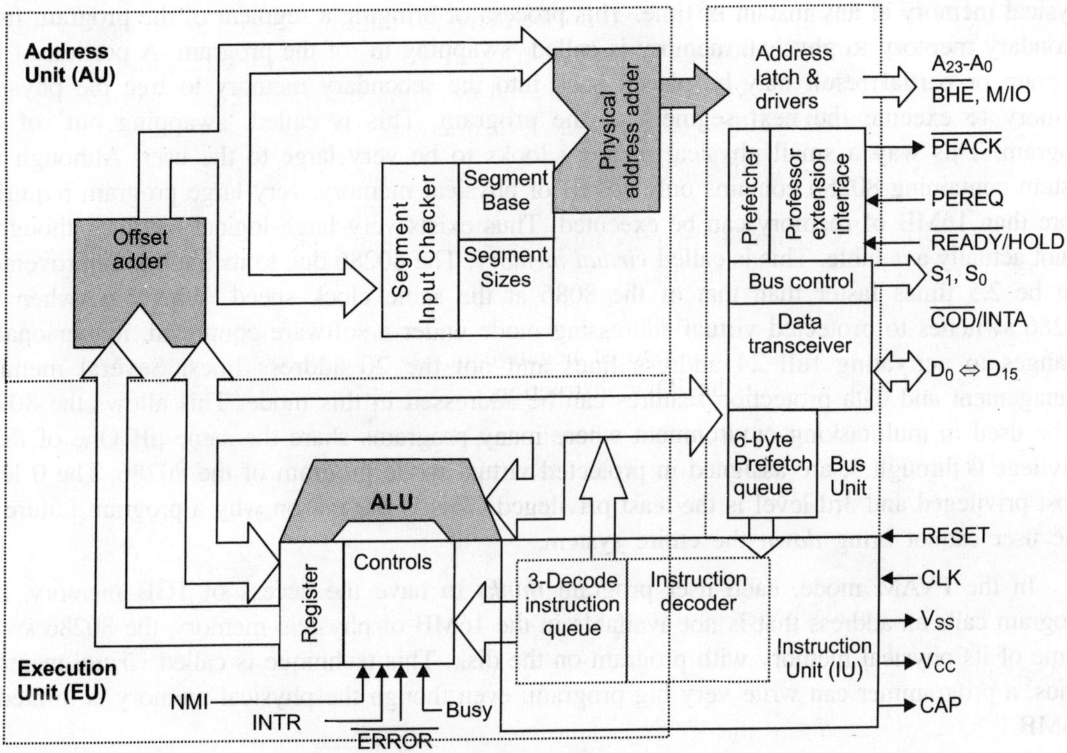

Fig. 2.14 *(a)* Functional block diagram of 80286

The bus unit (BU) interfaces the 80286 with outside world. It provides 16-bit data bus, 24-bit address bus, and control bus needed for bus interfacing. Unlike the 8086/8088, these buses are not multiplexed and the 80286 has separate 24-bit address, 16-bit data and control buses. This facility improves the performance of the 80286 w.r. to 8086.

The bus unit is responsible for performing all external bus operations. The bus unit also contains latch and drivers for the address bus, transceiver for the data bus, and control logic for generating control signals needed to perform the memory and I/O bus cycle. The address unit (AU) provides memory management and protection services. It discharges the duty of address generation, translation, and checking from the bus unit and hence further boost the performance of the 80286. It included the hardware for generating high-speed address calculation, virtual to actual address translation, and limit address rights attributes checks.

The maximum physical addressable memory of 16MB ($2^{24} = 2^4 \times 2^{20}$ = 16MB) by 80286 may not be sufficient to store the operating system and the set of application program to be executed by the microprocessors at the same time. This is achieved using the concept of memory management. The 80286 is the first microprocessor of Intel to support hardware for memory management. Let us consider a very big program or a set program is to be executed using 80286. The whole program is divided into number of segments (blocks). One segment of the program that is to be executed at that instant of time exists in the physical memory. Thus, the segment of the program that is to be executed is fetched from the secondary memory to the physical memory at any instant of time. This process of bringing a segment of the program from secondary memory to physical memory is called 'swapping in' of the program. A portion of the program or partial result may be placed back into the secondary memory to free the physical memory to execute the next segment of the program. This is called 'swapping out' of the program. This way a small physical memory looks to be very large to the user. Although the system containing 80286 contains only 16MB of physical memory, very large program requiring more than 16MB of memory can be executed. Thus exists very large logical memory, though it is not actually available. This is called *virtual memory*. The 80286 due to its internal improvement can be 2.5 times faster than that of the 8086 at the same clock speed. Where as, when the 80286 switches to protected virtual addressing mode under a software command, its personality changes to activating full 24 address lines and not the 20 address lines. Several memory management and data protection features can be addressed in this mode. This allows the 80286 to be used in multitasking environment where many programs share the same μP. One of the 4 privilege 0 through 3, are assigned in protected virtual mode program of the 80286. The 0 level most privileged and 3rd level is the least privileged. This is the reason why a program failure by one user cannot bring *down* the entire system.

In the PVAM mode, each user program *thinks* to have the access of 1GB memory. If a program calls an address that is not available in the 16MB of physical memory, the 80286 swaps some of its physical memory with program on the disk. This technique is called *virtual* memory. Thus, a programmer can write very big program, even though the physical memory is limited to 16MB.

The MS-DOS is even to-day used as an operating system for the 8086, 80186, 80286 etc. The MS-DOS neither supports multitasking nor virtual memory concept. The OS/2, the successor of the MS-DOS, is a multitasking operating system to allow many programs to run simultaneously. The OS/2 provides many window facilities to the user on the screen. Each window can run one program separately.

On the other hand, the 80286 in its PVAM has multi-tasking capability. The programmer can select one of these modes by loading appropriate data in the 16-bit *machine status word* (PSW) register by using load and store instructions. The examples of these instructions are

- LPSW ⇒ Load PSW register
- SPSW ⇒ Store PSW register.

The special instructions of the 80286 also support Operating System.

Architecture of 80286

Its architecture is divided into 4-separate processing units under dotted blocks as shown in Fig. 2.14 (*a*). These blocks are

- Bus Unit (BU),
- Instruction Unit (IU),
- Address Unit (AU), and
- Execution Unit (EU).

BU

The BU unit facilitates all memory and I/O read and write operations. The BU also performs data transfer between the 80286 and the coprocessor such as 8087. The prefetcher of BU prefetches instructions upto 6-bytes and places them in a queue.

IU

The IU translates or decodes upto 3-prefetched instructions and places them in the queue for execution by the execution unit.

AU

The AU calculates the 20-bit physical address based on the 16-bit contents of a segment register and a 16-bit offset just like the 8086. In this mode the 80286 address 1MB physical memory. The 80286 has 24-address pins. However, in real address mode, pins $A_{23} \Leftrightarrow A_{20}$ are ignored and $A_{19} \Leftrightarrow A_0$ pins are used to address 2^{20} = 1MB. In PVAM, however, the AU operates as a memory management unit (MMU) and utilizes all 24-address lines to provide $2^{24} = 2^{20} * 2^4$ = 16MB of physical memory.

The BU outputs memory or I/O address to devices connected to the 80286 after receiving them from AU. It is clear from the internal architecture that it has 2-queues

- a code queue,
- an instruction queue,
- and a separate set of logic for computing address.

The code queue feeds into the instruction unit that can supply the address unit with the advance address information. The instruction queue feeds into the execution unit that includes the μP registers and ALU. The inclusion of separate unit for calculating address and ability to compute address in advance significantly improves the performance of the 80286 over the 8086; and particularly useful when using the memory management feature of the PVAM.

Bit-0 of the status word determines the mode of the 80286. After a reset, bit-0 = 0 indicates that the 80286 is in the real address mode and starts executing at address FFFFF0H. The internal registers will be initialized as

Flag word = 0002H,	PSW = FFF0H,	IP = FFF0H
CS register = F000H,	DS register = 0000H	
ES register = 0000H,	SS register = 0000H.	

If PVAM operation is desired, this bit is set during system initialization. These 3-new bits have been defined in the PSW.

EU

The EU executes instructions from the IU sequentially. The EU has a flag register, general-purpose register, pointer register, index registers, and one 16-bit additional register called machine status word (PSW) register. The lower 4-bits of the PSW are used. One bit places the 80286 into protected virtual address mode (PVAM) while the other 3-bits control the processor extension (coprocessor) interface.

Registers of 80286

The 80286 microprocessor adds one major programmer-visible feature to the 8086 → protected mode operation. This text will not cover the 80286 protected mode of operation for a variety of reasons. First, the protected mode of the 80286 was poorly designed. Second, it is of interest only to programmers who are writing their own operating system or low-level systems programs for such operating systems. Even if we write software for a protected mode operating system like UNIX or OS/2, we would not use the protected mode features of the 80286. Nonetheless, it is worthwhile to point out the extra registers and status flags present on the 80286 just in case we come across them.

There are 3-additional bits present in the 80286 flags register. The I/O Privilege Level is a 2-bit value (bits 12 and 13). It specifies one of 4-different privilege levels necessary to perform I/O operations. These 2-bits generally contain 00B when operating in real mode on the 80286 (the 8086 emulation mode). The NT (*nested task*) flag controls the operation of an interrupt return (IRET) instruction. NT is normally zero for real-mode programs.

Besides the extra **3-bits in the flags register**, the 80286 also has **5-additional register** used by an operating system to support memory management and multiple processes. The registers are

- machine status word (MSW),
- global descriptor table register (GDTR),
- local descriptor table register (LDTR),
- interrupt descriptor table register (IDTR),
- task register (TR).

The 80286 is the 1st processor that supported the concept of virtual memory and memory management. The virtual memory means it does not physically (really) exists but it appears to the programmer that it is available.

About the only use a typical application program has for the protected mode on the 80286 is to access more than 1MB of RAM. However, as the 80286 is now virtually obsolete, and there are better ways to access more memory on later processors, programmers rarely use this form of protected mode.

Real address mode

Once the 80286 reset, it starts executing in the real address mode. This mode is referred to as real mode because physical memory address is produced directly by adding an offset to a segment base just as in the case of the 8086. In the real address mode, the 80286 addresses 2^{20} = 1MB memory locations. On the contrary, the 80286 calculates the address in the virtual mode in entirely different manner. The 80286 in the real address mode executes all the instructions of the 8086. The additional instructions of 80186 such as ENTER, LEAVE, and BOUND; and a few more instructions are used to switch the 80286 from the real address mode to the PVAM. The 80286 has several additional built-in interrupt types as listed in Table 2.14(*a*).

Many of these interrupts may not make much of the sense unless detail study of the 80286 is done. Similar to the 8086, the 80286 multiplies the interrupt type number by 4 and goes to the resulting address in the interrupt vector table to get the value of CS and IP to start the interrupt procedure. The interrupt vector table is in the 1st 1KB of memory in the real address mode. In the virtual mode, the interrupt vector table does not have fixed physical address in the memory.

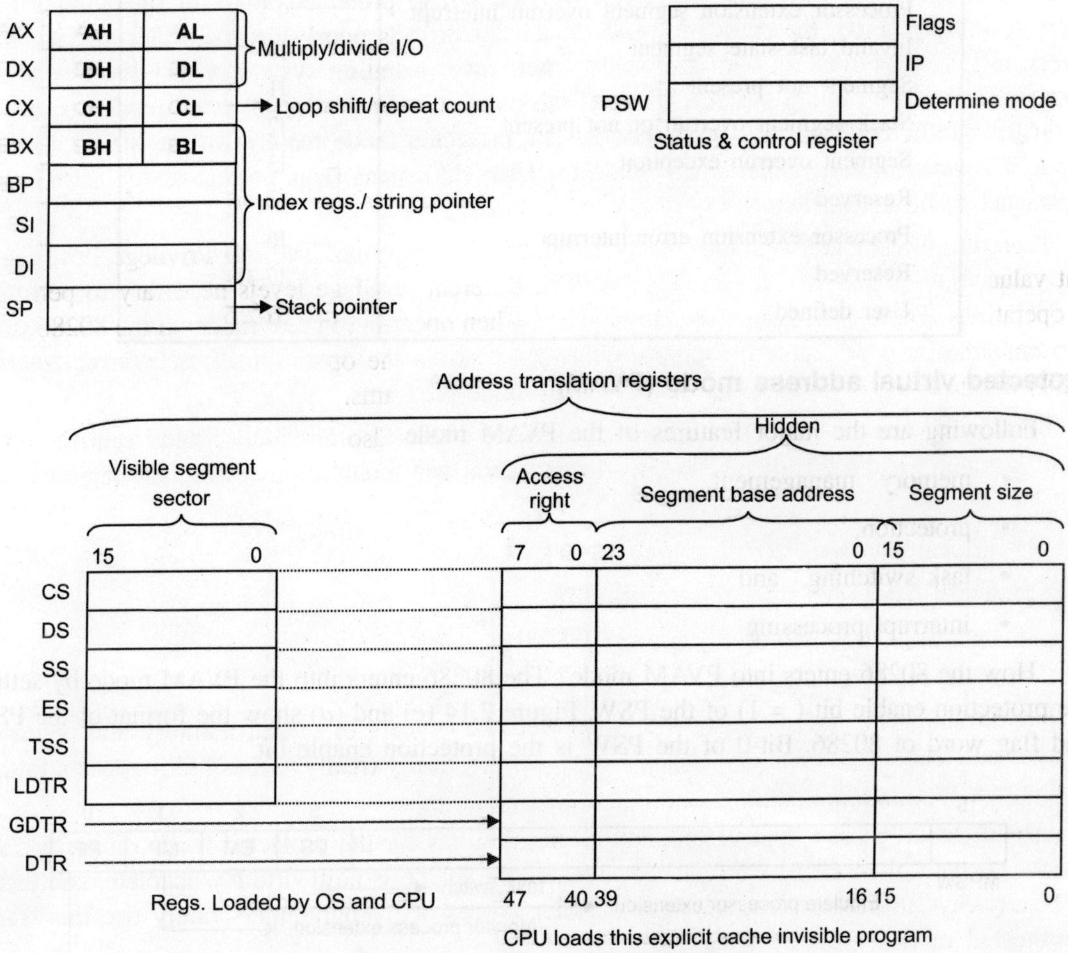

Fig. 2.14 (*b*) Registers of 80286

Table 2.14(*a*) Types of interrupts in 80286

Functions	*INTR No.*
Divide error exception	0
Single step interrupt	1
NMI interrupt	2
Break point interrupt	3
INT_0 detected overflow exception	4
BOUND range exceeded exception	5
Invalid opcode exception	6

(*Contd...*)

Processor extension not available exception	7
Interrupt table limit too small	8
Processor extension segment overrun interrupt	9
Invalid task-state segment	10
Segment not present	11
Stack segment overrun or not present	12
Segment overrun exception	13
Reserved	14, 15
Processor extension error interrupt	16
Reserved	17-31
User defined	32-255

Protected virtual address mode (PVAM)

Following are the major features of the PVAM mode

- memory management,
- protection,
- task switching, and
- interrupt processing

How the 80286 enters into PVAM mode? The 80286 enters into the PVAM mode by setting the protection enable bit (= 1) of the PSW. Figure 2.14 (*c*) and (*d*) show the format of the PSW and flag word of 80286. Bit-0 of the PSW is the protection enable bit.

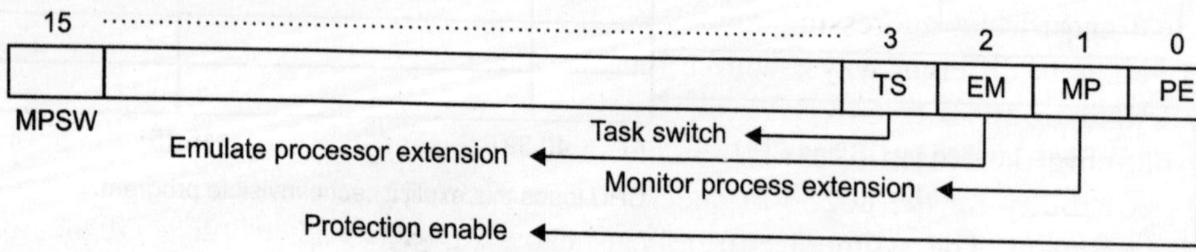

Fig. 2.14 (*c*) 80286 status word

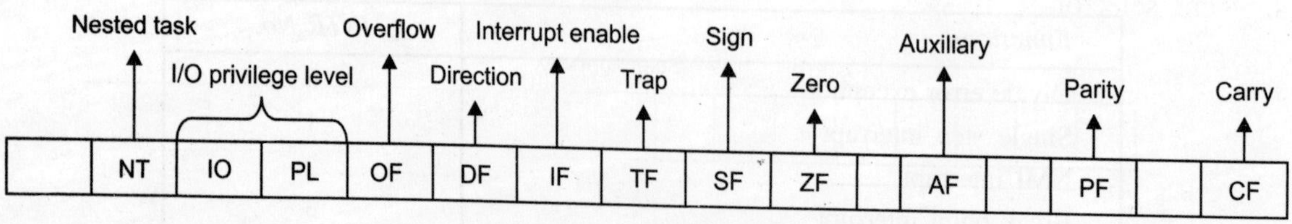

Fig. 2.14 (*d*) 80286 flag word

Bit-1, 2, and 3 are used to indicate whether a processor extension (coprocessor) is present or not. Bits are changed in the PSW by loading the desired word into register or memory location and executing the load machine status instruction (LMSW). Once executing the LMSW instruction enters the PVAM, the only way to get back to real address mode is by resetting the system.

Memory management

The memory management of the 80286 is based on address translation. The memory pointer of the 80286 includes two 16-bit word as

- one word is meant for a segment selection, and
- the other word as an offset onto the selected segment.

The 80286 is the 1st processor that supported the concept of virtual memory and memory management. The virtual memory means it does not physically (really) exists but it appears to the programmer that it is available in the system. The concept of virtual memory implemented using physical memory that access the secondary memory which is used as storage for data and program. The data and program are stored in the secondary memory initially. The segment of the program or data required for actual execution at that instant is fetched from the secondary memory into physical memory. After execution of this fetched segment the next segment required for further execution is again fetched from the secondary memory for further references. This continues until the complete program is executed.

The 80286 is capable of addressing virtual memory upto 1GB (2^{30} bytes) per task. Thus, complete virtual memory is mapped on the 16MB physical memory. In other words, if the program or data segments stored on the hard disk is larger than 16MB and is to be executed, it is fetched in terms of data or program segments of less than 16MB in size into the physical memory by swapping sequentially as sequence of execution.

Selectors

The real and virtual addressing modes compute the physical addresses from this selector and offset values in different ways. Unlike the 8086 segments, the segments for the 80286 can be of any length from 1B to 64KB. A size or limit is assigned to each segment when created. This way defining the length of the segments makes more efficient use of the memory. Also it provides possibility for the 80286 to check if a memory access accidentally goes outside the bounds of a segment. The segment currently used by a task is kept in physical memory. The 80286 computes the physical address from a 16-bit (selector) contents and a 16-bit offset just like the 8086/80186 in its real address mode. Figure 2.14 (*e*) shows the format of the 16-bit segment selector.

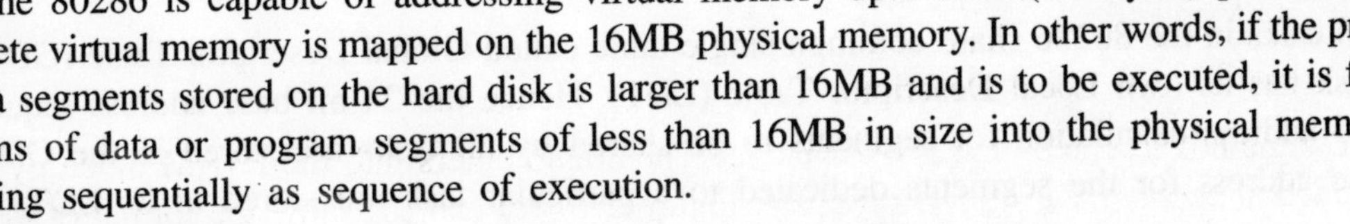

Fig. 2.14 (*e*) Segment selector format

The 16-bit selector is divided into a 2-bit Requested Privilege Level (RPL), a 1-bit Table Indicator (TI), and a 13-bit index. The functions of each bit of the segment selector are described in Table 2.14(*b*).

Table 2.14(*b*) Bit functions of segment selector

Bits	*Name*	*Functions*
1-0	Requested Privilege level	Indicates selectors privilege Level desired
2	Table Indicator	TI = 0, use global descriptor table TI = 1, use local descriptor table
3-15	Index	Select descriptor from table

In the descriptor table, the upper 13-bit (index) of a selector contains the number of the descriptor. This part of the selector is referred to as an index because the value of these bits, when internally multiplied by 8 points at the descriptor for that segment in a descriptor table. The index value is multiplied by 8 because each descriptor requires 8B in the descriptor table. Each descriptor is 8B wide and contains 24-bit base address required for physical address calculation. This address occupies only 3B out of the 8B. The composition of the rest of the bytes is

- 2-bytes for segment selectors,
- 1-byte for access rights,
- 2-bytes for segment size or limit.

All tasks in the 80286 share common single table called Global Descriptor Table (GDT) and each task has its own Local Descriptor Table (LDT). Hence the 24-bit base address required in physical address calculation for segments to be shared by all tasks are stored in the GDR and the base address for the segments dedicated to a particular task are stored in its LDT.

The Table Indicator (TI) bit selects one of the two tables, namely,

- Global Descriptor Table (GDT), or
- Local Descriptor Table (LDT).

With TI (bit-2) = 0, the GDT is selected as the lookup table by the upper 13-bit to index a segment descriptor. The LDT is used as the lookup table with TI = 1 by the upper 13-bits to index a segment descriptor in it. This is a form of indirect addressing. The selector points to the descriptor location in the descriptor table, and the descriptor at that location contains the actual base address and the other information about the desired segment. This seems to be the round about way to get to the start of a segment, but it is this indirect approach which allows several forms of protection to be built into the 80286.

The operating system uses the 2-bit Requested Privilege Level (RPL) for implementing the 80286's protection feature. The RPL is not used for physical address calculation. The contents of the 16-bit selector is shifted 4-times left and then the 6-bit offset is added to it to determine the physical address. In the virtual address mode (PVAM), the 32-bit address is called virtual address. Just like logical address, the virtual mode address also includes a 16-bit selector and a 16-bit offset. The 80286 determines the 24-bit physical address by 1st obtaining the 24-bit value from a table in the memory using segment value (selector) as an index rather than shifting the segment 4-times left as in the real address mode and then 16-bit offset is added to it. Figure 2.14 (*f*) shows the scheme of virtual address translation of the 80286.

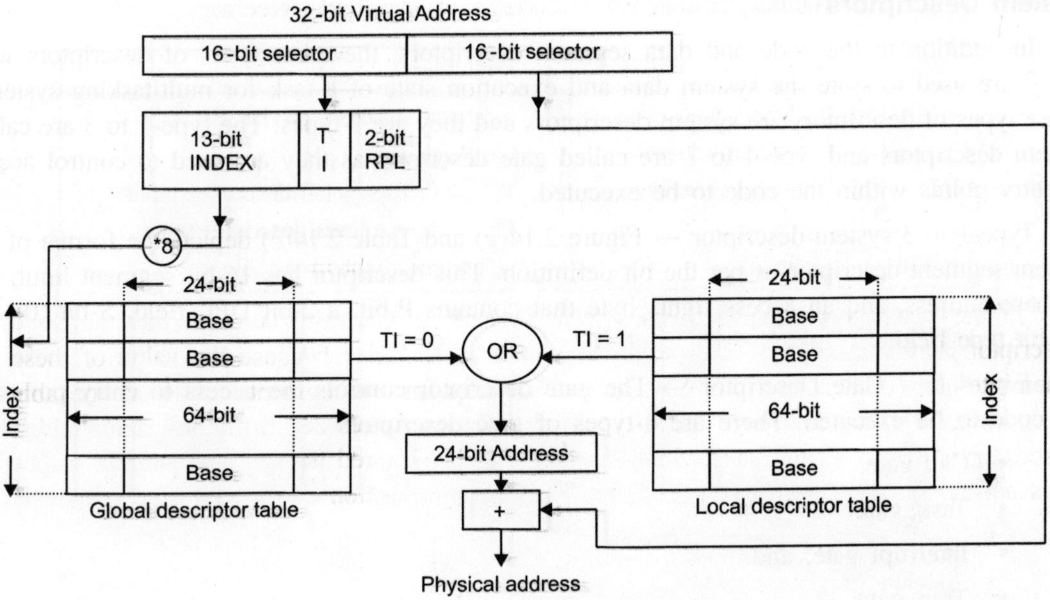

Fig. 2.14 *(f)* 80286 virtual address translation

Descriptors

The main frnotion of the descriptor is to keep the information about a segment and its access rights. In addition to this, the special types of descriptors are used to carry out additional functions such as **transfer of control** and **task switching**. The basic descriptors are for code, data, and stack segments in the 80286. The 80286 also has system control descriptor for special data segment and control transfer operations. Each descriptor is 8B long. The operating system uses the contents of descriptor (8B long) to support memory management operation.

A code or data segment descriptor contains 16-bit segment limit (2^{16} = 64KB block), 24-bit base address (physical address), 8-bit access right and remaining 16-bits are reserved for Intel upward compatibility.

Any segment access, violating the specified access rights in the descriptor, prevents memory cycle and generates an exception interrupt. The code segment descriptors are used to refer data segments.

S = 1, selects code/data/stack descriptor, S = 0, selects system control descriptor,

E = 1, indicates code segment, E = 0, indicates data segment

P = 1 (present), indicates whether segment is available in the physical memory

P = 0 (absent), indicates that the segment is not available in the physical memory

A = 1 (access), whether code/data/stack segment was accessed previously

A = 0(not access), the code/data/stack segment was not accessed previously

The μP will set A descriptor privilege level field indicating the accessibility of the descriptor.

System Descriptors

In addition to the code and data segment descriptors, the other types of descriptors with S = 0 are used to store the system data and execution state of a task for multitasking systems. These types of descriptors are system descriptors and they are 7-types. The type-1 to 3 are called system descriptors and type-4 to 7 are called gate descriptor as they are used to control access of entry points within the code to be executed.

Type-1 to 3 system descriptor → Figure 2.14(*g*) and Table 2.14(*c*) depicts the format of the system segment descriptor as per the bit definition. This descriptor has 16-bit segment limit, 24-bit base address, and an access rights byte that contains P-bit, a 2-bit DPL field, S-bit (0) and a 4-bit type field.

Type-4 to 7 Gate Descriptor ↪ The gate descriptor controls the access to entry points of the code to be executed. There are 4-types of gate descriptors.

- Call gate,
- Task gate
- Interrupt gate, and
- Trap gate.

The gate descriptor has the information about the destination of the control transfer, required stack manipulation, whether it is present in the memory or not, privilege level and its type. Gate descriptor provides a mechanism to keep track of the source and destination control transfer. Hence, the μP can perform the protection checks and control the entry point of destination codes.

Table 2.14(*c*) Format of system segment descriptor

Name	*Value*	*Description*
P	0	Descriptors contents not valid
	1	Descriptors contents valid
DPL	0 ↔ 3	Descriptor privilege level
TYPE	1	Available task state segment
	2	Local descriptor table
	3	Descriptor contents are valid
BASE	24-bit number	Base address of special system data segment in real memory
LIMIT	24-bit number	Offset of last byte in segment

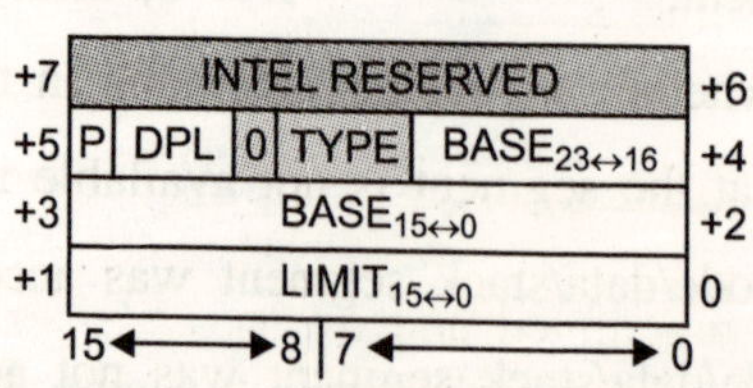

Fig. 2.14 (*g*) Descriptor structure of the 80286

Call gates are used to alter the privilege levels. Task gates are used to switch from one task to another. Interrupt and trap gates are used to specify corresponding service routine. Table 2.14(*d*) and (*e*) [Intel Corp] depicts the descriptor format and corresponding bit definition. If a destination selector does not refer to a correct descriptor type, exception-13 is generated. The task gate does not use destination-offset field. The task gate may only refer to a task state segment. The word count field is only used by a call gate descriptor to indicte the number of bytes to be transferred from the stack of the calling routine to the stack of the called routine, whcn a control transfer chagnes the priviliege levels. The access rights byte and its bit definition are the same as code or data segment descriptor.

Table 2.14(*d*) Format of system segment descriptor

Name	*Value*	*Description*
P	0	Descriptors contents not valid
	1	Descriptors contents valid
DPL	0 ↔ 3	Descriptor privilege level
TYPE	4	Call gate
	5	Task gate
	6	Interrupt gate
	7	Trap gate
Word count	0 ↔ 31	Number of words to be copied from callers stack to called proceeudre stack. Only used with call gate.
Destination selector	16-bit selector	Selector to target code segment (call interrupt of trap gate) selector to the target task segment (TSS gate)
Destination offset	16-bit offset	Entry point within the target code segment

Table 2.14 (*e*) Descriptor structure of the 80286

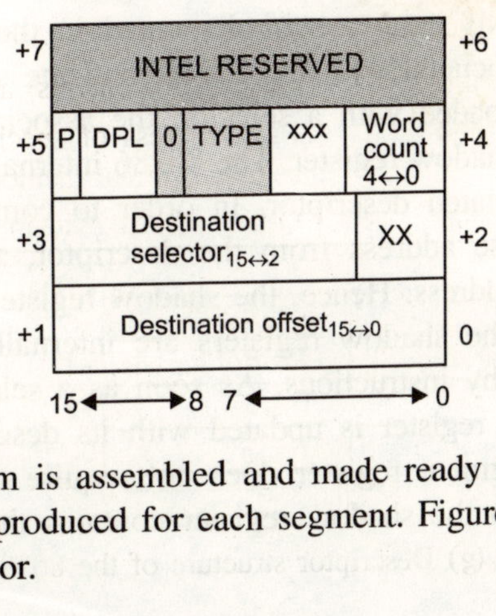

In PVAM when a program is assembled and made ready for execution, a unique, 8-byte quantity called a descriptor is produced for each segment. Figure 2.14 (*h*) shows the format for the segment and code descriptor.

Address translation registers

The virtual mode of address translation of the 80286 along with the 4-segment registers (CS, DS, SS, and ES) uses the Global Descriptor Table Registers (GDTR) and Local Descriptor Table Registers (LDTR). The composition of these registers is shown in Fig. 2.14 (*i*). The GDTR stores 24-bit base address and the length of the GDT (in byte) is one byte less than the capacity of the GDTR. The GDT is loaded during system initialization and is usually kept unchanged after that. The 80286 generates an interrupt if indexing beyond the GDT limit is attempted.

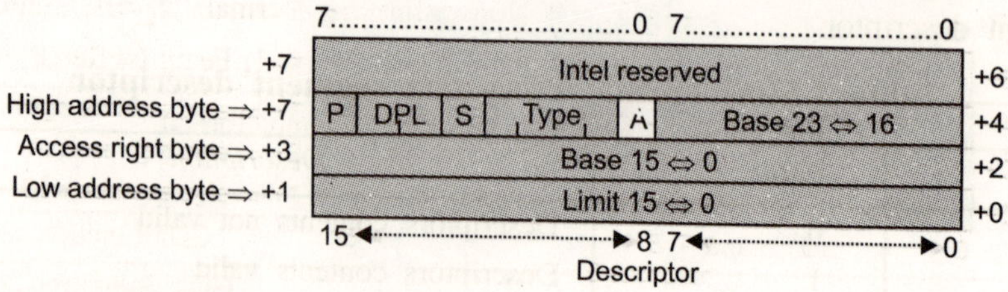

*Not used word should be filled with 0s

P ⇒ Present,

P = 1, means present and segment is mapped into physical memory,

P = 0, means not present. No mapping to physical memory exists. Base & limit are not used.

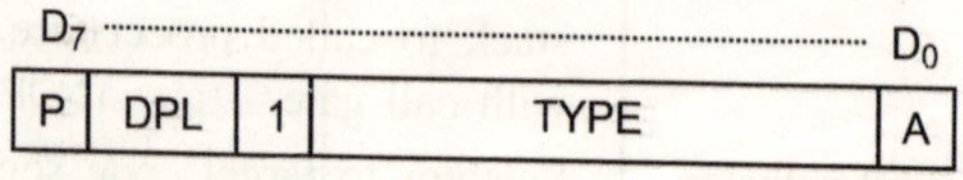

The LDTR stores a 16-bit selector for a selector in the present LDT. The GDT descriptor is indexed by this selector. The LDTR makes task switching very fast. As an example, the 80286 local address space can be modified during task switching by up-dating the LDTR.

The 80286 included improvement in the memory accessing speed by additional hardware. Inclusion of an internal register, the shadow register, with the LDTR and 4-shadow registers with the 4-segment registers does this. These shadow registers are internal to the 80286 and are not accessible via software. It is hardware implemented in the 80286. The shadow registers store respective descriptors associated with the LDTR and the segment register. Whenever the LDTR or segment register is loaded with a selector, the associated descriptor is automatically moved into the corresponding shadow register. The 80286 internally perform all steps to load the shadow register with the associated descriptor. In order to compute the physical address, the 80286 obtains the segment base address from the descriptor, adds to it the offset, and then access memory with physical address. Hence, the shadow registers have been included to speed up the memory referencing. The shadow registers are internally hardware arranged registers and hence cannot be accessed by instructions. As soon as a selector is moved into a segment register, the associated shadow register is updated with its descriptor automatically. Thus any memory accessed w.r.t. the segment register does not require referring to any look-up table, since the descriptor loaded into the shadow register contains the base address of the selected segment.

Access rights byte

DPL = Descriptor Privilege Level	$D_4 = 1$, Code or Data Segment
00 = Highest	$D_3 = 0$, Data Segment Descriptor, $D_3 = 1$, Code Segment Descriptor
01	$D_2 = 0$, Non-stack segment., $D_2 = 1$, Stack segment
10	$D_1 = 0$, Read only, $D_1 = 1$, Read/ Write
11 = Lowest	$D_0 = 0$, Segment not Accessed, $D_0 = 1$, Segment Accessed
	$D_2 = 0$, Nonconforming Normal, $D_2 = 1$, Conforming
	$D_1 = 0$, Fetch Only, $D_1 = 1$, Fetch or Read

Fig. 2.14 *(h)* Segment & code descriptor format

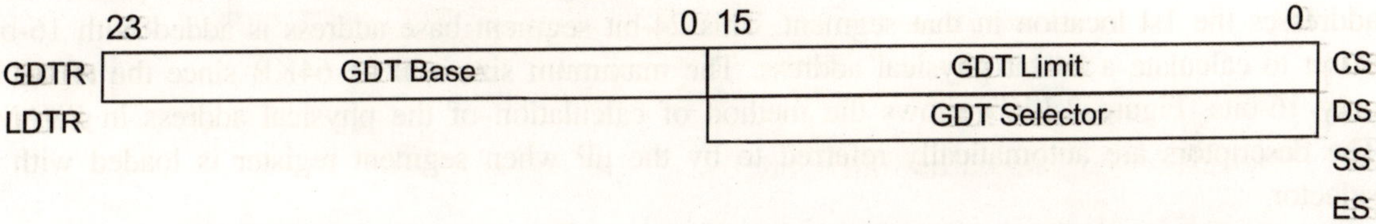

Fig. 2.14 *(i)* 80286 address translation registers segment register

Reset condition is shown in Fig. 2.14 *(j)*.

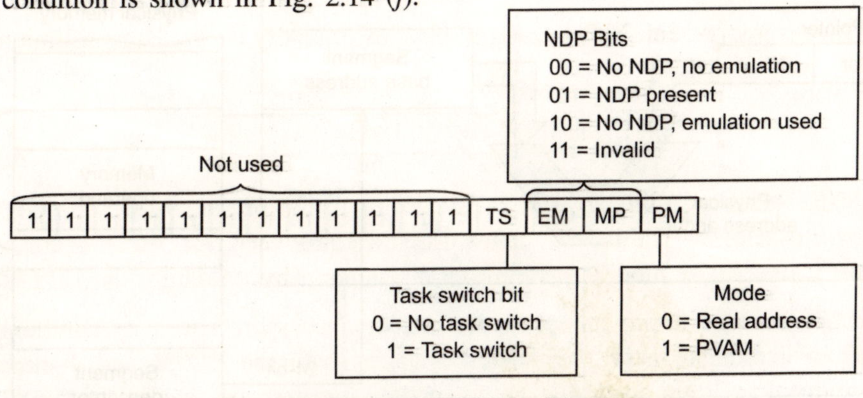

Fig. 2.14 *(j)* RESET condition

Physical address calculation in PVAM

The 80286 in PVAM uses the 16-bit contents of a segment register as a selector to address a descriptor stored in the physical memory. The descriptor is a block of continuous memory locations that contains information of a segment, like

- segment base address,
- segment limit,
- segment type,
- privilege level,
- segment availability in physical memory,
- descriptor type,
- segment used by another task.

The most important information about descriptor is the base address. This means starting address of a segment. The segment limit indicates the maximum size of a segment. Thus, the last location of a particular segment can determine using the base address of a segment and the segment limit. Similarly, each segment is assigned a type and its privilege level that indicate the importance of that segment. A segment with lower privilege level will not be allowed to access another segment with higher privilege level. This way it provides **protection** to segments from **unauthorized access.** Also a certain segment may or may not be present in the physical memory at any given instant of time. This information is also stored in a descriptor. Finally, whether any segment has been accessed by other task in the recent past, is also stored in the segment descriptor. This information helps in deciding whether the segment should be unswapped from the physical memory or not. A segment that has not been accessed in the recent passed may probably be unswapped from the main memory. The segment base address is 24-bit pointer that addresses the 1st location in that segment. This 24-bit segment base address is added with 16-bit offset to calculate a 24-bit physical address. The maximum size will be 64KB since the offset is only 16-bits. Figure 2.14(*k*) shows the method of calculation of the physical address in PVAM. The descriptors are automatically referred to by the μP when segment register is loaded with a selector.

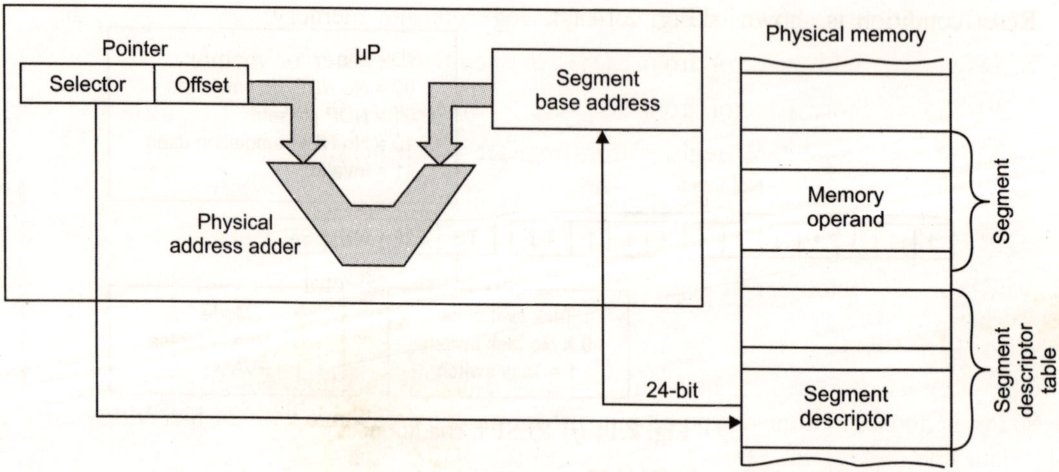

Fig.2.14 (*k*) Physical address Calculation for PVAM of 80286(Intel Corp.)

Instruction set

The instruction set of the 80286 is summarized as below. There are 16-new instructions that are in addition to the instruction for the 80186. These instructions are used for multitasking of the 80286 and most of these are for loading and storing the extra registers.

8086 Instructions : Instructions with restricted usage while in the PVAM are : HLT, IN, OUT, CLT, STI, and LOCK.

80186 Instruction : Instructions with restricted usage while in PVAM are; INS and OUTS.

80286 Instruction : Existing instructions with extended usage are; CALL, JMP, RET, IRET, and software interrupts.

Just like the 8086, the 80286 external memory is configured as odd ($\overline{BHE} = 0$) and even ($A_0 = 0$) memory banks. The 80286 operates in a mode similar to the maximum mode 8086. Some of the 80286 pins such as $M/\overline{IO}$, $\overline{S_0}$, $\overline{S_1}$, HOLD, HLDA, $\overline{READY}$, and $\overline{LOCK}$ have identical functions to that of the 8086. Two external pins NMI and INTR are provided on 80286 chip. The non-maskable interrupt (NMI) is serviced in the same way as in 8086. The INTR and COD/ $\overline{INTA}$ are used together to provide an interrupt type.

New instructions of 80286 for supporting the PVAM are

LGDT ⇒ Load global descriptor table registers from memory.

LIDT ⇒ Load interrupt descriptor table register from memory.

LLDT ⇒ Load selector and associated descriptor into LDTR (local descriptor table register)

LTR ⇒ Load task register with selector and descriptor for TSS(task state segment)

LPSW ⇒ Load PSW register from register or memory.

LAR ⇒ Load access right byte of descriptor into register or memory

LSL ⇒ Load segments limit from descriptor into register or memory.

CLST ⇒ Clear task switch flag to zero located in the PSW register.

SGDT ⇒ Store global descriptor table registers into memory.

SLDT ⇒ Store selector from LDTR in specified register or memory.

STR ⇒ Store selector from task register in register or memory

SPSW ⇒ Store PSW register from register or memory.

ARPL ⇒ Adjust requested privilege level of selector.

VERR ⇒ Determine if segment addressed by a selector is readable.

VERW ⇒ Determine if segment pointed to by the selector is writeable.

Function of pins

The pin outs of 80286 is shown in Fig. 2.14(*i*) and its functions are given in Table 2.14(*f*). The 80286 is houset in a quad type 68 pin surface mounted package with 24 bits address line and 16-bit data line.

Table 2.14(*f*) Pin functions of the 80286

Symbol	*In/Out*	*Description/ Z-state*
CLK	I	System clock. input pin. Internally divided by 2 to generate processor clock (16MHz for an 8MHz 80286).
$D_{15} \leftrightarrow D_0$	I/O–Z	Bidirectional Data Bus
$A_{23} \leftrightarrow A_0$	O–Z	Physical address output lines. Only $A_{15} \leftrightarrow A_0$ are used for I/O addressing. $A_{23} \leftrightarrow A_{16} = 0$ during I/O.
COD/ $\overline{INTA}$	O–Z	In combination with $M/\overline{IO}$ and $\overline{S_0}$, $\overline{S_1}$ distinguishes different memory, I/O and INTA cycles.

(Contd....)

$M/\overline{IO}$	O–Z	It distinguishes between memory and I/O operations.
$\overline{S_0}, \overline{S_1}$	O–Z	It indicates beginning of bus cycle and with $M/\overline{IO}$ and $COD/\overline{INTA}$ defines the type of bus cycle.
$\overline{LOCK}$	O–Z	Prevents other masters gaining the control of the bus cycle for current and following bus cycle.
$\overline{READY}$	I	Used to insert the wait state in the bus cycle for interfacing low speed I/O devices.
PEREQ	I	Processor Extension Request (means coprocessor here 80287).
$\overline{PEACK}$	O–Z	Acknowledge-PEREQ is used by the Coprocessor to request the 80286 to transfer a data operand and the 80286 to acknowledge that the requested operand is being transferred using $\overline{PEACK}$.
$\overline{BUSY}$	I	Processor Extension Busy. $\overline{ERROR}$ – $\overline{BUSY}$ is sent by the coprocessor to indicate it is busy. A low causes the 80286 to wait during a WAIT or ESC instructions.
$\overline{ERROR}$	I	The $\overline{ERROR}$ = 0 indicates to the 80286 that the processor extension has committed an error and it is reactivating the processor extension interrupt.
CAP	–	A capcitor 0.047μF + 20%, must be connected between this pin and ground to filter the ouput of the internal substrate and bias generator.
RESET	I	Reset input clears the internal logic and initializes 80286.

The 80286 chip has a new pin, called the CAP. The 80286 MOS substrate must be applied with a negative voltage for maximum speed. The negative voltage is obtained from the +5V supply by a bias generator on the 80286. An external capacitor can be connected to the CAP pin for filtering this bias voltage. Note the polarity of voltage of the capacitor. The 4-pins of the 80286 are used to interface it with the coprocessor. These pins are

- PEREQ
- $\overline{PEACK}$
- $\overline{BUSY}$
- $\overline{ERROR}$

The PEREQ (processor extension request) is the input pin to the processor to tell the 80286 to perform data transfer to or from memory for it. When 80286 is ready, it activates the $\overline{PEACK}$ (processor extension acknowledge) signal to inform the coprocessor of the start of the transfer. The 80286 $\overline{BUSY}$ signal inputs, when activated LOW by the coprocessor, stops 80286-program execution on WAIT and some ESC instructions until $\overline{BUSY}$ is HIGH. If a coprocessor finds some error during processing, it will activate the 80286 $\overline{ERROR}$ input pin. This will generate an interrupt. A service routine for this interrupt can be written to provide an indication to inform the user of the error.

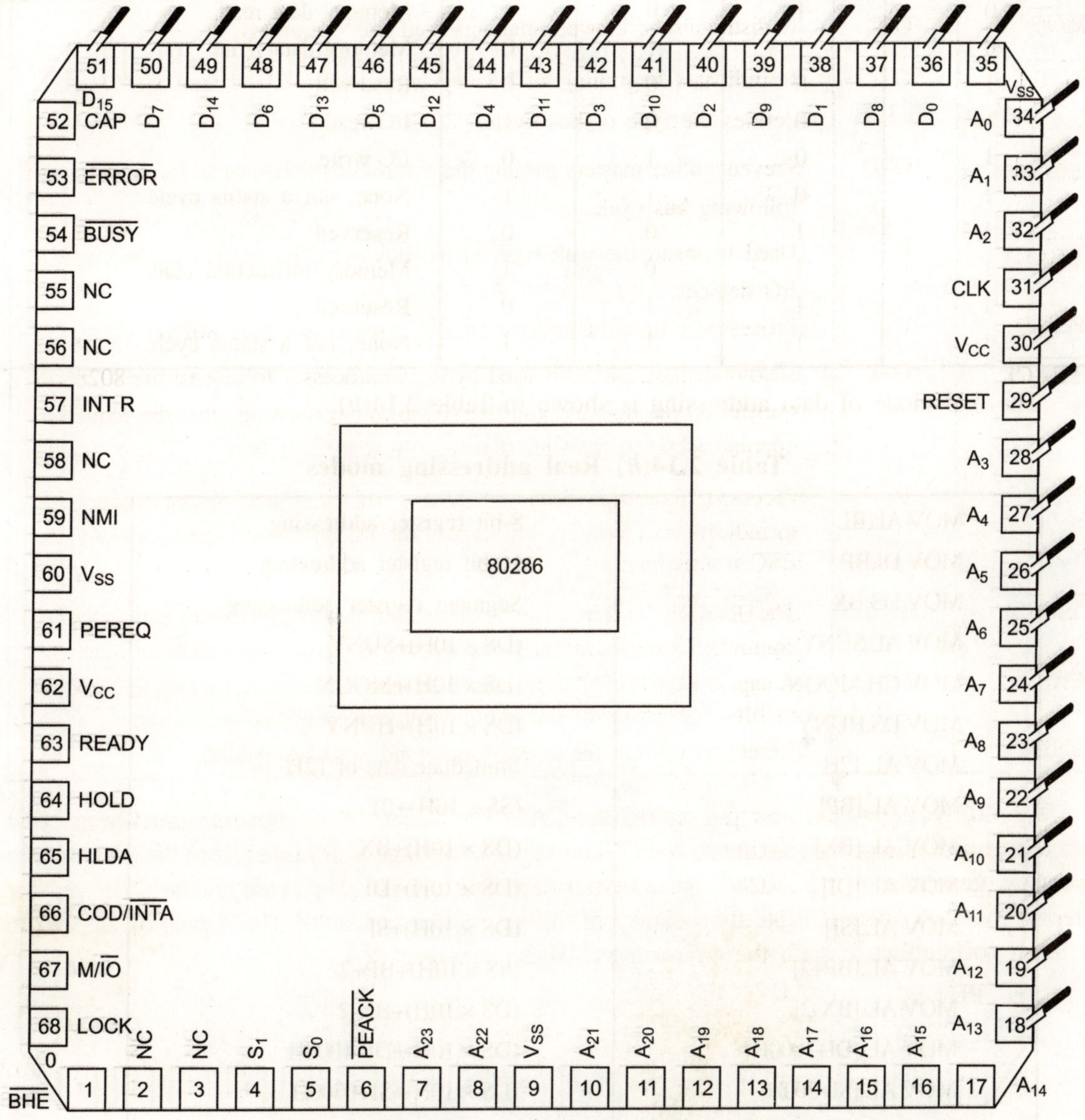

Fig. 2.14 (*l*) Pinouts of the 80286

Table 2.14(*g*) Definition of BUS cycle status of 80286

COD/$\overline{INTA}$	*M/*$\overline{IO}$	$\overline{S_1}$	$\overline{S_0}$	*BUS cycle Initiated*
0	0	0	0	Interrupt acknowledge
0	0	0	1	Reserved
0	0	1	0	Reserved
0	0	1	1	None; not a status cycle
0	1	0	0	If A_1 = 1 then halt; else shutdown

(*Contd....*)

0	1	0	1	Memory data read
0	1	1	0	Memory data write
1	0	0	0	Reserved
1	0	0	1	I/O read
1	0	1	0	I/O write
1	0	1	1	None; not a status cycle
1	1	0	0	Reserved
1	1	0	1	Memory instruction read.
1	1	1	0	Reserved
1	1	1	1	None; not a status cycle

The real mode of data addressing is shown in Table 2.14(*h*).

Table 2.14(*h*) Real addressing modes

MOV AL,BL	8-bit register addressing
MOV DI,BP	16-bit register addressing
MOV DS,BX	Segment register addressing
MOV AL,SUNY	(DS × 10H)+SUNY
MOV CH,MOON	(DS × 10H)+MOON
MOV DS,HUNY	(DS × 10H)+HUNY
MOV AL,12H	Immediate data of 12H
MOV AL,[BP]	(SS × 10H)+BP
MOV AL,[BX]	(DS × 10H)+BX
MOV AL,[DI]	(DS × 10H)+DI
MOV AL,[SI]	(DS × 10H)+SI
MOV AL,[BP+2]	(SS × 10H)+BP+2
MOV AL,[BX-2]	(DS × 10H)+BX-2
MOV AL,[DI+1000H]	(DS × 10H)+DI+1000H
MOV AL,[SI+300H]	(DS × 10H)+SI+0300H
MOV AL,SUNY[BP]	(SS × 10H)+BP+SUNY
MOV AL,SUNY[BX]	(DS × 10H)+BX+SUNY
MOV AL,SUNY[DI]	(DS × 10H)+DI+SUNY
MOV AL,SUNY[SI]	(DS × 10H)+SI+SUNY
MOV AL,SUNY[BP+2]	(SS × 10H)+BP+SUNY+2
MOV AL,SUNY[BP-2]	(SS × 10H)+BP+SUNY-2
MOV AL,SUNY[DI+100H]	(DS × 10H)+DI+SUNY+100H
MOV AL,SUNY[SI+100H]	(DS × 10H)+SI+SUNY+100H
MOV AL,[BP+DI]	(SS × 10H)+BP+DI

MOV AL,[BP+SI]	(SS × 10H)+BP+SI
MOV AL,[BX+DI]	(DS × 10H)+BX+DI
MOV AL,[BX+SI]	(DS × 10H)+BX+SI
MOV AL,[BP+DI+2]	(SS × 10H)+BP+DI+2
MOV AL,[BP+DI-2]	(SS × 10H)+BP+DI-2
MOV AL,[BX+DI+30H]	(DS × 10H)+BX+DI+30H
MOV AL,[BX+SI+30H]	(DS × 10H)+BX+SI+30H
MOV AL,SUNY[BP+DI]	(SS × 10H)+BP+DI+SUNY
MOV AL,SUNY[BP+SI]	(SS × 10H)+BP+SI+SUNY
MOV AL,SUNY[BX+DI]	(DS × 10H)+BX+DI+SUNY
MOV AL,SUNY[BX+SI]	(DS × 10H)+BX+SI+SUNY
MOV AL,SUNY[BP+DI+2]	(SS × 10H)+BP+DI+SUNY+2
MOV AL,SUNY[BP+SI-2]	(SS × 10H)+BP+SI+SUNY-2
MOV AL,SUNY[BX+DI-10H]	(DS × 10H)+BX+DI+SUNY-10H
MOV AL,SUNY[BX+SI+1AFH]	(DS × 10H)+BX+SI+SUNY+1AFH

REVIEW QUESTIONS AND ANSWERS

Q.1. Name the GPRs of the 8086/ 8088.

Q.2. What are different modes of addressing the 8086? Explain them.

Q.3. What is the purpose of the index registers?

Q.4. What is the purpose of the pointer registers?

Q.5. What is the purpose of the instruction pointers?

Q.6. How do the address/data buses differs in between the 8086 and 8088?

Q.7. How long the address is valid on lines $A_{19} \Leftrightarrow A_{16}$?

Q.8. What is the purpose of the queue?

Q.9. How many words do the queue stores in the 8086?

Q.10. How do we select the minimum versus maximum modes in the 8086?

Q.11. During what time period(s) of the bus cycle is $\overline{DEN}$ active?

Q.12. What is the state of $DT/\overline{R}$ for a *write* bus cycle?

Q.13. Name the hardware interrupts of the 8086.

Q.14. Which of the hardware interrupts of the 8086 are maskable?

Q.15. Which instructions are used to generate software interrupts?

Q.16. What signals control the direction of data in the 8086?

Q.17. Draw the block diagram of 8086 and explain the function of each block.

Q.18. What is a short jump?

Q.19. Which jump instruction is 5-byte long?

Q.20. What do you mean by the procedure?

Q.21. How does the near CALL instruction function?

Q.22. What is an interrupt?

Q.23. What instructions CALL an interrupt service procedure?

Q.24. How many different interrupts are available in 8086 microprocessor?

Q.25. What instructions control the function of the INTR pin?

Q.26. What type of jump is used when jumping anywhere in a segment?

Q.27. How much memory does the 8086 microprocessor address?

Q.28. How much memory does the 80286 microprocessor address?

Q.29. How many bytes do form a word of the 8086?

Q.30. How many bytes do form a word of the 80286?

Q.31. What are the differences between the logical and physical memory of the 8086?

Q.32. How many 16-bit general purpose registers are available in the 8086/8088? What are their designation?

Q.33. Why the CX register is called the count register?

Q.34. Why the DX register is called the data register?

Q.35. Name the 5-pointer and the index registers and explain their functions.

Q.36. What 2-pointers use the stack segment registers to address memory?

Q.37. How big is the flag of the 8086?

Q.38. Which base register is used to address the data in the stack segment?

Q.39. What is an overflow?

Q.40. What is the purpose of D-flag-bit?

Q.41. What do the following instruction do?

(*a*) MOV AX,BX, (*b*) MOV BX,AX (*c*) MOV BL,CL

(*d*) MOV SP,BP (*e*) MOV AX,CS

Ans. (*a*) Word size (BX) copied to word size (AX).

(*b*) Word size (AX) copied to word size (BX).

(*c*) Byte size (CL) copied to byte size (BL).

(*d*) Copies (BP) into (SP).

(*e*) Copies (CS) into (AX).

Q.42. Which of the followings are legitimate instructions? What do the legitimate instructions do?

(*a*) MOV ES,DS (*b*) MOV BL,BX (*c*) MOV CX,AX

(*d*) MOV DI,SI (*e*) MOV AX,[BX] (*f*) MOV CX,MOON

(*g*) MOV [BX+DI],CL (*h*) MOV AX,[BX+4] (*i*) MOV AX,SUNY[BP] (*j*) MOV AX,SUNY[BX+DI] (*k*) MOV BL,44H (*l*) MOV AX,44H

(*m*) MOV SI,0H (*n*) MOV AL, A (*o*) MOV AX, AB

Ans. (*a*) Segment to segment transfer is not allowed.

(*b*) Mixed sizes not allowed.

(*c*) Code segment register can not be used as destination register.

(*d*) Copies (SI) into (DI).

(*e*) Word size data from memory with the offset (BX) to register (AX).

(*f*) Word size contents from the memory Labeled MOON is copied to CX.

(*g*) Copies byte size data of (CL) into memory addressed by the (BX)+(DI).

(*h*) Copies word size data into (AX) from memory addressed by (BX)+4.

(*i*) Copies word size data from memory addressed by SUNY+[BP] into AX.

(*j*) Copies a word of data from memory addressed by SUNY+[BP+DI] into AX.

(*k*) BL filled with 44H.

(*l*) The (AH) = 00H and (AL) = 44H.

(*m*) (SI) = 0000H.

(*n*) Moves ASCII 'A(=41H)' to AL.

(*o*) Moves ASCII 'BA (= 4241H)' into AX.

Q.43. Which of the followings are legitimate instructions? What do the legitimate instructions do?

(*a*) MOV [BP],DL (*b*) MOV [DI],BH (*c*) MOV [DI],[BX]

(*d*) MOV DI,[DI] (*e*) MOV CH,2000H (*f*) MOV DATA,BP

Ans. (*a*) (DL) copied to memory addressed by BP in the stack segment.

(*b*) (BH) copied to memory addressed by DI in the data segment.

(*c*) Memory-to-memory moves not allowed except in the string case.

(*d*) Not allowed because the same register used to indirectly address memory may not be changed.

(*e*) Copies word size data into CH from memory addressed in the data segment by 2000H.

(*f*) (BP) copied into memory addressed by DATA in the data segment.

Q.44. What are 8-bit registers that have been used for register addressing?

Ans. (AX, BX, DX, CX)

Q.45. Can BX register be used as an index register in a data segment?

Ans. (No)

Q.46. Is the trap flag conditional one?

Ans. (No)

Q.47. Should the CALL instruction follow the RET instruction?

Ans. (No)

Q.48. What is the prerequisite for a conditional jump?

Ans. (one bit of the flag is tested)

Q.49. What does REP do? (It is a prefix instruction. It allows instructions to executed until CX=0).

Q.50. Which registers are used as the base location for all executable instruction and stack?

Ans. (CS and SS registers respectively).

Q.51. Does the execution of the program terminates with setting of zero flag?

Ans. (No. The program execution terminates when CX = 0).

Q.52. Explain the difference between JMP and JZ instructions.

Ans. (JMP instruction is allways taken as it is unconditional instruction. The JZ is conditional instruction and can be taken only when zero flag is set).

Q.53. What flags are used in signed comparison of data?

Ans. (zero, sign, and overflow).

Q.54. What is the difference between intrasegment and intersegment jumps?

Q.55. What is the difference between NEAR and FAR CALLs?

Q.56. Write a program to multiply the contents BX by 4 using shift instruction.

SHL BX, 0010H (*by 2)

SHL BX, 0001H (*by 4)

Q.57. Write a program to multiply the contents AX by 6 using shift instruction.

Ans.

```
PUSH AX,stack        ; save the contents of AX
MOV AX,double byte   ; 2-byte loaded in AX
SHL AX,0001H         ; AX × 2
MOV DX,AX            ; save AX × 2 in DX
SHL AX,0001H         ; AX × 4
ADD DX,AX            ; (DX) = AX × 2 + AX × 4 = AX × 6
POP AX, stack        ; retrieve the original value from the stack.
```

Q.58. Is that both 8086 and 8088 address 64KB of memory?

Ans. (No)

Q.59. What is the operation of 8086/ 8088 for $DT/\overline{R} = 1$?

Ans. (directs the bus buffer to transmit data from μP to memory and I/O devices).

Q.60. What is the purpose of the queue in BIU of the 8086?

Ans. (provides instantly the next instruction needed by the execution unit and improves the efficiency).

Q.61. What are conditions that causes BIU to suspend fetching instructions?

Ans. (Conditions are (*a*) current instruction requires access to memory or I/O port, (*b*) a transfer control (jump or call) instruction occurs, (*c*) instruction queue is full).

Q.62. What is information on the address/ data bus of the 8088 when ALE =1?

Ans. ($A_7 \Leftrightarrow A_0$).

Q.63. Specify the contents of the status flags after execution of ADD AL,1, if (AL) = FFH.

Ans. (ADD AL,1 = FF+01 = 100 H, CF =1, PF =1, AF = 1, ZF = 1, SF = 0, OF = 0)

Q.64. What would be the contents of AX and DX register after execution of following

(*a*) MOVDX,0000H	(*b*) MOV AX,2636H
MOV AX,2535H	MOV DX,2535H
MOV CX,3535H	MOV CX,3535H
MUL CX	DIV CX

Ans. (*a*) (DX) = 07BBH, (AX) = ACF9H, (*b*) (AL) = A0H ⇒ Quotient, (AH) = 15H ⇒ Remainder

Q.65. What do the following instructions do?

(*a*) PUSHF (*b*) POPF (*c*) PUSH AX

(*d*) POP BX (*e*) PUSH DS (*f*) PUSH 1234H

(*g*) POP CS (*h*) PUSH [BX] (*i*) PUSHA

(*j*) POPA

Ans. (*a*) Flag word stored on to the stack

(*b*) A word from stack loaded into the flags.

(*c*) (AX) copied on to the stack.

(*d*) A word from stack copied to BX.

(*e*) (DS) copied on to the stack.

(*f*) 1234H is stored on to the stack

(*g*) POP CS is illegal instruction.

(*h*) Copies a word from the data segment addressed by [BP] on to the stack.

(*i*) Copies all registers AX, CX, DX, BX, SP, BP, DI, & SI on to the stack

(*j*) Copies word data from the stack into SI, DI, BP, SP, BX, DX, CX & AX.

Q.66. What is wrong in MOV BL,CX and MOV DS,SS instructions?

Ans. Mixed byte size in MOV BL,CX is not allowed. Segment to segment moves not allowed

Q.67. What do the symbol [ ] indicate?

Ans. Indicates memory address.

Q.68. Determine the memory address accessed by each of the following instructions in real mode operation, if DS = 0200H, BX = 0300H, and DI = 400H.

(*a*) MOV AL,[1234] (*b*) MOV [DI],AL

Ans. (*a*) Effective address = 1234H, Physical address = 02000+1234 = 03234H

(*b*) Physical address = 02000+0400 = 02400H

Q.69. Determine the memory address accessed by each of the following instructions in real mode operation, if DS = 1000H, SS = 2000H, BP = 1000H, and DI = 0100H.

(*a*) MOV AL,[BP+DI] (*b*) MOV CX,[DI]

Ans. (*a*) Physical address = SS×10H+(BP+DI) = 20000+1000+100 = 21100H

(*b*) Physical address = DS×10H+DI = 10000+100 = 10100H

Q.70. Determine the memory address accessed by each of the following instructions in real mode operation, if DS = 1200H, BX = 0100H, and SI = 0500H.

(*a*) MOV [100H],DL (*b*) MOV DL,[BX+100H]

Ans. (*a*) Physical address = DS×10H+100 = 12000+100 = 12100H

(*b*) Physical address = DS×10H+0100+100 = 12000+0100 +100= 12200H

Q.71. Determine the memory address accessed by each of the following instructions in real mode operation, if DS = 1100H, BX = 0200H, SUNY = 0250H, and SI = 0500H.

(*a*) MOV CL,[BX+SI] (*b*) MOV CH,SUNY[BX+SI]

Ans. (*a*) Physical address = DS×10H+BX+SI = 11000+0200+0500 = 11700H

(*b*) Physical address = DS×10H+BX+SI+SUNY = 11000+0200 +0500+0250 = 11950H

Q.72. Determine the memory address accessed by each of the following instructions in real mode operation, if DS = 1300H, SS = 1400H, BP = 1500H, and SI = 0100H.

(*a*) MOV AL,[BP+SI-200H] (*b*) MOV AL,[SI-0100H]

Ans. (*a*) Physical address = SS×10H+BP+SI-200 = 14000+1500+0100-200 = 12400H

(*b*) Physical address = DS×10H+SI-100H = 13000+0100-100 = 13000H

Q.73. Indicate the type of jump, if the JMP THREE instruction is stored at memory address 10000H and the address of the THREE is

Ans. (*a*) 10020H (*b*) 11000H. (*c*) 0FFFFH (*d*) 30000H

(*a*) short, (*b*) near, (*c*) short, and (*d*) far

Q.74. Which are illegal instructions for 8086?

(*a*) MOV ES,4000H, (*b*) MOV ES,AX, (*c*) MOV DS,0000H

Ans. (*a*) and (*c*) are illegal instructions.

Q.75. What is the difference between 8088/8086?

Ans. Nearly internal processing of the 8088 and 8086 are the same, but there are few difference such as:

1. The 8088 access the external 8-bit data where as it's internal processing is done with 16-bit data.
2. The fetching and writing of 16-bit data by 8088 takes 2-consecutive bus cycle. Thus, 8088 is expected to take more time in performing a particular task than 8086.
3. The 8088 performance was optimized by reducing the prefetched code queue from 6 to 4-byte to avoid unnecessary prefetch operation for optimum use of the bus the BIU when prefetching the instruction.
4. The BIU of 8088 fetches a byte from the memory to load in the queue if at least one byte is free. In case of 8086 at least 2-byte should be free for the next fetch operation.
5. The over all execution time in 8088 are affected by 8-bit external data bus. All 16-bit operation requires additional 4-clock cycles. The μP speed is also limited by the speed of instruction fetches.

Pins available on 8088 also have some functional difference

- The pin $IO/\overline{M}$ of 8088 has been inverted from that of the pin the 8086 as $\overline{IO}/M$ (similar to 8085).
- Address bus $A_{15} - A_8$ are latched all the time have valid address only.
- $\overline{BHE}$ does not mean any thing for the 8088 as it fetches only 8-bit data/instruction.

3

32-BIT μPs (80386 TO 80486)

3.1 INTRODUCTION

Intel announced development of 32-bit microprocessor, the 80386 in 1985. The 80386 was a leap forward compared to 80286 with 32-bit data bus and 32-bit internal registers. The Intel-32 Architecture and the 80386 brought several welcome changes to Intel-based PCs, including the capability of processing data at twice the speed of earlier 16-bit Intel microprocessors and the ability to address up to 4GB of system memory as one linear address space. In real mode, an Intel microprocessor is limited to addressing 1MB of memory in 64KB segments. To keep compatibility with 80286, the 80386 included 3-operation modes : **real, protected,** and **virtual real**. The Intel 80286 microprocessor booted in real mode, and then it could be switched to protected mode by software command. However, it was not possible to switch the 80286 from protected mode back to real mode without resetting the microprocessor. The 80386 addressed this problem very well by permitting switching from either mode to the other via software commands, a welcome change for programmers. In protected mode, the 80386 is similar to 80286 with higher memory addressing capability. In virtual 8086 mode, the 80386 can simulate more than one 8086 at a time. This mode allows a single 80386 to divide its memory into many virtual machines. Because of the need to run software developed for its earlier 80x86 microprocessors, Intel included the option of switching the 80386 from protected mode back to the real mode. Resetting the microprocessor was no longer necessary as it had been with the 80286. This new mode of the 80386 was called virtual 8086 (V86) mode. In the V86 mode, an 80386 microprocessor could segment its memory into several virtual machines, each one performing like a separate computer with an 8086 microprocessor.

This 'virtual real mode' of the 8086 was used for many multitasking software such as Windows 3.1, OS/2 etc. In this mode the μP can run many real mode (8086/8088) software simultaneously in the system. In this mode multiple program runs in its protected memory. If one program crashes only that part of the memory can be reused, without disturbing the other program in other part of the memory. It is compatible with object code of the previous members of 80x86 family (80186, 80188, 80286, 8086, and 8088). This protects major investments in applications and operating system software developed for the 80x86 family, while offering a significant enhancement in performance. The architecture and performance of 80386 should allow it to be used in a wide range of demanding applications i.e., in engineering workstations, office system, robotic and control systems, and expert systems etc. The 1.5 mm CHMOS-III technology was used to manufacture the 80386 containing approximately 280,000 transistors in 2-layers of metallization but draws less current (400mA) than the 8086. This allows 16 MHz operation with low power consumption. The 80386 can operate on 12.5, 16, 20, 25, and 33MHz clocks also. This chip comes in 132-pins housed in a Pin Grid Array (PGA) package as indicated in Fig. 3.1 (*a*). It is designed using high speed CHMOS III technology that combines high speed of the HMOS and low power consumption of CMOS. The 80386 provides multitasking support, memory management, pipelined architecture, address translation cache, and a high-speed bus interface in a single chip. The 80386 is a true 32-bit microprocessor. The 80386 includes 32-bit address, data bus and all 8-internal general purpose registers of 32-bit wide. This allows access to 2^{32} = 4GB of physical memory. In the virtual memory mode, the 80386 supports 46-bit address. This allows the software to think that 64TB (64 million MB) memory exists. The processor can handle 8-, 16-, and 32-bit data types. It has separate data and address pins and generates a 32-bit physical address. Hence, it can address $2^{32} = 2^{30}2^{2}$ = 4GB of physical address and $2^{46} = 2^{40}2^{6}$ = 64TB of virtual memory.

The 80386 is highly pipelined and can perform instruction fetching, decoding, executing, and memory management functions in parallel. The on chip memory management and protection hardware translates logical address into physical addresses and provides the protection rules required in a multitasking environment. The 80386 has special hardware for task switching. A single instruction or interrupt is required for the 80386 to perform complete task switching. A 16MHz 80386 can save the state of one task (all registers), load the state of another task (all registers, segments, and paging registers if needed), and resume execution in less than 16ms.

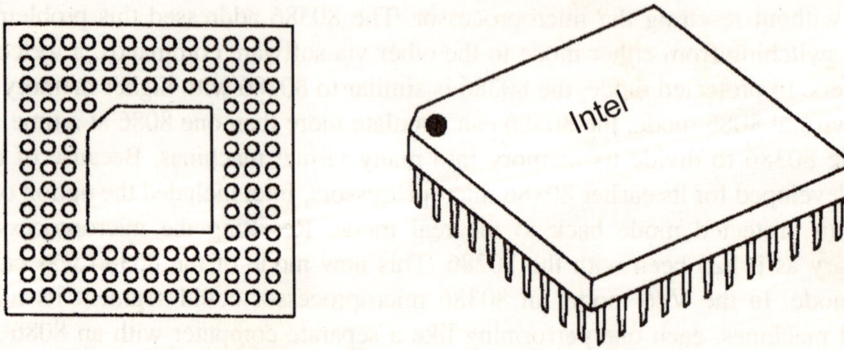

Fig. 3.1 (*a*) Look of pin grid array for 80386/80486

The 80386 has 129-instructions in all. The main **difference** between the 80286 and the 80386 are the 32-bit address and data type and data paging and memory management. The 80386 implements a full 32-bit architecture with a 32-bit internal data path including registers, ALU, and internal buses. It provides 32-bit instructions, addressing capability, data types, and a 32-bit external bus interface.

The 80386 are available in 80386SX, 80386DX, 80386SL versions. The 80386DX addresses 4GB of memory through 32-bit data and 32-bit address. The 80386SX (like 80286) addresses 16MB of memory with its 24-bit address bus via its 16-bit data bus. The 80386 was announced after 80386DX for applications that did not require full 32-bit bus version. The 80386SX is found in many personal computers that use basic motherboard design as the 80286. At that time, most of the applications required less than 16MB of memory and hence the 80386SX was fairly popular and less costly version of the 80386. The 25MHz version of the 80386 required 550mA average current, where as it required 500mA for 20MHz and 450mA for 16MHz. The 33MHz version of the 80386 required 600mA of power supply current. Thus, **80386** launched in 1985 **contained** 275,000 transistors. This chip included 32-bit with **multitasking** capabilities. This means it was capable of running multiple programs at the same time, thus increasing the efficiency.

Pin functions of 80386

Each output pin of the 80386 is capable of sinking (logic-0) 4.0mA (address and data buses) or 5.0mA (other connections). This represents significant improvement w.r.t. the earlier family sinking 2.0mA of current.

$A_{31} \Leftrightarrow A_2$: Address Bus Connections are used to address any of the 1GB × 32 memory locations found in the 80386-memory system. Note that A_1 and A_0 are encoded in the bus enable ($\overline{BE_3}$ and $\overline{BE_0}$) described elsewhere. (Note that the 80386SX contains address connections $A_{23} \Leftrightarrow A_1$).

$D_{31} \Leftrightarrow D_0$: Data Bus Connections are used to transfer data between the microprocessor and its memory and I/O system. (Note that the 80386SX contains data bus connections $D_{15} \Leftrightarrow D_0$)

$\overline{BE_3}$ - $\overline{BE_0}$: Bank Enable Signals are used to access a byte, word, or double word of data. These signals are generated internally of the microprocessor from address bits A_1 and A_0. (Note the 80386SX contains ($\overline{BE_3}$ and $\overline{BE_0}$) for bank selection).

$M/\overline{IO}$: Selects a memory device when a logic 1 or and I/O device when a logic-0. During the I/O operation the address bus contains a 16-bit I/O address.

$W/\overline{R}$: Indicates that the current bus cycle is a write for a logic-1 or a read when a logic-0.

$\overline{ADS}$ (Address Data Strobe) : Becomes active whenever the 80386 issues a valid memory or I/O address. This signal combined with the $W/\overline{R}$ signal to generate the separate read and write signals present in the earlier 8086/8088 system.

RESET : Reset initializes the 80386 causing it to begin executing software from memory location FFFFFFF0H. The 80386 is reset to the real mode and the leftmost 12-address connections remain logic 1's {FFFH} until a far jump or far call is executed when they are cleared to logic 0's {000H}.

CLK2 : Clock time 2 is driven by a clock signal that is twice the operating frequency of the 80386. For example, to operate the 80386 at 16 MHz, we apply a 32MHz clock to this pin.

$\overline{READY}$: Ready controls the number of wait states inserted into the timing to control memory accesses.

$\overline{LOCK}$: Lock becomes a logic 0 whenever an instruction is prefixed with the LOCK: prefix. This is most often used during DMA accesses.

$D/\overline{C}$: Data control indicates that the data bus contains data to/from memory or I/O when a logic-1. If it is logic 0, the microprocessor is halted or executing an interrupt acknowledge.

$\overline{BS_{16}}$: Bus size 16-bits selects either a 32-bit data bus or a 16-bit data bus. In most cases if an 80386 is operated on a 16-bit data bus, we use the 80386SX that has a 16-bit data bus.

$\overline{NA}$: Next Address causes the 80386 to output the address of the next instruction or data in the current bus cycle. This pin is often used for pipelining the address.

HOLD : Hold requests a DMA action as it did on the 8086/8088.

HLDA : Hold Acknowledge indicates that the 80386 is currently in a hold condition with the connection to the 80387 arithmetic coprocessor.

$\overline{PEREQ}$: The coprocessor asks the 80386 to relinquish control and is a direct connection to the 80387 arithmetic coprocessor.

$\overline{BUSY}$: BUSY is an input used by the WAIT or FWAIT instruction that waits for the coprocessor to become not busy. This is also a direct connection to the 80387 from the 80386. This pin functions as the TEST pin on the 8086/8088.

$\overline{ERROR}$: Error indicates to the microprocessor that the coprocessor has detected an error.

INTR : Interrupt Request is used by external circuitry to request an interrupt.

NMI : Non-maskable Interrupt requests a non-maskable interrupt as it did on the 8086 microprocessor.

3.2 REGISTERS OF 80386/80486

The 80386/80486 processor dramatically extended the 8086-register set. In addition to all the registers on the 80286 (and therefore, the 8086/486), the 80386 added several new registers extending the definition of the existing registers. The 80486 did not add any new registers to the 80386's basic register set, but it did define a few bits in some registers left undefined by the 80386.

From the programmer's point of view, the most important change in the 80386/486 was the introduction of a 32-bit register set. The AX, BX, CX, DX, SI, DI, BP, SP, FLAGS, and IP registers were all extended to 32-bits. The 80386 CALLs these new 32-bit versions EAX, EBX, ECX, EDX, ESI, EDI, EBP, ESP, EFLAGS, and EIP to differentiate them from their 16-bit versions (which are still available on the 80386). Besides the 32-bit registers, the 80386/486 also provides 2-new 16-bit segment registers, FS and GS. The FS and GS registers are additional extra segment registers that allows to access 6-different segments in memory without reloading a segment register. Note that all the segment registers on the 80386/486 are 16-bits. The 80386/486 did not extend the segment registers to 32-bits as it did for the other registers.

The 80386/486 did not make any change to the bits in the flag register. Instead, it extended the flags register to 32-bits (the EFLAGS register) and defined bits 16 and 17. Bit-16 is the debug resume flag (RF) used with the set of 80386 debug registers. Bit-17 is the Virtual 8086 mode flag (VM) which determines whether the processor is operating in virtual-86 mode (which simulates an 8086) or standard protected mode. The 80486 adds a 3rd bit to the EFLAGS register at position 18-the alignment check flag. Along with control register zero (CR_0) on the 80486, this flag forces a trap (program abort) whenever the processor accesses non-aligned data (e.g., a word on an odd address or a double word at an address which is not an even multiple of 4).

The 80386 added 4-control registers: CR_0-CR_3. These registers extend the MSW register of the 80286 (the 80386 emulates the 80286 MSW register for compatibility, but the information really appears in the CRx registers). On the 80386 and 80486 these registers control functions such as paged memory management, cache enable/disable/operation (80486 only), protected mode operation, and more.

The 80386/486 also adds 8-*debugging* registers. A debugging program like Microsoft Code view or the Turbo Debugger can use these registers to set breakpoints when we are trying to locate errors within a program. While we would not use these registers in an application program, we will often find that using such a debugger reduces the time it takes to eradicate bugs from our programs. Of course, a debugger that accesses these registers will only function properly on an 80386 or later processor.

Finally, the 80386/486 processors add a set of test registers to the system which tests the proper operation of the processor when the system powers up. Most likely, Intel put these registers on the chip to allow testing immediately after manufacture, but system designers can take advantage of these registers to do a power-on test.

For the most part, assembly language programmers need not concern themselves with the extra registers added to the 80386/80486/Pentium processors. However, the 32-bit extensions and the extra segment registers are quite useful. The programming model of the 80386/ 80486/ Pentium looks like that shown in Fig. 3.2 (*a*).

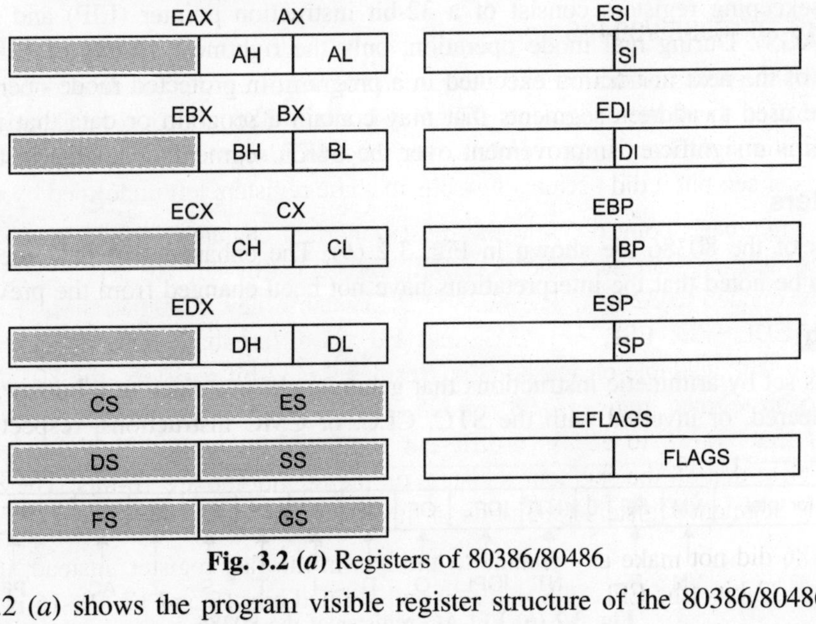

Fig. 3.2 (*a*) Registers of 80386/80486

Figure 3.2 (*a*) shows the program visible register structure of the 80386/80486 microprocessor. The registers are divided as

- General purpose,
  - 32-bit EAX, EBX, ECX, EDX, ESI, SDI, EBP, ESP.
- Special registers are:
  - segment (selector) registers 16-bit CS, DS, ES, SS, FS, GS.
- Housekeeping registers.
  - 32-BIT EIP
  - EFLAGs

The general-purpose registers are accessed through most instructions and are designated to hold 8, 16, or 32-bit data. The 8086 contains byte and word-sized registers, while the 80386 also contains double-word-sized or extended registers. The 8 and 16-bit registers are addressed using the same names as with the 8086. The new 32-bit registers are addressed as extended registers with the letter E augmenting the normal 16-bit designation. For example, EAX is the 32-bit extended version of AX. In general, it has 8-general purpose registers available for calculations and memory addressing, the flags register, and the instruction pointer. Other registers include control registers, 6-segment registers used to structure the 4GB address space and to facilitate system debug, and 6-debug registers used to control the setting upto 4-code or data breakpoints.

The 32-bit general-purpose registers are named : EAX, EBX, ECX, EDX, ESP, EBP, ESI, and EDI. The segment (selector) registers are CS, DS, ES, SS, FS, and GS. In real mode operation, segment registers contain a segment address and in protected mode operation they contain a selector just as in the 8086. We use prefix to address data using either MOV EAX, GS : DATA instruction. The prefix GS: or FS: is used to select these 2-new segments for any memory addressing mode.

The housekeeping registers consist of a 32-bit instruction pointer (EIP) and a 32-bit flag register (EFLAGS). During real mode operation, only the rightmost 16-bits of EIP contain the offset address of the next instruction executed in a program. In protected mode operation, all 32-bits of EIP are used to address segments that may contain a program or data that is up to 4GB in length. This is a significant improvement over the 64KB segments available in the 80286.

FLAG registers

The flags of the 80386 are shown in Fig. 3.2 (*b*). The enhancement over the 8086 is left vacant. It is to be noted that the interpretations have not been changed from the previous models.

C (Carry flag)

This bit is set by arithmetic instructions that generate either a carry or a borrow. This bit can also be set, cleared, or inverted with the STC, CLC, or CMC instructions, respectively.

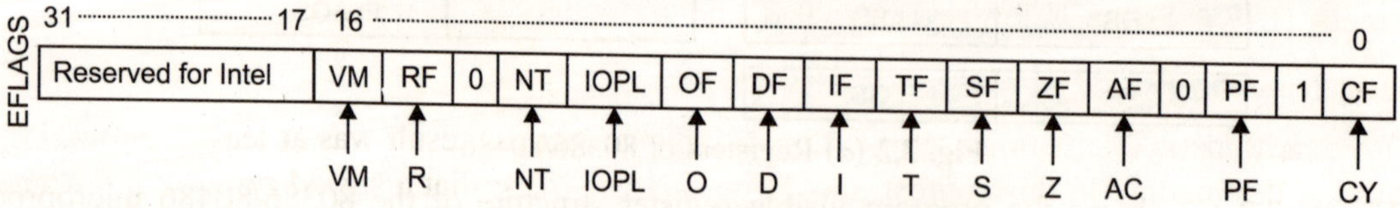

Fig. 3.2 (*b*) EFLAG register of the 80386

P (Parity flag)

This bit will be set by most instructions if the least significant 8-bits of the destination operand contain an even number of 1-bits. In other words this flag signals a byte with an even parity. Note that this flag represents the number of bits in the lower byte only, not in the entire operand.

AC (Auxiliary flag)

This is like half of the carry flag. Most comparison and arithmetic instructions set this flag if a carry or borrow occurred from the most significant nibble of the least significant byte of the accumulator (that is, bit-4 of the register EAX). Its seemingly perverse function exists to aid BCD arithmetic.

Z (Zero flag)

This flag is set by most instructions when the result of the operation is a binary zero. Comparison instructions set this flag without storing the result if the operands are equal, since comparisons are basically subtractions.

S (Sign flag)

Most comparison and arithmetic operations set SF equal to the most significant bit of the result. This will be bit-7, 15, or 31, depending on the magnitude of the operation. The 80386 correctly sets this bit whether the values involved are considered to be signed or unsigned. If they are signed, SF will tell you whether the result is positive or negative.

T (Trape flag)

This flag allows you to single-step through programs. When the processor detects that this flag is set it executes exactly one machine instruction and then automatically generates an internal exception 1. When the exception handler returns, the processor executes the next instruction and repeats the process. Single stepping continues until program code resets this flag.

I (Interrupt flag)

When this flag is set, the 80386 recognizes and handles external hardware interrupts on its INTR pin (B7 on the standard pin-grid-array package). If you clear this bit, the processor ignores any inputs on this pin. Note that this does not do the external non-maskable interrupt (NMI) input in any way, nor does it have any relationship with internally generated faults, exception, traps, and so forth. The IF is set and cleared with the STI and CLI instructions, respectively.

D (Direction flag)

This flag is set and cleared by the STD and CLD instructions. Its sole purpose is to determine whether string operations (LODS, STOS, CMPS, etc.) will automatically increment (DF = 0) or decrement (DF = 1) the index registers, ESI and EDI, after each iteration.

O (Overflow flag)

Most arithmetic instructions set this flag to indicate that the result was at least 1 bit too large to fit in the destination. In this context, too, large can also mean that a signed result was too small a negative number to be represented in the number of bits allowed in the destination.

IOPL (Input/Output privilege level flags)

This 2-bit field is peculiar to the protected mode of operation and so first appeared on the 80286. It holds the privilege level, from 0 to 3, at which your code must be running in order to execute any I/O-related instructions.

R (Resume flag)

This flag is the 1st-bit in the extended E FLAGS register. It has to do with new debug features and is related to registers DR_6 and DR_7. By setting it, you can selectively mask some exceptions while you are debugging code.

VM (Virtual 8086 mode flag)

The space allocated to this bit in the EFLAGS registers belies its importance. It has enormous impact on the way the 80386 behaves. When the flag is cleared the 80386 can operate in full 80386 Protected mode, 80286 Emulation mode, or Real Address mode. When it is set, the 80386 is essentially converted into a high speed 8086 until the bit is cleared again.

NT (Nested Task flag)

This bit is also an artifact of Protected mode. When set, it indicates that one system task has invoked another through a CALL instruction, as opposed to a JMP. By and large, you do not care about this bit, but the 80386 does, and the explanation on multitasking will show how you can manipulate it to your advantage.

3.3 OPERAND ADDRESSING OF 80386

The operands of the 80386 may reside in registers, main memory or I/O address space. Furthermore, an operand may be implied in the instruction or specified explicitly as a part of the instruction. Storing operands in registers generally provides the fastest method of processing data. Any arithmetic or logic operator can operate the contents of the 80386 general register on. Alternatively, 8, 16 or 32-bit constants (immediate) can be embedded directly in an instruction. The 16 and 32-bit operations may specify 8-bit sign-extended or zero-extended immediate. Table 3.3(*a*) includes sample instructions employing registers and immediate as operands. In general, register-to-register operations execute in two clocks on the 80386. At a clock rate of 16MHz, this translates to 125-nanos per operation.

Table 3.3(*a*) Examples of operand addressing in the 80386

Instruction	*CLKs*	*Semantics*
INC EAX	2	Increment contents of EAX by 1
IMUL EBX,-3	9	Multiply the integer in EBX by –3
CMP CX, 0	2	Compare contents of CX with 0 and set condition codes.
MOVSX EAX, SI	3	Sign extend the contents of the 16-bit register (SI) and move into EAX.
MOV DWORD PTR [56], –12445654	2	Assign –12445654 to the 32-bit integer at address 56
JMP Table[EBX * 4]	10	Jump to the address stored at entry EBX of jump table.
SUB DX, WORD PTR [EBP + EDI * 2 – 10]	7	Subtract from DX the 16-bit quantity at address [EBP + EDI*2 – 10].

Most operands are stored in main memory. The 80386 has a full complement of address generation mechanism for specifying the effective address of such operands. These mechanisms were developed in response to the storage paradigms present in high-level languages. The effective address can be obtained by summing the contents of 1 or 2-general-purpose registers and an optional immediate value or displacement. This register-based effective address scheme can be summarized as

[base register] + [index register] * (scale) + [displacement]

Here, the base register is any general-purpose register and the index register is any general register other than ESP. The scale specification is a constant value, either 2, 4, or 8. If specified, it scales the index register by the required amount, thus simplifying indexing into arrays of multi-byte elements. The displacement field is also a constant, its value ranging from –231 to +231 –1. Figure 3.3(*a*) shows all of the 80386's 32-bit memory addressing modes. The example of memory addressing appears in Table 3.3(*a*).

$$\begin{Bmatrix} \text{EAX} \\ \text{EBX} \\ \text{ECX} \\ \text{ESP} \\ \text{EBP} \\ \text{ESI} \\ \text{EDI} \end{Bmatrix} + \begin{Bmatrix} \text{EAX} \\ \text{EBX} \\ \text{ECX} \\ \text{ESP} \\ \text{EBP} \\ \text{ESI} \\ \text{EDI} \end{Bmatrix} * \begin{Bmatrix} 2 \\ 4 \\ 8 \end{Bmatrix} + \begin{Bmatrix} 0 \\ \text{8-bit displacement} \\ \text{16-bit displacement} \end{Bmatrix}$$

Fig. 3.3 (*a*) 32-bit memory addressing modes

Memory system

The physical memory system of the 80386DX is 4GB in size and can be addressed as such. In virtual addressing mode, 64TB are mapped into the 4GB of physical space by the memory management unit. Figure 3.3 (*b*) shows the organization of the 80386DX physical memory system.

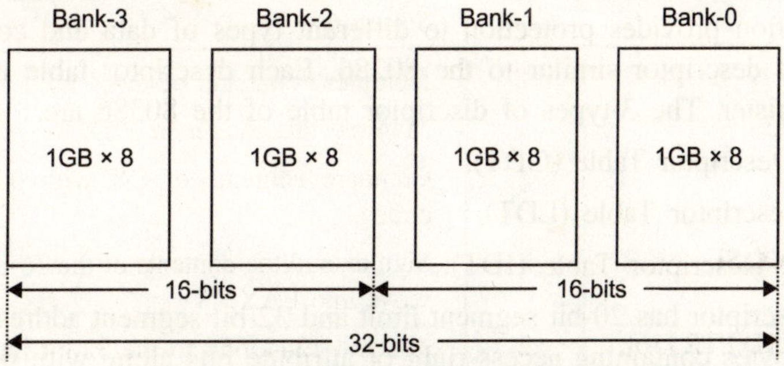

Fig. 3.3 (*b*) Memory system of the 80386

The memory is divided into 4-parts of 8-bit wide memory banks with each containing up to 1GB of memory data to be accessed directly. The 80386DX transfers up to a 32-bit wide number in a **single** memory **cycle**, where the early 8088 required **4-cycles** to accomplish the same transfer and the 80286 required **2-cycle**. Today, this data width is important, especially for single precision floating point numbers that are 32-bits in width and that are used in graphic programs. High level software normally uses floating point numbers for data storage, so 32-bit memory locations speeds up the execution of high level software if it is written to take advantage of this wider memory.

Each memory byte is numbered in hexadecimal as in the prior versions of the family. The difference is that the 80386DX used a 32-bit-wide memory address with memory bytes numbered from location 00000000H through FFFFFFFFH.

The 2-memory banks in the 8086 system are accessed via A_0 and $\overline{BHE}$. In the 80386DX, the memory banks are accessed via 4-bank enable signals $\overline{BE_3}$ - $\overline{BE_0}$. This arrangement allows a single byte to be accessed when 1-bank enable signal is activated by the microprocessor. It also allows a word to be addressed when 2-bank enable signal is activated. In most cases, a word is addressed in bank-0 and 1 or in bank-2 and 3. Memory location 00000000H is in bank-0, memory location 00000001H is in bank-1, 00000002H is in bank-2, and location 00000003H is in bank-3. The 80386DX does not contain address connections A_0 and A_1 because these have been encoded as the bank enable signals.

This is a direct mapped cache because only the leftmost 17-bits of the address are stored in the tag field. The remaining 13-bits of the address ($A_{14} \Leftrightarrow A_2$) are used to address one of the 8K data locations (4-bytes) and a location in the 8K-tag memory.

Cache

Each time that the microprocessor sends an address to the memory system, the cache controller checks the tag ($A_{31} \Leftrightarrow A_{15}$) to determine whether the location addressed by $A_{14} \Leftrightarrow A_2$ is stored in the cache. The comparison tests the address stored in the tag field against address bits $A_{31} \Leftrightarrow A_{15}$ to check for a match or hit. If the tag matches, the microprocessor fetches the data from the cache memory, stores it in the cache and also stores the tag in the tag memory. It may require up to 4-wait states to read data from the main memory, while a read from the cache requires no waits.

Descriptor Table

The segmentation provides protection to different types of data and code. The 80386 has 3-types of segment descriptor similar to the 80286. Each descriptor table is associated with a descriptor table register. The 3-types of discriptor table of the 80386 are :

- Global Descriptor Table (GDT),
- Local Descriptor Table (LDT),
- Interrupt Descriptor Table (IDT).

The 80386 descriptor has 20-bit segment limit and 32-bit segment address. The descriptor of the 80386 is of 8-bytes containing access right of attribute bits along with the base and limit of the segment.

Descriptor Attribute Bits

The access (**A**) attribute indicates whether the segment has been accessed by the μP or not. The TYPE field decides the type of the descriptor and hence the type of the segment. The S bit decides whether it is system descriptor (**S** = 0) or code/data segment descriptor (**S** = 1). The **DPL** field specifies the privilege level of the descriptor. The D bit decides the code segment operation size. With **D** = 1, segment operation size will be 32-bit operand segment else it is 16-bit operand segment. The P-bit (present) tell whether the segment is present in the physical address or not.

With **P** = 1, the segment is present in the physical memory, else it is not. The G-bit is called the **granularity** bit. This bit indicates whether the segment is page addressable or not. The zero bit (0) must remain zero for compatibility with future processor. The AVL field defines whether the descriptor is available for user or for operating system. The structure of the descriptor of the 80386 is shown in Table 3.3(*b*).

Table 3.3 (*b*) Descriptor structure of the 80386

<table>
<tr><td colspan="6">31</td><td colspan="6">0</td></tr>
<tr><td colspan="6">Segment base 15 ⇔ 0</td><td colspan="6">Segment base 15 ⇔ 0</td></tr>
<tr><td>BASE
31 ⇔ 24</td><td>G</td><td>D</td><td>0</td><td>AVL</td><td>LIMIT
19 ⇔ 16</td><td>P</td><td>DPL</td><td>S</td><td>TYPE</td><td>A</td><td>BASE
26 ⇔ 23</td></tr>
</table>

Base	*Base access of the segment*	*LIMIT*	*Length of the segment*
A	Access bit	AVL	Available field for user or OS
TYPE	Type of segment	D	1/0 segment size 32/16-bit
S	S = 0, segment descriptor, S = 1, code/data descriptor	0	0 for compatibility with future processor
DPL	Descriptor privilege level 0 to 3.	G	= segment length in page granular,
P	P = 1/0 segment present/absent		0 = segment length in byte granular
Note	In a maximum size segment (i.e., a segment with G = 1 and limit 19 ... 0 = FFFFFH), the lowest 12- bits of the segment base should be zero (i.e., segment base 11 ... 00 = 000H)		

The 80386 has 5-types of descriptors:

- Code or Data Segment Descriptors,
- System Descriptors,
- Local Descriptors,
- Task State Segment (TSS) Descriptors,
- GATE Descriptor.

3.4 80386 IMPLEMENTATION

In order to meet the performance objectives of the 80386, the chip's designers gave careful consideration to the internal implementation of the architecture i.e., the micro architecture. Performance enhancements in advanced microprocessor CPU designs can be obtained through

- Advances in technology and circuit design,
- Higher clock rates,
- Wider data paths,
- Advances in system and bus architecture and
- Advances in micro architecture.

The internal architecture of the 80386 is organized broadly in 3-sections, namely

- Processing unit,
- Memory management unit,
- Bus interface unit.

The processing unit is further sub-divided into

- Execution unit,
- Code prefetch unit,
- Instruction unit.

The execution unit has 8-general purpose registers and 8-special purpose registers. Each register can hold 32-bit data. The 32-bit multiplication is done in one microsecond by the multiply/divide logic available in the ALU.

The code prefetch unit is responsible for fetching instructions from memory when bus interface unit is not executing bus cycle. It uses an advance-instruction-fetch pointer to prefetch code from memory and stores it in a 16-bit temporary code queue. This queue also acts as a buffer between the prefetch unit and the instruction decodes unit. Since addresses generated by the prefetch unit are linear, they must be translated to physical addresses by the paging unit before the prefetch bus cycle request can be sent to the bus interface unit.

The instruction decode unit translates instructions from the prefetcher queue into micro-codes. The decoded instructions are then stored in an instruction queue (FIFO) for processing by execution unit. The instruction unit decodes the opcode byte received from the 16-byte instruction code queue and arranges them in 3-decoded instruction queue. The 3-decoded instruction queue is passed to the control unit for generating the necessary control signals. The barrel shifter enhances the speed of shift and rotates operations of the 80386.

The memory management unit is further sub-divided into

- Segmentation unit,
- Paging unit.

The segmentation unit uses 2-components to form the physical address such as,

- Segment,
- Offset.

The segmentation unit allows maximum segment of 4GB. The segmentation unit provides 4-levels of protection for protecting and isolating the system's code and data from application program.

The paging unit works under the control of the segmentation unit. In other words, each segment is further sub-divided into pages of 4KB size. The virtual memory is also divided into segments and pages. The paging unit converts linear address into physical address. The control and attribute PLA checks the privilege at the page level. Each page keeps the information about the task. The limit and attribute PLA checks the segment limits and attributes at segment level to avoid invalid access to code and data in the memory segments.

The BIU interfaces the 80386 with the memory and I/O. The bus interface unit has a prioritizer to resolve the priority of the various bus requests. The address drivers drives the bus enable and address signals $A_{31} \Leftrightarrow A_0$. The pipelining and dynamic bus sizing units handles the related control signals. The data bus buffers interface the internal data bus with the system bus.

Thus, there are 6-logical units that operate in parallel. Each unit is assigned a task or step in the fetching and execution of each instruction. This arrangement allows as much parallel execution of the instruction stream as possible. The parallel operation is known as pipeline processing. The units are pipelined and, for the most part, operate autonomously. The units and their interconnections are shown in Fig. 3.4 (*a*) in the functional block diagram of the 80386. Fetching, decoding, execution, memory management and bus access for several instructions are programmed simultaneously.

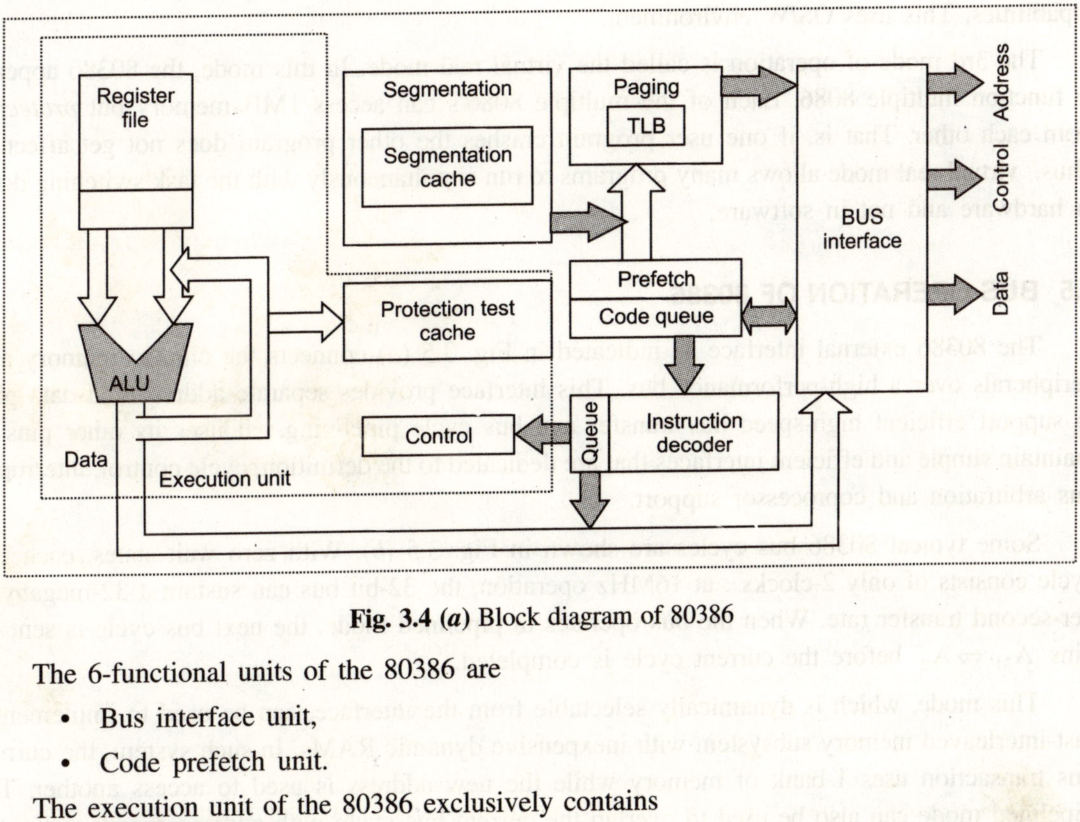

Fig. 3.4 (*a*) Block diagram of 80386

The 6-functional units of the 80386 are

- Bus interface unit,
- Code prefetch unit.

The execution unit of the 80386 exclusively contains

- The control unit,
- The data unit and
- Protection test unit.

The protection test unit checks for the segmentation violation under the control of microcode. Instructions requiring memory references send their requests to the segmentation unit for logical address computation, translation and segment protection violation checking. The segmentation unit produces a translated linear address that the paging unit inturn translates into a physical address. If paging is not enabled, the physical address is identical to the linear address and no translation is necessary. The paging unit also checks for paging violations before it sends a bus request and the address to the bus interface unit and external bus.

The 80386 uses 3-operating modes

- Real mode,
- Protected virtual mode and
- Virtual real mode.

In real mode, the 80386 is limited to 8086 performance and addresses 1MB of memory and one program may run at a time in MS-DOS environment.

In protected virtual mode, the 80386 functions as 80286 with virtual memory and multitasking capabilities. This uses OS/W environment.

The 3rd mode of operation is called the virtual real mode. In this mode, the 80386 appears to function multiple 8086. Each of the multiple 8086's can access 1MB memory but *protected* from each other. That is, if one user program crashes the other program does not get affected. Thus, virtual real mode allows many programs to run simultaneously with the task switching done in hardware and not in software.

3.5 BUS OPERATION OF 80386

The 80386 external interface as indicated in Fig. 3.5 (*a*) connects the chip to memory and peripherals over a high-performance bus. This interface provides separate address and data pins to support efficient high-speed bus transfer and bus cycle pipelining. It uses its other pins to maintain simple and efficient interfaces that are dedicated to the definition, cycle control, interrupts, bus arbitration and coprocessor support.

Some typical 80386 bus cycles are shown in Fig. 3.5 (*b*). With zero wait states, each bus cycle consists of only 2-clocks : at 16MHz operation, the 32-bit bus can sustain a 32-megabyte-per-second transfer rate. When the bus operates in pipelined mode, the next bus cycle is sent on pins $A_{31} \Leftrightarrow A_2$ before the current cycle is completed.

This mode, which is dynamically selectable from the interface, can be used to implement a fast-interleaved memory subsystem with inexpensive dynamic RAMs. In such system, the current bus transaction uses 1-bank of memory while the new address is used to access another. The pipelined mode can also be used to overlap the current bus cycle with address decode delays of the next cycle, making available longer system access times.

Slower memory or I/O systems use the READY pin to extend the current bus cycle as needed. Dynamic bus sizing allows software-transparent interfaces to 16 and 32-bit ports. The 80386 in protected mode can address 4GB of physical memory and 64TB of virtual memory per task. The 80386 in protected mode supports all software written for 80286 and 8086 under control of memory management and protection ability of the 80386. The protected mode supports additional instructions, addressing modes and additional capabilities.

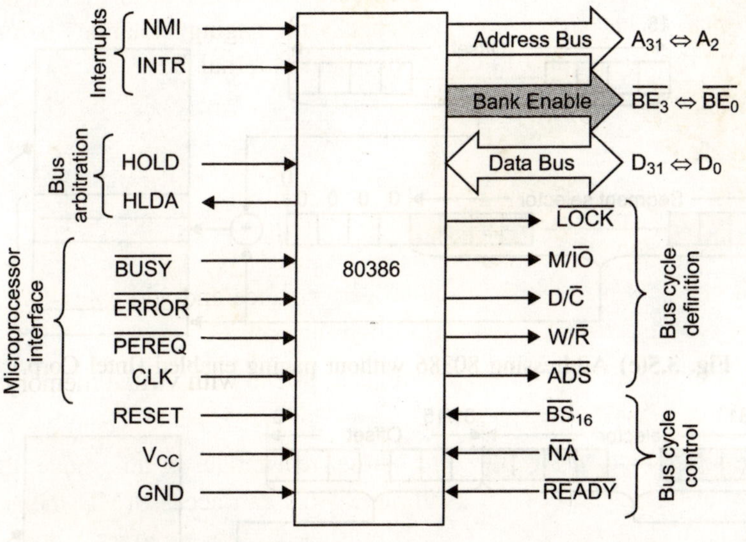

Fig. 3.5 (*a*) Interfacing signals with 80386

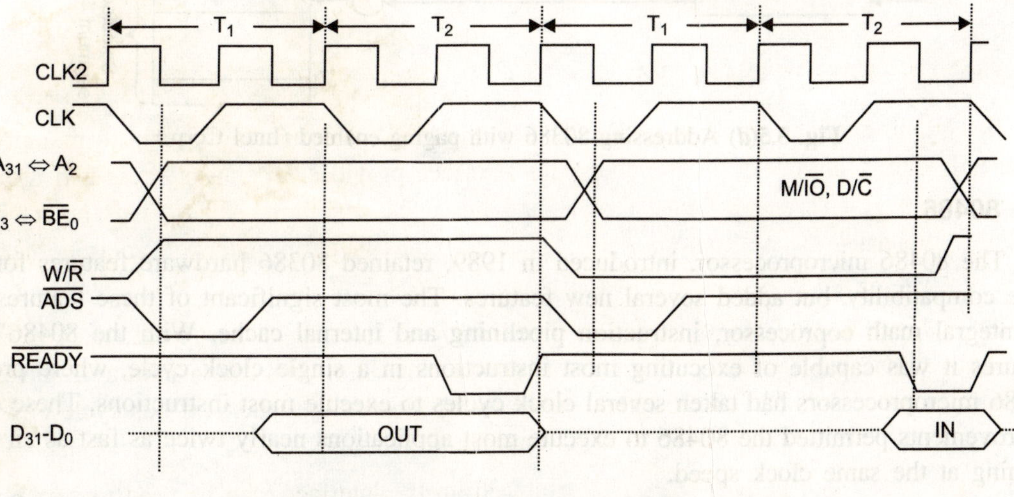

Fig. 3.5 (*b*) Bus operation of 80386

Addressing in Protected Mode

The contents of segment registers are used as selectors to address descriptor that contains the segment limits base address and access rights bytes of the segment. The effective address (EA) or the offset is added with the segment base address to calculate linear address. Thus, linear address is further used as physical address, if the paging unit is disabled, otherwise paging unit converts the linear address into physical address. The paging mechanism handles large segments of memory in terms of pages of 4KB size. The paging unit works under the control of segmentation unit. The paging unit in protected mode, if enabled, converts linear address into physical address. The address modes of 80386 in protected mode without and with paging unit enabled are shown in Figs. 3.5(*c*) and (*d*).

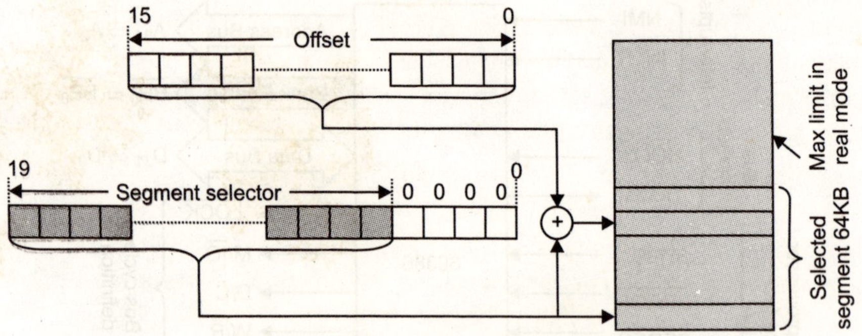

Fig. 3.5(*c*) Addressing 80386 without paging enabled (Intel Corp.)

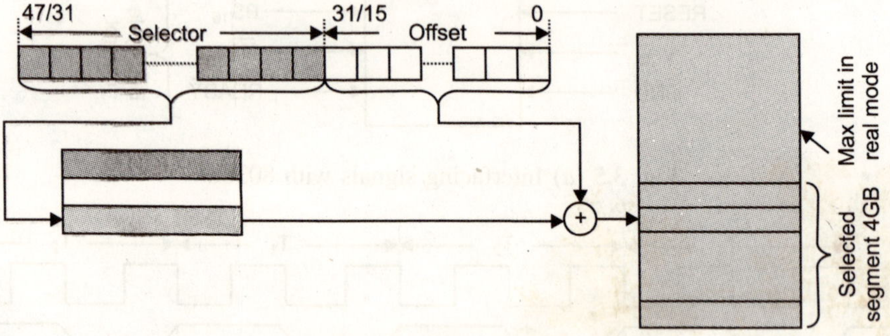

Fig. 3.5(*d*) Addressing 80386 with paging enabled (Intel Corp.)

3.6 80486

The 80486 microprocessor, introduced in 1989, retained 80386 hardware features for software compatibility, but added several new features. The most significant of these features were an integral math coprocessor, instruction pipelining and internal cache. With the 80486's new features it was capable of executing most instructions in a single clock cycle, where previous 80x86 microprocessors had taken several clock cycles to execute most instructions. These design improvements permitted the 80486 to execute most applications nearly twice as fast as an 80386 running at the same clock speed.

The 80486 is the 1st μP that had built-in math-coprocessor and an 8KB of internal cache. These 2-improvements have made the 80486 μP very fast compared to the 80386. The 80486DX chip has both internal and external 32-bit data bus, 32-address bus that can directly address ($2^{32} = 2^{30}2^2$ = 4GB) of physical memory and 64TB of virtual memory. Before introduction of the 80486, a math processor or floating point processor was provided separately for the 8086, 80286 and 80386 μPs. The math processor is a chip designed to perform mathematical operations at a very high speed compared to the μP. Whenever we add a processor 80287 to 80286 and a 80387 to 80386 based computer, the 80287 or 80387 takes over and does the mathematical calculations at a very high speed and gives result back to the μPs (either 80286 or 80386). Thus, this frees the main processor (μP) to do other works during that time and gives big boost to the system performance. The combination of 80386 and 80387 is equivalent to the 80486, but it provides twice as fast thruput because of high speed of the 80486DX and direct connection of the μP and math processor. Math coprocessor or Numeric coprocessor or Numeric Processor Unit (NPU) or Floating Point Unit (FPU) all mean the same thing. The math processor helps the main processor in mathematical operations. This uses over one million transistors and provides 2 to 4-times software performance to that of the 80386 with object code compatibility to 8086/ 8088. This is accomplished by streaming many of the frequently used instructions to a single clock cycle.

While still in the process of exploiting the full capabilities of the 80386, a new processor chip, the 80486, entered the PC market. Table 3.6(*a*) shows the entire Intel family tree. This took place because Intel began the development of the 80486 processor program shortly after it introduced the 80386 processor in 1985. This processor not only provided a very fast and improved 80386 but also came out in the form of EISA (Extended Industrial Standard Architecture) based system i.e., the system was designed around 32-bit bus without compromising for the software built on previous processors with 16-bit bus design.

From initial concept, the chip design team worked with the following μP goals

- ensure binary compatibility with the 80386 microprocessor and the 80387 math coprocessor,
- increase performance by 2-3 times over a 80386/387 processor system at the same clock rate;
- extend PC standard architecture of the 80386 μP with features suitable for minicomputers.

Table 3.6(*a*) Intel family performance

Chip	*Year of Introduction*	*Data*	*Address*
4004	1971	04	08
8008	1971	08	08
8080	1973	08	16
8085	1977	08	16
8086/ 88	1978	16/08	20
80186/ 188	1982	16/08	20
80286	1983	16	24
80386	1986	32	32
80486	1989	32	32
Pentium	1993	64	32

3.7 PERFORMANCE OF 80486

The 80486 microprocessor extended the 32-bit architecture of the 80386 microprocessor into a higher performance realm. The 1.5 mm technology was used to fabricate the 80386 μP that contains approximately 275,000 transistors. However, the rapid development in the semiconductor technology resulted in 1mm CHMOS-IV process for the manufacturer of the 80486 that provided more than 2-times the number of transistor i.e., 1,200,000 transistors. The CHMOS technology combines the advantages of high-speed characteristics of HMOS and low power consumption of CMOS process. The 80486 chip not only provides a powerful mini on a PC platform but also it is designed for workstations and file servers. The process also provided a dense static RAM cell suitable for an on-chip memory cache. The significant increase in the transistor budget provided the raw material for this performance improvement in 80x86 series.

Though the 80486 runs at similar clock speeds to that of the 80386 i.e. 25MHz, but the number of clock cycle required to execute an instruction stream has been reduced to more than half. In an average instruction mix, the 80386 takes slightly less than 2-clocks to execute an instruction. Thus, at the same CLK speed, the 80486 processes instruction twice as fast that of the 80386. This was achieved through a more *streamlined pipeline organization*, an *on-chip cache*, an *integrated floating-point unit* and a *burst mode* bus interface. Table 3.7(*a*) gives comparison of the 80486 w.r.t. that of the 80386 microprocessor.

The point-and-click era of computing came with the introduction of the 80486 in 1989. In other words, Graphical User Interface came into existence. The typical example of the graphical interface is the **Windows**. This enables users to simply point and select their requirements and eliminated the need to remember complex commands. It was also the 1st processor that speeded up the computing because it offloads complex math functions from the central coprocessor.

Table 3.7(*a*) Comparison between 80386 and 80486

Features	*80386*	*80486*
Technology used	CHMOS III	CHMOS IV
Transistors used	275,000	1,200,000
Pins available	32	68
MIPS at 25MHz	4.0 (many instructions in 2CLK)	15.0 (many instruction in 1CLK)
Pipeline capability	NO	YES
On Chip cache	NO	YES (8KB)
On Chip NDP	NO	YES (Arithmetic co-processor in DX version)
RISC care	NO	YES
Instruction set	—	6-Instructions more than 80386
Processing power	—	2-3 times more
CLK per instruction	4.4	1.6
Multiprocessing support	NO	YES
Prefetch queue size	16B	32B
Clock type	2X	X

3.8 BLOCK DIAGRAM OF 80486

Figure 3.8 (*a*) shows the block diagram of the 80486 processor. The chip contains 168-pins housed in a Pin Grid Array (PGA) package. The shaded blocks indicate functions that are partitioned onto separate chips in a system based on the 80386 µP but which appear on chip in the 80486 µP. These functions are

- FPU boosts performance of floating-point operations and
- 8KB Unified cache memory.

Although the remaining blocks look similar to the corresponding blocks in the 80386 µP, note that the underlying logic is quite different. Revamping has been done in the prefetch, instruction decode and control logic to optimally integrate the cache into the instruction execution pipeline. This critical improvement allowed the 80486 µP to execute the most frequent, simplest instruction in 1-clock cycle. The 80486 chip has all the hardware necessary for a complete memory management and protection mechanism. This mechanism provides 4-protection levels using a combination of segmentation and paging techniques. An on-chip paging cache or TLB (Translation Look aside Buffer) of 32-entries supports paging operations.

The design target of the 80486 was to design and produce a processor that can execute a sequence of instructions in less than half clocks of the 80386 while keeping the object code compatibility. This necessitated that many of the most used instructions execute in a single clock. In addition to executing the instructions in a single clock, the microprocessor must decode instructions at the same rate. The 80486 does this through a 4-part instruction decode/execution pipeline and the integrated cache. The number of clocks required, as an example, to execute some instructions are shown in Table 3.8(*a*).

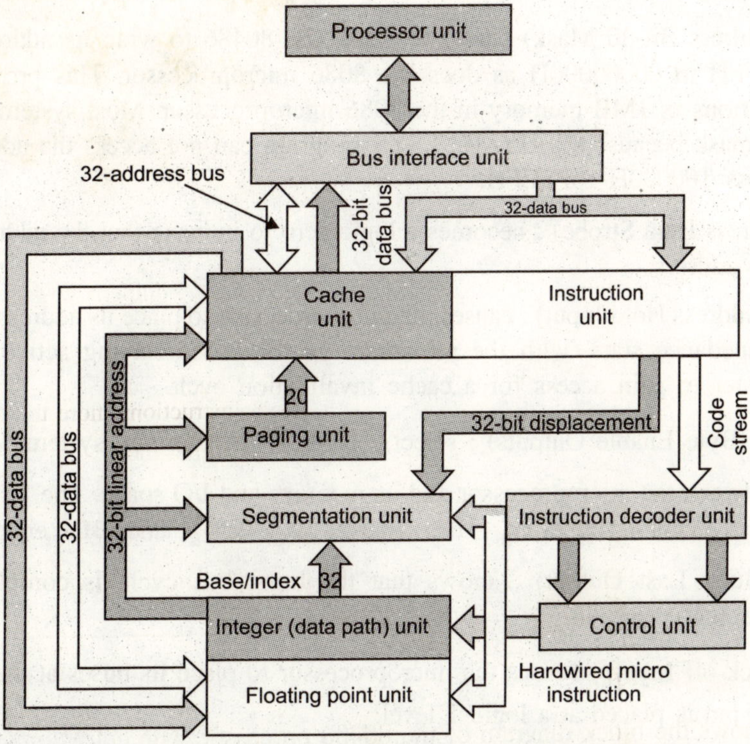

Fig. 3.8 (*a*) Block diagram of 80486

Table 3.8(a) Clock cycles required to execute instructions

Clock	*Instruction*
1	Memory load or store.
2	Register to register integer operation.
2	Memory to register integer operation
3	Register to memory integer operation.
1	Conditional branch not taken.
3	Branch taken.
1	Register PUSH or POP.
3	CALL
5	RETURN
13-42	32-bit register to register integer multiply
40	32-bit unsigned integer divide.

3.9 PIN FUNCTIONS OF 80486

$A_{31} \Leftrightarrow A_2$ (Address Outputs): provide the memory and I/O with the address during normal operation and during cache line invalidation, $A_{31} \Leftrightarrow A_4$ are used to drive the microprocessor.

$\overline{\text{A20M}}$ (Address Bit-20 Mask) : used to cause the 80486 to wrap its address around from location 000FFFFH to 00000000H as does the 8086 microprocessor. This provides a memory system that functions as 1MB memory in the 8086 microprocessor. Most systems do not use the address system mask because the HYMEM.SYS program can not access the additional memory located at address 100000H-10FFEFH.

$\overline{\text{ADS}}$ (Address Data Strobe) : becomes a logic zero to indicate that the address bus contains a valid memory address.

AHOLD (Address Hold Input) : causes the microprocessor to place its address bus connections at their high-impedance state, with the remainder of the buses staying active. Often used by another bus master to gain access for a cache invalidation cycle.

$\overline{\text{BE}_3}$ - $\overline{\text{BE}_1}$ (Byte Enable Outputs) : select a bank of the memory system when information is transferred between the microprocessor and its memory and I/O space. The $\overline{\text{BE}_3}$ signal enables $D_{31} \Leftrightarrow D_{24}$, $\overline{\text{BE}_2}$ enables $D_{23} \Leftrightarrow D_{16}$, $\overline{\text{BE}_1}$ enables $D_{15} \Leftrightarrow D_8$, and $\overline{\text{BE}_0}$ enables $D_7 \Leftrightarrow D_0$.

$\overline{\text{BLAST}}$ (Burst Last Output) : shows that the burst bus cycle is complete on the next activation of the $\overline{\text{BRDY}}$ signal.

$\overline{\text{BOFF}}$ (Back off Input) : causes the microprocessor to place its buses at their bus hold state until the $\overline{\text{BOFF}}$ pin is placed at a logic 1 level.

$\overline{\text{BRDY}}$ (Burst Ready Input) : used to signal the microprocessor that a burst cycle is complete.

BREQ (Bus Request Output) : indicates that the 80486 has generated an internal bus request.

$\overline{BS8}$ (Bus Size 8 Input) : causes the 80486 to structure itself with an 6-bit data bus to access bytewide memory and I/O components.

$\overline{BS16}$ (Bus Size 16 Input) : causes the 80486 to structure itself with a 16-bit data bus to access word wide memory and I/O components.

CLK (Clock Input) : provides the 80486 with its basic timing signal. The clock input is a TTL compatible input that is 25MHz to operate the 80486 at 25MHz.

$D_{31} \Leftrightarrow D_0$ (Data Bus) : transfers data between the microprocessor and its memory and I/O system. Data bus connections $D_7 \Leftrightarrow D_0$ are also used to accept the interrupt vector type number during an interrupt acknowledge cycle.

$DP_3 \Leftrightarrow DP_0$ (Data Parity I/O) : provide even parity for a write operation and check parity for a read operation. If a parity error is detected during a read, the $\overline{PCHK}$ output becomes a logic-0 to indicate parity error. If parity is not used in a system, these lines must be pulled-high to + 5V.

$D/\overline{T}$ (Data/$\overline{Control}$): indicates whether the current operation is a data transfer or control cycle. Table 3.10(*a*) illustrates function of $D/\overline{C}$, $M/\overline{IO}$), and $W/\overline{R}$.

$\overline{EADS}$ (External Address Strobe Input) : used with AHOLD to signal that an external address is used to perform a cache invalidation cycle.

$\overline{FERR}$ (Floating-Point Error Output) : indicates the floating-point has detected an error condition. Used to maintain compatibility with DOS software.

$\overline{FLUSH}$ (Cache Flush Input) : forces the microprocessor to erase the contents of its 8KB internal cache.

HLDA (Hold Acknowledge Output) : indicates that the HOLD input is active and that the microprocessor has placed its buses at their high-impedance state.

HOLD (Hold Input) : used to request a DMA action. It causes the address, data and control buses to be placed at their high-impedance state and also, once recognized, causes HLDA to become a logic-0.

$\overline{IGNNE}$ (Ignore Numeric Error Input) : causes the coprocessor to ignore floating-point errors and to continue processing data. This signal does not affect the state of the $\overline{FERR}$ pin.

INTR (Interrupt Request Input) : requests a maskable interrupt as it does on all other family members.

$\overline{KEN}$ (Cache Enable Input) : causes the current bus to be stored in the internal cache

$\overline{LOCK}$ (Lock Output) becomes a logic-0 for any instruction that is prefixed with the lock prefix.

$M/\overline{IO}$ (Memory/IO) : defines whether the address bus contains a memory address or an I/O port number. It is also combined with the $W/\overline{R}$ signal to generate memory address or I/O read and write control signal. The signals for bus cycle identification are in Table 3.9(*a*).

NMI (Non-Maskable Interrupt Input) : requests a type-2 interrupt.

Table 3.9(*a*) Bus cycle identification

$M/\overline{IO}$	$D/\overline{C}$	$W/\overline{R}$	*Bus Cycle Type*
0	0	0	Interrupt acknowledge
0	0	1	Halt/special
0	1	0	I/O read
0	1	1	I/O write
1	0	0	Code cycle (Op code Fetch)
1	0	1	Reserved
1	1	0	Memory read
1	1	1	Memory write

PCD (Page Cache Disable Output) : reflects the state of the PCD attribute bit in the page table entry or the page directory entry.

$\overline{\text{PCHK}}$ (Parity Check Output) : indicates that a parity error was detected during a read operation on the DP_3-DP_0 pins.

RESET (Reset Input) : initializes the 80486 as it does in other family members. Table 3.9(*b*) shows the effect of the RESET input on the 80486 microprocessor.

$\overline{\text{PLOCK}}$ (Pseudo-Lock Output) : indicates that the current operation requires more than one bus cycle to perform. This signal becomes a logic-0 for arithmetic co-processor operations that access 64- or 80-bit memory data.

PWT (Page Write Through Output) : indicates the state of the PWT attribute bit in the page table entry or the page directory entry.

$\overline{\text{RDY}}$ (Ready Input) : indicates that a non-burst bus cycle is complete. The $\overline{\text{RDY}}$ signal must be returned or the microprocessor places wait states into its timing until $\overline{\text{RDY}}$ is asserted.

$W/\overline{R}$ (Write/Read) : signals that the current bus cycle is either a read or a write.

Table 3.9(*b*) Effect of reset signal

Register	*Initial value with self test*	*Initial value without self test*
EAX	00000000H	?
EDX	00000400H+1D	00000400H
EFLAGs	00000002H	00000002H
EIP	0000FFF0H	0000FFF0H
ES	0000H	0000H
CS	F000H	F000H
DS	0000H	0000H
SS	0000H	0000H

(*Contd....*)

FS	0000H	0000H
GS	0000H	0000H
IDTR	Base = 0, limit = 3FFH	Base = 0, limit = 3FFH
CR_0	60000010H	60000010H
DR_7	00000000H	00000000H

3.10 BENCHMARK

The benchmarking is a technique for evaluating microprocessor performance. This technique is used in the computer industry to compare several computers for executing a set of programs (benchmark). The performance is compared in terms of *speed* and *memory* requirements. This technique has been adopted for comparing the microprocessors. The designer has to identify the 1st critical task to the specific application and then write a benchmark program that implements these task for each microprocessor. If the benchmark programs are written in assembly language, their execution speeds are determined by ascertaining the number of states required for execution of each instruction separately. The number of instructions in the benchmark then multiplies this number. This product is finally multiplied by the processor state time. This is the total execution time of benchmark. Computation of worst-case and typical execution time is helpful. The amount of memory required used by benchmark program is determined from the number of bytes required for each instruction. Its instructions at 25MHz clock frequency with 1.18 million transistors IC achieves a performance of > 37,000 Dhrystone/s as depicted in Fig. 3.10 (*a*).

It also provides 6.1-Double precision MWhetstone/s to meet the growing demands for floating-point performance in math-intensive applications viz. AUTOCAD and LOTUS1-2-3. The 33MHz 80486 provides 49,000 Dhrystone/s and 8.2 Double precision MWhetstone/s. The Dhrystrone benchmark is used to determine integer and memory performance of the processor while the Whetstone benchmark is used to determine the floating-point performance of the processor.

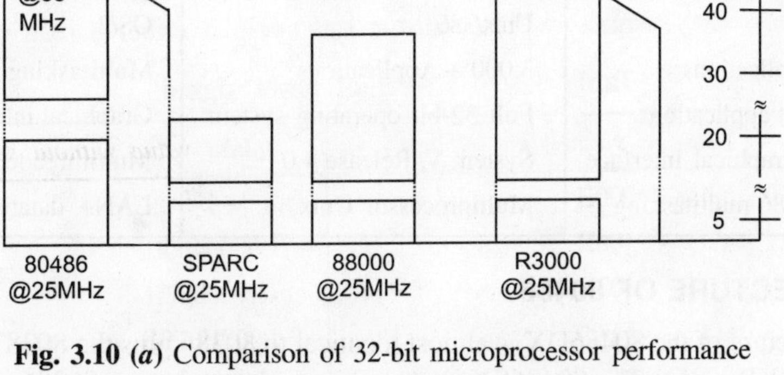

Fig. 3.10 (*a*) Comparison of 32-bit microprocessor performance

The 80486 architecture instructions have 3-parts

- The opcode specifier,
- The operand specifier; and
- Any immediate data or address offsets.

The 80486-microprocessor uses a 2-stages decodes pipeline to decode these 3-part instructions with a thruput of instruction per clock. In the 1st step, it determines the instruction type. In addition, it decides whether and what type of bus cycle needs to be run for that particular instruction and where the effective addressing components are to be found. In the 2nd part of the pipeline, the 80486 microprocessor calculates the actual address for any memory operand and fetches the 1st micro-code vector. In addition, the instruction decoder tells the register bank about the registers to be used in the instructions's 1st execution clock and how it should send those operands to the execution unit. Finally, the instruction decoder tells the ALU about type of arithmetic operation it needs to perform in the 1st execution clock. Both stages of decode pipeline work in parallel. This means that 2-instructions are being decoded in parallel as show in Fig. 3.8(*b*).

3.11 SOFTWARE COMPATIBILITY

With the evolution of new processor chip, a very important question arises in the mind of the user 'whether the existing software will work on the new processor or not'. In this context, the 80486 is fully compatible with its predecessors. Besides, the chip is especially suitable for Windows-98, OS/2, UNIX-V, and iRMK/ iRMX operating systems. One of the aims of the development of the 80486 was compatibility with the 80386 Family Architecture. Table 3.11(*a*) shows, a large base of installed software which runs on a processor that can execute its instruction set. This need to be compatible with an existing software base that influenced many of the design decisions during the development of the 80486 processor. In contrast to the clear choice for software compatibility, no strong point was in favour of hardware compatibility, that is to provide the same hardware interface (bus) used in the 80386 μP. Indeed, to achieve higher performance at the same clock rate, and to provide higher clock rate selections over time, the 80486 bus had to differ significantly from the 80386-processor bus. Intel's data book describes it in detail about the bus.

Table 3.11(*a*) 80386 family architecture software base

Operation System		
DOS	Unix/386	OS/2
10,000+ applications	3,000 + Applications	Multitasking
150 + 32-bit applications	Full 32-bit operating system	Graphical interface
Windows graphical interface	System V, Release 4.0	Multithreaded applications
Windows/386 multitasking	Multiprocessor Unix	LAN+ database

3.12 ARCHITECTURE OF 80486

The architecture of the 80486DX is almost identical to 80386 plus the 80387-math coprocessor and an internal 8KB cache. The 80486SX is also almost identical to an 80386 with an 8KB cache. Figure 3.8 (*a*) illustrate basic architecture of the 80486. The register architecture is similar to that of the 80386 shown in Fig. 3.2 (*a*). If this is compared to the architecture of the 80386 no differences are observed. The most prominent difference between the 80386 and the 80486 is that almost half of the 80486 instructions execute in *1-clock* instead of the *2-clocks* in the 80386.

As discussed, the 80386, the 80486 contains 8-numbers general-purpose 32-bit registers : EAX, EBX, ECX, EDX, EBP, EDI, ESI and ESP. These registers may be used as 8, 16, or 32-bit data registers or to address a location in the memory system. The 16-bit registers are the same set as found in the 80286 and are assigned: AX, BX, CX, DX, BP, DI, SI and SP. The 8-bit registers are AH, AL, BH, BL, CH, CL, DH and DL. In addition to the general-purpose registers, the 80486 also contains the same segment registers as the 80386 which are : CS, DS, ES, and SS in addition to the FS and GS. Each are 16-bit wide as in all earlier versions of the family. At any time, a program can thus access up to 6-segments in the linear address as indicated in Figure 3.12 (*a*).

The IP (Instruction Pointer) address the program located within the 1MB memory in combination with CS or as EIP (extended instruction pointer) to address a program at any location within the 4GB memory system. In protected mode operation, the segment registers function to hold selectors as they did in the 80286 microprocessors.

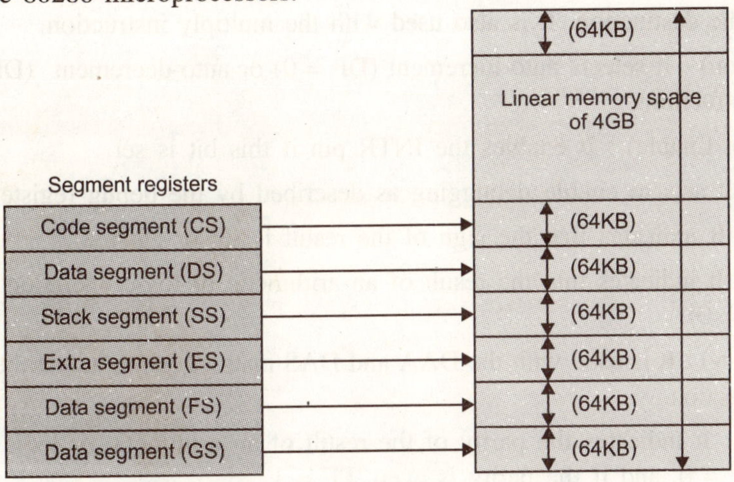

Fig. 3.12 (*a*) Segment representation of 80486 memory

The 80486 also contains the global, local and interrupt descriptor table registers and memory management unit as in the 80386. These registers are not illustrated in Fig. 3.8 (*a*) but are present as in the 80386. The extended flag register (EFLAGS) is illustrated in Fig. 3.12 (*b*).

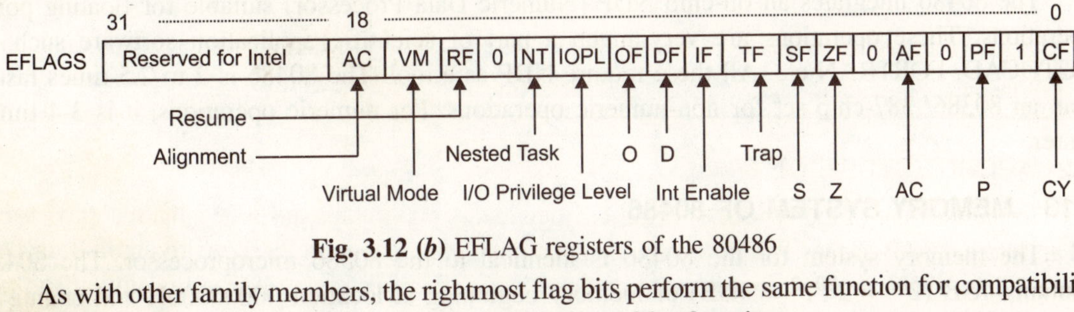

Fig. 3.12 (*b*) EFLAG registers of the 80486

As with other family members, the rightmost flag bits perform the same function for compatibility. Following is a list of each flag bit with a description of its function.

AC (Alignment Check) : It is new to the 80486 microprocessor and used to indicate that the microprocessor has accessed a word at an odd address or a double word stored at a non-double word boundary. Efficient software and execution require that data are stored at word or double word boundaries.

VM (Virtual Mode) : It enters in the Virtual mode by setting this bit while the 80486 is operated in the protected mode. This bit is always cleared by a PUSHF instruction, even if the 80486 is operating in the virtual mode.

RF (Resume) : It is used in conjunction with the debug registers as defined earlier in the 80386 microprocessor.

NT (Nested Task) : It sets to indicate that the 80486 is performing a task that is nested within another task.

IOPL (I/O Privilege Level) : It indicates the current maximum privilege level assigned to the I/O system.

OF (Overflow) : It indicates that the result of a signed arithmetic operation has overflowed the capacity of the destination. It is also used with the multiply instruction.

DF (Direction) : It selects auto-increment (DF = 0) or auto-decrement (DF = 1) operations for the string instructions.

IF (Interrupt Enable) : It enables the INTR pin if this bit is set.

TF (Trap): It sets to enable debugging as described by the debug registers.

SF (Sign) : It indicates that the sign of the result is set or cleared.

ZF (Zero) : It indicates that the result of an arithmetic or logic operation is zero (ZF = 1) or nonzero (ZF = 0)

AF (Auxiliary) : It is used with the DAA and DAS instructions to adjust the result of a BCD addition or subtraction.

PF (Parity) : It indicates the parity of the result of an arithmetic or logic operation. If the parity is odd, PF = 0, and if the parity is even, PF = 1.

CF (Carry) : It shows if a carry has occurred after an addition or a borrow after a subtraction.

NDP

The 80486 integrates an on-chip NDP (Numeric Data Processor) suitable for floating point operations. These operations are very much a part of scientific application software such as AUTOCAD, FORTRAN etc. All these require NDP as a tool. The 80486 is 2 to 2.5 times faster than an 80386/ 387-chip set for non-numeric operations. For numeric operations, it is 3-4 times faster.

3.13 MEMORY SYSTEM OF 80486

The memory system for the 80486 is identical to the 80386 microprocessor. The 80486 contains 4GB ($2^{32} = 2^{30}2^{2}$ = 4GB) of memory beginning at location 00000000H and ending at location FFFFFFFFH. The major change to the memory system is internal to the 80486 form of an 8KB cache memory that speeds the execution of instructions and the acquisition of data. Another addition is the checker/generator built into the 80486 microprocessor.

Parity checker/generator

Parity is often used to determine if data are correctly read from a memory location. Intel facilitates this by incorporating an internal parity generator/detector. The 80486 generate parity during each write cycle. Parity is generated an even parity and a parity bit is provided for each byte of memory. The parity check bits appear on pins $DP_3 \Leftrightarrow DP_0$ that is also parity inputs as well as outputs. These are typically stored in memory during each write cycle and read from memory during each read cycle.

On a read, the microprocessor checks parity and generates parity check error, if it occurs on the $\overline{PCHK}$ pin. A parity error causes no change in processing unless the user applies the $\overline{PCHK}$ signal to an interrupt input. Interrupts are often used to signal a parity error DOS-based computer system. Figure 3.13 (*a*) shows the organization of the 80486-memory system that includes parity storage. Note that this is the same as for the 80386, except for the parity bit storage. If parity is not used, Intel recommends that the $DP_3 \Leftrightarrow DP_0$ pins are pulled-up to +5V.

Cache memory

The performance of Von Neumann architecture gets considerably effected by the memory access time and the memory bandwidth (maximum memory transfer rate). These limitations have become of special interests for 32-bit and further advanced microprocessor with higher clock frequency. The static RAMs with 25ns access time were capable of keeping pace with the microprocessor rate. This precludes that they must be placed on the same board for the processor to minimize the delays, limiting the size of the high-speed memory available. The dynamic RAM has greater capacity per chip and the lower cost than the static counter part. But even the fastest dynamic RAM with 80ns access time are unable to keep pace with the high speed microprocessors, especially if they are located on the separate board attached to the memory bus. When the processor requires an instruction or data to/from the memory faster than the memory can read or write, the processor has to wait until the instruction/data is available. This considerably reduces the performance of the microprocessor.

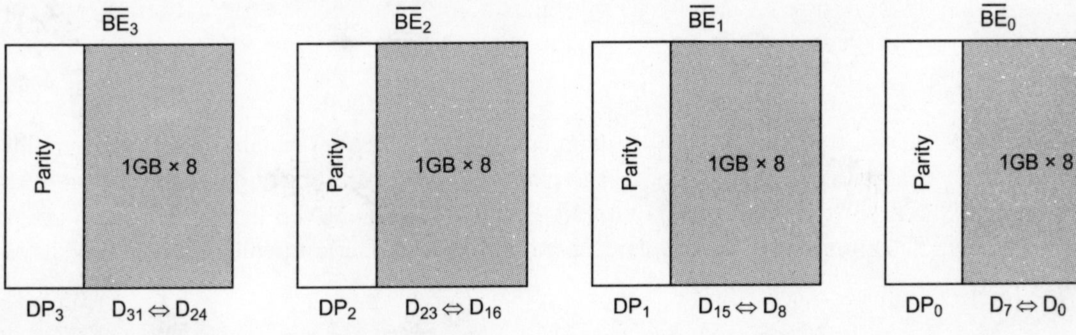

Fig. 3.13 (*a*) 80486 memory system with parity

A cache consisting of high-speed memory, either on the microprocessor chip or external to the chip but on the computer board, works as fast storage *buffer between the processor* and the slower *main memory*. The off-chip cache but on the board may require several memory cycles where the on-chip cache may only require one clock cycle. Even though the off-chip cache

Since the line size in the 80486 is of 16B, each line contains 16 × 8 = 128 bits. Thus, the entire 8KB cache is divided into 128 sets, each set will contain 8KB/128 = $2^3 \times 2^{10}/2^7 = 2^6$B = 64B. The total lines per set would be = 64/16 = 4-lines. Thus, cache organization is a 4-way set-associative cache. The 80486 high-end system can also use 2nd level cache to further enhance the processing speed. This added speed would make graphic-based window oriented program much more responsive. The row number (address) and block number refers the cache. The 16B row has been divided into 4B-lines. Any of the 4-lines can not be accessed partially.

Memory management unit (MMU)

What is the need of memory management unit? What are its functions? These are obvious questions that can be explained with the help of an example. A mini-computer has many terminals. One user for compiling his programs in FORTRAN shares the same compiler. Also that each user desires that his programs should not be used by some one else and should be secretly kept. Also that no one should interfere in between thereby interrupting not only him but also not interrupting the orderly working of the system such as shutting down the system all of a sudden.

These conditions precludes that the design of a mini-computer system should be such that the above mentioned problems are avoided. Therefore, a technique known as Memory Management was introduced. This technique incorporates hardware and software procedures for managing various programs residing in memory thereby presenting an efficient multiprogramming environment. The software procedures are taken care of by the operating system while hardware procedures are very much a part of the concerned central processor. The hardware aspects will only be discussed here.

Function of MMU

The basic function of MMU is to map the logical address into physical address space. This can be done by special concept known as segmentation and paging of memory. The segmentation unit allows management of the logical address space that may be larger than the physical address, but it may also be equal and some times smaller. Each segment provides easy data and code relocation and efficient sharing of global resources as the shared programs are placed in an unique segment in each user's logical address space so that a single physical copy can be shared.

The length of the segment can be allowed to expand or contract according to the program requirements. One way to achieve this is to associate with it a number of equal sized pages. Also, further, a field known as word field can give the specific word within the page. In the 80486, the memory can be organized into one or more variable length segments each up to 4GB. A segment has protection characteristics as well. Each task on an 80486 can have a maximum of 16,381 segments. Hence, total virtual memory for each task will be 64TB (Trillion Bytes), a vast amount of memory. The segment may also be divided into one or more 4KB pages as well. Figure 3.13 (a) illustrates the segmentation/paging incorporated in the 80486 microprocessor.

Fig. 3.13 (a) Segment and paging

Since the line size in the 80486 is of 16B, each line contains 16 × 8 = 128 bits. Thus, the entire 8KB cache is divided into 128-sets, each set will contain 8KB/128 = $2^3*2^{10}/2^7 = 2^6$B= 64B. The total lines per set would be = 64/16 = 4-lines. Thus, cache organization is a 4-way set-associative cache. The 80486 high-end system can also use 2nd level cache to further enhance the processing speed. This added speed would make graphic-based window oriented program much more responsive. The row number (address) and block number refers the cache. The 16B row has been divided into 4B-lines. Any of the 4-lines can not be accessed partially.

Memory management unit (MMU)

What is the need of memory management unit? What are its functions? These are obvious questions that can be explained with the help of an example. A mini-computer has many terminals. One user for compiling his programs in FORTRAN shares the same compiler. Also that each user desires that his programs should not be used by some one else and should be secretly kept. Also that no one should interfere in-between thereby interrupting not only him but also not interrupting the orderly working of the system such as shutting down the system all of a sudden.

These conditions precludes that the design of a mini-computer system should be such that the above mentioned problems are avoided. Therefore, a technique known as Memory Management was introduced. This technique incorporates hardware and software procedures for managing various programs residing in memory thereby presenting an efficient multiprogramming environment. The software procedures are taken care of by the operating system while hardware procedures are very much a part of the concerned central processor. The hardware aspects will only be discussed here.

Function of MMU

The basic function of MMU is to map the logical address into physical address space. This can be done by special concept known as segmentation and paging of memory. The segmentation unit allows management of the logical address space that may be larger than the physical address, but it may also be equal and some times smaller. Each segment provides easy data and code relocation and efficient sharing of global resources as the shared programs are placed in an unique segment in each user's logical address space so that a single physical copy can be shared.

The length of the segment can be allowed to expand or contract according to the program requirements. One way to achieve this is to associate with it a number of equal sized pages. Also, further, a field known as word field can give the specific word within the page. In the 80486, the memory can be organized into one or more variable length segments each up to 4GB. A segment has protection characteristics as well. Each task on an 80486 can have a maximum of 16,381 segments. Hence, total virtual memory for each task will be 64TB (Trillion Bytes), a vast amount of memory. The segment may also be divided into one or more 4KB pages as well. Figure 3.13 (*d*) illustrates the segmentation/paging incorporated in the 80486 microprocessor.

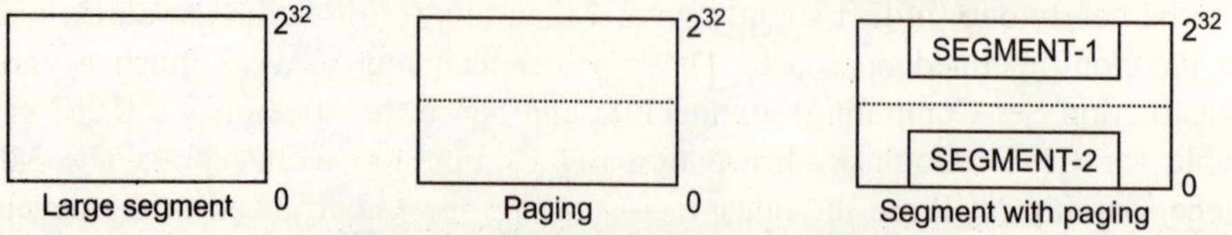

Fig. 3.13 (*d*) Segment and paging

3.14 PIPELINING

Modern microprocessors are structured and hence they contain many internal processing units. Each unit performs a specific task. In the real sense each of these processing units is actually a special purpose microprocessor. This means a modern microprocessor can often process several instructions simultaneously at various stages of execution. This ability is called pipelining. The 8085 is a typical 8-bit microprocessor that fetches an instructions during the read cycle. Figure 3.14 (*a*) illustrates the normal operation of the 8085. While the 8085 executes the instruction, the memory system is idle. The advanced microprocessors beginning with 8086/88 make use of this idle memory time by prefetching the next instruction while executing the current one. This process accelerates the overall execution of a program.

Figure 3.14 (*b*) illustrates the sequence of events for the 80486 microprocessor. The 80486 executes the instructions in pipeline fashion. Thus when one instruction is fetched, the other is being decoded, the third is being executed, and the 4th is being written back. All these take place within the same time duration thus giving over all execution rate of one instruction per clock cycle (in steady state). Considering the conventional approach which require 4-cycles to fetch, decode, execute, and write back for the processing of one instruction, the pipeline approach is far more superior.

Microprocessor	FETCH 1	Decode 1	Execute 1	FETCH 2	Decode 2	Execute 2	FETCH 3	Decode 3	Execute 3
BUS	BUSY	IDLE	BUSY	BUSY	IDLE	BUSY	BUSY	IDLE	BUSY

Fig. 3.14 (*a*) Non-pipelined execution (8085)

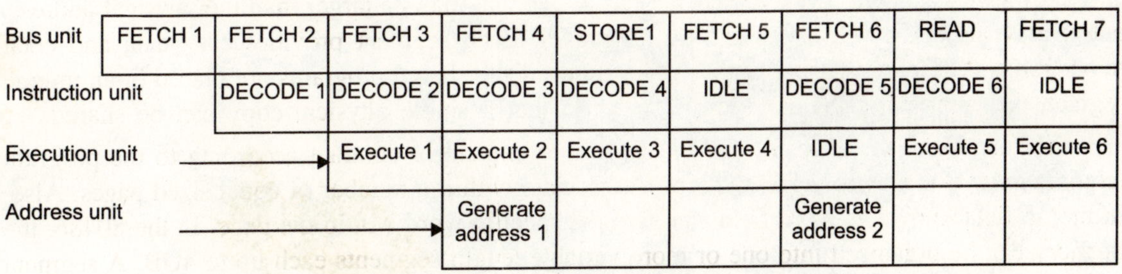

Fig. 3.14 (*b*) Pipelined operation of the 80486

If start and end times are also taken into account, the average (overall) rate of processing comes out to be nearly one (in fact slightly greater than unity) instruction per clock, which is also superior to the non-pipelined approach. The pipeline technique is very much a part of RISC (Reduced Instruction Set Computing) architecture and hence the 80486 has a RISC care besides being suitable for CISC (Complex Instruction Set Computing) architecture. The MMU, 8KB primary cache, on-chip NDP are the other factors which have contributed in the development of RISC on the 80486. The 80486 has 32-byte pipeline i.e., queue. Figure 3.14 (*c*) is another example of the pipelining.

that can result from DX2 or DX4 processors with different motherboard clock rate. Besides different clock speeds, the 80486 processors differ in pin-configurations also.

Table 3.16(a) Maximum clock speed of 80486

Processor	Clock Speeds in MHz									
	16	20	25	33	40	50	66	75	83	100
80486SX	YES	YES	YES	YES						
80486DX			YES	YES		YES				
80486DX2					YES	YES	YES			
80486DX4								YES	YES	YES

Table 3.16(b) Comparison of DX2 and DX4 speeds versus motherboard speeds

Motherboard Speed	16	20	25	33	50
DX2 processor speed	32 MHz	40 MHz	50 MHz	66 MHz	N.A.
DX4 (2X Mode) speed	32 MHz	40 MHz	50 MHz	66 MHz	100 MHz
DX4 (2.5X Mode) speed	40 MHz	50 MHz	75 MHz	83 MHz	N.A.
DX4 (3X Model speed	48 MHz	60 MHz	75 MHz	100 MHz	N.A.

The DX, DX2, and SX processors have identical 168-pin configuration whereas the Overdrive chips are available either in 168-pin or 169-pin configurations. The Overdrive chips are some times called 487SX. The motherboard has 2-sockets

- Primary socket for 168-pin configuration, and
- Secondary socket (Overdrive) for 169-pin configuration.

Most of the recent motherboards with single ZIF (zero insertion force) socket any of the 80486 processors except the DX4 can be inserted because it requires 3.3V instead of 5V for other chips. In upgrading the existing system DX4 should not be inserted in the 5V socket.

80486DX

The Intel introduced 80486DX on 10th April, 1989, the system with this chip appeared in 1990. The 1st 80486DX chip had maximum clock rate of 25MHz. Subsequent versions of 80486DX were available in 33MHz and 50MHz. The most important feature that distinguished 80486 from its predecessor 386 and 286 are:

- Integration, and
- Upgradability.

The 80486DX integrates function such as the math coprocessor, cache controller, cache memory etc. inside the chip. The 80486DX was designed with upgradability in mind. The double speed overdrive is upgraded to 80486DX and are available for almost all systems. Table 3.16(c) shows special feature of 80486DX processors.

that can result from DX2 or DX4 processors with different motherboard clock rate. Besides different clock speeds, the 80486 processors differ in pin configurations also.

Table 3.16(*a*) Maximum clock speed of 80486

Processor	*Clock Speeds in MHz*									
	16	20	25	33	40	50	66	75	83	100
80486SX	YES	YES	YES	YES						
80486DX			YES	YES		YES				
80486DX2					YES	YES	YES			
80486DX4								YES	YES	YES

Table 3.16(*b*)Comparison of DX2 and DX4 speeds versus motherboard speeds

Motherboard Speed	*16*	*20*	*25*	*33*	*50*
DX2 processor speed	32 MHz	40 MHz	50 MHz	66 MHz	N.A.
DX4 (2X Mode) speed	32 MHz	40 MHz	50 MHz	66 MHz	100 MHz
DX4 (2.5X Mode) speed	40 MHz	50 MHz	75 MHz	83 MHz	N.A.
DX4 (3X Mode) speed	48 MHz	60 MHz	75 MHz	100 MHz	N.A.

The DX, DX2, and SX processors have identical 168-pin configuration whereas the Overdrive chips are available either in 168-pin or 169-pin configurations. The Overdrive chips are some times called 487SX. The motherboard has 2-sockets

- Primary socket for 168-pin configuration, and
- Secondary socket (Overdrive) for 169-pin configuration.

Most of the recent motherboards with single ZIF (zero insertion force) socket any of the 80486 processors except the DX4 can be inserted because it requires 3.3V instead of 5V for other chips. In upgrading the existing system DX4 should not be inserted in the 5V socket.

80486DX

The Intel introduced 80486DX on 10th April, 1989, but system with this chip appeared in 1990. The 1st 80486DX chip had maximum clock rate of 25MHz. Subsequent versions of 80486DX were available in 33MHz and 50MHz. The most important features that distinguished 80486 from its predecessor 386 and 286 are

- Integration, and
- Upgradability.

The 80486DX integrates function such as the math coprocessor, cache controller, cache memory etc. inside the chip. The 80486DX was designed with upgradability in mind. The double speed overdrive is upgraded in 80486DX and are available for almost all systems. Table 3.16(*c*) shows special feature of 80486DX processors.

Table 3.16(*c*) Intel 80486DX processor features

Introduced	10th April,1989 (25MHz), 24th June, 1991 (50MHz)
Maximum speed	25 33 50MHz
µP clock multiplier	lx
Register size	32-bit
External data bus	32-bit
Address bus	32-bit
Maximum memory	4GB
Built-in cache type	4-way set associative, write-through.
Built-in cache size	8K
No. of transistors	1.2 million, 1.4 million (SL enhanced models)
Die size	1micron (2533MHz), 0.8 micron (50MHz and SL enhanced models)
Socket size	168-pin PGA, 196-pin PQFP, 208-pin SQFP.
Math processor	Integrated Floating-Point Unit (FPU)
Power management	SMM (System Management Mode)in SL enhanced models)
Burst-mode transfer	YES
Operating voltage	5V standard, 3.3V for 208-pin SQFP models
PGA = Pin Grid Array, PQFP = Plastic Quad Flat Pack, SQFP = Small Quad Flat Pack, Note : SQFP and PQFP are for SL enhanced models only.	

The standard 80486DX contains a processing unit, a FPU, (math processor), memory management unit, cache controller, 8K-built-in cache RAM. With built-in cache and more efficient processing unit, the 80486 family of processors can execute instruction on an average of 2-processor cycle as compared to 4.5-cycle in case of 386 and 286 and 12-cycles in case of 8086 and 8088. The built-in math processor, sometimes called FPU in 80486DX, is 100% software compatible external 387 math processor used with 386, but achieves double performance because it runs synchronously with the main processor and executes most instructions in half as many cycles as taken by the 80386.

80486SX

The 80486SX was introduced in April 1991 to be sold as low-cost version of the 80486. The 80486SX is identical to 80486DX processor with the exception that it does not include math processor (FPU). For very short time the 80486DX chip with FPU disabled due to some faults developed in it, was sold as 80486SX. The transistor count dropped from 1.2 million to 1.185 million in the mask of 80486SX. The 80486SX is 2-times faster than the 386DX with the same clock. Intel introduced 80486SX as an ideal chip for new computer buyers, as entry level users do not use much of the math processor functions. The 80486SX is available in 16, 20, 25, and 33MHz clock speeds. Table 3.16(*d*) shows specifications of 80486SX processor.

Die size	0.8 micron
Socket size	168-pin PGA, 196-pin PQFP, 208-pin SQFP
Math processor	Integrated Floating Point Unit (FPU)
Power management	SMM (System Management Mode) in SL enhanced models
Burst-mode transfer	YES
Operating voltage	5V standard, 3.3V for 208-pin SQFP models
PGA = Pin Grid Array, PQFP = Plastic Quad Flat Pack, SQFP = Small Quad Flat Pack	

Table 3.16(g) summarizes the cycles for most frequently executed instructions for the 386, 80486, Sun Microsystems Sparc, and Motorola 88100 processors. It is evident that 80486 achieves reduction in cycle counts between 2 to 4 times over that of the 385 processor. The 80486 does better than the RISC processors for data manipulation instructions but executes little slower on branches.

Register addressing modes of 80386/486

The 80386 (and later) processors provide 32-bit registers. All the eight general-purpose registers have 32-bit equivalents. They are EAX, EBX, ECX, EDX, ESI, EDI, EBP, and ESP. If we are using an 80386 or later processor we can use these registers as operands to several 80386 instructions.

Table 3.16(g) Clock cycle counts for basic instructions

Instruction type	80386	80486	Sparc	88100
Load	4	1	2	1-3
Store	2	1	3	1
ALU	2	1	1	1
Jump taken	9	3	1	2
Jump not taken	3	1	2	1
CALL	9	3	1	1

Memory addressing modes of 80386/486

The 80386 processor generalized the memory addressing modes, whereas the 8086 only allowed us to use BX or BP as base registers and SI or DI as index registers. The 80386 lets us use almost any general purpose 32-bit register as a base or index register. Furthermore, the 80386 introduced new scaled indexed addressing modes that simplify accessing elements of arrays. Beyond the increase to 32 bits, the new addressing modes on the 80386 are probably the biggest improvement to the chip over earlier processors.

Register indirect addressing modes

On the 80386 we may specify any general purpose 32-bit register when using the register indirect addressing mode. [EAX], [EBX], [ECX], [EDX], [ESI], and [EDI] all provide offsets, by

Die size	0.8 micron
Socket size	168-pin PGA, 196-pin PQFP, 208-pin SQFP
Math processor	Integrated Floating Point Unit (IFPU)
Power management	SMM (System Management Mode) in SL enhanced models
Burst-mode transfer	YES
Operating voltage	5V standard, 3.3V for 208-pin SQFP models
PGA = Pin Grid Array, PQFP = Plastic Quad Flat Pack, SQFP = Small Quad Flat Pack.	

Table 3.16(*g*) summarizes the cycles for most frequently executed instructions for the 386, 80486, Sun Microsystems Sparc, and Motorola 88100 processors. It is evident that 80486 achieves reduction in cycle counts between 2 to 4-times over that of the 386 processor. The 80486 does better than the RISC processors for data manipulation instructions but executes little slower on branches.

Register addressing modes of 80386/486

The 80386 (and later) processors provide 32-bit registers. All the eight general purpose registers have 32-bit equivalents. They are EAX, EBX, ECX, EDX, ESI, EDI, EBP, and ESP. If we are using an 80386 or later processor we can use these registers as operands to several 80386 instructions.

Table 3.16(*g*) Clock cycle counts for basic instructions

Instruction type	*80386*	*80486*	*Sparc*	*88100*
Load	4	1	2	1-3
Store	2	1	3	1
ALU	2	1	1	1
Jump taken	9	3	1	2
Jump not taken	3	1	2	1
CALL	9	3	3	1

Memory addressing modes of 80386/486

The 80386 processor generalized the memory addressing modes, whereas the 8086 only allowed us to use BX or BP as base registers and SI or DI as index registers. The 80386 lets us use almost any general purpose 32-bit register as a base or index register. Furthermore, the 80386 introduced new scaled indexed addressing modes that simplify accessing elements of arrays. Beyond the increase to 32-bits, the new addressing modes on the 80386 are probably the biggest improvement to the chip over earlier processors.

Register indirect addressing modes

On the 80386 we may specify any general purpose 32-bit register when using the register indirect addressing mode. [EAX], [EBX], [ECX], [EDX], [ESI], and [EDI] all provide offsets, by default, into the data segment. The [EBP] and [ESP] addressing modes use the stack segment by default.

Note that while running in 16 bit real mode on the 80386, offsets in these 32-bit registers must still be in the range 00000H to 0FFFFH. We cannot use values larger than this to access more than 64KB in a segment. Also note that we must use the 32-bit names of the registers. We cannot use the 16-bit names. The following instructions demonstrate all the legal forms

```
MOV AL,[EAX]
MOV AL,[EBX]
MOV AL,[ECX]
MOV AL,[EDX]
MOV AL,[ESI]
MOV AL,[EDI]
MOV AL,[EBP]            ;uses SS by default.
MOV AL,[ESP]            ;uses SS by default.
```

Indexed, base/indexed, and base/indexed/disp addressing modes of 80386/486

The indexed addressing modes (register indirect plus a displacement) allow us to mix a 32-bit register with a constant. The base/indexed addressing modes let us pair up two 32-bit registers. Finally, the base/indexed/displacement addressing modes allows us to combine a constant and 2-registers to form the effective address. Keep in mind that the offset produced by the effective address computation must still be 16 bits long when operating in real mode.

On the 80386 the terms base register and index register actually take on some meaning. When combining two 32 bit registers in an addressing mode, the 1st register is the base register and the 2nd register is the index register. This is true regardless of the register names. Note that the 80386 allows us to use the same register as both a base and index register, which is actually useful on occasion. The following instructions provide representative samples of the various base and indexed address modes along with syntactical variations

```
MOV AL, disp[EAX]       ;indexed addressing
MOV AL, [EBX+disp]      ; modes.
MOV AL, [ECX][disp]
MOV AL, disp[EDX]
MOV AL, disp[ESI]
MOV AL, disp[EDI]
MOV AL, disp[EBP]       ;uses SS by default.
MOV AL, disp[ESP]       ;uses SS by default.
```

The following instructions use the base + indexed addressing mode. The 1st register in the 2nd operand is the base register, the 2nd is the index register. If the base register is ESP or EBP the effective address is relative to the stack segment. Otherwise the effective address is relative

REVIEW QUESTIONS AND ANSWERS

Q.1. Name few 32-bit microprocessors?

Q.2. How much memory does the 80386 microprocessor address?

Q.3. How the 80486 microprocessor is faster than some of the earlier microprocessors?

Q.4. What is the difference between the logical and physical memory map of the 80486?

Q.5. How many 32-bit general purpose registers are available in 80386? What are their names and functions?

Q.6. How many 32-bit general purpose registers are available in 80486? What are their names and functions?

Q.7. What segment registers are added to the 80386 and 80486 microprocessors?

Q.8. How many pointer registers are available in the 80386 and 80486?

Q.9. What are index registers in 80386 and 80486? Explain their functions.

Q.10. What do the following instructions do?

(*a*) MOV ECX,EBX (*b*) MOV ESP,EDX (*c*) MOV EAX,SUNY[EBX + ECX]

(*d*) MOV AX,[ECX + 4] (*e*) MOV AX,SUNY[EBX] (*f*) MOV AL,[EAX + 4 × EBX]

Ans. (*a*) Copies (EBX) into (ECX).

(*b*) Copies (EDX) into (ESP).

(*c*) Copies 32-bit content of data segment of memory addressed by SUNY + EBX + ECX.

(*d*) Copies a word from the data segment addressed by ECX + 4 into AX.

(*e*) Copies a word from the data segment into AX addressed by SUNY + EBX.

(*f*) Copies a byte from the data segment into AL addressed by EAX + 4EBX.

Q.11. What do the following instructions do?

(*a*) MOV AL,[EDX] (*b*) MOV ECX,[EBX] (*c*) MOV [EAX + EBX],ECX

(*d*) MOV CL,[EDX + EDI]

Ans. (*a*) Copies a byte into AL from data segment or memory addressed by EDX.

(*b*) Copies double word into ECX from data segment of memory addressed by EBX.

(*c*) Copies double word contents of ECX in data segment of memory addressed by EAX + EBX.

(*d*) Copies a byte from the data segment into CL addressed by EDX + EDI.

Q.12. Determine the memory address accessed by each of the following instructions in real mode operation, if DS = 1000H, SS = 2000H, BX = 0200H, BP = 1000H, and DI = 0100H.

(*a*) MOV EDX,[BP] (*b*) MOV EAX,[BX] (*c*) MOV EAX,[BP + 200]

Ans. (*a*) Physical address = (SS × 10H) + 1000H = 20000 + 1000 = 21000H

(*b*) Physical address = (DS × 10H) + 0200H = 10000 + 0200 = 10200H

(*c*) Physical address = (SS × 10H) + 1000H + 200H = 20000 + 1200 = 21200H

Q.13. Determine the memory address accessed by each of the following instructions in real mode operation, if EAX = 00001000H, EBX = 00002000H, and DS = 0010H.

(*a*) MOV ECX,[EAX+EBX], (*b*) MOV [EAX + 2 × EBX],CL, (*c*) MOV DH,[EBX + 4 × EAX + 1000H]

4

PENTIUM

4.1 INTRODUCTION

The Pentium is the latest 64-bit Intel's family compatible microprocessors introduced in 1993. It was but natural in the sequel of the development of 8086/80186/80286/80386/80486 and also per the predictions projected in Table 4.1(*a*) to call the Pentium as 80586. But as the digits do not signify any real thing i.e., the generation, the word Pentium was coined from the combination of the Greek word '**pente**' and the popular word ending from the periodic table of elements '**ium**'. The word 'pente' meaning *five* associates Pentium with 5th generation microprocessor and the tail of the word 'ium' was used to make the word Pentium better sounding. Its design started in the early 1989 with the primary goal of maximizing the performance while preserving the software compatibility within the practical constraints of the available technology. The 60/66MHz Pentium processor was developed using +5V and 0.8 micron technology with 3.1-million transistors, while the 90/100MHz Pentium was designed using +3.3V and 0.6 micron BiCMOS technology with 3.3 million transistors. Increase in the number of transistors allowed to integrate more components on the chip such as math co-processor, cache, multiprocessor interrupt controllers etc. This increased the performance of the Pentium dramatically. Another way to increase the distance between components and increasing the speed at the same time at which they communicate to each other gave the idea of multiple layers of metal for interconnection between components and the processors. The Intel's 0.6 and 0.35 micron BiCMOS uses 4-layers of metal interconnections.

Pentium is the Intel's 64-bit processor having 32-bit address bus. All microprocessor manufacturers are pointing their intention all the time for better counts and system thruput by adding more and more on-chip functions to the existing architectures. The unprecedented upsurge in the semiconductor technology has been mainly responsible for many innovations in microprocessors' design. The development in the semiconductor technology has been so rapid but consistent, for so many decades, that industry started accepting the concocted '**laws**' to characterize the past progress and to project the future. A few samples of Moore's concocted laws are

- Number of transistors per chip doubles every 2-years (still strong since 1975),
- Feature size reduces to half after every 6-years,
- Number of charges needed to distinguish between 2-states reduces from millions to 10,000 electrons i.e., by a factor of 2-per generation.

Intel researchers have developed the world's fastest transistors that are nearly 1000% faster than those of today's fastest processor, i.e., the **Pentium® 4 processor.** This proves that there is no fundamental barrier to extending **Moore's Law** through the end of the decade.

At only 20nm in size, these transistors will lead to billion-transistor processors, running at speeds approaching 20GHz and operating at less than one volt, before the end of the decade. And despite the common belief that nanotechnology would replace silicon, these transistors are silicon-based. By using current materials and transistor structures to reach this level, Intel has shown that innovation can continue at its current pace.

At a speed of 20GHz, close to one billion calculations will be completed in the blink of a human eye. With all that processing power, it would be possible to create smarter computers that will understand "natural" language and handwriting, and will learn from user patterns to refine and simplify interaction.

These advancements are all part of Intel's goal to continue extending Moore's Law and shrink transistor size by 30% every 2 years. Due to extensive internal R & D, Intel's silicon technology continues to lead the industry. In fact, Intel's new Lab in Oregon is the 1st silicon research facility to focus its efforts on developing 300mm wafers.

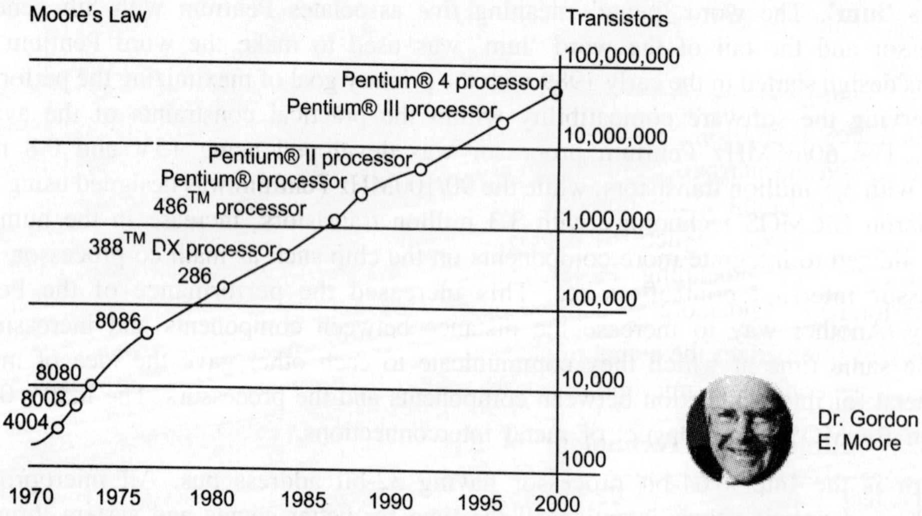

Dr. Gordon E. Moore, the Intel's Chairman, in 1989 in an IEEE Computer Society keynote address predicted that the semiconductor march promises to continue for another decade.

Intel's Director in June, 1990 itself announced the prediction of his company outlined as in Table 4.1(*a*). What a remarkable day in the history of microprocessors and microcomputers to have achieved a breakthrough in its architecture as the Pentium, announced on 22nd March, 1993. The announcement came out, with full of surprises having crossed the 100MIPS mark in the name of the Pentium.

In predicting the performance of the 80786, Gelsinger and his group at Intel's could foresee the parallel organization of 4-processors on the same chip, operating at 750MIPS (million instructions per second) each. In addition, the integration of 50-million transistors would be available for

- Floating point operation,
- Data cache,
- Instruction cache,
- Vector units,
- Graphics unit, and
- Various controllers.

Several factors contributed to the upsurge in the development of variety of new architectures. The most important among those factors is the integrated circuit technology. It evolved to a point to include hundred thousands of transistors on a single chip. The credit goes to the fast development in the integrated circuit technology that made 32-bit microprocessor possible on a single chip.

Table 4.1(*a*) Microprocessor performance prediction

Designation	*No. of transistors (Millions)*	*Availability*	*Clock rate (MHz)*
80486	1.2	1990	40
80586	4.5	1992	60
80686	22.0	1996	100
80786	50-100	2000	250

The continued search and rapid developments in the area of semiconductor technology provided innovation in microprocessor design at an incredible pace. Better technology resulted in higher level of integration made possible by reduced level of sizes and increased layer of interconnections, enabled designers to deploy additional hardware resources for more parallel computation and deeper pipelining. Faster device speed led to higher clock rates consistent to requirements for larger and more specialized on chip memory buffers.

Table 4.1(*b*) summarizes the actual implementation of technological improvements associated with the recent generations of microprocessor as against the projected prediction in Table 4.1(*a*).

Table 4.1(*b*) Technology of microprocessor development

μP	*Year*	*Technology*	*Transistors used*	*CLK (MHz)*
80386	1986	1.5μm CMOS	255K	16
80486	1989	1.0μm CMOS, 2-layer metal	1.2M	33
Pentium	1993	0.8μm BiCMOS, 3-layer metal/ 2-layer metal	3.1M	66

The 0.8 μm BiCMOS technology of the Pentium microprocessor enable 2.5-times the number transistor and twice the clock frequency of the original 80486 μP which was implemented on 1μm CMOS. The Intel on 19th October, 1992 announced the 5th generation microprocessor by the code name P5. Actually the Pentium was shipped on 22nd March, 1993 for the 1st time.

At the high end of the processing spectrum, the new generation chips would be coming up very fast. Millions of Pentium machines must have been sold in the market ranging clock frequencies 150MHz to 700MHz. People do think that as its internal register size is 32-bit wide, Intel is misleading by calling the Pentium a 64-bit processor. In fact 64-bit data transfer do take place. Hence, the Pentium is called 64-bit processor.

With the line of 80486 processors, Intel continued its aggressive upgrade path resulting into the Pentium line of chips. One of the most compelling reasons why Intel maintained an aggressive upgrade path has been the evolution of applications, more so, than the evolution of competing chips. For the development of increasingly sophisticated software, the suitable match of *bridegroom in the form of hardware* is imperative to run the application programs at reasonable speeds.

The cycle between the hardware and software works concurrently. Users define what they expect to foresee added to their existing applications. Applications developers, in turn, try to accommodate and even improve upon users expectations. Very often, the features built into state-of-the-art applications tax the capabilities of existing processors that generates an increased demand for more powerful processors and, consequently, more powerful PC systems. So software developers, by jumping through *hoops* whenever users say *boo,* actually have helped to create the trend towards more powerful hardware implementation.

Users think performance of the μP's in terms of the price paid for general-purpose desktop PCs. Both the Pentium and PowerPC systems provide top of the line of performance. We very often hear question from users, do we really need all that power? For most users, the answer today is *no*. But when users begin to examine and use software that were designed especially for these high-performance system, their answer changes to *yes* provided the *marriage* between *software* and *hardware* is negotiated at reasonable ***dowry***.

Intel produced Pentium keeping in view that it would go head-to-head with its own 80486 architecture. It was perceived by Intel strategists during the design phase of the Pentium (which at that time was code-named the P5) that the power of the Pentium would be valuable chiefly as a server for networks, and as a workstation. The Pentium would be able to run the new WindowsNT, an ideal operating system for a server that could support other Windows 3.x-based client machines, at *blinding* speeds.

Pentium could face significant competition from operating systems, thanks again to WindowsNT that provides symmetrical multiprocessing (SMP). The SMP essentially means that multiple μPs on a multiprocessor system can operate in parallel with little or no contention for memory and other resources. In other words, other system vendors can compete against the Pentium by designing low-cost servers that use multiple 80486 μPs. Since non-Intel 80486 μPs are only a fraction of the cost of a Pentium, this approach can be quite cost effective. It is worth mentioning here that the PowerPC architecture specifies that SMP support be built directly into the μP itself. Such techniques have weak storage models and locking features, which improve the ability of multiple processors to run in parallel without contending for specific memory and cache locations.

Although the Pentium and WindowsNT have been initially targeted for use in high-end systems (especially as network servers), history reveals that high-end μPs eventually have become the standard for desktop systems. With this in mind, Intel expected that the Pentium would become the desktop PC standard within a few years of its introduction ⇒ really to-day *all desktop computers use Pentium.*

Till recent past, Intel had almost no competition at the desktop level. Even superior RISC processors could not be marketed in mass because hardly any software support existed for them. Till date, Intel had no competition for its 80x86 line of chips. Cloned 80486 chips, available from such chipmakers as Cyrix and Advanced Micro Devices (AMD), have become increasingly popular in recent past and have begun eroding Intel's market share significantly.

The attack on Intel became severe with the evolution of its correctly guessed new opponent; the PowerPC. Apart from this, the 80x86 clonemakers are hard at work designing chips that could very well be Pentium killers. The Pentium, as a desktop standard, is seriously threatened.

Heat dissipation model

Designers of workstation, mini, and mainframe computers were forced to consider thermal models as part of system design. This simply means that system designers have to consider the amount of heat generated by a μP or bank of μPs and then designed a cooling system that can dissipate the heat successfully to prevent boards and other components from

- Overheating,
- Deforming, and
- Melting (possibly).

Thermal models for workstations and larger systems typically employ

- Sophisticated heat sinks,
- Cooling fans,
- Casing vents, and
- Other features designed to dissipate heat from the system boards.

The thermal modeling has been a relatively trivial issue for designers of most desktop PCs, with the exception of notebook systems. Systems that use 80286, 80386, or 80486 chips do generate heat but even a single small cooling fan is enough to prevent overheating. This is not the case with Pentium. As because Pentium typically generates 13W, it becomes too hot to be handled for many existing PC system designs. For this reason, Intel released thermal models of Pentium even while the Pentium was still in its pre-release stage. Intel wanted designers of 80486 systems to build Pentium capabilities into their systems to ensure an easy upgrade path for users. Although a few system manufacturers took this route, most did not. The cost to design Pentium-upgradable systems had to be passed on to consumers that would have led to a retail price tag too high for most buyers.

Intel later designed an easier upgrade path by creating a new OverDrive socket architecture that would allow the user of a 80486 to plug in a cooler Pentium OverDrive chip directly into the 80486 motherboard. The cost to design an 80486 with Pentium OverDrive capabilities was substantially less than the cost to design an 80486 that would support a 'true' Pentium upgrade. But herein lies the rub. An 80486 that is upgraded via a Pentium OverDrive is not a true Pentium.

The OverDrive chip can access a secondary (L2) cache only 32-bits at a time, while a true Pentium chip provides a 64-bit data to and from an L2 cache. This fact is especially important when we consider that virtually all-80486 systems have an L2 cache chip. Also, a Pentium

OverDrive chip is limited to the existing clock speed of the motherboard. The fine points of the Pentium OverDrive chip architecture bring yet another level of difficulty for system designers. The complex bus design of the Pentium OverDrive leaves little room for error in the design of a motherboard. *Caches*, *video bus*, *clock synchronicity*, and other system features all must be able to operate in synch with the Pentium chip. As a result the Pentium does not share the same easy upgrade path as earlier 80x86 chips. With Pentium upgrades are unavailable for many 80486 systems and for all 80386 systems, users will no doubt would like to chose other options when time comes to purchase more PC power. The smaller trace width technology allows to pack more transistors in lesser space. Intel's 0.8μ trace width technology is not state of the art, but it has proved aggressive enough to cram 3.1 million transistors on the Pentium die.

The alliance of IBM/Motorola/Apple PowerPC is pointing that the PowerPC 601 chip contains fewer transistors (2.8-million) than the Pentium, but supplies *much* more *processing power*, especially in floating-point performance. In all fairness, though, it is important to note that the 'core logic' transistor count for the PowerPC is about 1.2 million, about half the total transistor count. Here, the core logic just refers to the transistors that are devoted to handling the μP's instruction set. Other on-chip facilities, such as the cache, are not part of the core logic. By contrast, the Pentium uses about 2.3-million transistors for its core logic, about two-thirds the total 3.1-million transistor count. Most of the remaining transistors are included to support the on-chip data and instruction caches.

It is not the full story. Much of the core logic on the Pentium is devoted to support RISC features, many of which are not available on the PowerPC. Examples include *dual integer instruction units* and *aggressive branch prediction logic*. Any way, the PowerPC alliance tries to prove that the Pentium is transistor-heavy because it has to support the 80x86 architecture. This is wrong. The Pentium is transistor-heavy as because its RISC features are robust, although not as carefully packaged as they are on the PowerPC die. A mere 3% of the Pentium's die is devoted to microcode support for 80x86 complex instructions. Proper planning to reduce the trace width of future Pentium chip fabrications to 0.65 micron and less, Intel has provided room for the Pentium's RISC capabilities to grow. The point is that the Pentium is not limited by its microcode support for the 80x86 instruction set. However, it is severely limited.

The Pentium would have been considered an impressive design 2 to 3-years ago, but not today. The major problems with the Pentium are its manufacturing cost and its register architecture.

Performance improvement

Intel used to advertise that the performance improvements for the Pentium would be of the order of 4-times that possible with an 80486DX chip operating at a comparable clock speed during its development phase only. In reality, performance is only about 2-times that of a comparable 80486 and that assumes an application that has been recompiled to take advantage of the Pentium's superscalar architecture. For non-recompiled applications, performance improvement is 70% or less over a comparable 80486.

The 100% performance improvement seems pretty significant. But why the difference in expectation versus reality? In other words, it explains registers or more appropriately the lack of registers. The Pentium has only 8-general-purpose registers (GPRs) for use by both of its integer units, while the PowerPC has 32-GPRs for its single integer unit. Furthermore, one of those 8-GPRs of Pentium is the stack pointer that is not really general purpose at all. With the Pentium, Intel put most of its research and developmental dollars into the development of superscalar (multiple execution units) capabilities of the new chip and virtually zeros dollars into solving the problems that stem from maintaining a paltry 8-GPRs. This gives an indication that the next generations of Intel chips, the P6 and P7 might address this problem well.

This indication left a lot of people guessing about Intel and dragged Intel to compete from many sides. The most important foundations of the RISC technology is its support for a load/ store architecture that in turn relies on the proliferation of simple instructions that can be dedicated to moving values to and from registers. Also that all other instructions must operate on values stored in registers. The value of load/store architecture in increasing the thruput of multiple instructions during a single clock cycle is significantly reduced if the μP does not provide enough GPRs to support load/ store operations. The new RISC capabilities of the Pentium, then, were crippled by the chips small and outdated register file.

The Pentium was designed to increase the instruction thruput of the 80x86 architecture incorporating as many RISC concepts as possible while maintaining 80x86 instruction compatibility. The main RISC feature of the Pentium lies in *superscalar technology*. Most of the processors execute 1-instruction at a time, but the Pentium processor has twin data pipeline that feeds 2-instructions together to the processor. This enhances the processors' ability to execute 2-instructions simultaneously that is called *superscalar architecture*. The use of superscalar technology by the processor to execute more than one instruction at the same time is also called multithreading. The Pentium can execute more than one instruction at a time. This is different from the way Windows, OS/2 or any other operating system performs multiple works at the same time. When a multitasking OS executes more than one job at a time, it does not executes all of them at a time inside the μP, instead what the processor does it allots a fixed time piece for each process and executes each of them for a very short period of time one by one. The change between different process is so fast that this gives impression as if the processor is executing all the process at the same time. Multithreading is different from multitasking in the sense that in the multithreading the processor actually performs 2-process at the same time whereas in multitasking the processor makes use as if it is performing 2 or more jobs at the same time.

The common software environment permits the 80x86 architecture to exercise several operating modes. The performance of any microprocessor is the complex function of many parameters that vary among

- Applications,
- Compilers, and
- Hardware systems.

The design team of Pentium gave due considerations to these aspects for each of the popular software environments. As a result, Pentium features tuned compilers and cache memory.

The Pentium roughly achieves 2-times speed on integer code and up to 5-times on the floating point vector code as compared to an 80486 of identical clock frequency.

4.2 BLOCK DIAGRAM

Figure 4.2 (*a*) is the simplified block diagram organization of the Pentium microprocessor. The core execution units have 2-integer pipelines and a floating-point pipeline with dedicated adder, multiplier and divider. Separate on-chip instruction code and data caches supply the memory demands of the execution units, with a branch target buffer augmenting the instruction cache for dynamic branch prediction. The external interface includes separate address and 64-bit data buses.

Integer Pipeline

A critical feature of the Pentium is that the 2-integer pipelines; the U-pipe and the V-pipe. They are not identical. Users who are only familiar with the Pentium in a very general sense understand that the Pentium has 2-integer units and therefore assume that 2-integer instructions can typically execute in parallel. That is certainly the goal of any superscalar processor to keep as many of its execution units busy simultaneously to increase instruction thruput, but the Pentium imposes some formidable limitations on a program's ability to execute 2-integer instructions in parallel.

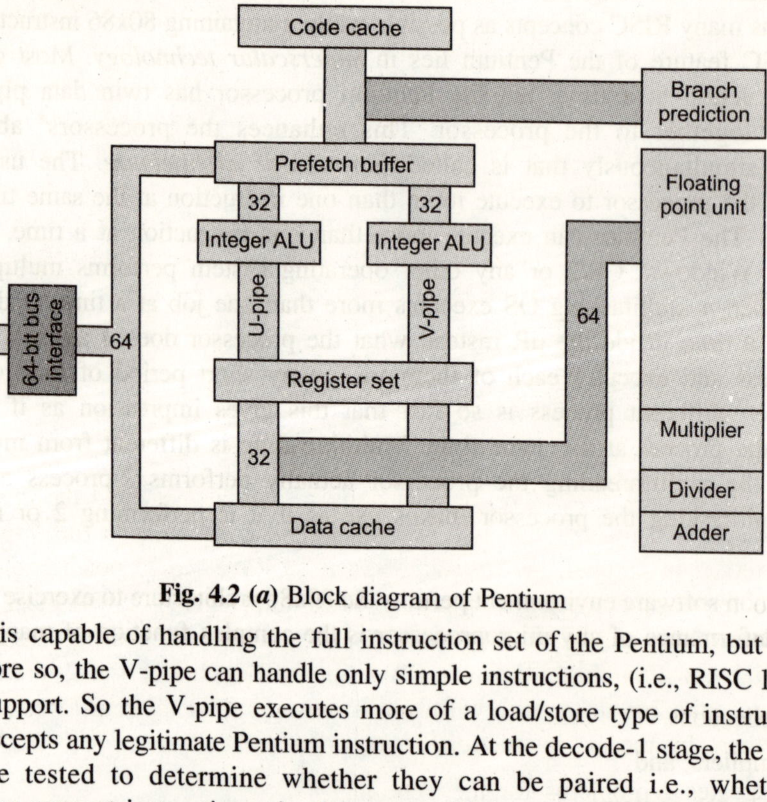

Fig. 4.2 (*a*) Block diagram of Pentium

The U-pipe is capable of handling the full instruction set of the Pentium, but the V-pipe is more limited. More so, the V-pipe can handle only simple instructions, (i.e., RISC like) requiring no micro-code support. So the V-pipe executes more of a load/store type of instruction set than the U-pipe that accepts any legitimate Pentium instruction. At the decode-1 stage, the 2-instructions in each pipe are tested to determine whether they can be paired i.e., whether both the U-and V-pipes can support instructions simultaneously. If the instruction in the V-pipe does not qualify as a simple instruction, or if the μP determines that some dependency limitation prevents the V-pipe instruction from being paired with the instruction in the U-pipe, the V-pipe instruction is transferred to the U-pipe. The V-pipe remains empty until an instruction placed in the V-pipe's decode-1 stage can be paired with the current instruction in the U-pipe's decode-1 stage.

This limitation of pipelining capabilities can lead to significant processing bottlenecks because only simple instructions can be paired and because both the U-pipe and V-pipe must operate in parallel. When Pentium is able to pair 2-integer instructions, the instruction in the U-pipe exits in the decode-2 stage at the same time the instruction in the V-pipe also exits in its decode-2 stage. Both instructions enter the execute stage at the same time. However, if the instruction in the V-pipe stalls, or if it requires more than one clock cycle to execute or depends on some result stored by the instruction in the U-pipe, the instruction in the U-pipe is allowed to continue in the write back stage. That is the result is allowed to be written back to the cache. On the other hand, if the instruction in the U-pipe stalls, the instruction in the V-pipe also halts (i.e., the μP actually forces the V-pipe to stall) until the instruction in the execution stage of the U-pipe has been moved to the write back stage.

The explanation in the previous paragraph basically points to 2-nagging limitations. The first is that the Pentium does not support out-of-order execution and dispatch of instructions. This feature is available in the PowerPC in a limited way. Since the Pentium can't handle instructions out of order, the instruction in the U-pipe gets completed always before the instruction in the V-pipe.

The paired instructions must finish together partly to enforce the pairing feature itself. For 2-instructions to pair, both instructions must enter the decode-1 stage of their respective pipes at the same time. The paired instructions exit the pipes at the same time that enables the prefetcher to place 2-new instructions in the U-and V-pipes at the same time. This limitation can lead to numerous pipeline bubbles (and consequently waste processing cycles) if a program has not been optimized to ensure that paired instructions both execute in one clock cycle. (A few load/store type instructions that execute in 2 or 3-clock cycles can be successfully paired with single-cycle instructions. The Pentium uses sequencing logic to stall one pipe if the other pipe needs an additional cycle or 2 to 'catch up to' the other pipe).

The 2nd major problem of the Pentium's lockstep approach to its dual pipeline architecture stems from data dependencies. Instructions can not be paired if either the instruction in the U-pipe or in the V-pipe attempt to access the same register.

Basically, this type of register contention comes in 2-flavours

RAW : The read-after-write (RAW) register contention would occur if the instruction in the U-pipe writes a result to a register that will be read by the instruction in the V-pipe. Since the U-and V-pipes execute in parallel, the instruction in the U-pipe would not be able to make the result available to the V-pipe when the V-pipe accesses the register. When the decode-1 stage of the Pentium detects this condition, it prevents 2-instructions from pairing; instead, they must both execute, serially, within the U-pipe. A RAW can still occur for the 2-instructions in the U-pipe, in which case the 2nd instruction will stall while waiting for the register to make available the result stored there by the 1st instruction.

So a RAW creates 2-problems for the Pentium

- The inability of the μP to pair integer instructions, and
- A pipeline bubble in the U-pipe.

By now you might have realized that RAWs can occur in any heavily pipelined architecture, including one that provides only one integer unit (such as the 80486 and the PowerPC).

WAW : A write-after-write (WAW) condition occurs when an instruction tries to write to the same register where the previous instruction writes too. If the Pentium μP identifies a WAW with the 2-instructions in the decode-1 stage of the integer pipes, it dose not allow the 2-instructions to pair, even if other pairing requirements have been met. Instead, both instructions must execute serially in the U-Pipe. The WAW condition stems directly from a problem called register starvation, probably the single *biggest drawback* of the Pentium.

Register starvation

The Pentium maintains backward compatibility with the Intel 80x86 architecture by providing only 8-general-purpose registers to be used by its 2-integer units. This suggests that Intel had no choice but to preserve this architecture in whole cloth. It will become clear that it did have a choice if you go through the Cyrix's innovative M1 chip.

Intel did maintain the backward compatibility with the existing 80x86 instruction set. This makes sense for its survival. But less sensible was Intel's reluctance to support backward compatibility while encouraging a more flexible use of RISC features-specifically, a *larger register file* that could be exploited without sacrificing the 8-GPRs requirement of existing 80x86 programs.

Probably Intel *misjudged* both the quality and the quantity of its *competition*. Intel never really believed (until very recently) that a RISC architecture, like the PowerPC, would infringe significantly on its market share. For this reason, it has not shown adequate respect for the R&D capabilities of rival 80x86 chipmakers like Cyrix and AMD. The result is that the Pentium is starved of registers by comparison with other advanced processors. Intel, by exhibiting the kind of cockiness long associated with IBM, became vulnerable to assault not only from IBM but also from all other PowerPC and 80x86 chipmakers.

Intel's strategy in the short time is likely to amount to a 'quick fix' i.e., improve manufacturing technology so that faster Pentium chips can be produced. By doing so, Intel can then brag that its Pentium chips are faster than anything from the PowerPC alliance. In fact, Intel shipped millions of 90MHz and 100MHz P54C chips, a Pentium design that is faster and architecturally smaller than the original Pentium. The P54C is not cheap, but it is cool enough to be used in notebook PCs, something that could not be done away with the original Pentium chips.

The 'faster is better' strategy might work well. But in the long term, Intel has to be able to **pit** the Pentium against PowerPC chips in terms of **price**. Faster may mean better, but it almost never means cheaper. The PowerPC alliance is poised to design faster and cheaper chips. It is not just an opinion. IBM sources have confirmed that 100MHz Pentium will not be able to approach it in terms of speed. Here, Intel has probably lost the hardware war, but it can still keep up its market share by ensuring that the PowerPC never wins the software war. The onus lies on Intel to ensure that Microsoft and other major vendors of 80x86 software do not migrate their applications to the PowerPC platform.

The Cyrix estimates that the 80x86 based application sales in more than $50 billion, a number that continues to climb. Clearly, PC users will be relying on existing 80x86 applications for many years. It is wrong to say that the Intel 80x86 architectural standard will disappear with the advent of the PowerPC and other μP technologies. Intel will continue to prevail, but the question is for how long?

We have explained in some detail why the Pentium's architecture poses hazards for applications designed with previous 80x86 processors. The best way to improve the performance of existing 80x86 applications on the Pentium is to recompile them. A Pentium optimizing compiler can rearrange code to reduce or even eliminate register contention and to improve the number of instructions that can be paired in the U- and V-integer pipes, but 2-problems emerge. The 1st and the most important is the Pentium's sticky superscalar architecture makes it difficult for even the most sophisticated Pentium-based compilers to rearrange an existing application's code to achieve an ideal level of pairing in the Pentiums 2-integer pipes. The truth is, an application can be optimized best by programmers who can handcraft the code in assembly language; a time-consuming and elaborate process that requires a lot of expensive programmer savvy Compilers to make 'educated guesses' when recompiling an existing application for the Pentium. Also sometimes those guesses are less than great. Human programmers can do a much better job at hand optimizing code-provided they understand all the nuances of the processor, in this case, the Pentium.

The 2nd problem involves economy of the scale. Few software development vendors have the time or budget to allow their programmers to handcraft code for the Pentium. Intel has to convince vendors of 80x86 programs that the Pentium architecture is innovative enough to substantiate the cost involved in recompiling and reselling Pentium-specific applications. But vendors monitor their customers to gain information about trends; they do not rely on chipmakers to provide this news.

Once again, we are back to the old price/performance issue. The average PC consumer including the average corporate information systems (IS) manager responsible for buying PC does not care much about the nuts-and-bolts of RISC μP design. Consumers are willing to pay more for new technology only if the higher price tag translates into a corresponding additional value, a much bigger bang for relatively few added bucks.

Let us analyze the Pentium with a question. Does it provide substantial price/performance payoff? Not at all. Sales of Pentium-based systems have been far from stellar, and the major reason is easy to identify. The availability of native Pentium software (it is the software that has been specifically recompiled to take advantage of Pentium's specific features and thus cannot run on any other 80x86 μP) is slight, at best. Intel seems to be gaining only modest ground in its attempt to climb a very slippery slope. Applications designed for more traditional 80x86 chips do run faster on the Pentium, but not fast enough for buyers to justify relatively high expense involved in purchasing new Pentium systems. Users benefit from Pentium systems only when the majority of their applications have been Pentium-tuned.

As users have been slow to purchase Pentium systems, application vendors became slower to port their existing 80x86 programs to optimize Pentium versions. This fact is more than ironic and positively sadistic for Intel. Intel has been shouting to anybody willing to listen that the PowerPC architecture would not proliferate because the native software base for PowerPC systems will never grow larger than an ant-hill. Actually, the commitment on the part of vendors to design native PowerPC systems grows exponentially on a weekly basis, while commitments to recompile applications for the Pentium have been relatively few and far between.

Intel claims that the PowerPC can not compete with the Pentium because the installed base for 80x86 applications is too overwhelming. According to Intel, such PowerPC players as IBM,

Apple, Motorola, and other partners would not be able to convince software developers to abandon the 80x86 platform in favour of the PowerPC platform. But the bottom line is that most users do not care about platforms, they do not care about RISC versus CISC issues, and they do not care about the shrill debate taking place between Intel and PowerPC alliance. They do care about the cost, performance, and time.

A number of corporate players in the PowerPC stadium have encouraged 3rd-party developers to write native PowerPC versions of their existing software that helps to ensure that numerous native PowerPC applications are available for existing PowerPC systems; at or shortly after the availability of the systems themselves. The current availability of native software for Power Mac systems is the proof of this trend.

On the contrary, Intel has to rely only on itself for the continued success and proliferation of the 80x86 architecture. Intel is literally starving for help. The starvation is a good metaphor for Intel, since the feeding frenzy for application developers might very well move to the PowerPC platform or even to other CISC chips. The next few sections explain why.

The Pentium processor's integer pipeline is similar to that of the 80486 μP. The pipeline has 5-stages.

4.3 PREFETCHER

The 2-integer units and one floating-point unit of the Pentium is just like the one shown in Fig. 4.3 (*a*). The 2-integer units, each one, having the same basic 5-stage pipeline that is available on the 80486. They are

- Prefetch,
- Decode-1,
- Decode-2(called the segment or address generation stage in the 80486),
- Execute, and
- Writeback.

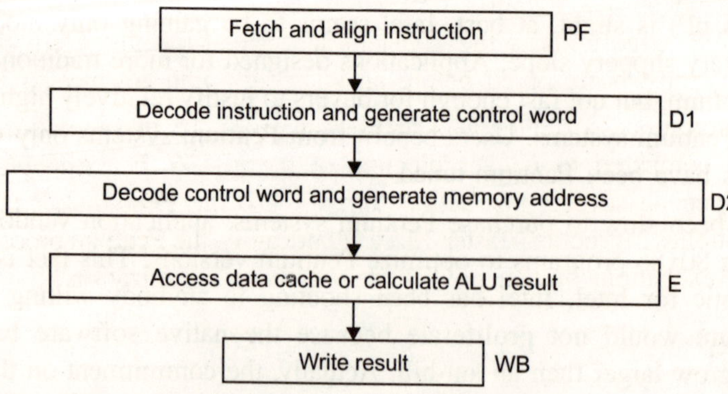

Fig. 4.3 (*a*) Integer pipeline

However, each stage on the Pentium works a bit differently than on the 80486. The differences is involved in the need to support the Pentium's superscalar (multiple execution units). The prefetcher on the Pentium contains 2-instruction queues, but only one queue is used at a time, depending on the type of instruction currently being executed. When instructions are being executed in sequence, the pipelines get their instructions from the 1st instruction queue.

The 2nd pre-fetch queue comes into use only when a branch instruction enters the execution stage in the V-pipe and only if the Pentium's branch-prediction logic determines that the branch should be taken. When that happens, the μP identifies the target location for the branch (a form of speculative execution), loads instructions that begin at the branch target into the second queue, and the μP then executes instructions from the 2nd queue. The μP will switch back to the 1st pre-fetch queue if the branch is not taken (in which case the 2nd queue and both pipelines have to be flushed). This switching from one queue to another continues whenever a branch instruction is taken. In other words, when the μP predicts that the branch will be taken and therefore speculatively executes instructions starting at the branch target address. During the PF stage, the μP prefetches code from the instruction cache and aligns the code to the initial byte of the next instruction to be decoded. Because instructions are of variable length, this stage includes buffers to hold both the lines containing the instruction being decoded and the next consecutive line.

First decode

The μP in the D1 stage decodes the instruction to generate a control word. A single control word executes instructions directly; more complex instructions require micro-coded control sequencing in D1.

Second decode

The μP in the D2 stage decodes the control word from D1 for use in the E-stage. In addition, the μP generates address for data memory references.

Execute

The μP in the E-stage either accesses the data cache or calculates results in the ALU, barrel shifter, or other functional units in the data path.

Write back

The μP in the WB stage updates the registers and flags with the instruction's results. All exceptional conditions must be resolved before an instruction can advance to WB.

The Pentium microprocessor integrates additional hardware in several stages to speedup instruction execution compared to integer pipeline of the 80486. For *example*, the 80486 μP requires 2-clocks to decode several instruction formats, but the Pentium μP takes 1-clock and executes shift and multiplies instructions faster. More significantly, the Pentium processor substantially enhances superscalar execution, branch prediction and cache organization.

4.4 SUPERSCALAR

The Pentium has superscalar organizations. It enables 2-instructions to be executed in parallel. Figure 4.4 (*a*) shows that the resources for address generation and ALU functions have been replicated in independent integer pipeline, called U and V. (The pipeline names were selected because U and V were the 1st 2-consecutive letters of the alphabet neither of that was the initial of a functional unit in the design partitioning).

The μP in the PF and D1 stages can fetch and decode 2-simple instructions in parallel and issue them to the U and V-pipelines. Additionally, the μP for complex instructions in D1 can generate microcoded sequences that control both U and V-pipelines. Several techniques are used to resolve dependencies between instructions that might be executed in parallel. Most of the logic is contained in the instruction issue algorithm as indicated in Fig. 4.4 (*b*) of D1.

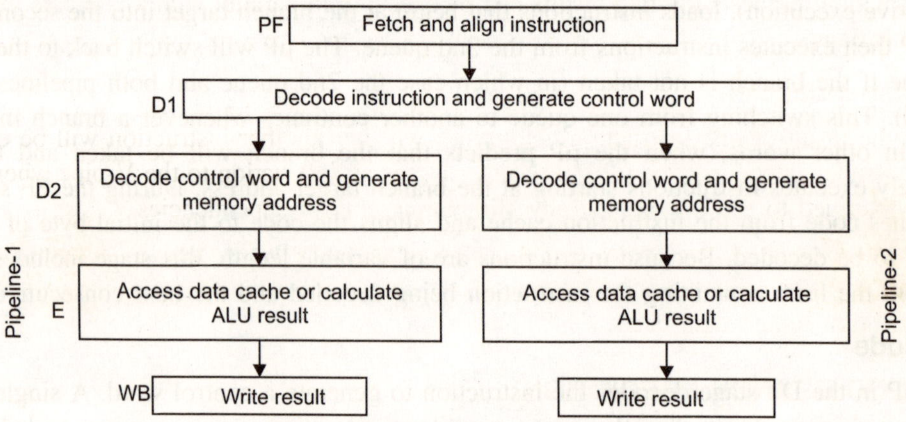

Fig. 4.4 (*a*) Superscalar execution

Decode two consecutive instructions: I1 and I2
If the following are all true
I1 is a 'simple' instruction
I2 is a 'simple' instruction
I1 is not a jump instruction
Destination of I1 ≠ source of I2
Destination of I1 ≠ destination of I2
Then issue I1 to U-pipe and I2 to V-pipe
Else issue I1 to U-pipe

Fig. 4.4 (*b*) Instruction issue algorithm

Resource dependencies

When 2-instructions require a single functional unit or data path, a resource dependency occurs. The μP during the D1 stage issues 2-instructions for parallel execution if both belong to the class of 'simple' instructions, thereby eliminating most resource dependencies. The instructions must be directly executed that does not require micro-coded sequencing. The instruction being issued to the V-pipe can be an ALU operation, memory referencing or a jump. The instruction being issued to the U-pipe can be from the same categories or from an additional set that uses a functional unit available only in the U-pipe, such as the barrel shifter. Although, the set of instructions identified as 'simple' might seem restrictive, more than 90% of the instructions executed in the integer SPEC benchmark suite are simple.

Data dependencies

When one instruction writes a result that is read or written by another instruction, a data dependency occurs. Logic in D1 ensures that the source and the destination registers of the instruction issued to the V-pipe differ from the destination register of instruction issued to the U-pipe. This arrangement eliminates read after write (RAW) and write after write (WAW) dependencies. Write after read (WAR) dependencies need not be checked because reads occur in an earlier stage of the pipelines than writes. The design includes logic that enables instruction with certain special types of data dependency to be executed in parallel. For example, a conditional branch instruction that tests the flag results can be executed in parallel with a compare instruction that sets the flags.

Control dependencies

When the result of one instruction determines whether another instruction will be executed, a control dependency occurs. The μP in D1 never issues an instruction to the V-pipe when a jump instruction is issued to the U-pipe, thereby eliminating control dependencies.

Note that resource dependencies and data dependencies between memory references are not resolved in D1. Dependent memory references can be issued to the two pipelines; as explained in the description of the data cache.

4.5 DYNAMIC BRANCH PREDICTION

The Pentium uses pipeline to increase the performance of μP. The pipeline should always be full for getting maximum throughput. For the pipeline, the μP always fetches some instruction ahead of the actual processing location. Thus, most likely set of instructions to be executed is predetermined and the pipeline is kept full accordingly. This technique has been borrowed from the main frame for improving the performance. The branch prediction unit forecasts the branch to be taken by the software based on the assumption that the previous branch that was taken will be used again. The Pentium uses its Branch Target Buffer (BTB) to do so. The Pentium uses 2-prefetch buffers to function efficiently. One of the prefetch buffers fetches the code in a linear fashion (for the next execution step) while the other pre-fetches instructions based on the address in the BTB. Thus, the needed code is pre-fetched before it is required for execution.

The prediction in a loop is based on the assumption that the instruction in the loop is likely to be executed in the next instruction. There are 2-types of branch

- Unconditional, and
- Conditional.

In unconditional branch, the processor is sure that the branch will take place and the execution will start from a specific location. On the contrary, in conditional branch, the processor is not any way able to know if the branching will take place or not until the instruction is processed completely. If the processor waits for the instruction to finish, to decide from where to fetch the next instruction, then the pipeline may miss a lot of opportunity to fetch the instruction and the pipeline may become empty.

The Pentium excels over the PowerPC in the branch prediction area of the μP design. You might recall that the PowerPC uses a static branch prediction method. Under this approach, when a branch instruction enters the branch-processing unit, the BPU determines whether the branch instruction's target address is in front of or behind the current address being processed in the integer unit. The BPU speculatively executes instructions at the target address only if the address is behind the current address. This logic is considered to be static because it behaves the same way for all branch instructions.

The branching logic of the Pentium is more aggressive and therefore, is called dynamic branch prediction. The Pentium uses a 4-state model

- Strongly taken,
- Weakly taken,
- Weakly not taken, and
- Strongly not taken.

based on the history of a particular branch instruction to determine whether to speculatively execute instructions at the branch target address.

The Pentium essentially tracks the history of branch instructions to determine when and how often they have been taken in the past. The Pentium analyzes this history to determine whether a branch should be taken (speculatively executed) on its current pass through the μP. The Pentium's superior branching logic helps to reduce pipeline flushes (for incorrectly executed branches), although it does not really affect the Pentium's rate of instruction pairing.

The major benefit of the Pentium's dynamic branching logic is that the successful branch prediction is less dependent on compiler optimization. With the static branch prediction used in the PowerPC, it is important that the compiler optimize loops and other instructions that rely heavily on branching. Optimization is critical here because the branch processor is relatively limited in its ability to correctly guess when a branch will be taken. (It simply guesses based on whether the branch target address is in front of or behind the address of the currently executing sequence of instructions).

The μP does most of the work in determining when a branch will be taken with dynamic branch prediction. Thus, optimization of loops (using techniques like loop unrolling) and other code structures that incorporate branches is relatively unimportant. In fact, loop unrolling is often counter-productive for Pentium optimized applications. The fact that dynamic branch prediction supports non-optimized applications should raise a few eyebrows. As we have already explained, Intel stresses that applications will take full advantage of the Pentium's dual integer pipes and other capabilities only when a Pentium compiler has optimized them.

The 80486 μP has a simpler technique for handling branches. When a branch instruction is executed, the pipeline continues to fetch and decode instructions along the sequential path until the branch reaches the E stage. In E, the μP fetches the branch destination and the pipeline resolves whether or not a conditional branch is taken. If the branch is not taken, the μP discards the fetched destination and the execution proceed along the sequential path without delay. If the branch is taken, the fetched destination is used to begin decoding along the target path without clocks of delay. Taken branches are found to be 15% to 20% of the instructions executed, representing an obvious area for improvement by the Pentium processor.

The Pentium microprocessor employs a branch target buffer (BTB) that is an associative memory used to improve performance of taken branch instructions as indicated in Fig. 4.5 (*a*). When a branch instruction is 1st taken, the μP allocates an entry in the branch target buffer to associate the branch instruction's address with its destination address and to initialize the history used in predicted algorithm. As instructions are decoded, the μP searches the BTB to determine whether it holds an entry for a corresponding branch instruction. When there is a hit, the μP uses the history to determine whether the branch should be taken. If it should, the microprocessor uses the target address to begin fetching and decoding instructions from the target path.

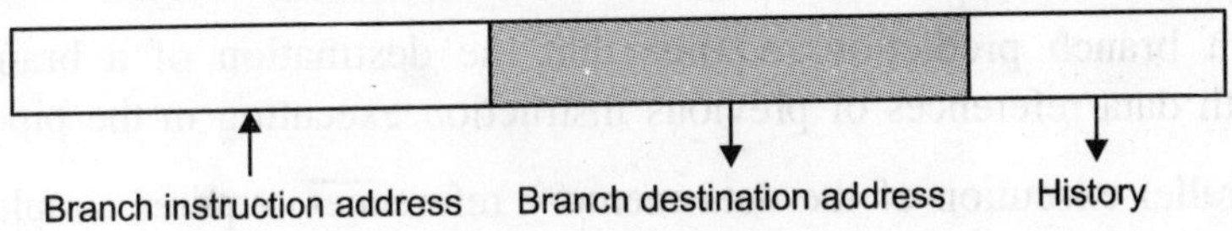

Fig. 4.5 (*a*) Branch target buffer

The branch is resolved early in the WB stage and if the prediction was incorrect, the μP flushes the pipeline and resumes fetching along the correct path. The μP updates the dual-ported history in the WB-stage. The branch target buffer holds the entries for predicting the 256 branches in a 4-way associative organization.

The Pentium μP using these techniques executes correctly predicting branches without delay. In addition, conditional branches can be executed in the V-pipe paired with compare or other instruction that sets the flags in the U-pipe. Branching executes with full compatibility and no modification to the existing software. (WE explain aspects of interactions between branch prediction and self-modifying code later).

4.6 CODE AND DATA CACHE

On-line cache on the Pentium enhances the performance by acting as a temporary storage places for commonly used instructions and data avoiding the need to go off-chip to the system's memory to fetch information. The Pentium has separate *code* and *data* caches, each one of 8KB. This provides the option to use one cache for data and the other for the program instruction. The 80486 μP employs a single 8KB on-chip cache to handle both data and instructions. This creates bus conflicts between the instructions prefetch and the data access. Approximately 90% of time is required by the data in the cache itself without accessing the main memory. Both data and code caches are 2-way set associative. The data cache of the Pentium uses 2 important techniques : *write-back* method to transfer data to the cache without going out to the main memory. The data is written to the main memory only when it is removed from the cache. The previous generation used 'write-through' cache implementation to transfer the data to the external memory each time the processor writes data to the cache. Thus, the write-back technique increased the performance by avoiding the bus utilization for write-through process. In order to maintain the consistency between the data in the cache and the main memory, the data cache implements a protocol called MESI.

As just explained, branch targets are pre-fetched in the E-stage, effectively using the same hardware as data memory references. There are potential advantages of such an organization over that separates code and data.

- For a given size of cache memory, a unified cache has a higher hit rate than separate cache because it balances the total allocation of code and data lines automatically.
- Only one cache needs to be designed.
- Handling self-modified code can be simpler.

Despite of these advantages of a unified cache, all of which apply to the 80486 μP, the Pentium microprocessor uses *separate codes and data catches of 8KB each*. The reason is that the superscalar design and branch prediction demand more bandwidth than a unified cache similar to that of the 80486 μP can provide. They are described as

1stly, efficient branch prediction requires that the destination of a branch be accessed simultaneously with data references of previous instruction executing in the pipeline.

2ndly, the parallel execution of the data memory references requires simultaneous accesses too for loads and stores.

3rdly, in the context of the overall Pentium microprocessor design, handling self-modifying code for separate code and data cache is only marginally more complex than for a unified cache.

Both the instruction cache and data cache, each, are 8KB, two way associative with 32B-lines.

The *cache memory* is used for temporary storage of currently used *codes and data* copied from system's (main) memory. If the processor is executing a loop, the codes within the loop are copied into the cache. This eliminates processor from going off the chip to access the main memory during loop execution and thus improves the performance.

The programs executing on the 80486 μP typically generate more data references than when executing on RISC microprocessors. Measurements on integer SPEC benchmarks show 0.4 to 0.6 data references per instruction for the 80486 microprocessor and only 0.17 to 0.33 for the MIPS processor. These difference results directly from the limited number (8) of registers for the 80x86 architecture, as well as produce calling conventions that require passing all parameters in memory. A small data cache is adequate to capture the locality of the additional references. (After all, additional references have sufficient locality to fit in the register file of the RISC microprocessors). The Pentium microprocessor implements a data cache that supports dual accesses by the U-pipe and V-pipe to provide additional bandwidth and simplify compiler instructions scheduling algorithm.

Figure 4.6 (*a*) depicts that the address path to the translation look-aside buffer and data cache tags is a fully dual-ported structure.

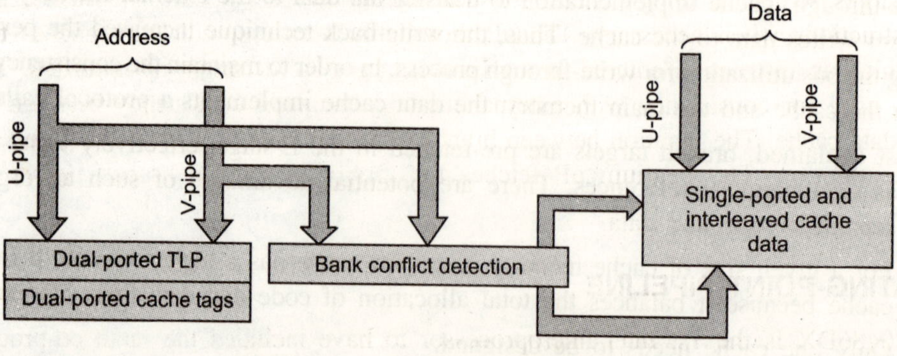

Fig. 4.6 (*a*) Dual-access data cache

However, the data path is single ported with 8-way interleaving of 32-bit-wide banks. When a bank conflict occurs, the U-pipe assumes priority and the V-pipe stalls for a clock cycle. The bank conflict logic also serves to eliminate data dependencies between parallel memory references to a single location. For memory references to double-precision floating-point data, the μP accesses consecutive banks in parallel, forming a single 64-bit path.

The Intel design team considered a fully dual-ported structure for the data cache. But feasibility studies and performance simulations showed that the interleaved structure is more effective. The dual-ported structure eliminated bank conflicts, but the SRAM cell would have been larger than the cell used in the interleaved scheme resulting in a smaller cache and lower hit ratio for the allotted area. Additionally, the handling of data dependencies would have been more complex.

The 80486DX2 μP with a write-through cache-consistency protocol and 32-bit data bus uses buses 80% of the time; 85% of all bus cycles are writes cycles. The 80486DX2 μP has a core pipeline that operates at twice the bus clock frequency. With higher performance core pipelines and 64-bit data bus of the Pentium microprocessor, using the write-back protocol for the cache consistency was an obvious enhancement. The write-back protocol uses 4-states

- Modified,
- Exclusive,
- Shared and
- Invalid (MESI).

Self-modifying code

One of the challenging aspects of the Pentium microprocessor's design was supporting self-modifying code compatibility. Compatibility requires that when an instruction is modified followed by execution of a taken branch instruction, subsequent executions of the modified instruction must use the updated value. This is a special form of dependency between data stores and instruction fetches.

The interaction between branch predictions and self-modifying code requires the utmost attention. The Pentium μP fetches the target of a taken branch before previous instruction have completed stores, so dedicated logic checks for such conditions in the pipeline and flushes incorrectly fetched instructions when necessary. The μP thoroughly verifies predicted branches to handle cases in which an instruction entered in the branch target buffer might be modified. The same mechanism used for consistency with external memory maintain consistency between the code cache and data cache. The iteration between branch predictions and self-modifying code requires the utmost attention. The Pentium μP fetches the target of a taken branch before previous instructions.

4.7 FLOATING-POINT PIPELINE

The 80486DX is the 1st Intel microprocessor to have included the math co-processor on chip. previous generations used off-chip math co-processors. The on-chip floating-point unit incorporates sophisticated 8-stage pipelining, Fig. 4.7 (*a*), and hardwired functions. A *3-stage* floating point instruction pipelines is appended to the integer pipelines. Most floating-point instructions begin execution in one of the integer pipelines, then moved to the floating-point pipeline. In addition, common floating-point functions such as add, multiply, divide etc. are hardwired for faster execution.

Integer pipe	PF	D1	D2	E				
Floating-point pipe	PF	D1	D2	E	X1	X2	WB	ER

Fig. 4.7 (*a*) Floating-point pipeline

Comparing the Pentium's FPU performance against the 80486's FPU performance is a bit like comparing *fast to a passenger train*. Do we doubt fast train runs faster w.r.t. the *passenger trains*? But there are so many other vehicles that outperform the fast ones. A better comparison of floating-point performance can be made by pitting the Pentium's FPU against the PowerPC 601. In this arena, the Pentium punts. The SPECfp92 rating for the PowerPC 601 is 72. Apple has published a SPECfp92 rating of 80 for the PowerPC 601. Against this comparison, the Pentium's 57 SPECfp92 rating is not impressive. A brief tour of the Pentium's FPU can be useful to show both its improvements over the 80486 and its limitations by comparison with the PowerPC chip's floating-point capabilities.

Unlike the FPU on the 80486, the Pentium's FPU contains dedicated

- Adder,
- Multiplier and
- Division units.

Independent circuitries perform the above functions (addition, multiplication, and division) to handle math function within less clock cycle.

The 2nd major enhancement is a more super-pipelined FPU that allows multiple FPU instructions to move through the pipe during a single clock cycle. Functionally, the Pentium's FPU has these 7-pipeline stages

- Prefetch,
- Decode-1,
- Decode-2,
- Execute (Dispatch),
- FP Execute-1,
- FP Execute-2 and
- Writeback.
- Error reporting.

The 1st 4-pipelining stages look similar to stages of the prefetcher. In fact, the 1st 4-stages of the FPU actually are performed in the U-integer pipe. Only the last 3-stages are dedicated to floating-point execution. This approach, although offers some benefits in its ability to fetch 64-bit operands by using both the U and V-pipes, makes it impossible to pair an integer instruction with a floating-point instruction.

The PowerPC 601 simply devotes more silicon to its FPU, allowing it to feature a richer set of floating-point instructions. However, in the real world of floating-point intensive applications, the inability of the Pentium to execute floating-point instructions in parallel with integer instructions is a problem. Most floating-point routines are a mixed bag of floating-point and integer instructions. The PowerPC's architecture, in this regard, is more suited t o pounding its way through floating-routines with brute force and at blinding speeds.

The 80486 μP integrated the floating-point unit (FPU) on chip. This eliminates overhead of the communication protocol that resulted from using a coprocessor. Importing the FPU on chip substantially boosted performance in the 80486 μP. Nevertheless, due to limited devices available for the FPU, its micro-architecture was based on a partial multiplier array and a shift and add data path controlled by micro-code. Floating-point operations could not be pipelined with any other floating-point operations i.e., once a floating-point instruction is invoked, all other floating-point instructions stall until its completion.

The larger transistor budget available for the Pentium microprocessor permits a completely new approach in the design of the floating-point micro-architecture. The aggressive performance goals for the FPU presented an exciting challenge for the designers, even with more silicon resources available.

Furthermore, maintaining full compatibility with previous products and with the IEEE standard for floating-point arithmetic was an uncompromising requirement.

The Pentium's floating-point pipeline has 8-stages. The first 2-stages are processed by the common (integer pipeline) resources for prefetch and decode. In the 3rd stage, the floating-point hardware begins activating logic for instruction execution. The entire first 5-stages for pipeline sequencing and synchronization is indicated in Fig. 4.7 (*b*).

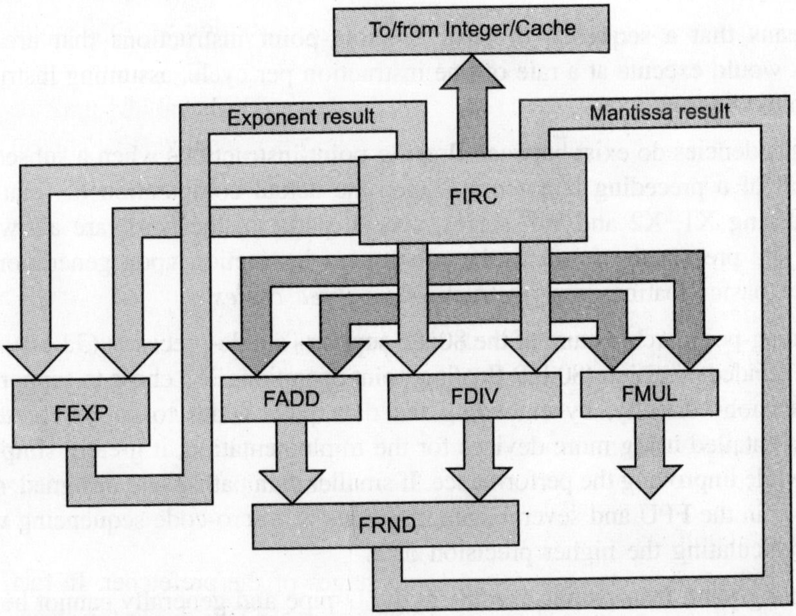

Fig. 4.7 (*b*) Block diagram of floating-point unit

Pre-fetch : The PF stage is the same as in the integer pipeline.

1st decode : The D1 stage is the same as in the integer pipelines.

2nd decode : The D2 stage is as in the integer pipeline.

Operand fetch : In this E stage the FPU accesses both the data cache and the floating-point register file to fetch the operands necessary for the operation. When floating-point data is to be written to the data cache, the FPU converts internal data format into the appropriate memory representation. This stage matches the E stage of the internal pipeline.

1st Execute : In the X1 stage the FPU executes the 1st step of the floating-point computation. When floating-point data is read from the data cache, the FPU writes the incoming data into the floating-point register file.

2nd Execute : In the X2 stage the FPU continues to execute the floating-point computation.

Write float : In the WF stage the FPU completes the execution of the floating-point computation and writes the result into the floating-point register file.

Error reporting : The FPU in the ER stage reports internal special situations that might require additional processing to complete execution and updates the floating-point status word.

The 8-stage pipeline in the FPU allows a single cycle thruput for most of the **basic** floating-point instructions such as floating-point

- Add,
- Subtract,
- Multiply, and
- Compare.

This means that a sequence of basic floating-point instructions that are free from data dependencies would execute at a rate of one instruction per cycle, assuming instruction cache and data cache hits.

Data dependencies do exist between floating-point instructions when a subsequent instruction uses the result of a preceding instruction. Since the actual computation of floating-point results takes place during X1, X2 and WF stages, special paths in the hardware allow other stages to be bypassed and present the result to the subsequent instruction upon generation. Consequently, the *latency* of basic floating-point instructions is *three cycles*.

The floating-point architecture of the 80x86 supports single-precision (32-bit), double precision (64-bit) and extended precision (80-bit) floating-point operations. We chose to support all computation for the 3-precisions directly, by extending the data path width to support extended precision. Although this entailed using more devices for the implementation, it greatly simplified the micro-architecture while improving the performance. If smaller data paths were designed, special enrouting of the data within the FPU and several state machines or micro-code sequencing would have been required for calculating the higher precision data.

The floating-point instructions execute in the U-pipe and generally cannot be paired with any other integer or floating-point instructions (the one exception will be explained later). The design was tuned for instructions that use one 64-bit operand in memory with the other operand residing in the floating-point register file. Thus, these operations may execute at the maximum thruput rate, since a full stage (E stage) in the pipeline is dedicated to operand fetching. Although floating-point instructions use the U-pipe during the E stage, the 2-ports to the data cache (which are used by the U-pipe and V-pipe for integer operations) are used to bring 64-bit data to the FPU. Consequently, during intensive floating-point computation programs, the data cache access the ports of the U-pipe and V-pipes to operate concurrently with the floating-point computation. This behavior is similar to superscalar load/ store RISC designs where load instructions execute in parallel with floating-point operations.

The overview of the floating-point of the Pentium microprocessor consists of 6-functional sections as depicted from Fig. 4.7 (*b*). The *floating-point interface*, *register files* and *control section* (FIRC) is the only interface between the FPU and the rest of μP. Since the function of floating-point operations is usually self-contained within the floating-point computation core, concentrating all the interface logic in one section helped to create a modular design of the other sections. The FIRC section also contains most of the common *floating-point resources*; *register file*, *centralized control logic*, and *safe instruction recognition logic* (described later). The FIRC can complete execution of instructions that do not need arithmetic computation. It dispatches the instructions requiring arithmetic computation to the arithmetic sections.

The floating-point exponent section (FEXP) calculates the exponent and the sign results for all the floating-point arithmetic operations. It interfaces with all other arithmetic sections for all the necessary adjustments between the mantissa and the sign-and-exponent fields in the computation of floating-point results

The floating-point multiplier section (FMUL) includes a full multiplier array to support single-precision (24-bit mantissa), double precision (53-bit mantissa) and extended precision (64-bit mantissa) multiplication and rounding within three cycles. FMUL executes all the floating-point multiplication operations. It is also used for integer multiplication, which is implemented through micro-code control.

The floating-point adder section (FADD) executes all 'add' floating-point instructions, such as floating-point add, subtract and compare. FADD also executes a large set of micro-operations that are used by micro-code sequences in the calculation of complex instructions, such as binary coded decimal (BCD) operations, format conversions and transcendental functions. The ADD section operates during the X1 and X2 stages of the floating-point pipeline and employs several wide adders and shifters to support high-speed arithmetic algorithms while maintaining maximum performance for all data precision. The μP achieves a latency of three cycles with a throughput of one cycle for all the operations directly executed by the FADD section for single-precision, double precision and extended precision data.

The floating-point divide (FDIV) section executes the floating-point divide, remainder and square-root instructions. It operates during the X1 and X2 pipeline stages and calculates two bits of the divide quotient every cycle. The overall instruction latency depends on the precision of the operation. FDIV uses its own sequencer for iterative computation during the X1 stage. The results are fully accurate in accordance with IEEE standard 754 and ready for rounding at the end of the X2 stage.

The processors in between 66MHz to 120MHz reportedly had FDIV bug. It effects division and division related other calculations also. Testing for this FDIV bug is straightforward using some test numbers.

Example

$$z = x - \frac{x}{y}y$$

Naturally for any value of x and y, z must be equal to zero. But the Pentium in the range of 66MHz to 120MHz for $x = 4,195,835$ and $y = 3,145,727$ yields

$$z = 4195835 - \frac{4195835}{3145727}3145727 = 0 \text{ (correct answer)}$$

$$z = 4195835 - \frac{4195835}{3145727}3145727 = 256 \text{ (flawed Pentiums' Answer)}$$

This FDIV effects MSD arbitrarily that becomes evident from other examples as

Example

$\frac{5,505,001}{294,911} = 18.66665197$ is correct answer but flawed Pentium reports 18.666000093 i.e., error in the 6th MSD.

Example

$\frac{962,306,957,033}{11,010,046} = 87,402.6282027341$ is correct but flawed Pentium reports 87,399.5805831329 i.e., error starts from 3rd MSD itself.

The above calculations can be seen on the computer using spreadsheets (Lotus or Excel) that handles 13 to 15-significant digits. Calculators, normally, do not show this flaw, as it can not handle high numbers. While calculating FDIV, the processor must be enabled for FPU.

Note

The bug observed in the Pentium processor could not be confirmed until when mathematics Professor publicly came out on the Internet. The flaw was found in the floating-point unit of the math coprocessor. The Intel claimed that an error would occur only every 26,000 years in purely random fashion. However, the main customer of Intel microprocessors, the IBM contradicted it saying that the typical spreadsheet user could incur an error every 24-days and ultimately halted its shipment of the Pentium based PCs. The other group of Intel sympathizer Gateway, Packard Bell, Dell and others have continued the shipment telling that the problem is not as bad as the IBM claims. The Microsoft hurriedly came out with a 'patch' that bypasses the faulty area. It has been predicted that more than 4-million flawed Pentium would have been sold by end of 1995. The Intel's President, Mr. Andy Grove apologized for the flaw and promised to send the replacement processor to PC users who choose to go for the replacement themselves and will offered telephonic technical assistance as well.

The floating-point rounder (FRND) section rounds the results delivered from the FADD and FDIV sections. It operates during the WF stage of the floating-point pipeline and delivers a rounded result according to the precision control and the rounding control, which are specified in the floating-point control word.

Safe instruction recognition

Floating-point computation requires longer execution time than integer computation. Pentium's floating-point pipeline uses 8-stages, while the integer pipeline uses only 5-stages. Compatibility requires in order instruction execution as well as precise exception reporting. To meet these requirements in the Pentium processor, floating-point instructions should not proceed beyond the X1 stage, that is, allow subsequent instructions to proceed beyond the E stage, unless the floating-point instruction is guaranteed to complete without causing an exception. Otherwise, an instruction may change the state of the µP, while an earlier floating-point instruction (which has not yet completed) might cause an exception that requires a trap to a software exception handler.

In order to avoid a substantial performance loss due to stalling instructions until the exception status of a previous floating-point instruction is known, Pentium's floating-point unit employs a mechanism called safe instruction recognition (SIR). This logic determines whether a floating-point instruction is guaranteed to complete without creating an exception and therefore is considered 'safe'. If an instruction is safe, there is no need to stall the pipeline, and the maximum thruput can be obtained. However, if the instruction is not safe, the pipeline stalls for 3-cycles until the unsafe instruction reaches the ER stage and a final determination of the exception status is made.

There are 6-possible exceptions that can occur on the Pentium microprocessor's floating-point operations, namely,

- Invalid operation,
- Divide by zero,
- Denormal operand,
- Overflow,
- Underflow, and
- Inexact.

The XIR logic determines whether the instruction is guaranteed to be exception free (safe) or not (unsafe) early in the floating pipeline-in the X1 stage-before any computation takes place. The 1st 3-of the 6-exceptions can be detected without any floating-point calculation. From latter 3- exceptions, the inexact exception is usually 'masked' by the operating system or the software application (using the precision mask, or PM, bit in the floating-point control work), otherwise, a trap will occur whenever rounding of the result is necessary. When the precision (inexact) exception is masked, the pipeline delivers the correctly rounded result directly. For overflow and underflow exceptions SIR logic uses an algorithm that monitors the exponent fields of the input operands to conclude the exception status (safe or unsafe).

The architecture the 80x86 μP stores floating-point operands in the floating-point register file with an extended-precision exponent, regardless of the precision control in the floating-point control word. The extended precision exponent supports much greater range than the double-precision format. Overflow and underflow exceptions caused by converting the data into double precision or single precision formats occur only when storing the data into external memory. These characteristics of the 80x86 floating-point architecture give a unique advantage to the effectiveness of the SIR mechanism in the Pentium μP, since the SIR algorithm can use the internal (extended precision) exponent range. Thus, the occurrence of unsafe operations is extremely rare. Our evaluation of the SIR algorithm for the FPU design found no unsafe instructions in simulated execution of the SPEC89 floating-point benchmarks.

4.8 REGISTER STACK MANIPULATION

The floating-point instruction set of the 80x86 uses the register file as a stack of 8-registers in which the top of stack (TOS) acts as an accumulator of the results. Therefore, the top of the stack is used for the majority of the instructions as one of the source operands and, usually, as the destination register.

The floating-point pipeline performance can be improved by optimizing the use of the floating-point register file. The Pentium's FPU can execute the **FXCH** instruction in parallel with any basic floating-point operation. The FXCH instruction 'swaps' the contents of the **TOS** register with another register in the floating-point register file. All the basic floating-point instructions may be paired with FXCH in the V-pipe. The pair executes in parallel, even when the data dependency between the 2-instructions in the pair exists. The use of parallel FXCH redirects the result of a floating-point operation to any selected register in the register file, while bringing a new operand to the top of the stack for immediate use by the next floating-point operation. The example shown in Fig. 4.8 (*a*) illustrates the use of parallel FXCH.

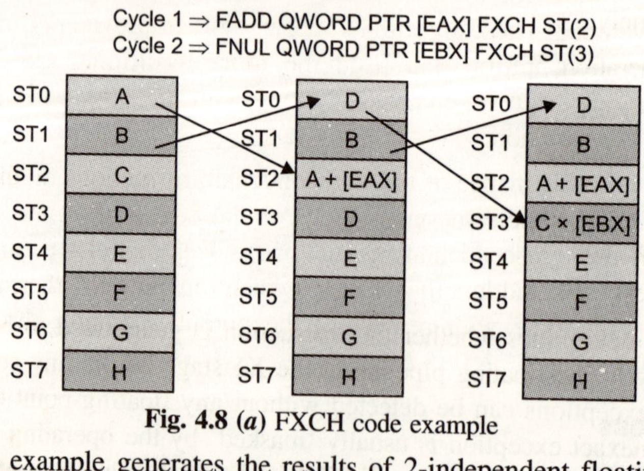

Fig. 4.8 (*a*) FXCH code example

The code in the example generates the results of 2-independent floating-point calculations. The floating-point register file contains initial value prior to code execution

- Register ST0 (TOS) contains the value A,
- Register ST1 contains the value B,
- Register ST2 contains value C, and so on.

The 2-operations are

- Floating-point addition of value A with the 64-bit floating- point operand addressed by the general register EAX, and
- Floating-point multiplication of value C by the 64-bit floating-point operand addressed by the general register EBX.

When the floating-point pipeline is fully loaded and these 2-operations are part of the code sequence, the parallel FXCH allows the calculation to maintain maximum throughput of one cycle per operation. Within one cycle the Pentium μP writes the result of the addition to ST2, while the operand for the next operation moves to the top of the stack in the next cycle, the processor writes the result of the multiplication to ST3, while the top of the stack contains value D, which may be used for a subsequent operation.

Transcendental instructions

The μP supports all 8-transcendental instructions that are defined in the instruction set through direct execution of micro-code sequences. These transcendental instructions are given in Table 4.8(*a*).

Table 4.8(*a*) Table of transcendental equations

FSIN	*Sine*	*FPATAN*	*Arctangent*
FCOS	Cosine	F2XM1	2**X – 1
FSINCOS	Sine and Cosine	FYL2X	Y*Log2(X)
FPTAN	Tangent	FYL2XP	Y*Log2(X + 1)

A table driven algorithms for the transcendental functions using polynomial approximation techniques has been developed. These algorithms substantially improved performance and accuracy over the 80486 μP implementation that used the more traditional Cordic algorithms. The approximation table resides in an on-chip ROM along with the other special constants that are used for floating-point computation.

The improvement of performance of the transcendental instructions on the Pentium processor ranges from 2 to 3 times over the same instructions on the 80486 μP at the same frequency. The worst-case error for all the transcendental instructions is lees than 1μ1p (unit in the last place) when rounding to nearest even and less than 1.5μ1ps when rounding in other modes. The functions are guaranteed to be monotonic with respect to the output operands, throughout the domain supported by the instruction.

Development process

The development of a highly integrated microprocessor involves close interaction between numerous teams having diverse technical specialization and working under the discipline of well-defined methodologies. A small team of architects and **VLSI** designers developed the initial concepts of the design. This group conducted feasibility studies of parallel instruction decoding and options for branch prediction techniques. Simultaneously, it evaluated performance by hand for short benchmarks and compiler optimizations. As initial directions were established, additional engineers participated, and a sub-team focused on the following areas

- Behavioral modeling of the micro-architecture,
- Circuit feasibility design for caches, decoding programmable logic arrays (PLAs), floating-point data path, and other critical functions;
- A flexible, trace driven simulator of instruction timing for performance evaluation;
- A prototype compiler; and
- Enhancements to existing instruction tracing tools.

The Pentium was refined using both top-down and bottom-up methods throughout the design. Architects using the performance simulator accomplished top-down refinement through comprehensive characterization of executing benchmark workloads on the 80486 μP and trace-driven experiments concerning alternative machine organizations conducted.

The VLSI design engineers' were constantly evaluating its features critical to the targeted area and frequency and refined the design from the bottom-up. On 2-occasions in the design, the accumulation of changes from the bottom-up refinement caused the need for substantial restructuring of the microprocessors' global chip plan, or '**die diets**'. On those occasions, interdisciplinary teams of specialists had several brainstorming sessions and evaluated ideas that could satisfy the global or local design constraints. In one instance, they found it necessary to refine the sets of instructions that could be executed in parallel. Constraints have been assigned to the area and speed of the decoder PLAs. The VLSI designers identified combinations of instruction formats that would feasibly decode in parallel, and the compiler writers determined the optimal selection.

The measured performance of the Pentium in production systems is within 2% of that predicted before the design was completed. The logic validation of the Pentium processor design presented a major challenge to the design team. A comprehensive test base from the validation of previous 80×86 microprocessors was available. However, the Pentium processor micro-architecture introduced several new fundamental techniques, such as superscalar, write-back cache, and floating-point algorithms, that required a more rigorous verification technology.

Different validation approaches were used in pre-silicon testing of the Pentium microprocessor

- Architecture verification looked at the 'black box' functionality from the programmer's point of view. It was designed for the comprehensive tests to cover all possible aspects of the programming model and all the Pentium processor user-visible features.
- Design verification checked the internal functionality from the point of view of a logic designer who would understand the behavior of every internal signal. This testing approach is considered a 'white box' technique, in which tests are written to exercise all the internal logic and verify its correct behavior.
- Random instruction testing was a valuable tool to cover all those situations that are rarely covered by the more traditional, hand written tests. Running finally tuned random tests let us verify correct functionality by comparing the results generated by a logic design description of the Pentium processor to the results generated by a software-emulated model.

- A logic design hardware (Quick Turn) enabling increased testing coverage capacity by allowing a much larger software base to run on the processor model before the first silicon was available. We ported the logic model of the Pentium processor onto a Quick Turn setup, which was capable of handling the complete design, and tested major operating systems and application programs before finalizing the design.

A special effort was dedicated to verify the new algorithms employed by the FPU in addition to the general validation approach. A high-level software simulator was developed to evaluate the intricacies of the specific *add*, *multiply*, and *divide* algorithms used in the design. The simulator then evolved into a testing environment, allowing the verification of the FPU logic design model independently from the rest Pentium processor. Also, the new algorithms used for the floating-point transcendental functions required an extensive test strategy that verified the accuracy and monotonicity of the results through out the development process, comparing the results to a 'super accurate' software model. Eventually, when the first silicon of the Pentium processor was available for testing, we used automatic testing techniques to assure the correctness of the transcendental instructions.

The compiler technology developed with the Pentium includes machine-independent optimizations common to current high performance compilers, such as inlining, onrolling, and other loop transformations. In addition, the specially developed techniques for the 80×86 architecture were used and tuned them for the Pentium processor's micro-architecture. The 80×86 architecture has certain characteristics different from those for RISC architectures. The architecture supports a variety of instruction formats for equivalent operations.

Why Pentium will exist

Applications on Windows and DOS will run under emulation on the PowerPC. Macintosh applications on Apple's own PowerPC machines have to run under emulation. Emulation is slow and necessitates unnecessary layer of complexity. Thus, Emulation is an evil.

- Probably it was not a good marketing decision to call the technology behind the PowerPC 'RISC'.
- Intel stock ensures instant delivery.
- The ability to run DOS and Windows applications faster on 80486 machine and Intel wins on this front.
- Intel is favourite.
- Computer buyers value compatibility and versatility.
- The Austin factor is a reliability measure and hence the question 'Can we be absolutely sure that Motorola would not withdraw the PowerPC from the market?'

The Pentium chip is fully compatible with its predecessors, but it also differs from them in many ways. The twin data pipeline enables it to execute 2-instructions simultaneously. The U-pipe (called primary pipe) executes all integer and floating-point instructions. the V-pipe (secondary pipe) can execute only simple integer instructions and certain floating-point instructions. Execution of two instructions simultaneously in different pipe is called pairing. When pairing is not possible only U-pipe is used. The Pentium is 100% software compatible with its 386 and 80486. The

Pentium processor includes a Branch Target Buffer to employ branch prediction technique. This technique avoids, more appropriately, minimizes the stalling in one or more pipes caused by delays in fetching instruction that branch to non-linear memory location. The BTB decides whether a program branch will be taken or not and only then fetches the next instruction. Thus BTB enables Pentium to allow both U- and V-pipes operate at full speed. The 2nd generation Pentium uses 0.6 microm BiMOS technology to reduce the physical die size as well as power consumption. The 100MHz Pentium consumes 3.25A at 3.3V that equals only 10.725W. The 90MHz takes 2.95A at 3.3V which equals only 9.735W. Table 4.8(b) short lists important feature of Pentium.

Table 4.8(*b*) Pentium characteristics

Introduced	March22, 1993(P5), March 07,1994(2nd generation).
µP clock multiplier	1x(1st generation), 1.5x, 2x(2nd generation)
Maximum rate speeds	60/66MHz (1st generation), 90/100MHz (2nd generation)
Register size	32-bit
Memory address bus	32-bit
External data bus	64-bit
Maximum memory	4GB
On-chip cache	8K code and 8K data
On-chip cache type	2-way associative, write-back data.
Burst mode transfer	Yes
Transistors used	3.1 million (60/66MHz), 3.3million (75MHz and more)
Size of the chip	0.8micron (60/66MHz), 0.6micron (75MHz and above)
Package size	273-pin PGA (Pin Grid Array), 296-pin SPGA (Staggered Pin Grid Array).
Math coprocessor	Built-in floating-point unit
Power management	SMM enhanced in 2nd generation
Operating voltage	5V(1st generation), 3.3V (2nd generation).
Maximum current	3.25A at 3.3V

4.9 GENERATIONS OF PENTIUM

The Pentium is available in many designs such as

- 1st generation, and
- 2nd generation.

1st Generation Pentium

The 1st generation Pentium was manufactured on 0.8µ BiCMOS technology with 273-SPGA pins. It runs at 60 and 66MHz speeds with +5V. The 1st generation Pentium runs at the speed of the motherboard. In other words, it uses 1∗X clock speed. This generation of Pentium has no FDIV bug and power management bug due to absence of SMM (sleeping management mode) function as needed in green PCs.

The process combined with 3.1 million transistors resulted in a large die size and complicated manufacturing process. As a result, Intel could not manufacture as much as required. The large die size with +5V operating voltage caused 66MHz Pentium to draw as much current as 3.2A resulting in the consumption of 16W power. This created problem of tremendous heating. Hence, separate fan was used for cooling it. It was but natural to receive severe criticism from other manufacturers such as IBM and Motorola that used 0.6 micron technology.

2nd Generation Pentium

Intel received the criticism with good test and announced the 2nd generation Pentium on 7th March, 1994 that ran at 90MHz and 100MHz speeds. It used 0.6μ BiCMOS technology to shrink the die size and power consumption. The 2nd generation Pentium operates on 3.3V. The 100MHz Pentium draws a maximum of 3.25A at 3.3V consuming 10.725W. Its slower version 90MHz draws only 2.95A at 3.3V consuming 9.375W. The 75MHz version probably consumes only 5W. The 2nd generation Pentium contains 3.3 million transistors i.e., it uses extra transistors w.r.t. the 1st generation. The extra number of transistors was used to implement additional clock-control SL such as Advanced Programmable Interrupt Controller (APIC) and dual processor interface.

The APIC and dual processor interface facilitate the dual processor configuration in which two 2nd generation Pentium chips can process on the same motherboard simultaneously. The 2nd generation Pentium uses clock-multiplier circuitry that allows the processor to operate faster than the bus. The 90MHz Pentium can operate at 1.5 times bus frequency that is normally 66MHz. Similarly, the 100MHz Pentium processor can run at 1.5 × clock in a system using 66MHz bus speed or 2 × clock on motherboard that is running at 50MHz.

In general, the motherboard faster than 66MHz is if not impractical, then very difficult because of memory and local-bus performance constraints. The faster Pentium system could combine the 66MHz motherboard operation with 1.5 × clock to result in 100MHz processor operation. One might think that 1.5 × 66MHz = 99MHz, but 66MHz is always 66.6666MHz which approximately makes it 100MHz.

We must know that a Pentium motherboard has 2-numbers of 320-pin processor sockets that fully meet the Intel 320-pin Socket 5 specification. This facilitates us to add a 2nd Pentium processor to take advantage of SMP (Symmetric Multi-Processing) support in the newer operating systems. Also the motherboard should have facility to reconfigure it for 66MHz operation. This will enable to take advantage of future Pentium Overdrive processor that will support the higher motherboard clock speeds.

Cache of Pentium

We have already discussed in brief in article 3.13 regarding cache and its function. Here, once again we talk about the cache in slightly more detail. The program performance is squeezed between two pincers

- Instruction performance, and
- Memory bottleneck.

The machine instructions have been made relentlessly faster over the years in the field of microprocessors development to the point where many machine instructions (and nearly all of the most frequently used instructions) now execute in only one system clock cycle. The newer µPs, both RISC and CISC types, execute more than one instruction in a single cycle due to the presence of multiple EUs. This is called **superscalar execution**.

The memory access bottleneck is the far worse problem. DRAMs are slower as compared to modern µPs and making it faster is not the easy job. Faster SRAMs become bulkier and also become much more expensive per byte than DRAM, especially in MB quantities required by modern desktop computing systems.

The system designers have tried to make compromise between the SRAM and DRAM and create a memory system using a compromise combination of both. The result is called the **cache memory system,** and is perhaps the single most important computer performance feature present in modern desktop computers.

External cache

The earlier microprocessor had relatively small amount of high-speed SRAM on the motherboard connected by a local bus. A **local bus** is a short and direct extension of the µPs address and data lines, without the extensive buffering and times synchronization logic required for traditional expansion bus to support hardware expansion cards. Early cache systems used 32KB or 64KB of SRAM that could be included at the time without blowing the price of the system beyond affordability. The price of SRAM has come down lately to the point where 256KB and 512KB SRAM caches are common and even higher are not out of question.

The cache, quite simply, holds the instructions or data at rate of the µP processing. When the µP tries to read data or instructions from the main memory, 32B or 128B block of main memory are copied into the cache memory, from which the µP can access it very quickly, if not quite instantly. The blocks of main memory that are moved into cache are commonly called cache lines. In some respects a cache is like a very large prefetch queue or perhaps a postfetch queue, since the cache is filled with information after the µP requests it, not before.

It is but natural to think that the data movement delays would cancel the additional speed of the SRAM in the cache as because the information in the cache has to be moved twice (once from main memory to the cache and once from cache into the µP). In fact, because an entire cache line is moved into cache at once (and because DRAM is still obnoxiously slow) the hit taken due to moving the same bytes twice is relatively minor.

The *real magic* of the cache does not become apparent until we realize that a great deal of computing is done through the use of program loops in which the same short sequence of instructions is executed repeatedly, once for each count of a counter variable, or perhaps repeatedly until some flag indicates that the job at hand has been completed. Either way, if a program loop is completely loaded into the cache, execution will happen at full speed, without any additional delay for loading the cache, until execution completes that loop and run beyond the end of what is in the cache. The cache performance depends on its access time and the hit ratio that is dependent on size of the cache and the number of bytes brought into the cache on any fetch from main memory (line size). Increasing the line size increases the chances that there will be a cache hit on the next memory reference.

If a 4KB cache with 4B-line size has a hit ratio of 80%, the doubling the line size might increase the hit ratio to 85% but doubling the line size again might only increase the hit to 87%.

The overall memory performance is a function of cache access time, cache hit ratio and main memory access time for cache misses. Any system with 85% cache hit ratio and 120ns cache access time may access main memory 20% of the time with an access time of 600ns. The average access time would be

$$[0.8\times120+\{0.2\times(600\times120)\}\,] = 240 \text{ ns}$$

The reduction in bus bandwidth required for processor-to-memory transfers results from the use of a cache calculated as

$$\text{Bus bandwidth (MB)} = \frac{\text{Bus utilization} \times \text{Bytes transferred}}{(\text{Clock cycles/transfer}) \times \text{Cycle time in ns}}$$

Example

What would be the bus bandwidth, if each memory access transfers 4-bytes of data taking 3-clock cycles, each cycle of 60ns, and utilizes the bus for 82%?

$$\text{Bus bandwidth (MB)} = \frac{0.82\times4}{3\times60 \text{ ns}} = 18.2\text{MB/s}$$

If the cache hit ratio is 100% and bus utilization in above example is 60%, what is bus bandwidth.

$$\text{Bus bandwidth (MB)} = \frac{0.60\times4}{3\times60 \text{ ns}} = 13.3\text{MB/s}$$

Hit and miss

When the μP sends out an address for a desired memory location onto its data bus, the cache logic checks to see whether the address being requested is already loaded onto the cache. If it is, that is called a cache hit, and the address is passed along to cache memory without further delay. The effectiveness of a cache system is measured by the % of cache hits measured against total memory accesses. In tightly written code that uses a lot of program loops, the cache hit rate can exceed 90%.

If the μP sends an address out onto the bus and that address is not present in the cache that is considered as cache miss. Then the cache logic must go out to main memory and bring in the next cache line (or perhaps a number of cache lines in more sophisticated caching schemes) to satisfy the request. This scheme is shown in Fig. 4.9 (*a*).

Least recently used

Each time the cache logic detects the cache miss, the cache line containing the requested address is fetched from main memory and written into the cache. Now it is not likely that any empty space is there in 256KB cache even a few seconds after a program begins running. This new cache line has to go somewhere, so the cache logic determines which cache line in the cache was least recently accessed by the μP. The newly loaded cache line overwrites that cache line, and if the code in the overwritten cache line is to be executed later, it must be fetched again from the main memory. A system like this is called the least recently used (LRU) cache.

The special purpose logic (sometimes huge specialized cache-controller ICs nearly as complex as the μP itself) has been designed to do all the work such that the μP does not have to work of detecting cache hits and misses and determining where to load cache lines onto cache memory. The μP simply has to ask and wait if necessary and requested code or data will appear in cache memory without any further fuss on the part of the μP.

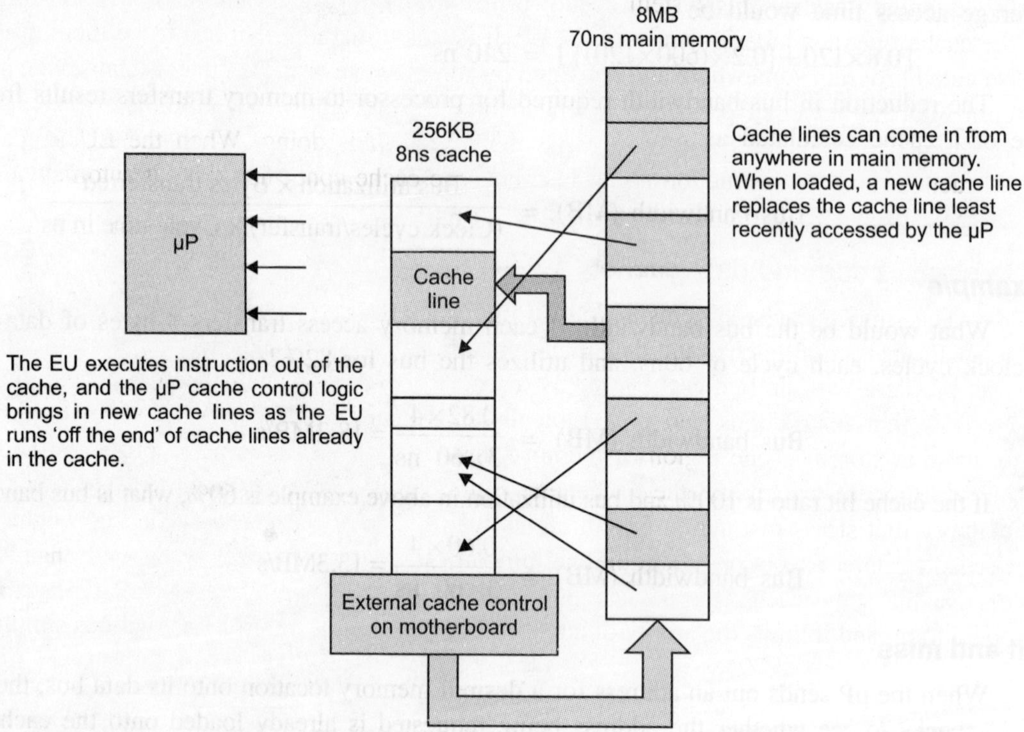

Fig. 4.9 (*a*) Single level least recently used cache

Primary and secondary caches

The single-level cache described in the previous section improves code performance nicely. But nicely is generally not good enough. As μPs grow faster and faster, even the scorchingly fast 8ns SRAM forming the cache begins to drag on code performance. When the system clock is ripping along at 50MHz, even 8-billionths of a second becomes a **long** time. So the μP designers took bold step of putting cache memory actually **inside** the μP chip itself. As we can imagine a lot of cache inside the μP will cost much as the chip-size will increase. The 80486 has only 8,192B of cache and the Pentium has just twice (16,384B). The PowerPC only doubles Pentium cache to 32,768B. Still there is very little delay involved in accessing cache memory that is within the μP chip, and when a tight program loop fits entirely within the μP cache, the μP finally has the chance to run about as fast as it can run, unconstrained by the memory access bottleneck.

The problem with a system like this lies in keeping the small internal cache full by loading cache lines from slow DRAM. (This again is analogous to the problem of a fast μP continually draining the prefetch queue). A *voracious* μP like the PowerPC *eats cache lines* like *corn flakes* and the delay getting cache lines in from DRAM cancels much of the speed advancing of executing code from the internal cache.

In order to avoid this problem, a 2nd cache is placed between the μP's internal cache and the slow main memory. The on-chip cache is called the primary (L1) cache, and the intermediate cache is called the secondary (L2) cache. The secondary cache built from fast SRAM in the 8ns to 10ns range. There are 2-cache controllers : one inside the μP to manage the on-chip primary cache, and an external cache controller (usually a large custom chip on the motherboard) to handle the secondary cache. This system is shown in Fig. 4.9 (*b*). The *main points* to remember is that the cache operates independently of the μP. It is not something that has to be monitored by a μP executed program somewhere and activated by subroutine calls. Even the on-chip cache controllers built into the newer μP's operate independently of instruction execution. The *right hand* of the μP almost literally does not know what the *left hand* is doing. When the EU needs the next instruction, and the instruction is not in cache, the cache controller kicks it automatically to load the next cache line from memory.

Separate code and data cache

The simpler cache systems are called unified cache i.e., a single cache that handles memory for both instruction fetches and data manipulation. The 80486 is a good example of unified cache. One problem with unified cache systems (especially relatively small ones, like that one in 80486 chip) is that instructions and data typically exist in widely different parts of memory. If an instruction stored in one region of memory must read or modify a piece of data stored in a far-away region of a memory, the cache line containing the data bytes must be loaded from main memory that slows instruction execution and can overlay code in the cache that might in fact be needed for later execution. An additional problem with unified caches is that with instructions and data in the same cache, the μP cannot read code and data simultaneously. This is called cache contention, and it limits the ability of advanced μPs to execute multiple instructions simultaneously.

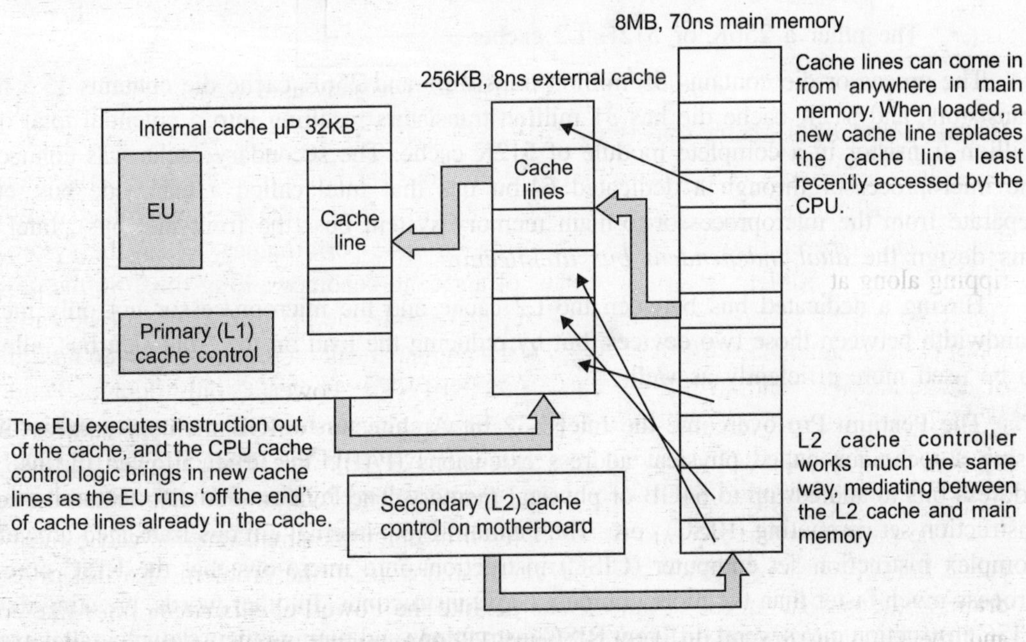

Fig. 4.9 (*b*) Two-level least-recently used cache

The problem of the cache contention prompted the designers to come out with newer μP like the Pentium to implement a separate cache for code and data, called the split cache. The code cache contains only machine instructions and the data cache only the data being acted upon by those instructions. This allows 2-different areas that the μP is dealing to fetch instructions data on the same machine cycle, further speeding up program execution.

The PowerPC actually followed the older idea of a unified cache, but with a fairly large (32K) on-chip cache and with some extremely clever tricks to get around cache contention. The extremely powerful caching systems of newer processors such as the Pentium and PowerPC have almost eliminated memory access as program-performance bottleneck. If memory can in fact supply instructions as quickly as the μP requires them, improving performance further becomes a matter of making instructions execute in as little time as possible. Two concepts have emerged as important point in this context. Both are ways of achieving this in parallelism to save time.

Pipelining : It is method of splitting instructions into pieces and processing them in parallel.

Superscalar Execution : It divides the μP into multiple EUs that then executes multiple instructions in parallel.

Pentium Pro

The Pentium Pro was released by Intel in September 1995 but became widely available in October 1996. It was no longer a single chip, but was what Intel called a multi-chip module (MCM) physical format in the form of Dual Cavity PGA (Pin Grid Array) package. The 387-pin chip carrier of Pentium Pro contains 2-die

- One containing the actual Pentium Pro processor, and
- The other a 256K or 512K L2 cache.

The processor die contains 55 million transistors and 256K cache die contains 15.5 million transistors, and 512K cache die has 31 million transistors resulting into a potential total of 36.5 million transistor in a complete module of 512K cache. The secondary cache was connected to the microprocessor through a dedicated 64-bit bus that Intel called a back-side bus, entirely separate from the microprocessor-to-main memory system bus (the front-side bus). Intel called this design the *dual independent bus architecture*.

Having a dedicated bus between the L2 cache and the microprocessor not only increased bandwidth between those two devices, but by reducing the load on the front side bus, allowed it to be used more efficiently as well.

The Pentium Pro overcome the Intels'-32 bit Architecture's 4GB memory address limit by using a technique called physical address extensions (PAE). The PAE allowed the use of 36 address bits to address up to 64GB of physical memory. The Pentium Pro also featured a reduced instruction set computing (RISC) core. The Pentium Pro's internal circuits translated conventional complex instruction set computer (CISC) instructions into micro-ops that the RISC core could process much faster than the more complex CISC instructions. In other words, breaking down the CISC instruction into several different RISC instructions and running them down parallel execution pipelines, the overall performance is increased.

Other significant architectural change in the Pentium Pro included 11 execution units and a 12-stage pipeline, allowing it to execute 32-bit instructions much faster than previous Intel microprocessors. However, because it was designed primarily for 32-bit instructions, it showed little or no performance improvement when executing 16-bit instructions.

The Pentium Pro architecture contains 3-internal instruction pipelines that execute multiple instruction in one cycie. The main processor die includes 16K-split L1 cache with 8K two-way associative cache for primary instruction, and an 8K four-way associative cache for data. The Pentium Pro can execute instruction out of order and has dynamic branch prediction and speculative capabilities.

Compared to the Pentium, the Pentium Pro is faster as long as it runs on 32-bit software. The Pro's Dynamic Execution is maximized for performance primarily when it runs on 32-bit software such as Window NT. If it runs on 16-bit software such as Windows 95 (it partly operates on 16-bit environment), the Pentium Pro will not provide a marked performance improvement over similar speed rated Pentium and Pentium-MMX processor. For this reason, Windows NT is often regarded as the mandatory operating system with Pentium Pro. Tables 4.9 (*a*) and (*b*) give details of Pentium Pro processor.

Table 4.9(*a*) Pentium Pro specification

Introduced	September 1995
Maximum rated speed	150 166 180 200MHz.
µP clock multiplier	2.5x to 3x
Register size	32-bit
External data bus	64-bit
Integrated-cache bus	64-bit
Memory address bus	32-bit
Maximum memory	4GB
Integrated-cache size	8K code, 8K data
Integral-cache type	Non-blocking, L1 cache
Number of transistors	5.5 million
Transistors in L2 cache	15.5 million (256K cache), 31 million (512K cache)
Circuit size	0.35micron
External package	387-pin Dual Cavity PGA
Math coprocessor	Built-in FPU
Power management	SMM (system management mode)
Operating voltage	3.3V

The speeds of Pentium Pro and its motherboard is shown in Table 4.9(*b*).

Table 4.9(*b*) Pentium Pro and its motherboard speed

μP type/speed	μP	Motherboard Speed
Pentium Pro 150	2.5x	60
Pentium Pro 166	2.5x	66
Pentium Pro 180	3x	60
Pentium Pro 200	3x	66

Pentium MMX™

The MMX is called **multimedia extension.** This technology was incorporated to improve 2D and 3D imaging for multimedia applications. All algorithms developed for multimedia applications involve operations on several pixel (picture cell) at a time. For an example, a colour pixel consists of three component **red, green,** and **blue.** Each one of these three components is 8-bit integer. Thus, the intensity of the pixel varies from 0 to $2^8 - 1 = 255$ in each component of colour. Now a pixel of an image can be represented by 24-bit quantity. The black and white picture requires only 8-bit for pixel representation. The Intel developed MMX technology as a direct consequence of the growing importance and increasing demands of multimedia and communication applications. Many such applications run repetitive loops of instructions that take large amount of time to execute it. The Intel incorporated MMX technology, a Process called Single Instruction Multiple Data (SIMD) in the MMX technology to solve this problem. The SIMD process allows one instruction to perform the same function on many pieces of data. Furthermore, 57-new instructions have been added on the chips that were designed, specifically, to handle video, audio music synthesis, speech recognition, speech compression, image processing, image compression, games, video conferencing, and graphics data etc.

This 57-new instruction facilitates the programmer to write efficient program for image enhancement, image filtering, coding algorithms etc. The conventional PCs allows us to operate on 2-pixels concurrently, at the most. Where as the MMX technology allows us to load 8 -pixels simultaneously and perform concurrent operation on them?

Many such applications run repetitive loops of instructions that take large amount of time to execute it. In order to solve this problem, the Intel incorporates MMX technology, a Process called Single Instruction Multiple Data (SIMD) that allows one instruction to perform the same function on many pieces of data.

All image processing algorithm and image compression techniques require multimedia applications involving matrix multiplication and matrix convolution including multiple number of pixels in the operation. Thus, most of the multimedia operations require Single Instruction Stream Multiple Data Stream (SIMD) type of architecture. Intel accepted this by providing 57 MMX instructions. This instruction facilitates the programmer to write efficient program for image enhancement, image filtering, coding algorithms etc. The conventional PCs allows us to operate on 2-pixels concurrently, at the most. Where as the MMX technology allows us to load 8-pixels simultaneously and perform concurrent operation on them?

In January 1997, Intel introduced its first multimedia extensions (MMX)-enhanced Pentium microprocessor. The MMX Pentium incorporated a larger L1 cache plus other features that permitted it to recognize and rapidly execute MMX commands. An MMX-enabled Pentium was capable of executing programs written to take advantage of this architectural difference faster than a conventional Pentium; however, it offered little or no speed advantage with conventional programs. It is available for clock rates of 66/166MHz, 66/200MHz, 66/233MHz. Apart from these features available for conventional Pentium, it includes new features such as pipelined MMX unit, 16KB code and write-back cache (8KB in earlier Pentium) and 4.5 million transistors. Pentium- MMX chips are produced on an enhanced 0.35µP BiCMOS silicon process that allows for a lower voltage of 2.5V.

Intel MMX architecture

We know that there are 8-numbers of floating-point registers in the floating-point unit of the Pentium. Each of these floating point registers is 80-bit wide. Out of the 80-bit, 64-bits are used for **mantissa** and the rest 16-bits for **exponent** for floating point operations. The MMX instructions use these floating point registers as MMX registers and use 64-bits mantissa portion only to store MMX operands. Thus, MMX programmers virtually get 8-new MMX register, each of 64 bit wide. Though it is possible to use same set of register as floating point registers and the MMX register in the same program, but it is not used. After the sequence of execution of the MMX instruction, these registers are cleared by EMMS instruction that indicates that the stack of the MMX instruction is now empty. It is advisable that the multimedia program developer should partition the MMX instructions in separate library routine. Since, the MMX registers are 64-bit wide; it can pack a total of 8-pixel values (each 8-bit) in a single register. Thus, *while 80×86 processor can manipulate only one pixel at a time, using MMX architecture, we can manipulate 8 such pixels in a single 64-bit register concurrently.*

Multimedia registers

Figure 4.9 (*c*) indicates register file of multimedia instruction MM7 through MM0. These registers are mapped into lower 64-bit of the 80-bit floating-point register stack. *The multiplexed use of the floatingpoint register must not intermix the instructions that process floating point data and MMX data. Therefore, it should be cleared after execution of either floating-point or MMX computation. The execution of the EMMS instruction invalidates the data in all MMX registers.*

MMX data types

Figure 4.9 (*d*) shows packed SIMD data types: packed byte, packed word, packed double word, packed quad word. This way, all the source and destination operands of multimedia instructions are 64-bit in width.

The *MMX instruction has been so designed that they can perform parallel operations on multiple data elements*. They can operate on packed data as mentioned above. For an instance, it can add 2-groups of 8-packed bytes using PADDB instruction. Similarly, PADDW denotes adding of 2-group of 4-packed words.

Wraparound and saturation arithmetic

We know that in 8086 arithmetic of 16 bit number results into 17 bit number, overflowing the most significant bit. This is known as wraparound with storing the lower 16-bit number in the 16-bit register. In saturation arithmetic, instead of losing the 17th bit, the result is clamped to the largest possible unsigned number that can be represented in 16-bit register, i.e., FFFF (16-bit).

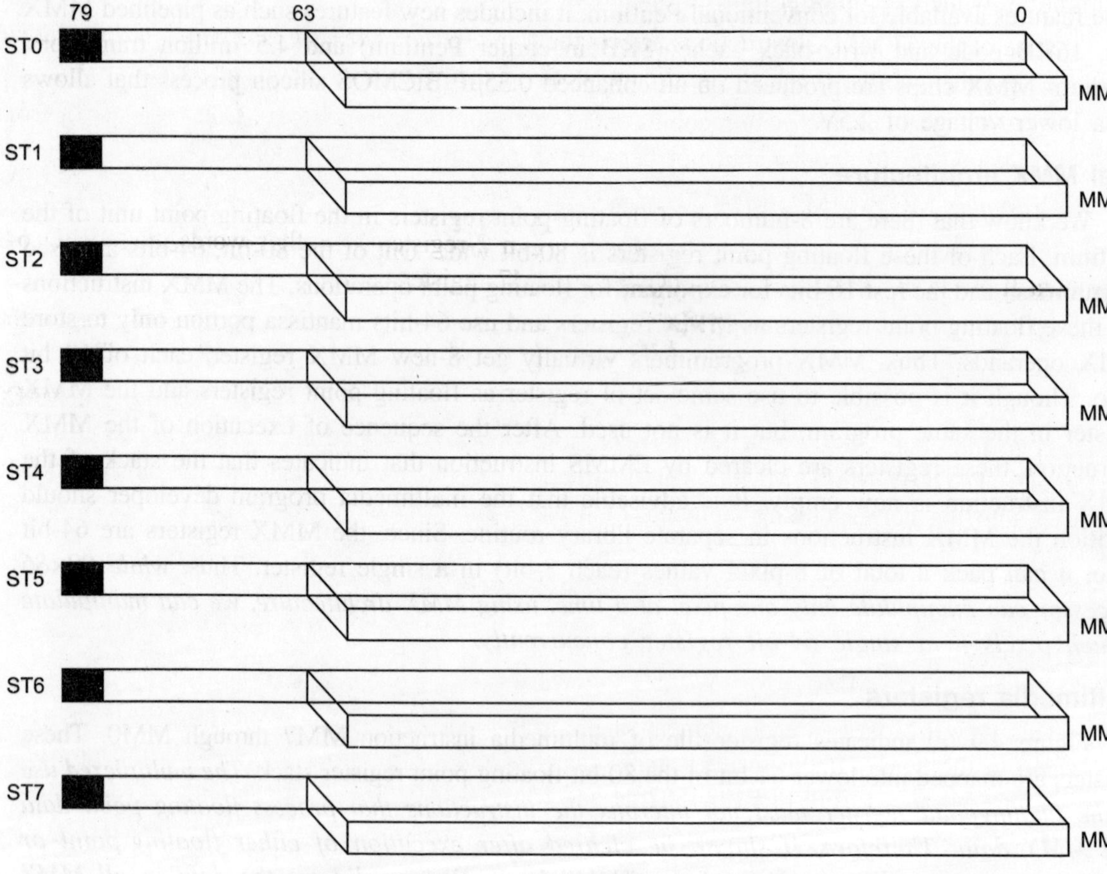

Fig. 4.9 (*c*) MMX and FPU register aliasing

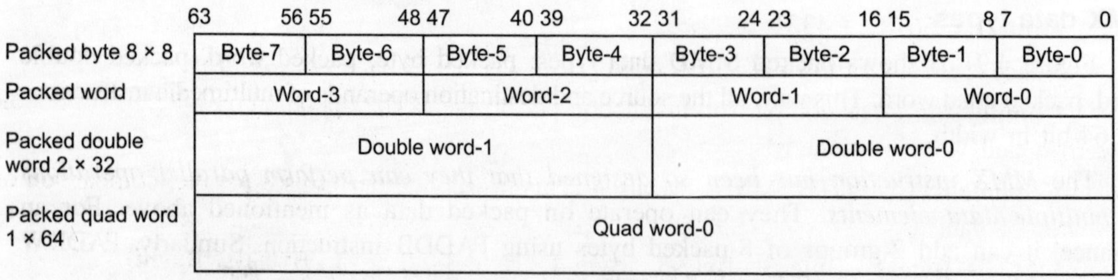

Fig. 4.9 (*d*) MMX technology packed data type

MMX instructions

There are 57-new MMX instructions. In all MMX instructions, the source operand may be found in either, MMX register or in memory. The destination operand, however, resides in the

MMX register. In the data transfer group of instruction, both the source and destination operands may reside in either in MMX registers or in memory.

The suffix S of an instruction implies signed saturation whereas US indicates unsigned saturation, while executing arithmetic operation.

None of the 57 MMX instructions affects the flag register. The byte is ordered in little endian format. In other words, the less significant word of data stored at lower address.

MOV(D, Q) : This instruction copies data from memory location to a MMX register or from one MMX register to another, but not between 2-locations of the memory.

Example: MOV MM7, MM0

This instruction performs data transfer between 2-registers. In other words, the SIMD 64-bit data from MM0 register is copied into the MM7 register.

MOV MM3, SI : This instruction copied the data from the source storage pointed by the contents of the source index register (SI) into the MM3 register.

We have seen how to move data from one memory location to a MMX register or one MMX register to the other. Now, we examine operations of some instructions that process SIMD data. Example : PADDW MM7, MM0 : This instruction adds 4-word packed data elements of MM0 register into the corresponding elements of 4-word packed data of MM7 register as explained below.

MM7 =	Word-3	Word-2	Word-1	Word-0
	+	+	+	+
MM0 =	Word-3	Word-2	Word-1	Word-0

This way it processes each element of the MMX operands in parallel. The word-0 of MM0 is added to the MM7. If the result is more than 16-bits, wraparound is done by truncating it to the 16-bits. MM7 = FFFFFFFF12345678H and MM0 = 0002000187654321H

Word-3 = FFFF + 0002 = 10001 $\rightarrow$ FFFFH $\rightarrow$ due to saturation

Word-2 = FFFF + 0001 = 10000 $\rightarrow$ FFFFH $\rightarrow$ due to saturation

Word-1 = 1234 + 8765 = 9999H

Word-0 = 5678 + 4321 = 9999H

Hence, MM7 = FFFFFFFF99999999H

Comparison : The comparison instruction performs parallel operation on byte, word, double word, Quad word element wise.

Example : PCMPGTB MM7 , MM0 $\rightarrow$ This instruction performs 8-parallel comparison on 8-bytes of data as;

MM7 =	Byte-7	Byte-6	Byte-5	Byte-4	Byte-3	Byte-2	Byte-1	Byte-0
	>	>	>	>	>	>	>	>
MM0 =	Byte-7	Byte-6	Byte-5	Byte-4	Byte-3	Byte-2	Byte-1	Byte-0

If the specific byte of the MM7 is greater than the corresponding byte of the MM0, comparison is true and the destination register MM7 is set to FFH, otherwise it is reset to 00H.

Example: MM7 = FFFF000012345678H and MM0 = FFFF000087654321H. Here the Double word-1 of MM7 and MM0 (equal) = FFFF0000H → true → FFFFFFFFH

Double word-0 of MM7(12345678H) < MM0 (87654321H) → false → 00000000H

Hence, MM7 = FFFFFFFF00000000H

Pack and unpack : The pack and unpack conversion instructions are used for storing data after converting the size of the element as per the size of the destination operand.

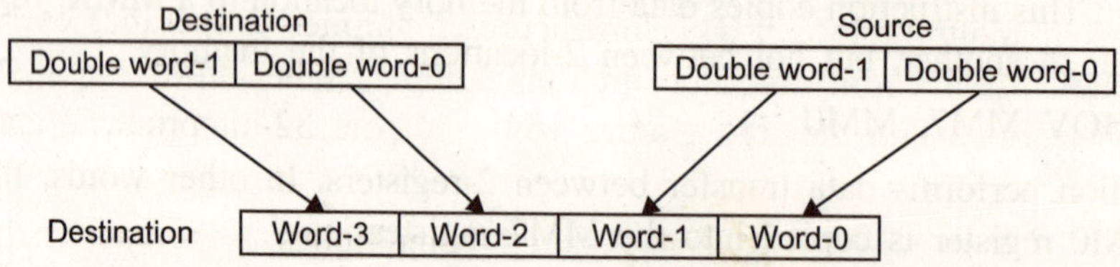

Example : PACKUSDW MM1, MM0 → This instruction packs the 4-double word elements of the source and destination MM0 and MM1 linto 4-words and packs them into the destination MM1

If MM1 = 0000AA0000004321H and MM0 = 000100FF80000123H

MM0 Word-3 = 000100FFH → FFFFH → unsined saturation overflow
Word-2 = 80000123H → FFFFH → unpack saturation overflow

Now we see how to arrange the unpacked instruction to realign the data. Figure below is the realignment process of unpacked data.

Example : PUNPCKLWB MM4, MM3

MM4 = 0F0F0F0F07050301H and MM3 = FFFF000006040200H

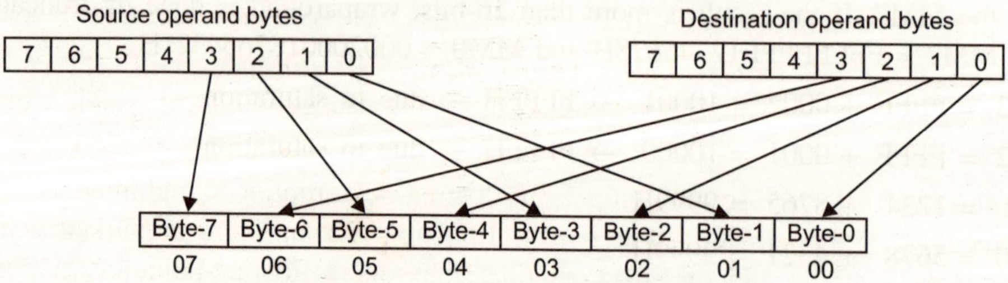

After execution of instruction MM4 = 0706050403020100H

PADD (B, W, D) : This instruction performs addition on 2-packed byte or word and double word format.

PSUB (B, W, D) : This instruction performs subtraction on 2-packed byte or word and double word format.

PCMPEQ (B, W, D) : This instruction compares the data elements of 2-packed data types in parallel.

If the result of comparison is success (true) a 1s is generated, if not then 0s is generated in the destination operand.

PCMPGT (B, W, D) : This instruction is similar to the PCMPEQ instruction and is used to check greater than condition. The result of the PCMPGT on 2-packed doubled word looks like as below

45	35	22	FF
41	30	23	OB
1....1	1....1	0....0	1.....1

PMULLW (Packed multiply low) : This instruction multiplies 2-operand; each 4-packed signed word in parallel using 16-bit precision multiplication. In other words, four 16 × 16-bit multiplications are performed and the lower order 16-bits of the 32-bit products are stored in destination operand.

PMULHW (Packed multiply high) : This instruction multiplies 2-operand; each 4-packed signed word in parallel using 16-bit precision multiplication. In other words, four 16 × 16-bit multiplications are performed and the higher order 16-bits of the 32-bit products are stored in destination operand.

PMADDWD (Packed multiplies and adds) : This instruction multiplies 4-signed word of destination operand with 4-signed word of source operand that results in two 32-bit double words. The higher order words are added and results are stored in high order double word of the destination operand. Similarly, two lower order words are added and stored in lower double word of the destination operand.

PAND, POR, PXOR : These instructions perform bit wise logical AND, OR, and XOR operation on 64-bit source and destination operands stored in packed format and the result is stored in the destination operand.

MOV (D, Q) : This is the data transfer instruction that transfers 32-bit double word or 64-bit quad word between memory and MMX registers but not between 2-locations of the memory.

PSRA (W, B), PSLL (W, D, Q) : These instructions perform arithmetic shift right or logical shift left/right in a single cycle. This is meant for shifting only packed word and double word data types.

EMMS : This instruction is used to switch from floating point routine to multimedia and vice-versa. After execution of the MMX instruction, the floating-point tag word is marked valid i.e., we do not use EMMS at the end of MMX routine, subsequent floating-point instruction will produce erratic result.

Pentium II

With the introduction of the Pentium II in 1997, Intel moved the L2 cache and the back-side bus to the microprocessor die that resulted in a dramatic reduction in cache latency. In the Pentium II, the back-side bus runs at half the processor clock speed. In 1998, the Pentium II Xeon™ microprocessor, designed for high-speed workstations and servers, increased the back-side bus speed to full processor clock speed. The full-speed back-side bus significantly increases microprocessor-to-cache bandwidth. The Pentium II chip is characterized by its Single Edge Contact (SEC) cartridge and large heat-sink. The processor mounts to its own little board along with L2 cache that is then plugged into the motherboard through an edge connector, much like a PCI I/O card. The Intel Pentium II specification are offered as in Tables 4.9(*c*) and (*d*). The Pentium II processors are still manufactured with 0.35micron technology and not the speculated technology of 0.25 micron. The maximum current drawn for the 233MHz Pentium II processor is 11.8A; the 266MHz chip draws 12.7A, and the 300MHz unit takes 14.2A. At 2.8V all Pentium II processor runs in excess of 30W. Table 4.9(*d*) illustrates Pentium II characteristics.

Table 4.9(*c*) Pentium II and its motherboard speed

µP type/ Speed	µP	Motherboard Speed
Pentium II 233	3.5x	66
Pentium II 266	4x	66
Pentium II 300	4.5x	66

Table 4.9(*d*) Pentium II processor specification

Introduced	May 1997
Maximum rated speed	233 266 300MHz.
µP clock multiplier	3.5x, 4x to 4.5x
Internal bus width	300-bit
External data bus	64-bit
Integrated-cache bus	64-bit
Memory address bus	32-bit
Maximum memory	64 GB
Integrated-cache size	16 K code, 16K data
Integral cache type	Non-blocking, L1 cache
L2 cache size	512 K
Number of transistors	7.5 million
Transistors in L2 cache	31 million
Circuit size	0.35 micron
External package	242-pin single ended cartridge
Math coprocessor	Built-in FPU
Power management	SMM (system management mode)
Operating voltage	2.8V

Pentium III and KNIS

Moving towards faster multimedia, the Pentium III was launched in 1999. It featured 70-new Internet Streaming SIMD extension which dramatically improves the performance of advanced imaging, 3D, streaming audio, video and speech recognition applications. The processor incorporates 9.5 million transistors, and was introduced using 0.25 micron technology.

The Pentium III (code-named Katmai) introduced several significant architectural changes and a new instruction set called the Katmai New Instruction Set (KNIS). Because KNIS instructions are of a type called single instruction multiple data (SIMD), KNIS is now called streaming SIMD extensions (SSE). With SSE, several bytes, or words, are packed into a 128-bit package that can be manipulated with a single instruction.

The SSE adds a total of 68 instructions to the standard Intel instruction set, including:

- 45 new floating-point instructions.
- 11 SIMD integer instructions (MMX extensions).
- 5 new cache management instructions.

Among the architectural changes introduced in the Pentium III are an enhanced floating-point execution unit and eight new registers. The Pentium III also offers a front side bus speed of 133 MHz when used with the Intel 810e chip set. Its new instructions enhanced multimedia processing and video streaming. The SIMD technology provided enhanced performance of advanced imaging, 3D simulation, animation, streaming audio, video and speech recognition applications. Table 4.9(*e*) gives more details.

Table 4.9(*e*) Comparison of P4, PIII, and Celeron processors

	P4	*PIII*	*Celeron*
Comparison as desktop processor	Ideal for Internet audio and streaming video, image processing, video content creation, speech, 3D, CAD, games, multimedia, multitasking and power for future applications	Ideal for today's usages like shopping and surfing on the Internet, e-mail, and productivity applications	Ideal for Internet, educational programs, general multimedia, productivity applications
Processor core speed	2GHz, 1.8GHz, 1.7GHz, 1.5GHz, 1.4GHz, and 1.3GHz	1.2GHz (0.13 micron), 1.1GHz, 1.0GHz, 933MHz, 866MHz, 850MHz, 800MHz, 750MHz, 733 MHz, 700MHz, 667 MHz, and 650MHz	1.2GHz, 1.1GHz, 1GHz, 950MHz, 900MHz, 850MHz, 800MHz, 766MHz, 733MHz, 700MHz, 667MHz, 633MHz, 600MHz, 566MHz, 533MHz, 533AMHz, and 500MHz
System bus	400 MHz	100MHz and 133MHz	100 MHz or 66 MHz
Level 1 cache	Execution Trace Cache	32K (16K for infrastructure and 16K for data)	32K (16K for infrastructure and 16K for data)
Level 2 cache	Advance Transfer Cache	512KB unified, non-blocking, or an integrated 256 KB Advance transfer cache	128 KB unified, non-blocking
Motherboard	Intel® D850GB, 850MD, 850MV, D845WNL, D845WN, D845HVL, and D845HV	Intel® D820LP, D815EEA, D810EMO, VC820, CA810E, SE440BX-2, JN440BX, and others	Intel® D810E2CA3, D810E2CB, D815BN, D815EEA2, D815EPEA2

(*Contd....*)

RAM	Dual channel RDRAM and PC133 SDRAM	SDRAM and RDRAM	SDRAM
Micro-architecture	Intel NetBurst Micro-architecture	P6 dynamic execution	P6 micro-architecture
Chipset	Intel® 850 and 845 chipsets	Intel® 815E, 815, 815EP, 815P, 815G, 815EG, 820, and 820E chipsets	Intel® 810, 810E, 810E2, 815E, 815P, 815EP, 815G, 815EG, 440BX, 440ZX, 440ZX-66
Micron technology	0.18 micron technology	0.13, 0.18 and 0.25 micron technology	0.18 micron technology
Power management	Yes	Yes	Yes

Does Moore's law still apply? The Katmai P3 has about 8.2 million transistors and had initial launch speeds in the 500MHz range. The current high-end Coppermine P3 processors have over 28 million transistors and speeds have hit the 1GHz mark.

The Coppermine's 3X transistor count increase is a bit misleading since the chip now includes 256KB of L2 cache on the core. The Katmai P3 had kept its L2 cache off-die on discrete SRAM chips. Nonetheless, you can't argue with the 2X clock speed increase. We could pick up Intel's 'limited availability' disclaimer, but the high-speed chips do exist and are available from OEM system builders. Moore's Law seems to be holding up, but will it continue?

Thus Pentium launched in 1993, created a new industry standard around the world with its **Superscalar architecture**. It proved to be 80% faster in DOS applications as compared to the 80486. It was the 1st processor that truly catered heavy requirements of the multimedia. The Pentium is the most successful Intel processor.

Intel-32 architecture (IA-32)

Intel's roadmaps show plans to continue improving the performance of the IA-32 architecture with newer and faster 32-bit microprocessors. New microprocessors shown on the current Intel roadmap include

- Coppermine, a 0.18 micron Pentium III microprocessor, introduced in October 1999. Initially Coppermine microprocessors operates at 700MHz and features a 256KB or larger on-chip secondary cache. In approximately the same time frame, Intel had planned to introduce the Camino chip set, its 1st to support *Rambus memory*. However, recent problems with Rambus have delayed the Camino chip set.
- Cascades, a slot 2 microprocessor introduced in late 1999, featured a 133MHz front side-bus (FSB) and a choice of 256KB or larger on-chip L2 cache. According to Intel literature, Cascades will offer several new reliability, availability, and serviceability (RAS) features.
- Foster, the most advanced IA-32 microprocessor on Intel's current roadmap introduced sometime in 2000. Intended for use in high-end workstations and servers, Foster will be introduced with a core speed of over 1GHz and features a new 32-bit

micro-architecture with a longer pipeline and larger on-chip L1 and L2 caches. Intel recently announced that Foster included a new feature called a trace cache. The trace cache is said to eliminate the need to decompress instructions, thus reducing instruction processing time and significantly increasing the microprocessor's performance.

- Willamette, a 32-bit microprocessor that was introduced at approximately the same time as Foster, is intended primarily for the high-end desktop PC market. Willamette is likely to include features similar to Foster but scaled down to reduce its cost.

CISC versus RISC

Although a significant number of microprocessors are based on RISC technology, RISC never achieved the market penetration that its early proponents hoped for. In part, this limited acceptance was because performance improvement offered by RISC were offset by a very large installed base of 80×86-compatible CISC computers. With large investments in software for CISC computers, corporate decision makers could not justify switching to RISC in many cases.

The CISC μPs are based on a complex instruction set in which instructions are executed by micro-code—one or more small programs that carry out the logical instructions required to execute each CISC instruction. Micro-code allows developers to change hardware designs and still maintain backward compatibility with instructions for earlier computers by changing only the micro-code. Although CISC designs allow a lot of hardware flexibility, the supporting micro-code slows microprocessor performance because of the number of operations that must be performed to execute each CISC instruction. A CISC instruction set typically includes more instructions than a RISC instruction set and CISC instructions are harder to pipeline.

In contrast, RISC is based on the premise that 20% of a computer's instructions do 80% of the work. A RISC instruction set includes fewer and simpler instructions that require little or no micro-code to execute, permitting easier pipelining and faster execution. With RISC microprocessors, performance depends heavily on the compiler, which is responsible for analyzing the program code and rearranging instructions for parallel execution to optimize the RISC processor's performance. Maximum performance is achieved by keeping the microprocessor's pipelines full of instructions and keeping the instructions flowing smoothly.

Although Intel microprocessors use a CISC instruction set and are considered CISC chips, the internal architecture has gradually migrated to RISC. Beginning with the Pentium Pro, Intel used a RISC core, converting CISC instructions to RISC like instructions that Intel calls micro-ops. The micro-ops overcome much of the speed penalty of CISC architecture by converting all instructions to the same length before they are processed. Micro-ops also eliminate arithmetic operations that directly change memory by loading memory data into registers before processing.

With the addition of RISC core technology, MMX, and SSE, CISC performance has become very competitive with that of RISC computers. Future CISC performance enhancements are likely to blur the differences between CISC and RISC even further.

Intel Celeron

Facing tough competition from formidable competitors like Advanced Micro Devices (AMD), Cyrix, etc., Intel was forced to develop a low-price Pentium III based on Pentium

Pro architecture in the name of **Celeron** in 1999. The clock speed of the Celeron varies from 266MHz to 700MHz. The earlier versions like 300MHz did not include L2 level on-chip cache. The higher version starting from 333MHz included 128KB L2 level cache. They have separate 16KB code and 16KB data cache on chip. These chips integrated 19 million transistors. The Celeron processor has an extra MMX pipeline. This pipeline is in addition to the 2-integer pipelines and one floating point pipelines. It provides multiple branch prediction that predicts program execution through several branches, accelerating the thruput. It includes 75-new instructions for multimedia uses. It uses SIMD instruction. It contains 8-numbers of 64-bit MMX registers. Its special functions are

- 8-bytes can be packed in one 64-bit.
- 4-numbers of 16-bit word can be packed in one 64-bit.
- 2-numbers of 32-bit double word can be packed in one 64-bit.
- One quad word in 64-bit.

The above features are especially advantageous in graphic applications. Each pixel in graphic is represented by 8-bit integer or byte. With the availability of MMX 64-bit wide register, 8-pixels can be packed together in a MMX register. All 8-pixels can be taken together in parallel for arithmetic and logical execution and can be transferred to another.

Intel compatible processors

Several manufacturers, such as AMD (Advanced Micro Design), Cyrix, MIPS etc, have developed processors that are compatible with Pentium processors. Most of the chip are pin-compatible, others requires custom designed motherboards. Any hardware or software that works on Intel-based PCs works on PCs made with any of these third-party μP chips.

AMD (Advanced Micro Devices)

AMD developed 32-bit microprocessors in the trade name of K5, K6, K6-MMX, and K7. The K5 is the competitor of the Pentium. The K6-MMX has been developed especially for multimedia systems. The K6-2 was realized with 9.3 million transistors on 81 mm die using 0.25 micron technology operating at 400MHz clock. It operates on all operating systems such as MS-DOS, Unix, OS/2, Windows 95/NT, Novell, Netware etc. K7 is called *Athelon* operating at 1.1GHz. It is a competitor of Celeron.

CYRIX

CYRIX, the competitor of Intel, developed 586 and 686 as 32-bit processors. The IBM manufactures 6X86MX microprocessor developed by CYRIX only. It is better than Pentium II running at 233MHz.

SUN SPARC

SUN SPARC is a 64-bit superscalar RISC type microprocessor having 3.8 million transistors. It can issue 4-instruction per clock cycle. It includes 9 independent functional blocks. It has 5-way set-associative instruction cache and 16KB four-way set-associative data cache. The Indian Supercomputer, PARAM 10000, uses ULTRASPARC III processor.

MIPS

The MIPS R4000, R5000, MIPS10000, MIPS12000 series are popular RISC type 64-bit microprocessors. They are suitable for workstations. Its low-end R4400 microprocessor is suitable for desktop computers. The MIPS R4400 processor is a 0.35 micron, CMOS-based, 250MHz, 64-bit processor. The chip contains a floating point unit, an arithmetic logic unit, an integer unit, cache control, separate 16KB instruction and 16KB data caches and a 64-bit bus interface. The MIPS10000 integrates 6.4 million transistors and operates at 200MHz. It has 5-functional units, 32-number of general purpose and 32 floating-point registers. It can issue 4-instructions in one clock.

4.10 PENTIUM 4

It is believed that the brand new Pentium 4 processor introduced in November 2000, shall make Intel the provider of the fastest, most advanced and thus simply best microprocessor in the world. Intel wants to get back to the top leaving the AMD far behind. The mysterious Pentium 4, codename 'Willamette', is supposed to be powerful enough to ensure that Intel reaches this juicy goal.

Architecture of Pentium 4

I am quite sure that most of us must have already heard about Pentium 4's funky 'NetBurst-Architecture'. However, I would still like to dedicate a major amount of time of this article to a detailed explanation of what is really behind Pentium 4 and all its fancy new features. I would like to start discussing the block diagrams of Pentium 4, Pentium III and AMD's latest Athlon processor for a comparative look. Let me once again take few terms to be used later here in this article.

These terms are

- Branch Target Buffer (BTB) : This table gives us all the addresses to where a branch will or could be made. *Athlon* is also using a BHT = Branch History Table that stores the addresses where branches were made to. A software program uses branches to make decisions. The program asks a question and according to the answer a branch is made or not.
- Micro-Operation/Operand (μOP) : As per Intel this is like an instruction that can be directly understood by the execution units of the microprocessor. The AMD calls them μOPs, because they are a bit advanced and can contain more information than Intel's μOPs. Both 'OPs' have one important thing in common. They represent very simple instructions that can be quickly carried out by the processor. Unlike 80×86-instructions, those 'OPs' are of a defined size and can thus easily be fed into the execution pipeline. The decoder translates an 80×86-instruction into one or many more 'OPs', unless the 80×86-instruction was so complex (and rare) that the 'Micro Instruction Sequencer' has to produce a sometimes rather longish sequence of 'OPs', using the 'Micro Code ROM' found in any modern superscalar microprocessor. In average, most 80×86-instructions get decoded in about two 'OPs'. Some extremely simple instructions like i.e., an 'AND', 'OR', 'XOR' or 'ADD' are often producing only one 'OP', while a

'DIV' or 'MUL', or an indirect addressed operand will produce more. Complex instructions like e.g. trigonometric commands can easily produce up to hundreds of 'OPs', coming out of the 'Micro Instruction Sequencer'.

- Arithmetic Logic Unit (ALU) : This is the name of what we call the Integer-unit. Arithmetic operations like adding, multiplying and dividing as well as logic operations such as 'OR', 'AND', 'ASL', 'ROL', ... are carried out by the 'ALUs'. Those operations represent the vast majority of program code in most software programs.
- Address Generation Unit (AGU) : This unit is just as important as the 'ALU', because it is responsible for the data from or to the correct address to either be loaded or stored. Absolute addressing in programs is only used in rare exceptions. As soon as we get arrays of data, the program code uses indirect addressing, keeping the 'AGUs' busy.

The diagram of the good old Pentium III is shown in Fig. 4.10 (*a*). We can see that it is a lot less complex than the Pentium 4 diagram shown in Fig. 4.10 (*b*). Still, we might spot the few advantages that P III has over P 4. This will become clear further down in the text.

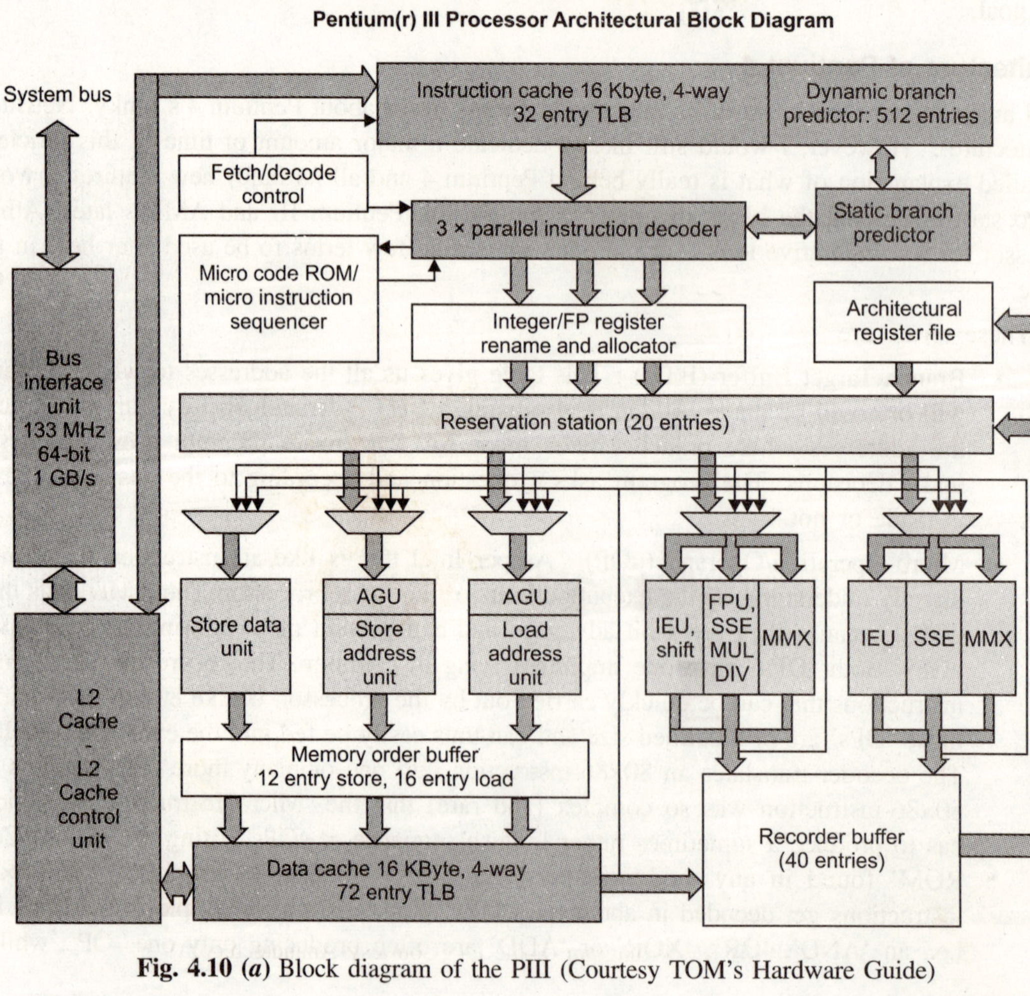

Fig. 4.10 (*a*) Block diagram of the PIII (Courtesy TOM's Hardware Guide)

NetBurst architecture

It starts bursting with Net. Don't you love marketing? NetBurst is the new name for the Pentium 4 processor's new architecture. The architecture design should improve net and multimedia related performance in areas such as imaging, streaming video, speech, and 3D graphics. There are several main parts of the NetBurst architecture which are listed below

- Hyper Pipelined Technology
- Execution Trace Cache
- Rapid Execution Engine
- Advanced Dynamic Execution
- Intel i850 Chipset

Our web pages would not pop up any faster without NetBurst. The downloading will just take long and the Internet would not 'burst' either. However, Intel is trying hard to be trendy and since the Internet is still hip, it is a perfect vehicle to market Pentium 4.

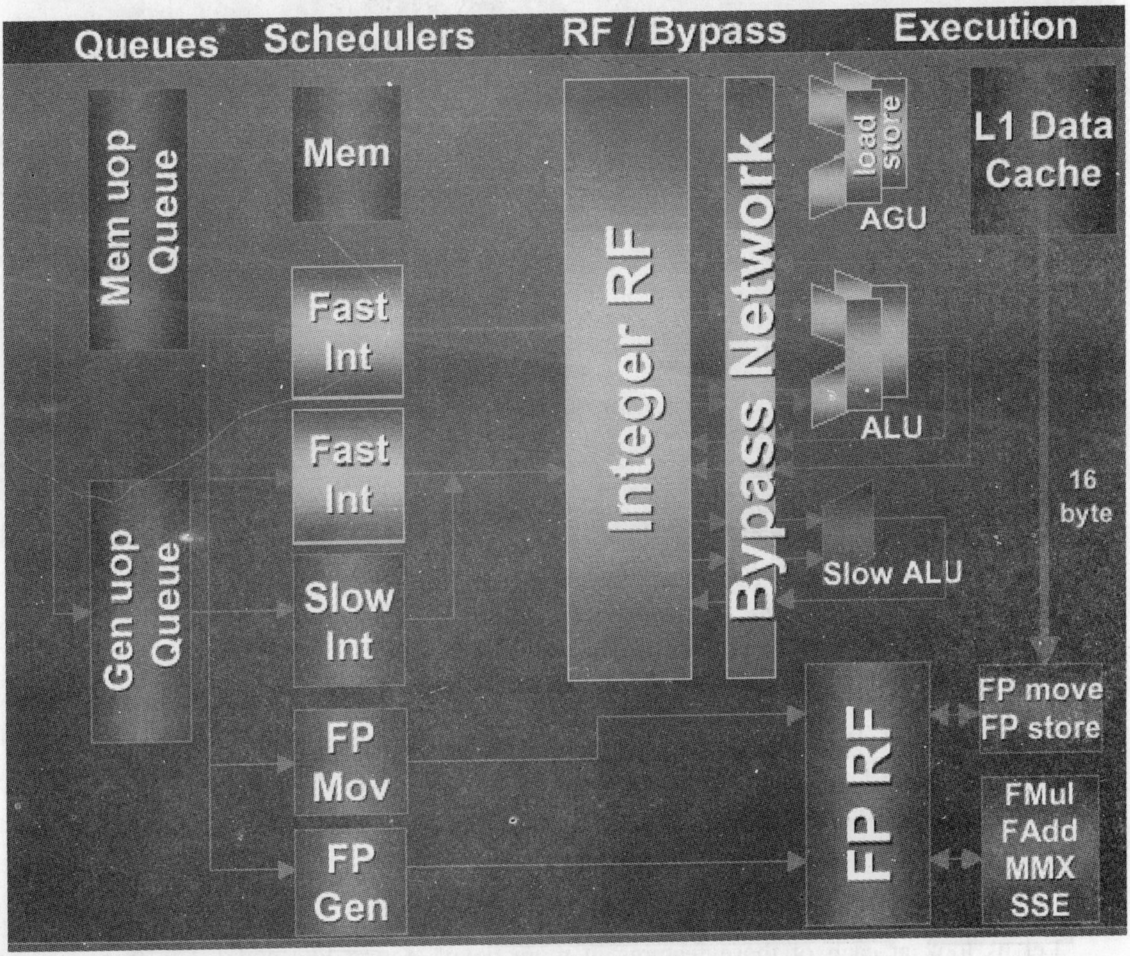

Fig. 4.10 (*b*) Block diagram of the P4 (Courtesy Anandtech.com)

The name 'NetBurst' could also be a hint towards Pentium 4's performance characteristics. Professional PC-users might not care quite as much for an 'Internet-accelerating' processor, but more for a product that makes them get their work done as fast as possible. Looking at the benchmark results, the Pentium 4 shines a lot more at recreational software than in professional applications.

Another big issue with Pentium 4's NetBurst-Architecture is its obvious focus to deliver highest clock rates. Intel wants clock rate at almost any cost. The Intel also wants to prove that the Pentium 4 design is perfect to deliver exactly that. The GHz clock rate concludes that this alone is already good enough to make Pentium 4, the fastest processor in the universe. That is all about why the name 'NetBurst' seems to be given.

Figure 4.10 (*c*) gives more details of the P4 architecture.

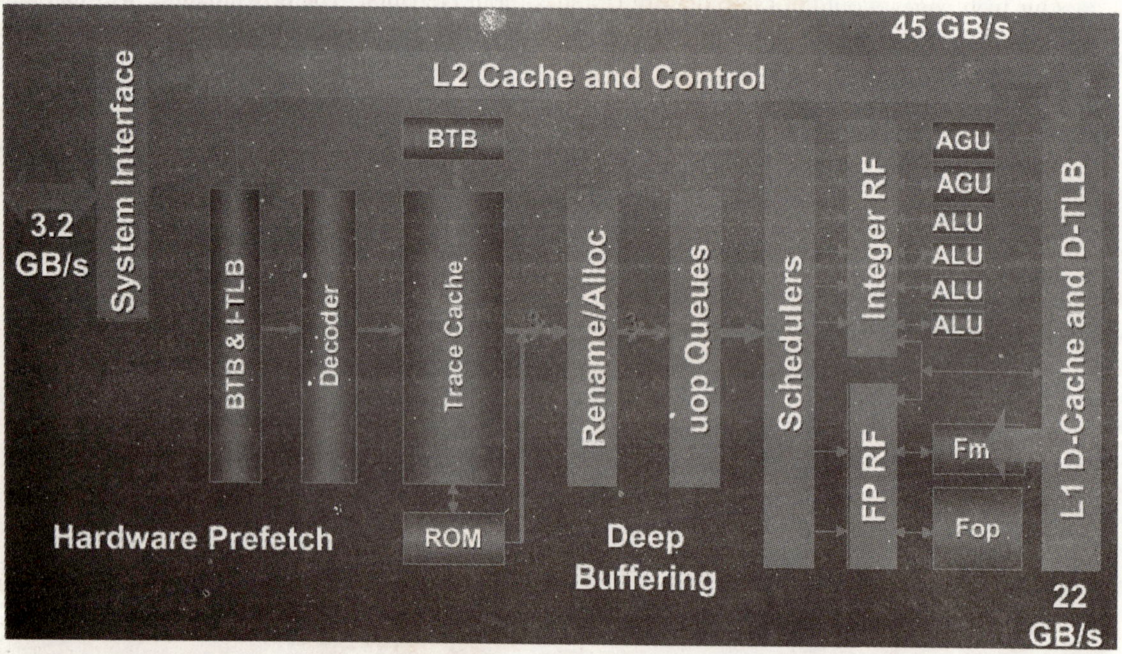

Fig. 4.10 (*c*) Detail block diagram of the P4 (Courtesy TOM's Hardware Guide)

The NetBurst includes the following goodies that have been implemented into the new Pentium 4 design as shown in Fig. 4.10 (*d*).

- Faster System Bus
- Advanced Transfer Cache (already known from Pentium III)
- Advanced Dynamic Execution (Execution Trace Cache, Enhanced Branch Prediction)
- Hyper Pipelined Technology
- Rapid Execution Engine
- Enhanced Floating Point and Multi-Media (SSE2)

Let us look at each of those features in more detail, acting like code/ data that is being processed by Pentium 4.

New processor bus

The 1st new feature seen by code or data as it enters Pentium 4 is the new system bus as drawn in Fig. 4.10 (*e*). The well-known FSB of Pentium III is clocked at 133 MHz and able to transfer 64-bit of data per clock, offering a data bandwidth of 8 byte × 133 millions/s = 1.066 GB/s (133MHz is actually 133.3MHz). Pentium 4's system bus is only clocked at 100MHz and is also 64-bit wide, but it is quad-pumped using the same principle as AGP4x. Thus, it can transfer 8 byte × 100 millions/s × 4 = 3.2 GB/s. This is obviously a tremendous improvement that even leaves AMD's recently 'upgraded' EV6-bus quite far behind. The bus of the most recent Athlon's is clocked at 133MHz, 64-bit wide and 'double-pumped', offering 8 byte × 133 millions/s × 2 = 2.133 GB/s.

The new bus of Pentium 4 enables it to exchange data with the rest of the system faster than any other 80×86-processor, thus removing one important bottleneck that Pentium III was suffering from. However, the fastest processor bus does not help much unless the system's main memory can deliver data at an matching pace.

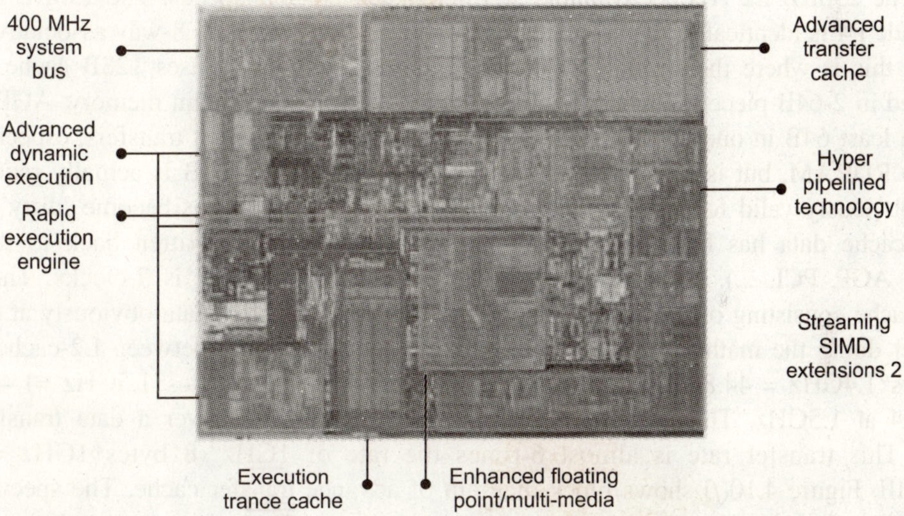

Fig. 4.10 (*d*) NetBurst architecture layout (Courtesy TOM's Hardware Guide)

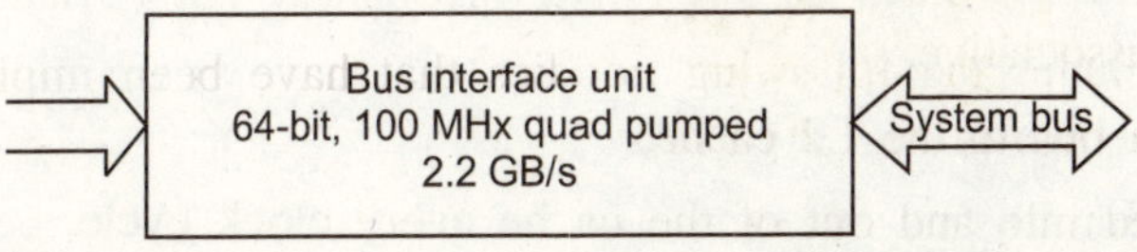

Fig. 4.10 (*e*) New processor bus (Courtesy TOM's Hardware Guide)

Intel's new 850 chipset for Pentium 4 that currently represents the only chipset for this new μP, uses 2-Rambus channels and therefore the expensive and unpopular RDRAM. However, these 2-RDRAM channels are able to deliver the same data bandwidth as Pentium 4's new bus (3.2GB/s), making them a perfect match at least on paper. This constellation enables Pentium 4-systems to have the highest data transfer rates between processor, system and main memory, which is a clear benefit. At the same time system cost is impacted by the high price of RDRAM plus the fact that a Pentium 4-system always requires two or even four RDRAM-RIMMs of the same size and spec. One, three or mixed RIMMs are not an option.

RAMBUS versus DDR

As we have known, the power of RAMBUS could not be realized without the Pentium 4 and in fact, slows down Pentium III's. Is it the utopian RAM that has been promised to be? Well the speed of RAMBUS can have a serious effect on the Pentium 4. The difference between using 600 and 800MHz RAMBUS RAM can make up to a 10% real performance difference. Yes, the Pentium 4 gets a boost from RAMBUS, however, the Athlon gets just as big of a boost from the much cheaper DDR-SDRAM so basically it is just like everyone has been saying. The RAMBUS was more of a marketing item than it was a real competitive technology.

Advance transfer cache

The 2nd new thing that most of the data has to pass is Pentium 4's on-die L2-cache. Since the days of **Pentium III 'Coppermine'**, the Intel calls the L2-cache 'Advanced Transfer Cache' (ATC). The 256KB L2 ATC is available at 1.8GHz and 2.0GHz speeds. The 256KB size of L2 cache inside P4 is identical to the L2-cache of Pentium III and both are 8-way associative as well. However, this is where the similarities ends. Pentium 4's L2-cache uses 128B cache lines that are divided in 2-64B pieces. When it fetches data from the system (main memory, AGP, PCI, ...), it reads at least 64B in one go that ensures great performance for burst transfers, especially when talking to RDRAM, but is rather bad if only one byte out of that 64B is actually required. The same is obviously valid for write operations in case the cache line has become 'dirty', meaning that the cache data has been altered and therefore needs to be written back to the system (memory, AGP, PCI, ...). The read latency of Pentium 4's L2-cache is 7-clocks. The advance transfer cache consisting of 256 bit (32-bytes) interface that transfers data obviously at core clock rate. After doing the math we get to an impressive data bandwidth between L2-cache and core of 8 bytes*1.4GHz = 44.8GB/s for Pentiume 4 @ 1.4GHz and (8bytes*1.5GHz =) 48GB/s for Pentium 4 at 1.5GHz. Thus, 1.5GHz Pentium 4 processor can deliver a data transfer rate of 48GB/s. This transfer rate is almost 6-times the rate of 1GHz (8 bytes*1GHz =) 8GB/s. Pentium III. Figure 4.10(*f*) shows block diagram of advance transfer cache. The special features of the ATC includes

- Non-blocking, full speed, on-die level 3 cache,
- 8-way set associative,
- 256-bit data bus to the L2 cache,
- Data clocked into and out of the cache every clock cycle.

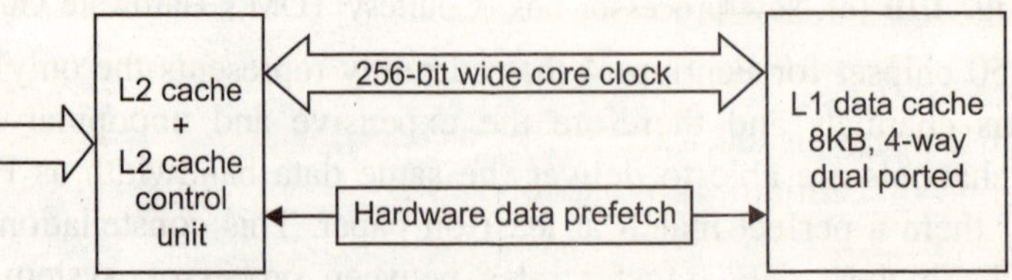

Fig. 4.10 (*f*) Advance transfer cache

Pentium 4's L1 cache

With the phenomenal increase in the clock rate, the design of the cache has to be improved to let Pentium 4 to have the most ideal performance. The Pentium III is equipped with a 16KB L1 cache for instructions and a 16KB L1 cache for data. One special thing about the Pentium 4's L1-cache is that the size has been reduced to 8KB only. This is only half the size of the Pentium III's L1-cache and only an eighth of Athlon's. Intel was probably forced to reduce the size of the L1 data cache down to only 8KB to enable its extremely low latency of only 2-clock cycles. It results in an overall read latency of less than half of Pentium III's L1 data cache already in the Pentium 4 at 1.4GHz.

While the L2-cache uses 8-way set associative, the L1-cache of Pentium 4 uses 4-way set associative. The L1 data cache of Pentium 4 is 4-way set associative and uses 64B cache-lines. The dual-port architecture allows one load and one store operation per clock.

Hyper pipeline

One of the most well known features of the new Pentium 4 is its extremely long pipeline. While the pipeline of Pentium III (P6-microarchitecture) has 10-stages and one of Athlon 11, Pentium 4 has no less than 20-stages as shown in Fig. 4.10 (*g*). The pipeline is a new 20-stage branch prediction/ recovery pipeline. The reason for the longer pipeline is Intel's wish of Pentium 4 to deliver highest clock rates. Once a pipeline has more stages, it allows for the µP to do less per clock cycle and increase the speed of the clock rate and allow for more processing. If more stages can increase the speed of clock rate, why not make it 50-stages instead of 20. The reason is that as soon as it turns out at the end of the pipeline that the software will branch to an address that was not predicted, the whole pipeline needs to be flushed and refilled. The longer the pipeline the more 'in-flight' instructions will be lost and the longer it takes until the pipeline is filled again. This is the main disadvantage of one long pipeline.

Intel was proud to announce that the Pentium 4 pipeline can keep up to 126 instructions 'in-flight', amongst them up to 48-load and 24-store operations. The improved trace cache branch prediction unit described above is supposed to ensure that flushes of this long pipeline are only rare occasions.

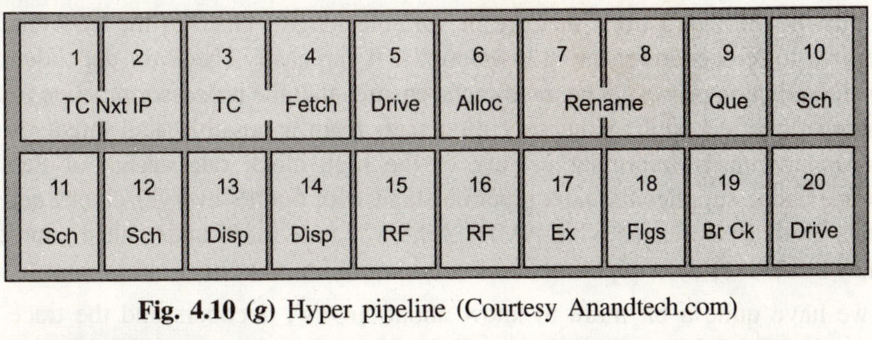

Fig. 4.10 (*g*) Hyper pipeline (Courtesy Anandtech.com)

The stuff that happens in the trace cache, as mentioned above, only represents the first 5-stages of the pipeline of Pentium 4. What follows is

- Allocate resources
- Register renaming.

- Write into the 'μOP' queue.
- Write into the schedulers and compute dependencies.
- Dispatch 'μOPs' to their execution units.
- Read register file (to ensure that the correct ones of the 128 all-purpose register files are used as the register(s) for the actual instruction)

After that comes the actual execution of the μOP, which will be discussed in more detail in the next paragraph. Of the above-mentioned previous stages the schedulers as well as the register file read are the most interesting.

Hardware prefetch

Intel has added another nifty feature that I want to bring to our attention in the L1/L2 cache context. If we think of the Pentium III launched in February 1999, we might remember Intel's introduction of the 'streaming' SIMD Extensions. The 'streaming' bit of 'SSE' is actually represented by the prefetch-instructions of Pentium III that enable software to load data into the caches before it is requested by the processor core.

Those instructions still exist in Pentium 4's instruction set, but with the new hardware prefetch feature of Pentium 4 a lot of this is done automatically. This new unit is able to recognize data access patterns of the software executed by Pentium 4, so that it 'guesses' which data will be needed next and 'pre-fetches' it into the cache.

The procedure might sound familiar to us from the complex hard drive cache algorithms and we might also be aware how much this can speed up hard disk accesses under certain circumstances. Pentium 4's hardware prefetch is probably able to significantly accelerate the execution of software that is using a lot of large data arrays.

Entering execution pipeline-Pentium 4's trace

As already mentioned in description of the term 'μOPs', those simple instructions are the language understood by the execution units. They are of a defined size and thus easier to be sequenced than 80x86-instructions that are of variable length. Once in the trace cache, Pentium 4 saves the time to re-decode repeating instructions. It can easily check for dependencies required for the branch prediction process. The trace cache ensures that the processor pipeline is continuously fed with instructions, decoupling the execution path from a possible stall-threat of the decoder units. This is particularly important in case of the high clock rate design of Pentium 4. The execution trace cache supplies the next pipeline stage with 6μOPs every 2-clocks and thus 3μOPs per-clock that is about as fast as what AMDs Athlon is able to do under ideal conditions. Figure 4.10 (*h*) shows the execution pipeline.

Now we have quite a bit more to know about 'μOPs', decoders and the trace cache. First of all, those 'μOPs' are not exactly small. In fact they are considerably larger than an 80x86-instruction although they contain less information (most 80x86-instructions are represented by more than one μOP). The 'μOPs' of Pentium III are known to be as large as 118-bit. Intel never reported the physical size of the execution trace cache and neither the size of Pentium 4's 'μOPs'. We only know that the trace cache is supposed to contain about 12K decoded 'μOPs' in the order of program execution. This increases the performance by removing the decoder from the main execution loop and makes more efficient use of the cache storage space since the instructions that are branched around are not stored. The result is to deliver high volumes of instructions to the processor's execution units and a reduction in the overall time required to recover from branches that have been mis-predicted.

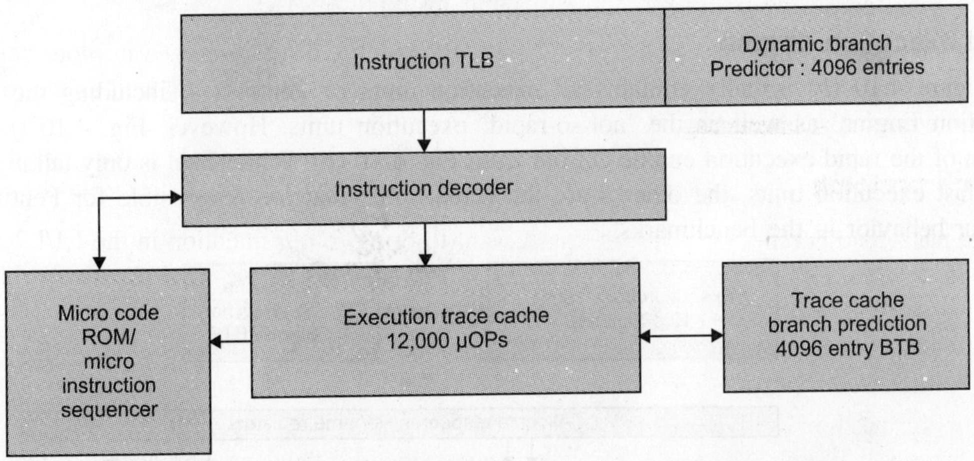

Fig. 4.10 (*h*) Execution pipeline (Courtesy TOM's Hardware Guide)

Looking at the Pentium 4 die chips that was supplied by Intel to the press at Comdex and comparing the trace cache area with the L2-cache area it looks as if the trace cache is about 92-96KB in size. It therefore seems to be a good guess estimating the size of a Pentium 4 µOP in the neighborhood of 64-bit.

The 96KB (8KB L1 cache and 12KB *ETC*) is quite a considerable size and 6-times larger than Pentium III's 16KB L1 instruction cache. However, Intel was careful enough not to waste the space offered by Pentium 4's execution trace cache. The decoder units that feed the trace cache ensure that only those 'µOPs' are stored in the trace cache that will actually be executed.

From the description of 'µOPs' we might remember the case when an 80x86-instruction is rather complex. The decoder required in the micro code ROM of the processor must produce a sometimes very long chain of 'µOPs'. In this case the trace cache does not get filled up with all of those 'µOPs'. As a placeholder it only contains some kind of flag that signalizes that the micro-instruction sequencer is supposed to supply the 'µOPs' to the next pipeline stage. It is not known how many 'µOPs' per clock the micro-instruction sequencer is able to deliver, but it would not be surprising if it is less than the 3 'µOPs' per clock that the trace cache can send to the next pipeline stage. This can obviously have an important performance impact on the Pentium 4 µP that has been tuned for simple instructions, but it seems to suffer from complex ones, as we will see later.

As already mentioned, the trace cache can also be of significant benefit, in case of a mis-predicted branch. In this case the alternative code could already be found in the trace cache. To check if certain code already resides in the trace cache, it has a rather complex structure of tags, indices and cache lines.

Trace cache branch prediction unit

Intel is very proud on the branch prediction unit that aids the execution trace cache. Its branch target buffer is 8-times as large as the one found in Pentium III and its new algorithm is supposed to be way better than AMD's latest G-share algorithm used in Thunderbird and Spitfire. Intel claims that this unit can eliminate 33% of the mispredictions of Pentium III. Figure 4.10 (*i*) depicts the arrangement of trace cache prediction unit.

Rapid execution engine

Figure 4.10 (*b*) actually contains all execution units of Pentium 4, including the 'Rapid Execution Engine' as well as the 'not-so-rapid' execution units. However, Fig. 4.10 (*j*) is the portion of the rapid execution engine cut out from Fig. 4.10 (*b*). While Intel is only talking about the 4-fast execution units, the other 4-are the actual units that are responsible for Pentium 4's peculiar behavior in the benchmarks.

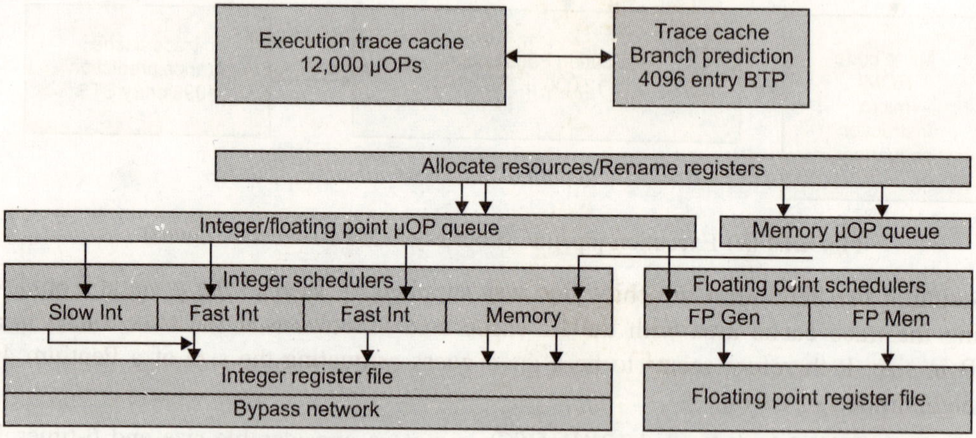

Fig. 4.10 (*i*) Trace cache prediction (Courtesy TOM's Hardware Guide)

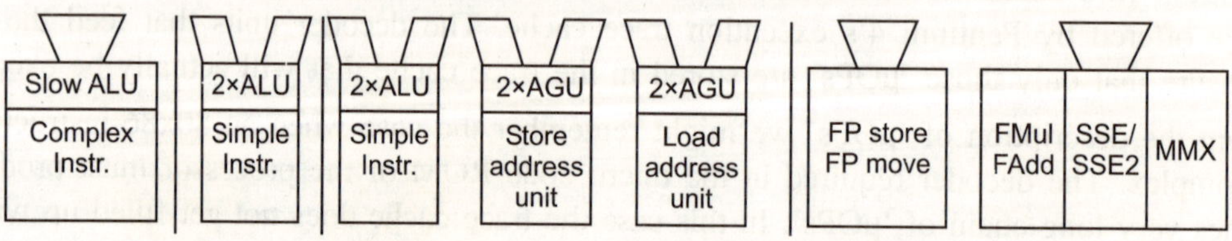

Fig. 4.10 (*j*) Rapid execution engine (Courtesy TOM's Hardware Guide)

Basic part of the rapid execution engine are the 2 'double-pumped' ALUs and AGUs. Each of the 4-is said to be clocked with double the processors clock, because they can receive a µOP every half clock. Intel never disclosed if those units are now indeed clocked with twice the processor clock or if each of those units is in reality consisting of two identical sub-units running at normal clock that can merely receive the µOPs alternately every half clock. It doesn't really matter which of the two is actually true, because the result is the same. Simple µOPs that can be processed by the Rapid Execution Engine are executed in half a clock, which is obviously a very good thing. The Pentium 4's 2-ALUs run at twice the frequency of the processor core, so effectively 3GHz for a 1.5GHz Pentium 4. This allows two things. It allows the processor to execute certain instruction at ½ the frequency of the processor core and allows for higher execution speed and reduced latency. It permits the processor to re-calculate bad cycles in a timely fashion that would not allow it to lose execution time. If it miscalculates during a cycle it still has the 2nd half cycle to retrace its steps and correct the error.

The story looks a lot different for the instructions that cannot be processed by the rapid execution units. Those instructions or 'μOPs' need to use the one and only 'Slow ALU', which is not 'double pumped'. The majority of instructions need to use this path that obviously sounds scary. However, the majority of code is in actual fact consisting of the most simple 'AND', 'OR', 'XOR', 'ADD', ... Instructions, making Intel's 'Rapid Execution Engine'-design sensible though not particularly amazing.

Things look worse if we look into the FPU-part of Pentium 4. Please take the time and compare this part to the Pentium III block diagram. We will see that Intel has actually castrated quite a bit of the SSE/MMX part of Pentium 4. Pentium III used to have two MMX and two SSE units, but Pentium 4 has only got one of each. Intel claims that additional units would not have improved the SSE/SSE2, MMX or FPU performance. However, our benchmark results speak a different language.

However, this does not help in integer base applications, such as business applications. Even with all these new features to help out with miscalculated cycles and keep the processing flowing a pace that is higher than that of the Athlon. When dealing with integer applications they generally cannot be predicted, meaning the cycles on the Pentium 4 will have to start often. Even with Rapid Execution Engine to help speed up the processing time of instructions the Pentium 4 has to pay very high penalties for a mis-prediction on any one cycle. Like everything else out there, nothing is perfect and we must pay a price for success, in Pentium 4's case a longer pipeline to achieve a higher clock speed.

SSE2-new double precision streaming SIMD extensions

I need not forget to mention SSE2 to conclude this epic piece about Pentium 4's internal architecture. The 144 new instructions are finally enabling everything that SSE was expected to be in the 1st place. The 128-bit of packed data that could only be in form of 4-single-precision floating-point values under SSE can now be operated in all of the following options:

- 4 single precision FP values (SSE)
- 2 double precision FP values (SSE2)
- 16 byte values (SSE2)
- 8 word values (SSE2)
- 4 double word values (SSE2)
- 2 quad word values (SSE2)
- 1 128-bit integer value (SSE2)

The options are vast and the usefulness are undoubted. Intel hopes that software developers will soon replace the old ×87-FPU-instructions with the double-precision FP instructions of SSE2, so that Intel's currently false claim that Pentium 4 has the most powerful FPU finally becomes reality. The AMD is very impressed with SSE2 as well that is why it announced privately only to a few of us that the upcoming Hammer-line of 80×86-64 processors will include SSE2 as well.

We doubt if SSE2 will be able to replace 80x87-instructions in scientific software. We should not forget that the original FPU is using 80-bit FP-values, not the less exact 64-bit FP-values offered by SSE2.

Die size/package/socket

Here we see a die size comparison of Pentium 4, Pentium III, Athlon and Duron in Fig. 4.10 (*k*). The 217mm^2 of Pentium 4 are more than double the size of Pentium III and almost double the size of Athlon. This seems rather surprising when we realize that Pentium 4 has got 42 million transistors and Athlon 37 million. Each of the four is manufactured in 0.18μ-process. Bottom line: Pentium 4 is big!

Intel has reacted to the complaints about damaged flip chips of Pentium III and Celeron. The recent flip chip package of the 2-processors exposes the die to direct forces from outside. A badly mounted heat sink can easily damage the sensitive silicon and destroy the processor. Intel equipped Pentium 4 with a protective metal cover that saves the die from any damage.

Pentium 4's package and Socket is a bit larger than Pentium III's, but the difference isn't much. When we see a Pentium 4 on its own we mistakenly estimate it's of the same size as a Pentium III.

Fig. 4.10 (*k*) Package and socket of P4

This is the new Socket423. As we never would have guessed it comes with 423-pins, thus 53 pins more than Pentium III. The majority of those new pins are required to supply Pentium 4 with the vast amount of power it requires.

i850 Chipset

With development of any new processor, a new chipset is must to go along with it as indicated in Fig. 4.10 (*l*). The Pentium 4 features a 400MHz NetBurst system bus that provides nearly 3-times the system bandwidth over other platforms. With 3.2GB/s of system bandwidth, compared to 1.06GB/s of bandwidth from Pentium III running at 133MHz system bus. This gives the processor tons of processing headroom for complex applications, such as e-commerce applications which demand bandwidth of highly complex applications.

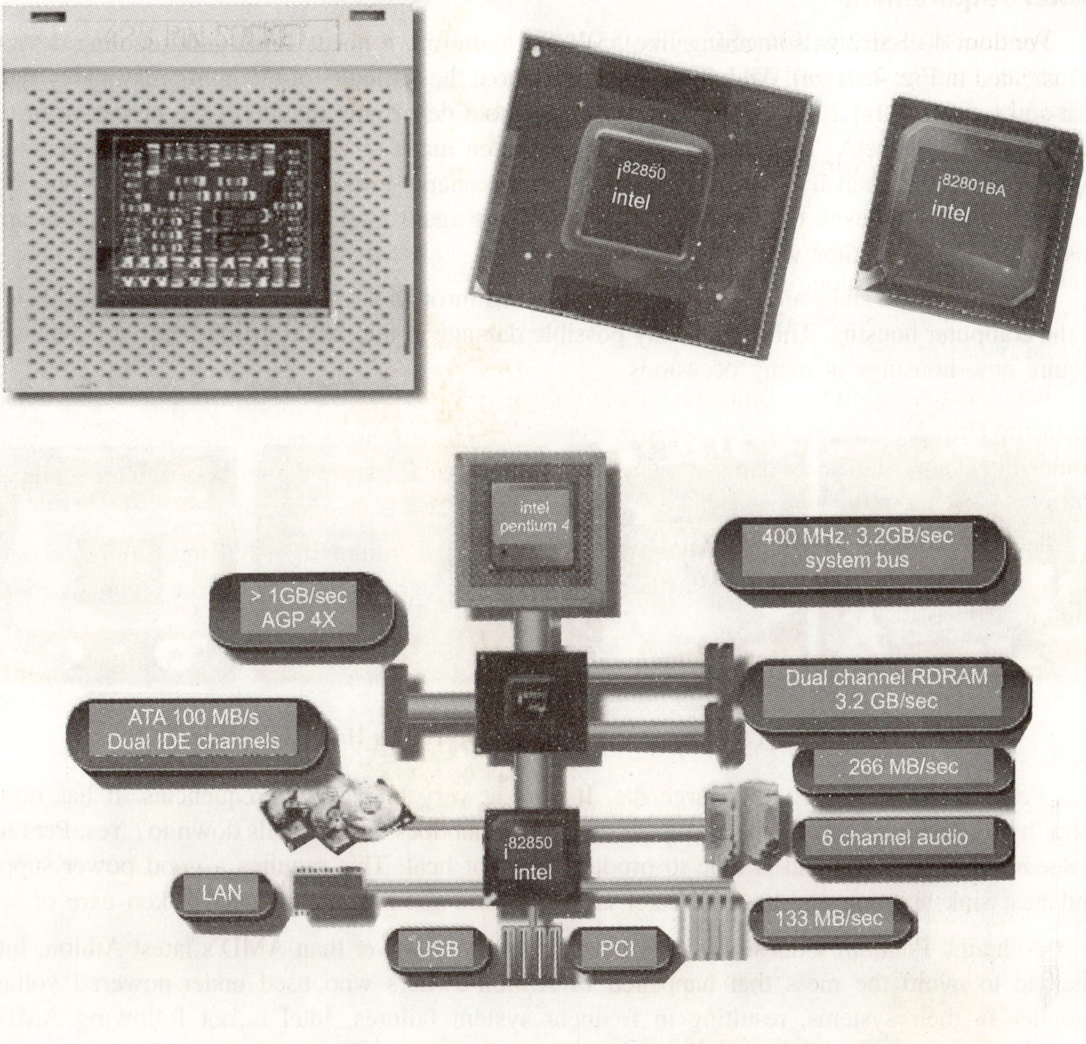

Fig. 4.10 (*l*) Chipset (Courtesy TOM's Hardware Guide)

Furthermore, the 850-chipset features enhanced I/O controller Hub (ICH2). The ICH2 features such as an additional USB controller for 4-ports and twice as much bandwidth for USB than any other bridge architecture; 24Mbps. Also the 850 chipset features ATA100 support for the best performance in hard drives and allows for a cost effective solution features of the HDD technology without losing performance from cost-effective HDD solutions.

There is not really that much to say about the one and only chipset available for Pentium 4 right now. Basically, i850 has a lot of similarities with i840E. Both chipsets are using a dual-Rambus channel memory architecture, both share Intel's 'hub architecture', both are using the 82801BA aka ICH2 chip for I/O and PCI and both are pretty expensive.

The only difference is the bus to the processor that is a normal 133MHz bus for Pentium III in case of i840 and the quad-pumped 100MHz bus for Pentium 4 in i850.

Power requirement

Pentium 4 dissipates something like 55W and therefore it needs some good cooling devices as indicated in Fig. 4.10 (*m*). Athlon owners know about the difficulty of mounting reasonably sized heat-sinks on their 1-1.2GHz Athlon processors without destroying the die of the processor or the notches of the SocketA. Top-notch heat sinks are often made of copper and can easily weigh up to 0.500kg (more than a pound). In a worst-case scenario heat-sink could rip either only the socket, but possibly even the motherboard apart. Once again Intel decided not to make the same mistake as its competitor with the green logo.

Pentium 4 heat sinks are supposed to be mounted through 4-holes around the Socket423 right to the computer housing. This avoids any possible damage to the socket or motherboard. This will require new housings in many occasions.

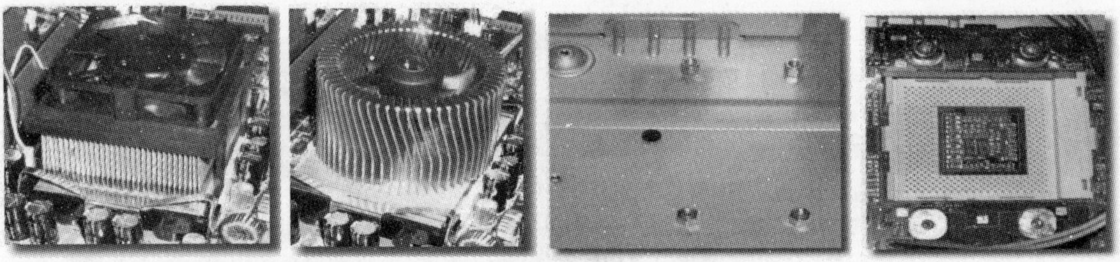

Fig. 4.10 (*m*) Cooling devices (Courtesy TOM's Hardware Guide)

Pentium 4 has got a rather large die. It runs at very high clock frequencies. It has rather long 'hyper-pipeline' and a supply voltage of 1.7V. What does that all boils down to? Yes, Pentium 4 needs a lot of power and is able to produce a lot of heat. This requires a good power supply and heat sink solution and because Intel is Intel, these things were properly taken care of.

Although Pentium 4 does not really need much more power than AMD's latest Athlon, Intel decided to avoid the mess that happened to Athlon-owners who used under-powered voltage supplies in their systems, resulting in frequent system failures. Intel is not following AMD's basically ignored compatibility list that is hardly worth the paper to be written on. This was proved by Asus' A7V motherboard which was not officially supported by AMD for a long time while AMD shipped review systems with exactly this board to the press. Intel is well-known for going ahead and establishing new industry standards, and as much as this may bug a lot of us who don't own the hardware required by the new standard, it assures that systems which accord to this guidelines will actually work without any glitches.

Pentium 4 requires a new kind of power supply that ensures the delivery of 10-12A from the 12V line of the power supply. This results in the need for additional connectors that can carry this current. Although we managed to run all of our Pentium 4 test motherboards with a normal power supply as well, we encourage every Pentium 4 owner to ensure he's got one of the new power supplies that come with two additional connectors that need to be plugged into the motherboard. The two new power supplies available to us were from AOpen and Delta Electronics.

Power consumption

Increasing the clock frequency will increase the power, a processor consumes and thus the heat load, a processor must dissipate. Different cooling techniques will be required on a system level to allow the P4 to achieve dramatically higher clock frequencies. In the near term cooling can be accomplished using forced convection with a plenum directing air over the processor and memory. There is already interest in vapor compression cooling for high performance desktop computers?

Reducing the clock frequency reduces the power consumption. Although this negates some of the advantages of the P4, it could be useful for portable computers. For example, when the computer is plugged into the wall it would operate at a higher clock frequency than when it is operating off battery. Intel already has a technology called SpeedStep that does this.

Another way to reduce power consumption is to reduce the voltage levels. The 1.5GHz P4 runs at 1.7V to 1.75V. We have found nothing to indicate that Intel is planning a lower voltage version of the P4 at this time.

Processor placement

Intel even defined the place where the processor is supposed to be placed on the motherboard (right at the edge), but many motherboard makers preferred to go their own ways. Asus for example, considered it to be technically better to have the processor in the middle of the board. In return Asus ships those board with a metal back plate, which makes the motherboard boxes surprisingly heavy. The heat sink is mounted through the board to this back plate, thus ensuring stability. Asus ensures that its P4T motherboard can still be mounted into any kind of computer case, but we have our doubts if all of the el-cheapo cases will indeed be able to host it. Figure 4.10 (*n*) depicts the placement of processor on motherboard.

Fig. 4.10 (*n*) Processor placement (Courtesy TOM's Hardware Guide)

Gigabyte's beautifully blue GA-8TX is placing the Socket423 where Intel wants it, but it ships the board with heat sink mounts that are fixed to the motherboard with some plastic nipples.

Future Pentium 4 processors that run at 1.6GHz and more require their heat sinks to be grounded (connected to V_{SS}) to avoid EMI. To ensure this, board makers have to place ground pads around the mounting holes of the heat sink, as we can see in this picture.

Price versus performance

It is logical to assume that one development trend for the P4 processors will be toward higher clock frequencies. Increase in clock frequency for the P4 produce a proportionately greater increase in performance than increases in clock frequency for the PIII. Some critics argue (Mihocka, 2001) that the P4 is less efficient then the PIII in terms of clock cycles to complete various benchmarks. Intel claims that the P4 was designed to operate at higher clock frequencies and compares the P4 to the PIII at dissimilar clock frequencies.

Future versions of the P4 will have better price versus performance than current versions due to their higher clock rate. When a processor has been designed to run at a higher clock frequency the cost of increasing the clock frequency is low compared to the increased cost of a new micro architecture.

Further performance gains will come from the software side. Software developers can take advantage of the P4 architecture by using NetBurst optimized libraries and optimizing compilers. Microsoft Visual C++ with the processor pack, Intel C++ 5.0 and Intel FORTRAN 5.0 support the P4 architecture.

One enhancement to reduce the cost of a complete P4 system is using the SDRAM memory instead of the RDRAM memory. The new Intel i845 chipset allows the use of SDRAM or DDRAM. SDRAM support was expected in the second half of 2001 and DDRAM is expected to be supported by first quarter of 2002 (Altavilla, 2001).

Overclocking

Many of the overclockers of this world were afraid that Pentium 4's quad-pumped 100MHz bus would make bus overclocking of this processor as difficult and restrictive as with Athlon and its dual-pumped 133/100MHz-bus. Surprisingly we can see that the Pentium 4 is as overclockable as Intel's other processors. We can imagine that the multiplier of official Pentium 4 processors will be locked, but with a good P4-motherboard we would not have any problems in overclocking the bus.

The jumperless-mode of the Asus P4T-motherboard could be managed to let 2-different P4 processors run up to 125MHz bus clock. Even the 1.4GHz Pentium4 was overclocked to 14 × 115MHz = 1610MHz as well as the 1.5GHz Pentium 4 overclocked to 16 × 108MHz = 1728MHz in the benchmark results. It requires only raising the voltage from 1.7V to 1.8V. There was no thermal issue, as Pentium 4 heat sinks are already designed for much higher heat dissipation than what current Pentium 4 processors are actually able to produce.

Benchmark

After all what we have learnt in the architectural part of this article, we should expect Pentium 4 to show excellent performance in all benchmarks that are heavily integer based and the ones that take great advantage of the new high-speed bus between processor, system and memory. SSE2-optimized software should obviously run very fast on Pentium 4 as well. Although Intel claims that Pentium 4 has the worlds best floating-point performance, in reality the normal FPU of Pentium 4 is hardly even able to live up to Pentium III standards. Only floating-point applications that use SSE2 could possibly support Intel's bold claim. Today's standard software is obviously not yet SSE2-optimized, so that standard FPU-intensive software will probably run rather slow on Pentium 4 systems.

This benchmark discussion starts with a big bang. Only a heavily overclocked Pentium 4 at 1728MHz is able to surpass AMD's 'older' Athlon with SDRAM and 100MHz FSB. The 'newer' Athlon with DDR-SDRAM and 133MHz FSB is simply untouchable in this benchmark right now. Even Pentium III is able to leave both official Pentium 4 models at 1.4 and 1.5GHz behind!

Whatever it may be, Pentium 4 cannot handle current office software very well. One of the reasons may be its small 8KB L1 data cache, but that's certainly not all. In some way that Intel has shown guts in releasing a processor, they call, 'the world's best performing microprocessor' although Intel is well aware of the fact that one of the major standard benchmarks shows this processor in a pretty bad light. What does this result mean to us? Not much really. Office applications have not been a challenge to any of the processors that were released.

Clock for clock

The P4-sample provided by Intel was not multiplier-locked, so that it could run at 10 × 100MHz = 1GHz without any need to alter the system bus clock.

Clock-for-clock Pentium 4 gets beaten by Pentium III in all but the Quake 3 and the MPEG4 benchmarks! If the Pentium III at 1.13GHz were available and Intel would have launched Pentium 4 at 1.3GHz, the difference between the 2-would have been hardly noticeable. Nevertheless, please do not forget that Pentium III is much pretty at the end of its days. It cannot get much faster. Pentium 4 however has a long way to go. Clock speed will soon equalize this nasty defeat.

What was done with Pentium III can obviously be done with Athlon as well? In this run an under-clocked Pentium 4 at 1.2GHz (12 × 100MHz) had to show how it performs against AMD's current flagship. Compared to Athlon clock-by-clock Pentium 4 is looking even worse. It is barely able to beat Athlon in MPEG4 and the lead in Quake 3 Arena Demo1 is very thin as well. Clock speed and compiler optimization is the only way Pentium 4 will keep the Athlon at arms length.

Impact of the memory speed

Finally let us see the excellent data thruput of Pentium 4's system bus and memory interface that makes it perform so well in Quake 3 and the MPEG4-compression. The Pentium 4 ran twice at 1.6GHz and 100MHz system bus, once with the RDRAM clocked at 400MHz, once clocked at only 300MHz, reducing its bandwidth from 3,200MB/s down to 2,400 MB/s. Then it was 'overclocked' at 1.4GHz Pentium 4 to 1.610GHz by increasing the system bus clock to 115MHz and thus the Rambus block to 460MHz, increasing the memory bandwidth to 3,680MB/s.

We can see that the memory speed does indeed have a major impact on all the benchmark results except of the 3D Studio Max scores. In some cases the difference between the slowest and the fastest score is more than 10%! This proves clearly that Pentium 4 lives from the high memory bandwidth that RDRAM is finally able to deliver. Keep that in mind in case someone wants to sell us PC600 RDRAM!

Physical specifications of P4

- 217 square mm.
- 0.18 micron manufacturing process.
- 42 million transistors.
- 423 pins.

Size matters

As we mentioned earlier, the Coppermine Pentium III has 28-million transistors. The Willamette Pentium 4 has 42-million transistors and a larger die size, 217 square mm as compared to the PIII's 104 square mm die (95 square mm for C-0 PIII processors).

In the case of die size, bigger is not better. The Pentium 4 is roughly twice as large as the Pentium III. The P4's larger size cuts the number of chips Intel can fit on a single wafer in half. Larger die sizes also lowers yields since a defect is more likely to occur in a chip with a larger area.

Even though the P4 initially used the 0.18-micron manufacturing process, we most likely could not see volume P4 production until Intel moved to the 0.13-micron Northwood P4. Intel still can't keep up with PIII demand, and dedicating manufacturing facilities to the larger Pentium 4 chips only result in lower chip production.

Intel also plans on switching from 200-mm diameter wafers to larger 300-mm wafers. This should double the number of chips that can fit on each wafer.

Few comments

We have learnt that Pentium 4 has got a rather exciting and interesting brand new design that comes with a whole lot of potential. However, the benchmark results might seem a bit sobering to the majority of us. Whatever Pentium 4 is right now, it is certainly not the greatest and best performing processor in the world. It is not a bad performer as well.

Intel seems very determined to make Pentium 4 a success and we feel that it will succeed. The implementation of SSE2-instructions into future software as well as the uses of code-optimizing compilers for Pentium 4 will make sure that Pentium 4 will be standing in a much better light very soon. However, we believe that Pentium 4's strongest side is its clock speed potential. Just realize that I overclocked this brand new 1.5GHz Pentium 4 to beyond 1.7GHz without any problems. I do not care whatever the latest roadmap of Intel may be saying. I am certain that Intel will deliver very fast Pentium 4 processors very soon. Intel has finally won back the ability to make AMD's life a lot harder.

What do we think of the components around Pentium 4? We must admit it, but with Pentium 4 Rambus is finally able to deliver for the 1st time. If we look at Pentium 4's design closely enough, we can see that it is engineered to live with RDRAM in perfect harmony. The memory benchmarks from above show that Pentium 4 really requires the 3,200 MB/s of data bandwidth supplied by the two Rambus channels. We are forced to doubt that it will perform as well with DDR-SDRAM, unless 2-channels will be used. One DDR-SDRAM channel offers 'only' 2,122 MB/s of data bandwidth, which might make quite a difference with Pentium 4.

The new power supply and housing requirements for Pentium 4 might be a nuisance to some, but they make perfect sense. I hope that AMD will follow Intel's example and come up with some solid new specifications for Athlon-platforms as well.

Pentium 4 at 1.4GHz goes for $644, Pentium 4 at 1.5GHz costs $819. It is not exactly a bargain, but who really cares about price if it really is all about style? Please do not forget to read the **Intel roadmap update** from October 13, 2000 to understand Intel's strategy with Pentium 4.

Our code has now passed the system bus, L1 and L2-cache, so that it is finally time to enter the execution path of Pentium 4. We remember that Pentium 4 is not using an L1 instruction cache, but a much niftier thing instead. Let us 1st explain what is bad about an L1 instruction cache.

With Pentium III or Athlon, both having an L1 instruction cache, code is fetched by this cache and stored until it is about time to enter the execution path. This is done by code entering the decoder unit, which e.g. in case of Athlon consists of 3 'direct path' and 3 'vector path' decoders, which alternatively produce the 'OPs' (as explained above) that can get executed by the execution units of the processor. This situation has a few glitches. First of all, some 80x86-instructions are rather complex, taking a lot of time to be decoded by the slow or 'vector path' decoders. In the worst case all decoder units are busy decoding complex instructions, thus stalling the execution pipeline of the processor. Another problem is the fact that 80x86-instructions that are supposed to be executed repeatedly (e.g. in small loops) need to be decoded each time they enter the execution path, thus wasting a lot of time. Software branches are another wasteful situation for a processor with L1 instruction cache that starts its pipeline at the decoder level.

Pentium 4's fancy Execution Trace Cache does not suffer from the above-described problems. Once we understood it, the idea of the trace cache is actually rather simple, but it takes quite a bit more silicon resources and design skill to replace the good old L1 instruction cache with something like Pentium 4's trace cache. Basically, the 'Execution Trace Cache' is nothing but a L1 instruction cache that lies BEHIND the decoders. Obviously it's quite bit more complex, but once we understood this basic fact we start to realize the benefits of the trace cache.

4.11 ITANIUM

The itanius is a new processor in the family of Pentium μP. Intel and Hewlett Packard designed its architecture jointly. The itanium is

- Big (powerful processor),
- Complex (it has many functional units and many instructions),
- Expensive (more costly as it is more powerful processor),
- Provides massive internal resources.

The Itanium is cutting edge processor design. It uses a new form of instruction called Explicit Parallel Instruction Computing (EPIC). This brings in unseen level of parallel processing in it. The Itanium gains performance over older processor architecture in many ways as described below:

- Through 64-bit address,
- EPIC (Explict Parallel Instruction Computing) architecutre,
- Wide parallel emulation core,
- 25.4 million transistors
- 0.18 micron CMOS process

- 6 metal layers
- C4 (flip-chip) assembly technology
- 1012-pad organic land grid array
- 733MHz and 800MHz initial release clock speeds
- 2 bundle dispersal windows
- 3 instructions per bundle
- 9 function unit slots
- 2 integer slots
- 2 floating point slots
- 2 memory prediction slots
- 3 branch prediction slots
- Maximum of 6 instructions issued per cycle
- 2 extended and double precision FMACs (Floating-point Multiply Add Calculators)
- 4 double or single precision operations per clock maximum
- 3.2 GFLOPS of peak double precision floating point performance at 800MHz
- 2 additional single precision FMACs
- 4 single precision operations per clock maximum
- 6.4 GFLOPS of peak single precision floating point performance total at 800MHz
- 4 single cycle integer ALUs
- 4 MMX units
- 3 branch units

The Itanium will be fully 80x86 compatible in hardware. Applications and operating systems should be able to run without any changes. A decoder internal to the μP decodes 80x86 instructions into EPIC instructions, then dynamically schedules them to run with increased parallelism. While the Itanium will be compatible with 80x86 software, it is not expected to do it quickly. Compatibility is being included to ease the transition from 80x86 code to EPIC code.

According to Intel, the EPIC architecture was designed with about 25 years of headroom for future development in mind. The 1st Itanium-family processor, originally codenamed Merced, has been delayed but looked like it would hit in the 1st half of 2001. McKinley announced to follow the original Itanium and would integrate its L3 cache onto the μP die. McKinley will arrive in the 2nd half of 2002. Madison may also arrive in 2002 on a 0.13 micron process. Deerfield will arrive not long after, also on a 0.13 micron, at a lower price and performance level but with more performance for the dollar than Madison.

Itanium roadmap

In addition to pure clock speed boosts, future Itanium μPs are sure to have more functional units (FMACs, ALUs, etc.), more cache, and a wider instruction dispersal window than the original Itanium. A deeper pipeline may also be in the works.

The Itanium will be Intel's 1st bid into the 64-bit enterprise computing market. Primary competition will come from Sun, IBM, HP, Compaq, and SGI, all of whom have their own 64-bit processor solutions. In addition to the promise of performance, Intel offers their ability to standardize the 64-bit market onto one processor family and instruction set. IBM, HP, Compaq, and SGI will all offer Itanium solutions alongside their own. Sun may remain as the main holdout, and we expect a large and long battle between Sun and Intel. AMD may also enter the market at a later date. For now, AMD is beginning to lay the groundwork of their move to the 64-bit arena with their announcement of 80x86-64, a 64-bit extension to 80x86.

Off-die in 2 or 4-chips

- 2MB or 4MB
- Runs at core clock
- 4-way set associative
- Up to 294.8 million transistors
- 128-bit bus
- 21+ cycle latency

On-die

- 96KB of full-speed cache
- 6-way set associative
- 256-bit bus
- 6-cycle + latency

On-die

- 16KB instruction cache
- 4-way set associative
- 16KB integer only data cache
- 4-way set associative
- 2-cycle + latency
- ECC or parity checking on most caches and busses inside and near processor
- Data poisoning to recover from bad data without a reboot
- Error logging
- Designed to correct, recover from, or contain machine errors
- Hardware decoder turns 80x86 instructions into EPIC instructions
- Dynamic scheduler optimizes 80x86 for EPIC micro-architecture
- Shared cache
- Shared execution core

The Itanium has a complex, bleeding edge, forward looking processor family that holds promise for huge gains in processing power. The processor uses the entirely new EPIC architecture that has the potential to deliver large improvements in processor parallelism. The EPIC processors are capable of addressing a 64-bit memory space. In comparison, 32-bit 80x86 processors access a relatively small 32-bit address space, or up to 4GB of memory. The concept of a 64-bit memory space is immense and therefore difficult to get your mind around. At some point in the future, a 64-bit memory space may be a limiting factor to performance, but I believe that day is well off in the future. This gives the Itanium the memory addressing ability needed to meet current and foreseeable future high-end processing needs. Through bank switching, 80x86 processors, such as the Intel Pentium III Xeon and the AMD Athlon, can address more than 4GB of memory. Unfortunately, there is hardware and software overhead to bank switching that harms performance and increases complexity.

The 1st generation of Itanium systems, using the 460GX chipset, will be expandable with up to 64GB of memory. Generations beyond that will be able to take more memory. Higher end Itanium systems designed by the likes of SGI, IBM and HP should eventually be able to take far more than 64GB. While it may be hard to imagine 4GB or even 64GB of memory being a bottleneck to performance, when you consider SGI has mentioned plans to eventually build machines using 512 Itanium processors accessing more than a terabyte of data in main memory, 64GB of memory, let alone 4GB, begins to look rather small.

When we have a massive amount of data to process we need to access and process in real-time, a RAID attached by FCAL may not cut it. The EPIC architecture and Itanium are intended in part for just such situations, where in order to get acceptable performance, a system needs more than 4GB or 64GB of main memory.

The Itanium uses the Explicitly Parallel Instruction Computing (EPIC) architecture. The EPIC is a description of the Itanium architecture, like CISC describes the Athlon and Pentium, RISC describes the PowerPC, and VLIW describes Transmeta's Crusoe. The EPIC architecture uses complex instruction wording that, in addition to the basic instruction, contains information on how to run the instruction in parallel with other instructions.

The EPIC instructions are put together by the compiler into a threesome called a 'bundle'. Bundles instructions are sent to the μP together. The 'bundles', or their parts, are put together in an 'instruction group' with other instructions. The rule for instructions to share a group, quoted from an Intel presentation, is that they do 'not have read-after-write or write-after-write dependencies between them and may execute in parallel'. Translation: instructions in a bundle do not affect each other with the data they are working on, so they can run together without getting in each other's way. There is no limit to the size of an instruction group, and an instruction group can begin or end in the middle of a bundle.

The instructions are actually bundled and grouped together when software is compiled. This simplifies the process of running multiple instructions at once on an Itanium μP, allowing it to make greater use of multiple execution units without having to rely on complex on-die logic to determine what operations can run in parallel. The Itanium will still use on-die logic to improve upon instruction level parallelism, but EPIC instructions, at the minimum, provide a parallel blueprint for the Itanium processor. Of course, for this reason, compiler technology and programming algorithms will have a massive impact on Itanium performance.

In addition to grouping operations into instructions, the compiler handles several other important tasks that improve efficiency, parallelism and speed. CISC puts most of the burden of scheduling instructions onto the μP hardware. RISC gives some of this responsibility to the compiler. VLIW removes even more load of the μP and gives more importance to the compiler. The EPIC takes things much, much further.

The compiler adds branch hints, register stack and rotation, data and control speculation, and memory hints into EPIC instructions. The compiler also uses predication. We will have more on some of these features later on, but at this point, it is enough to know that these EPIC instructions and features contribute to an increase in parallelism. These extra features are why we believe, in addition to marketing reasons, Intel coined a new term for their style of architecture, EPIC, instead of calling it VLIW (Very Long Instruction Word), which it closely resembles.

Predication is a compiling technique used in the Itanium that optimizes or removes branching code by working it so that much of the code runs in parallel. By designing software and compilers to rework branches into parallel code with fewer or no branches, and by running this code on a 'wide' processor that can process this code in parallel, which the Itanium is intended to be able to do, the number of cycles it takes to complete a task drops.

What predication does to minimize the time it takes to run if-then-else situation and uses processor width to run both the 'then' and 'else' in parallel. When the 'if' branch is determined, the incorrect branch's result is discarded. By removing branches and making code more parallel, predication reduces the number of cycles it takes to complete a task while making better use of a wide processor.

Jerry Huck of HP gave a great metaphor for predication in a presentation back in 1997 at the Microprocessor Forum. Here is a reworking of his example. Imagine we walk into a bank. We will make either a deposit or withdrawal. The teller may predict that we will make a withdrawal as they know that we usually do, so they fill out the withdrawal form as we get in line. If we get to the front and make a withdrawal, all is well, but if we are there to make a deposit, the teller then has to fill out the deposit slip and the time it takes to complete the transaction increases.

With predication, the teller is ambidextrous and when we get in line, they fill out both a withdrawal and deposit slip, so that when we get to the front, no matter what task we intend on doing, the process will run without a hitch.

In the metaphor, predication is the teller's knowledge that they should fill out both the deposit and withdrawal forms before they know exactly what we want. The teller's ambidexterity, the ability to fill out both forms at once, is akin to the ability of an EPIC processor to run instructions in parallel. Predication removes or lessens the penalty of if-then and allows the if-then-else process to run with as few steps as possible.

Another side benefit of predication is that, by removing branches, there are less-branch mis-predicts. Branch mis-predicts necessitate that the pipeline be flushed, a very cycle expensive procedure, so by lessening the number of branches, predication can greatly reduce wasted processor time.

At the heart of the Itanium lies a parallel execution core. Itanium family processors are intended to be very 'wide'. That is to say, they are able to run multiple instructions and operations at once in parallel. The Itanium will also be somewhat deep, with a ten-stage pipeline. The 1st-generation Itanium processor issues up to 6-EPIC instructions in parallel every clock cycle. We can compare this to the Pentium III and Athlon architectures that are able to issue up to 3- 80×86 instructions per clock.

The 6-issue (2-bundle) scheduler disperses instructions into 9-functional slots, 2-integer slots, 2-floating-point slots, 2-memory slots, and 3-branch slots, giving a total of 9-dispersal slots. This limits the number of each type of instruction that can be assigned in a single clock cycle. If an instruction or instructions cannot be executed because too many slots of one type are filled, the instructions are delayed until the next cycle. Proper compiler design should be able to handle most situations without overloading any type of functional slot.

Itanium block diagram

Backing up the Itanium's 6-issue scheduler have 11 execution units: 4-integer, 2-floating point, 3- branch, and 2-load/ store units. These help support the various EPIC instructions that can launch more than one operation in a single instruction, such as SIMD floating point operations. Combined with the EPIC instruction set, the Itanium can execute up to 20 operations in a single cycle when doing some floating point- intensive tasks.

The Itanium may not consistently run 20 operations per cycle, but the potential is there and proper coding and compiling should yield efficient usage of the μP. Either way, in concert with the EPIC instruction set and predication, this width will allow tasks to be completed in relatively few clock cycles. For more integer-oriented tasks, where there are few instructions with multiple operations, running eight operations per clock is the theoretical maximum.

The Itanium contains 4-pipelined FMAC units (Floating-point Multiply Add Calculator). The primary 2 are each capable of processing 2-single-precision, 2-double precision, or 2-double-extended-precision floating-point operations per clock that yields up to 3.2GFLOPS of highly precise floating point processing.

There are an additional 2-FMACs tuned for 3D applications. They are each capable of processing up to 2-single-precision floating-point operations per clock. That yields another 3.2GFLOPS of single-precision processing power. All together, the Itanium has a theoretical max of 6.4GLOPS of single-precision floating point processing power.

There are 4-pipelined ALUs (Arithmetic Logic Unit) in the original Itanium. Each can process one integer calculation per cycle. They can also process MMX type instructions. While the Itanium has the potential to be a massive floating-point powerhouse, its integer performance also has tremendous potential.

The Itanium will come with 128 floating point and 128 integer registers. When processing up to 20 operations in a single clock, the registers give plenty of room for data inside the processor. This reduces the chances of the execution of an instruction being delayed because data could not be held locally. This is especially important since the Itanium can process up to eight floating-point operations in a single clock. With the possibility of eight operations running in a single clock, having too few registers could be a serious bottleneck.

The registers also have the ability to rotate. Rotating registers allows the processor to perform an operation on multiple software accessible registers in turn. This increases µP pipeline utilization and efficiency when dealing with streams of data to process.

When a processor is waiting for data or instructions, time is wasted. The longer it takes for data and instructions to get to the µP, the worse it gets. When data and instructions are in cache, the processor can grab them much quicker than when having to go to slow main memory. Not only is cache latency much lower than DRAM latency, the bandwidth is much higher.

There are some trick programming techniques in use to keep often-used data and instructions in cache and they are not the kind of techniques we learn in our high school BASIC course. Still, the easiest way to keep data and instructions in cache is to have a lot of cache to keep them in. Intel knew that when they designed the Itanium.

The Itanium has 3-levels of cache. L1 and L2 are on-die while L3 is on cartridge. According to Intel, the L3 cache weighs in at 2MB or 4MB of 4-way set associative cache on 2 or 4-1MB chips. IDC reports that the L2 cache size is 96KB in size, and the L1 cache, which does not deal with floating point data, has a 16KB integer data and a 16KB instruction cache.

The 294.8 million transistors of (4MB) level 3-cache runs at the full processor speed, giving 12.8GBps of memory bandwidth at 800MHz. With 2MB or 4MB of L3 cache on the Itanium, the chances of the required data and instructions being in cache are quite good, bus traffic can be reduced, and performance increases. With six pipelines hungry for instructions and data, the Itanium needs all cache it can get.

In order to make caching even more effective, Intel uses data speculation and cache hints. Data speculation is caching and calling for data that may be needed or may be changed before it is needed, so that, in the case that the data is needed and it has not changed, the µP does not have to take a latency impact from calling for the data. The processor, with the help of compiled instructions, looks ahead, anticipates what info it may need, and then brings it to cache or into the processor. This helps hide memory latency. Cache hints are 2-bit markers for memory loads set by the compiler that help the µP find data in cache. This improves the speed of retrieving data from cache.

A major link in the food delivery system for the Itanium is the system bus. The Itanium will use a 2.1GBps multi-drop system bus to keep well fed with data and instructions. We expect it to have a 128-bit 133MHz bus. The memory subsystem and I/O will be determined by the chipset used. First generation systems should use dual-memory ported SDRAM giving 4.2GBps of memory bandwidth. Later generations will have the option to use DDR SDRAM or RDRAM. Eventually, Intel plans on moving server platforms to DDR II. 64bit, 66MHz PCI and AGP Pro (4x) should be common on Itanium motherboards and support will be included in Intel's 460GX chipset.

The 1st generation of Itanium processors was expected to come in the 1st half of 2001 at 733MHz and 800MHz. The 1st generation's clock speed may not be particularly quick, but Intel has several generations ahead of the Itanium already in the works that should increase performance. Intel claims they have plenty of clock headroom in the Itanium design and are aiming for a greater than 1GHz clock speed with their 2nd generation Itanium processor, McKinley, which will have the L3 cache on-die.

The Itanium was not designed for small systems, it is intended for 1 to 4000 processor workstations and servers. There are several Itanium features designed to help with hardware scalability, a full-μP-speed Level 2 bus, a large L3 cache, deferred-transaction support and flexible page sizes.

The full-μP-speed Level 3 bus provides quick communication between μPs. The large L2 cache reduces Intel-μP bus traffic by keeping data close to the μP that needs it. Deferred-transaction support can stop one μP from getting in the way of another. Flexible page sizes, from 4KB to 256MB, give the Itanium family the flexibility to access small amounts of memory in small chunks and massive amounts of memory in massive chunks without the overhead of smaller page sizes.

The 1st generation Itanium chipset, the 460GX, will support up to 4-processors, and OEMs will be able to build eight-way and larger systems. Successive generations of chipsets should be successively more scalable. The 3rd party solutions should also increase scalability.

The Itanium will have extensive error handling capabilities. It features ECC and parity error checking on most processor caches and busses. If a machine error occurs and a piece of data becomes corrupted, the ECC or parity checking will allow the machine to recognize the error, fix it if possible, or flag it as corrupted. The processor also has the capability to kill an application or thread that has experienced a machine error without having to reboot.

Chipset, OS, and system designers, which will include the likes of HP, IBM, Compaq, SGI, Microsoft and Intel, will bring out their own error handling and reliability processes that should further enhance Itanium-based server uptime to 99.9% and beyond.

The Itanium will be fully 80x86 compatible in hardware. Applications and operating systems should be able to run without any changes. A decoder internal to the μP decodes 80x86 instructions into EPIC instructions, than dynamically schedules them to run with increased parallelism. While the Itanium will be compatible with 80x86 software, it is not expected to do it quickly. Compatibility is being included to ease the transition from 80x86 code to EPIC code.

Analysts generally agree the new chip will shine on benchmarks and be popular in areas such as biosciences that require heavy weight numbers crunching. Unfortunately for Intel, the chip is arriving at a time when the entire industry is in a deep freeze. Corporate customers have drastically curtailed IT budgets and are less inclined to adopt or even test new technologies, a process that can consume thousands of dollars and hundreds of engineering hours. Though, the Itanium 2 processor marks a significant step in Intel's strategy to penetrate the upper echelons of the computing market, yet analysts and others are asking whether it will thrive in the current economic ice age.

Some Intel design teams advocated getting to 64-bit chips by merely stretching the existing 32-bit architecture, sources said, a relatively easy leap in terms of engineering. And as an added bonus, both chip families would be able to handle the same software. Intel performed this sort of stretching magic when going from 4-bit to 8-bit chips in the 1970s and moving from 16-bit to 32-bit chips in the early 1980s.

But for the 64-bit chips, the company decided to work with HP on building processors around the EPIC idea. A new software base would have to be created for EPIC chips, but the switch would also let Intel break free of many of the inherited twists in the **IA** architecture, the design behind all the other the Intel processors.

A quick glance at Intel's new server chip

Code name : McKinley, Originally expected : 2nd half of 2001, Number of transistors : 221 million, Size : 42mm^2, Speeds : 900MHz, 1GHz, Cache size: 3MB or 1.5MB level three cache, The **Successors** of Itanium 2 are : Madison and Deerfield (2003), Montecito (2004), Chivano (2006).

The future EPIC chips by the mid-1990s, known by the code name Merced, loomed as a danger to Sun Microsystems, Digital Equipment, IBM and other server makers who also made their own chips. With its mass manufacturing capabilities, Intel would be able to pump out low-cost, high-speed 64-bit chips faster than these individual manufacturers, analysts predicted at the time.

Making Merced, however, proved difficult in practice and, after several lengthy delays, the chip went from being a giant killer to a conceptually interesting test vehicle. The chip finally came out commercially in May 2001. Performance was relatively anemic, and sales were even worse. In the third quarter of 2001, less than 500 Itanium servers were shipped, according to IDC.

What's new about Itanium?

But although the 1st Itanium was a flop, newer versions could succeed. For one thing, Intel still wields a larger manufacturing base than any other chipmaker, giving the company an inherent advantage when it comes to lowering costs, accelerating speeds or developing multiple products simultaneously.

In the long run, even in the short run, our capabilities in manufacturing will keep us ahead, said enterprise-chip marketing chief Hambrick.

A 4-processor Itanium 2 server with 8GB of memory will cost around $41,000, less than competing Sun servers. In contrast, a Sun Fire 3800 with four 900MHz UltraSparc III chips with 4GB of memory goes for $133,000, according to Sun's Web site. Hambrick says that on benchmarks the Itanium 2 system will outperform this machine.

Lower prices, combined with improved performance, will certainly make Itanium 2 more attractive than its predecessor, said TimesTen's Shetler.

It looks more promising, Shetler said. 'With the recession, it becomes even more of a viable option.' Sun disagrees, saying Itanium will be the more expensive option.

'The migration costs are going to be very big, there is very little software available and it is an unproven architecture,' said Martin Chorich, a Sun spokesman. 'This is the 2nd time out with a demonstration vehicle.'

Another factor is that Intel has put tremendous amounts of energy and money into building a world around Itanium. The IA-64 fund invested approximately $250 million in start-ups to get them to develop Itanium applications and worked extensively with established hardware and software manufacturers to get them to make Itanium-based projects. Consulting and testing labs were set-up around the world so large corporate buyers could tune their applications for the new chip.

Intel will even build complete servers for manufacturers who cannot spare the engineers. This option is expected to be popular with manufacturers in developing nations, such as Russia's Kraftway, said Phil Brace, a marketing manager for Intel.

The company has also straightened out many of the kinks that held back the original Itanium. The photos of the core of the original Itanium looked like a 'collage' of chip components rather than a tightly integrated processor, Krewell said.

'They slapped the 1st one together to get it out the door after it slipped so badly,' Krewell said. Itanium 2 is vastly improved. It contains a 400MHz bus i.e., 128-bits wide, compared with the 64-bit wide bus on the original Itanium, a change that allows for greater data transfer rates. The level three cache, a reservoir of memory for rapid data access, goes up to 3MB.

The pipeline that serves as a kind of assembly line inside the processor, is also shorter. In addition, improvements have been made to the compiler, software that organizes and schedules how different resources on the chip will get deployed, Krewell said.

Programming of Pentium

It is time to stop programming using SASM and get to the real stuff: Pentium assembly language.

Notes ahead of time

The assembly language details presented are not specific to just the Pentium architecture. They are applicable to the 80486 and to the Pentium II and the Pentium Pro. Many details also work for the earlier Intel architectures.

We are only presenting a subset of the Pentium instruction set. The real instruction set is really large, and contains many instructions that no one really uses anymore. We do not cover those 'obsolete' instructions. We also do not cover the instructions that only the operating system would use.

- It is not a load/ store architecture.
- The instruction set is huge! We go over only a fraction of the instruction set.
- There are lots of restrictions on how instructions/operands are put together, but there is also an amazing amount of flexibility.

Registers

The Intel architectures as a set just do not have enough registers to satisfy most assembly language programmers. Still, the processors have been around for a LONG time, and they have a sufficient number of registers to do whatever is necessary. For our (mostly) general-purpose use, we get

32-bit	*16-bit*	*8-bit (MSBs of 16-bit)*	*8-bit (LSBs of 16-bit)*
EAX	AX	AH	AL
EBX	BX	BH	BL
ECX	CX	CH	CL
EDX	DX	DH	DL

and

EBP	BP
ESI	SI
EDI	DI
ESP	SP

There are a few more, but we would not use or discuss them. They are only used for memory accessibility in the segmented memory model.

Use of registers

As an operand, just use the name (upper case and lower case both work interchangeably)

- EBP is a frame pointer
- ESP is a stack pointer

This is the only architecture that we know of where the programmer can designate part of a register as an operand. On all other machines, the whole register is designated and used.

Status register

Many bits used for controlling the action of the processor and setting state are in the register called EFLAGS. This register contains the condition codes

OF	Overflow flag
SF	Sign flag
ZF	Zero flag
PF	Parity flag
CF	Carry flag

The settings of these flags are checked in conditional control instructions. Many instructions set one or more of the flags. (Note that we only utilized SF and ZF in SASM.)

There are many other bits in the EFLAGS register:

The use of the EFLAGS register is implied (rather than explicit) in instructions.

Accessing memory

There are 2 memory models supported in the Pentium architecture. (Actually it is the 486 and more recent models that support 2 models).

In both models, memory is accessed using an address. It is the way that addresses are formed (within the processor) that differs in the 2 models.

Flat memory model

This is the memory model that we use and the memory model that every other manufacturer's processors also use.

Segmented memory model

Different parts of a program are assumed to be in their own, set-aside portions of memory. These portions are called segments.

An address is formed from 2 pieces: a segment location and an offset within a segment.

Note that each of these pieces can be shorter (contain fewer bits) than a whole address. This is much of the reason that Intel chose this forn of memory model for its earliest single-chip processors.

There are segments for

- code
- data
- stack
- extra

Which segment contains something can be implied by the memory access involved. An instruction fetch will always be looking in the code segment. A push instruction always accesses the stack segment etc.

Addressing modes

Some would say that the Intel architectures only support 1 addressing mode. It looks (something like) this

Effective address = base reg + (index reg × scaling factor) + displacement

where,

- Base reg is EAX, EBX, ECX, EDX or ESP or EBP
- Index reg is EDI or ESI
- Scaling factor is 1, 2, 4 or 8

The syntax of using this (very general) addressing mode will vary from system to system. It depends on the preprocessor and the syntax accepted by the assembler.

For our implementation, an operand within an instruction that uses this addressing mode could look like

```
[EAX][EDI*2 + 80]
```

The effective address calculated will be the contents of register EDI multiplied 2 times, added to the constant 80, added to the contents of register EAX.

There are extremely few times where a high-level language compiler can utilize such a complex addressing mode. It is much more likely that simplified versions of this mode will be used.

Register mode

The operand is in a register. The effective address is the register.

Example

MOV EAX, ECX

Both operands use register mode. The contents of register ECX is copied to register EAX.

Immediate mode

The operand is in the instruction. The effective address is within the instruction.

Example

MOV EAX, 26

The second operand uses immediate mode. Within the instruction is the operand. It is copied to register EAX

Register direct mode

The effective address is in a register.

Example

MOV EAX, [ESP]

The second operand uses register direct mode. The contents of register ESP is the effective address. The contents of memory at the effective address are copied into register EAX.

Direct mode

The effective address is in the instruction.

Example

MOV EAX, var_name

The second operand uses direct mode. The instruction contains the effective address. The contents of memory at the effective address are copied into register EAX.

Base displacement mode

The effective address is the sum of a constant and the contents of a register.

Example

MOV EAX, [ESP + 4]

The second operand uses base displacement mode. The instruction contains a constant. That constant is added to the contents of register ESP to form an effective address. The contents of memory at the effective address are copied into register EAX.

Base-indexed mode (Intel's name)

The effective address is the sum of the contents of two registers.

Example

MOV EAX, [ESP][ESI]

The contents of registers ESP and ESI are added to form an effective address. The contents of memory at the effective address are copied into register EAX.

Note that there are restrictions on the combinations of registers that can be used in this addressing mode.

PC relative mode

The effective address is the sum of the contents of the PC and a constant contained within the instruction.

Example

JMP a_label

The contents of the program counter is added to an offset that is within the machine code for the instruction. The resulting sum is placed back into the program counter. Note that from the assembly language it is not clear that a PC relative addressing mode is used. It is the assembler that generates the offset to place in the instruction.

Instruction set

Many (most) of the instructions have exactly 2 operands. If there are 2 operands, then one of them will be required to use register mode and the other will have no restrictions on its addressing mode.

There are most often ways of specifying the same instruction for 8, 16 or 32-bit operands. We left out the 16-bit ones to reduce presentation of the instruction set. Note that on a 32-bit machine, with newly written code, the 16-bit form will never be used.

Operand specifications

reg -	register mode operand, 32-bit register
reg8 -	register mode operand, 8-bit register
r/m -	general addressing mode, 32-bit
r/m8 -	general addressing mode, 8-bit
immed -	32-bit immediate is in the instruction
immed8 -	8-bit immediate is in the instruction
m -	symbol (label) in the instruction is the effective address

Data movement

```
MOV     reg, r/m     ; copy data
        r/m, reg
```

```
                reg, immed
                r/m, immed
        MOVSX   reg, r/m8       ; sign extend and copy data
        MOVZX   reg, r/m8       ; zero extend and copy data
        LEA     reg, m          ; get effective address
```

Examples

```
MOV EAX, 23        ; places 32-bit 2's complement immediate 23 into register EAX
MOVSX ECX, AL      ; sign extends the 8-bit quantity in register AL to 32 bits, and places
                     it in ECX
MOV [ESP], -1      ; places value -1 into memory, address given by contents of ESP
LEA EBX, loop_top  ; put the address assigned (by the assembler) to label loop_top into
                     register EBX
```

Integer arithmetic

```
ADD    reg, r/m       ; two's complement addition
       r/m, reg
       reg, immed
       r/m, immed
INC    reg            ; add 1 to operand
       r/m
SUB    reg, r/m       ; two's complement subtraction
       r/m, reg
       reg, immed
       r/m, immed
DEC    reg            ; subtract 1 from operand
       r/m
NEG    r/m            ; get additive inverse of operand
MUL    EAX, r/m       ; unsigned multiplication  (EDX||EAX ← EAX*r/m)
IMUL   r/m            ; 2's comp. Multiplication (EDX||EAX ← EAX*r/m)
       reg, r/m       ; reg ← reg*r/m
       reg, immed     ; reg ← reg*immed
DIV    r/m            ; unsigned division (does EDX||EAX/r/m, EAX ← quotient, EDX ←
                        remainder)
IDIV   r/m            ; 2's complement division (does EDX||EAX/r/m, EAX ← quotient,
                        EDX ← remainder
```

```
CMP   reg, r/m          ; sets EFLAGS based on
      r/m, immed        ; second operand - first operand
      r/m8, immed8
      r/m, immed8       ; sign extends immed8 before subtract
```

Examples

```
NEG [EAX + 4]   ; takes double-word at address EAX+4 and finds its additive inverse, then
                ; places the additive inverse back at that address the instruction should
                ; probably be NEG dword ptr [EAX + 4]
INC ECX         ; adds one to contents of register ECX, and result goes back to ECX
```

Logical instructions

```
NOT   r/m               ; logical NOT
AND   reg, r/m          ; logical AND
      reg8, r/m8
      r/m, reg
      r/m8, reg8
      r/m, immed
      r/m8, immed8
OR    reg, r/m          ; logical OR
      reg8, r/m8
      r/m, reg
      r/m8, reg8
      r/m, immed
      r/m8, immed8
XOR   reg, r/m          ; logical exclusive OR
      reg8, r/m8
      r/m, reg
      r/m8, reg8
      r/m, immed
      r/m8, immed8
TEST  r/m, reg          ; logical and to set EFLAGS
      r/m8, reg8
      r/m, immed
      r/m8, immed8
```

Examples

```
AND EDX, 00330000H    ; logical and of contents of register EDX (bitwise) with 0x00330000,
                      ; result goes back to EDX
```

Floating-point arithmetic

Since the newer architectures have room for floating point hardware on chip, Intel defined a simple-to-implement extension to the architecture to do floating-point arithmetic. In their usual zeal, they have included MANY instructions to do floating-point operations.

The mechanism is simple. A set of 8-registers is organized and maintained (by hardware) as a stack of floating point values. ST refers to the stack top. ST(1) refers to the register within the stack that is next to ST. ST and ST(0) are synonyms.

There are separate instructions to test and compare the values of floating point variables.

```
FINIT                  ; initialize the FPU
FLD     m32            ; load floating point value
        m64
        ST(i)
FLDZ                   ; load floating point value 0.0
FST     m32            ; store floating point value
        m64
        ST(i)
FSTP    m32            ; store floating point value and pop ST
        m64            ;
        ST(i)
FADD    m32            ; floating point addition
        m64
        ST, ST(i)
        ST(i), ST
FADDP   ST(i), ST      ; floating point addition and pop ST
```

I/O

The only instructions that actually allow the reading and writing of I/O devices are privileged. The OS must handle these things. But, in writing programs that do something useful, we need input and output. Therefore, there are some simple macros defined to help us do I/O.

These are used just like instructions.

```
put_ch  r/m    ; print character in the least significant byte of 32-bit operand
get_ch  r/m    ; character will be in AL
put_str m      ; print null terminated string given by label m
```

Control instructions

These are the same control instructions that all started with the character 'b' in SASM.

JMP	m	; unconditional jump
JG	m	; jump if greater than 0
JGE	m	; jump if greater than or equal to 0
JL	m	; jump if less than 0
JLE	m	; jump if less than or equal to 0

REVIEW QUESTIONS AND ANSWERS

Q. 1. What is DDR memory?

Ans. DDR memory, or Double Data Rate memory, does exactly what it sounds like—it doubles the rate of speed at which standard SDRAM can process data. That means DDR is roughly twice as fast as standard SDRAM.

Q. 2. How many address and data lines are available in Pentium?

Ans. 32 = address lines and 64 = data lines.

Q. 3. What is Moore's law?

Ans. Number of transistor per chip doubles every 2-years.

Q. 4. Why it was called Pentium?

Ans. To signify the 5th generation μP, the word Pentium was coined using the Greek word Pente, meaning five and the tail of the word 'ium' in the periodic table for better sounding.

Q. 5. What does the thermal model of the Pentium do?

Ans. The thermal model included many feature to dissipate heat from its mother board along with

- Sophisticated heat sink,
- Casing vent,
- Cooling fans.

Q. 6. How many general-purpose registers are available in the Pentium?

Ans. It has only 8-general-purpose registers including the stack pointer that is not considered as the general-purpose register.

Q. 7. What is superscalar technology?

Ans. The superscalar technology stands for the multiple execution units working in parallel.

Q. 8. How many integer and floating-point units the Pentium contains?

Ans. The Pentium contains 2-integer and 1-floatig-point units.

Q. 9. Why the Pentium has 2-integer units?

Ans. The 2-integer units of the Pentium are known as U and V-pipes. The V-pipe handles only the simple instruction such as load/ store etc. that does not require micro-code support. The U-pipe, on the other hand, accepts all legitimate instructions.

Q. 10. Who decides which pipe will handle which instruction?

Ans. The decode-1 stage decides whether 2-instructions in 2-pipelines can be executed simultaneously or not. If an instruction in the V-pipe does not qualify as a simple instruction, or the μP determines some dependency and prevents V-pipe instruction being paired with U-pipe. The V-pipe instruction then transferred to U-pipe. The V-pipe remains empty unless decode-1 stage determines that the 2-instructions in U and V-pipe can be paired.

Q. 11. What is register starvation in Pentium?

Ans. The Pentium has only 8-general purpose registers. This is too less and register contention usually occurs in the form of RAW (read after write) and WAW (write after write).

Q. 12. What is RAW in Pentium?

Ans. The RAW contention of register occurs when U-pipe writes result in the register that will be read by the instruction in the V-pipe. Since, U- and V-pipes executes in parallel, the instruction in the U-pipe would not be able to make the result available to the V-pipe when V-pipe access the register. When decode-1 stage of the Pentium detects this condition, it prevents 2-instructions from pairing, instead they must be accessed serially within U-pipe.

The RAW can still occur for 2-instructions in U-pipe in case when the 2nd instruction will stall that waiting for the register to make available the result stored by the 1st instruction.

Q. 13. What problem occurs due to RAW in Pentium?

Ans. This creates 2-problems

- Inability of the Pentium to pair integer instructions, and
- Pipeline bubble in U-pipe.

Q. 14. How WAW occurs in Pentium?

Ans. The WAW contention occurs due to limited number of registers in the Pentium. Perhaps, this is the biggest drawback of the Pentium. When an instruction tries to write to the same register where the previous instruction too was written, the WAW contention occurs. If the Pentium identifies a WAW with the 2-instructions in the decode-1 stage of the integer pipeline, it does not allow the 2-instructions to be paired, even if other pairing requirements have been met. Instead both instructions must execute serially in the U-pipe.

Q. 15. What is the Intel Pentium 4 processor?

Ans. The Pentium 4 processor is Intel's highest-performing desktop processor. Containing the fastest and smallest (60-nanometers) CMOS transistors in the mass production and many advanced memory, graphics and processing capabilities, it maximizes the performance of current and future Web, productivity and multimedia applications.

Q. 16. At what speeds are boxed Pentium 4 processors available?

Ans. Pentium 4 processors are currently available at operating frequencies from 1.30GHz to 2.53GHz.

Q. 17. Who is the Pentium 4 processor designed for?

Ans. Pentium 4 processors are designed for mainstream to high-end performance and are ideally suited for desktop users who want to be on the leading edge of technology, both today and tomorrow. Professionals, PC gamers, enthusiasts and entry-level workstation consumers alike will benefit from systems equipped with Pentium 4 processors.

Q. 18. What new features does the Intel Pentium 4 processor offer?

Ans. The Pentium 4 processor offers Intel NetBurst micro-architecture, including hyper-pipelined technology, rapid-execution engine and 533MHz system bus. The Pentium 4 processor also offers such newly improved features as execution trace cache, advanced transfer cache, enhanced floating point and multimedia unit, streaming SIMD extensions 2 and advanced dynamic execution.

Q. 19. What makes the Pentium 4 processor different from the Pentium® III processor?

Ans. The Pentium 4 processor is based on Intel NetBurst micro-architecture. The Pentium III processor is based on Intel's P6 micro-architecture. The Pentium 4 processor has become the highest volume shipping microprocessor in the world with the fastest and smallest (60-nanometers) CMOS transistors in mass production.

Q. 20. What is the difference between Pentium 4 processors based on 0.18 micron technology and on 0.13 micron technology?

Ans. Micron technology refers to the manufacturing process of a microprocessor. Utilizing Intel's 0.13 micron manufacturing process, the new Pentium 4 processor increases the performance scalability of the Intel NetBurst micro-architecture by delivering higher clock speeds with lower thermal power and doubling the Level 2 cache to 512K. 0.13 micron technology uses the fastest circuits in the industry, ensuring further performance and speed improvements for NetBurst micro-architecture in the future.

Q. 21. What is the advantage of the 400MHz bus system?

Ans. The 400 MHz system bus provides a 3.2 gigabyte per second transfer speed between the Pentium® 4 processor and the memory controller and is the highest bandwidth desktop system bus available, increasing overall CPU performance and raising the ceiling for clock speeds.

Q. 22. What is Streaming SIMD Extensions 2 (SSE2)?

Ans. Another key change in the Pentium 4 is the integration of 76 new SIMD instructions and improvements to 68 integer SIMD instructions (together called Streaming SIMD Extensions 2 or SSE2). Like the SSE instructions in the Pentium III, the new SIMD instructions target CPU-intensive encoding and decoding operations - streaming video, speech, 3D rendering, and other multimedia procedures.

Q. 23. What is hyper pipelined technology?

Ans. This is a 20-stage pipeline compared to the 10-stage pipeline in the PIII.

Q. 24. What is NetBurst technology?

Ans. NetBurst incorporates a number of elements that increase the processor's ability to handle instructions more efficiently. NetBurst also features a 400MHz system bus, which means the processor communicates more quickly with other system components, plus additional instructions for multimedia called Streaming SIMD (single instruction multiple data) Extension 2 (SSE2). SSE2 builds on the original SSE technology by adding 144 new instructions, including support for 128bit arithmetic.

Q. 25. What's new about the Intel® Pentium® 4 processor?

Ans. The Intel® Pentium® 4 processor with Intel® NetBurst micro-architecture, currently available at 1.30, 1.40, and 1.50GHz, maximizes the performance of cutting-edge technologies such as digital video and online gaming. A new Hyper Pipelined Technology doubles the pipeline depth to 20 stages, significantly increasing processor performance and frequency capability. A Rapid Execution Engine pushes the processor's Arithmetic Logic Units to twice the core frequency, resulting in higher execution throughput and reduced latency of execution. The 400MHz System Bus, the improved Advanced Dynamic Execution, and an improved floating point pushes data efficiently through the pipeline for lifelike video and 3D graphics. Along with Streaming SIMD Extensions 2 (SSE2) that extends MMX™ technology and SSE technology, and the addition of 144 new instructions, the Pentium® 4 processor gives us the advanced technology to get the most out of today's and tomorrow's Internet.

Q. 26. Why would we want to buy a high-performance PC based on a high-performance Pentium® 4 processor?

Ans. A high-performance Pentium® 4 processor-based system provides an extremely powerful computing experience, whether we have a broadband connection to the Web, we are playing cutting edge online games, watching or creating videos or running other performance-intensive applications. It intensifies the 3D action of our favourite games and enables clear and smooth audio and video streaming. Pentium® 4 processors provide the performance to power the connected home, which means linking all our digital devices in order to extend their capabilities, from our digital camera and MP3 player to our entire home entertainment system. Increased performance and headroom allows us to take advantage of the emerging Internet and computer technologies enabling the connected home. Pentium® 4 processors also deliver high-performance when networking multiple PCs, or when attaching our PC to home consumer electronic systems and new peripherals.

Q. 27. Can we upgrade our current Pentium III processor to a Pentium 4 processor?

Ans. The Pentium 4 processor introduces a new generation of processing power with Intel NetBurst micro-architecture (the first new micro-architecture in five years) that requires a new chipset and motherboard. As a result the processor socket pin out is different, making it impossible to upgrade. However, the Pentium 4 processor is backward compatible with all IA-32 software and works with industry standard memory and peripherals.

Q. 28. How many cache levels are present in the P4 architecture?

Ans. The Level 1 cache is of 8KB data cache and 8KB instruction cache, both closely coupled to pipeline. The 2nd level cache is of 256KB, 512KB, or 1MB static RAM that is coupled to the core processor through a full clock-speed 64-bit cache bus.

Q. 29. What is dynamic execution in P4 processor micro-architecture?

Ans. The P4 micro-architecture has an out-of-order execution mechanism called dynamic execution. The dynamic execution incorporates 3-data-processing concepts

- Deep branch prediction,
- Dynamic data flow analysis,
- Speculative execution.

Q. 30. What does the branch prediction unit do?

Ans. The branch prediction is a modern technique to deliver high performance in pipelined micro-architectures. It allows the processor to decode instructions beyond branches to keep the instruction pipeline full. The P6 processor family implements highly optimized branch prediction algorithm to predict the direction of the instruction stream through multiple levels of branches, procedure calls and returns.

Q. 31. What is a front end pipeline?

Ans. The front end pipeline supplies instructions in program order to the out-of-order core that has very high execution bandwidth and can execute basic integer operations with 1/2 clock cycle latency. The front end fetches and decodes IA-32 instructions and breaks them down into simple operations called micro-ops (μops). It can issue multiple μops per cycle, in original program order, to the out-of-order.

Q. 32. What basic function does the front end pipeline do?

Ans. The front end performs following basic functions

- Prefetch IA-32 instructions that are likely to be executed.
- Fetch instructions that have not already been prefetched.
- Decodes IA-32 instructions into μops
- Generate microcode for complex instructions and special-purpose code.
- Predict branches using highly advanced algorithm.

Q. 33. What are major source of delays in front end pipeline?

Ans. Following are responsible for the major source of delays in the front end

- When a μop completes and writes its result into the destination, it is retired. Up to 3-μops may be retired per cycle. The Reorder Buffer (ROB) is the unit in the processor that buffers completed μops, updates the architecture state in order and manages the ordering of exceptions.
- The retirement action also keeps track of branches and sends updated branch target information to the BTB to update branch history. In this manner, traces that are no longer needed can be purged from the trace cache and new branch paths can be fetched, based on updated branch history information.

Q. 34. What is hyper-threading technology?

Ans. The hyper-threading technology enables a single physical processor to execute 2 or more separate code streams (called threads) concurrently. This technology improves the performance of IA-32 processors when executing multi-threaded operating system and application code or single-threaded application under multi-tasking environments.

Q. 35. Is a Pentium 4 Machine better than an AMD Athlon machine?

Ans. There is no clear-cut of which one is better. Pentium 4 is better at running bandwidth intensive applications, while AMD's latency is lower than Pentium 4 in most cases. The determining factor to decide which one is better for us really depend what kind combination of applications we will use most.

Q. 36. How does hyper-threading technology work?

Ans. Figure below shows a comparison of an IA-32 processor with Hyper-Threading Technology (implemented with 2-logical processors) and a traditional multiple processor (MP) system.

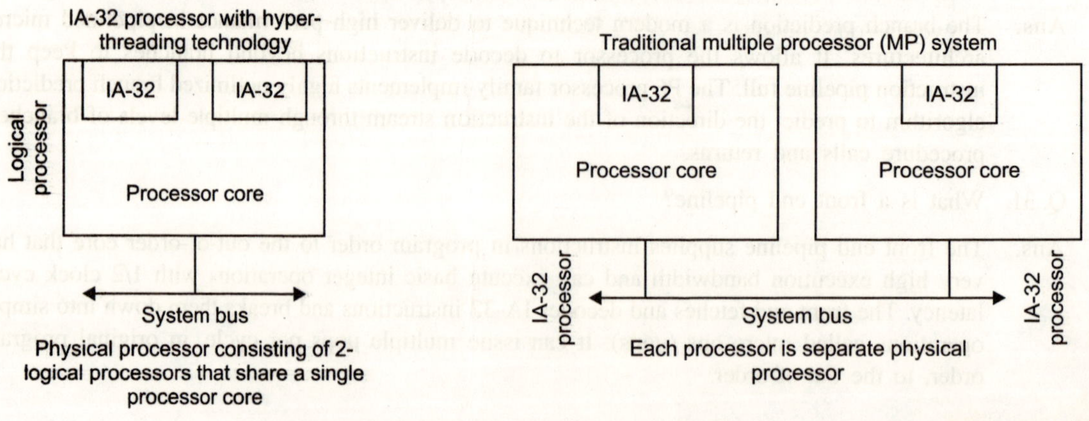

Q. 37. How large is Pentium 4's L1 cache, and how is it compared to AMD Athlon?

Ans. Pentium 4's L1 cache is 8KB, while AMD Athlon has 64KB of L1 cache.

Q. 38. Is Pentium 4 faster than AMD Athlon?

Ans. So far, Pentium 4 is faster in pure clock speed, but that does not guarantee Pentium 4 will run applications faster than AMD Athlon. How fast a machine runs the applications depends on the combination of clock speed, cache latency, bandwidth and bus speed.

Q. 39. How large is Pentium 4's L2 cache, and how is it compared to AMD Athlon?

Ans. Pentium 4's L2 cache is 256KB, while AMD Athlon also has 256KB of L2 cache.

Q. 40. What is the name of the 0.13-micron Pentium 4?

Ans. Northward.

Q. 41. Will the 0.13-micron Pentium 4 implement a 512KB of L2?

Ans. Most likely, since that will lessen the effects of RDRAM's relatively high latency.

Q. 42. What kind of memory buses Pentium 4 and Athlon use?

Ans. Pentium 4 uses RDRAM, and Athlon uses DDR SDRAM

Q. 43. What is Revamped Pipeline?

Ans. It is another name for Hyper-Pipelined Technology.

Q. 44. How large is the L2 cache of the Pentium 4/Northwood processor?

Ans. The Intel S478 Pentium 4's L2 cache size is 256KB while Northwood's L2 cache size is 512KB.

Q. 45. At what clock speeds will the Intel S478 Pentium 4 and Northwood processors operate?

Ans. Initial versions of the Intel Pentium 4 processor are planned to operate at clock speeds of 2.0GHz and 1.9GHz, based on Intel's 0.18-micron process technology. Later on, there will be 1.8GHz, 1.7GHz or lower speed versions for value PC segment. Northwood's speeds will range from 2.0GHz to 2.4GHz and beyond.

Q. 46. How well does the Intel S478 Pentium 4 processor perform relative to the S423 Pentium 4 processor?

Ans. Intel S478 Pentium 4 based system has better performance than all existing PC systems, and it fully utilizes the most advanced process technology to stretch its computing power into 2.4GHz and beyond. Most importantly, the price for such an advanced, next-generation CPU could priced affordably for everyone. S423 Pentium 4 platform might be our interest, but due to lack of higher CPU speed support (limited to 2.0GHz), if future upgradeability and longer system life cycle is our concern, then S478 Pentium 4 platform will be our only choice.

5

POWER PC

5.1 INTRODUCTION

The warfare between RISC and CISC continued for long and these technologies are converging towards a 3rd that might be entirely different from both RISC and CISC. The RISC is the acronym of Reduced Instruction Set Computer. It does not mean really that the instruction set has been reduced. The earliest laboratory experiments in high-performance computing needed space on their experimental μP chip to build new performance feature like cache and greatly expanded instruction sets. Hence, manufacturers made room for these new features by shoveling instructions off the chip wholesale. It will not be out of place to mention that one of the earlier experimental RISC chips had as low as only 39-instructions.

It is well known that complex instructions take a lot of space on a μP w.r.t. the simple one. One answer to this problem was to imitate the complex instructions by staggering together a series of simple instructions. Also that, if the simplified chip becomes fast enough such that these emulated complex instructions run as fast as on the simplified chip as the built-in complex instructions that more traditional μP had. This was one of the methods for cutting the occupied space of a μP without sacrificing functionality.

The RISC manufacturers also took advantage of the philosophy that the rich instruction set adds enormous power to a μP and added all instruction back those instructions that were removed to call it the RISC chip. In addition some more were added apart from leaving the room for cache and registers as well. Thus, today the RISC μP has as many instructions as CISC. The PowerPC 601 μP has more instructions than the Pentium with which it is invariably compared. More so the instruction set of some PowerPC are complex indeed.

The Apple, IBM, and Motorola are jointly responsible for production of the RISC series of microprocessors. It was hoped that it would work as an engine to bring radical changes in main stream of computing systems in terms of

- improvement of power,
- price,
- performance, and
- functionality.

The 601 chip was the 1st PowerPC chip produced and is used in Macintosh and IBM desktop systems, as well as IBM's Power250 RS/6000 workstation.

The RISC is posing to be very strong competitor of CISC architecture. For years together, the semiconductor technologists have been forecasting that the balance would soon be shifting towards the RISC because of its simplicity. As it is simple at its fundamental level, it is cheaper and easier to design and manufacture. The PowerPC within a very small span has shown its potential to make its presence felt in the desktop systems too. The high performance coupled with its unprecedented industry backing of the RISC technology, is expected to give a good fight to its most powerful long old CISC counterpart, even the Pentium. The Apple would have incorporated the PowerPC processors into its entire Macintosh line by the end of 1994 amounting something like 700,000 units. The IBM were producing the PowerPC in the trademark of RS/6000 engineering workstation. This workstation will directly compete with DOS/ Widows PCs. Thus, the PowerPC would be able to run most of the today's PC and Macintosh software using a variety of emulation techniques.

It was forecasted that the PowerPC would easily outrun Intel's acclaimed Pentium microprocessor, but could do so at a fractional price. With superior RISC technology, the PowerPC requires half the number of logic transistors as that of the Pentium. Thus, fewer transistors culminate into smaller, cheaper and more energy-efficient chips. The MS-DOS and Windows users would be able to run all their programs on the PowerPC platform. The PowerPC processor, 1st introduced into several mainstream and high-end Macintosh models, provides 2-to-4-times the performance of Motorola's 68040 or Intel's 80486 chip using comparable applications designed for PowerPC.

The 601 PowerPC is the 1st implementation of this series of architecture developed jointly by IBM and Motorola. The alliance of the IBM, Motorola, and Apple is developing a family of processors that would serve the purpose of computer markets requiring high performance, low power consumption, and outstanding price/performance.

Goals of 601

The 1st goal was to capture the marketplace quickly. The IBM's POWER (Performance Optimized with Enhanced RISC) architecture formed the basis of the PowerPC architecture and the RISC Single Chip Vector (RSV) processor as the base for the 601 microprocessor. Motorola considered various architectures that helped to form the PowerPC architecture. In addition, the bus interface from the Motorola 88110 microprocessor served as a basis for the PowerPC 601-bus interface.

The 2nd goal was to offer competitive price/performance cost. The basic RISC design was enhanced to allow more concurrent instruction execution and higher frequency operation. Integrating the processor onto a relatively small die and selecting an economical package minimized the cost.

The 3rd goal was to offer capabilities suitable for wide range of system design points. This was achieved by extending the versatile interface of the Motorola 88110 to support the PowerPC architecture, advanced protocol operations, and enhanced multiprocessing support.

5.2 ARCHITECTURE

The 3rd generation RISC architecture were designed to accommodate diverse computing environment. The result is a powerful architecture that embraces fundamental concepts of simplicity and general applicability. It is also extensible to advance technology that will allow it to be used in future generations of PowerPC microprocessor.

The application of binary interface (ABI) of IBM's POWER (Performance Optimization with Enhanced RISC) architecture was maintained in this design. Such compatibility allowed PowerPC-base machines to exploit the existing application base of the RISC system/6000 machines. The power user-level instruction set and programming model was adopted as a starting point for the architecture. Ultimately, some instructions were added and others were deleted. The compilers and operating systems could manage these changes.

A 3rd objective was to provide support for a wide range of uni-processor and multiprocessor system configurations. The architecture also allows either a Big-Endian or a Little-Endian byte-ordering convention to support different operating system needs.

The final architecture goal was to allow upward compatibility of 32-bit applications after defining 64-bit extension. The 64-bit instruction operation and 64-bit memory management was defined as a logical extension to the 32-bit execution model. Each implementation can either comply with the base 32-bit PowerPC architecture or the extended 64-bit architecture.

Bi-Endian architecture

A single instruction or data unit is called a **scalar**. In turn, the smallest addressable unit of storage is 1-byte (8-bits). A scalar is typically composed of multiple bytes. A half-word scalar is 2-bytes, a full word scalar is 4-bytes long, and a double word scalar is 8-byte long.

A μP may require to use multiple bytes of storage (for half-word and larger scalars). A μP can either assign bytes from a low address to a higher address, or from higher address to a lower address. If a μP assigns the bytes such that the most significant 8-bits start at a low address, and successive bytes are stored at higher addresses, it is said to operate in Big-Endian mode. On the other hand, if a μP assign the bytes in the reverse order, it operates in Little-Endian mode. Figure 5.2 (*a*) shows the difference in byte ordering and addressing for both modes.

It is important to define the Endian-ism because most μPs only support one of the two-Endian modes. The μPs in RS/6000 workstations and in Macintosh computers (the 6800X0 family) operate in Big-Endian mode. But the Intel family of 80x86 μP all operate in Little-Endian mode. The PowerPC chips, on the other hand, are **Bi-Endian**. In other words, they support both Little-Endian/Big-Endian byte ordering. This feature made it possible to develop PowerPC systems that conform to the Endian mode assumed by existing system software (such as DOS, Windows, AIX, and System7).

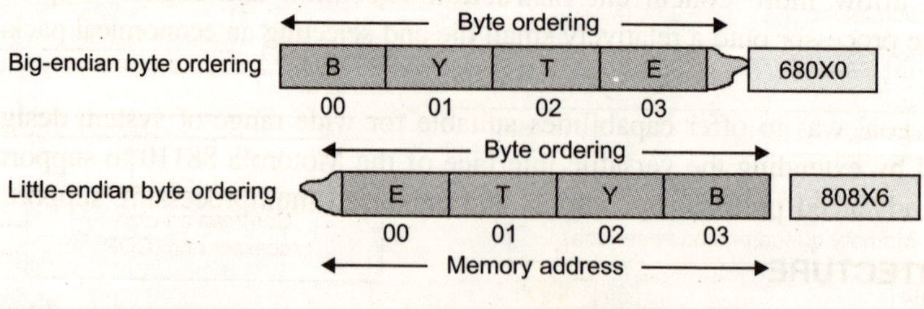

Fig. 5.2 (*a*) Big-Endian and little-Endian byte ordering

By default, all PowerPC run in Big-Endian mode. However, operating system can easily change the byte ordering to Little-Endian during boot-up. The byte ordering does have significant effect on software emulation.

Storage control

It contains a robust storage control structure that includes definitions for memory management, caching, and memory operations. The 32-bit architecture supports a 52-bit virtual address and 32-bit real address. The 64-bit architecture supports an 80-bit virtual address and a 64-bit real address.

Address translation

A well known recognition process translates the effective address into the virtual address, then a paging process translates the virtual address into a real address. The page translation is implemented through a memory-resident hashed page table. The page table is commonly cached in a translation look aside buffer, so the architecture provides instructions to control the TLB. The page size was fixed to be 4KB. In addition to the segmentation and paging mechanism, the architecture also defines a block address translation mechanism that allows translation of blocks that can vary in size from 12KB to 256MB only.

5.3 MACHINE ORGANIZATION

The 601 PowerPC is a 32-bit implementation of the PowerPC architecture. The superscalar machine organization can dispatch up to 3-instructions each cycle. The 601 μP also benefits from instruction pipelining that helps to improve its clock rate. Figure 5.3 (*a*) illustrates the block diagram of the 601 microprocessor and Fig. 5.3 (*b*) shows the pipeline structure.

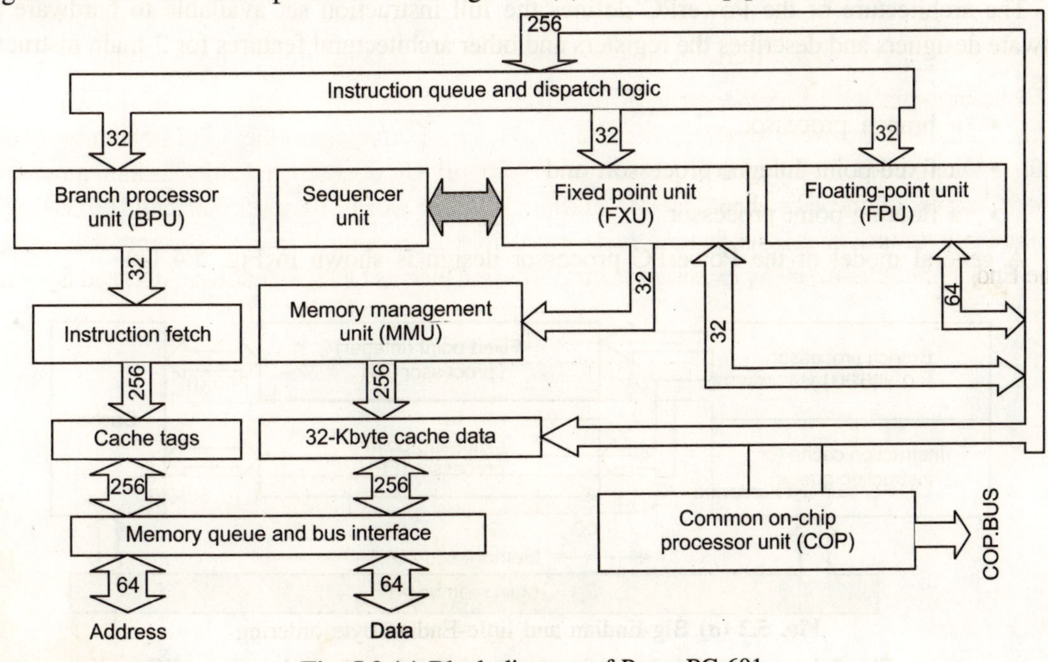

Fig. 5.3 (*a*) Block diagram of PowerPC 601

The instruction queue and dispatch logic buffer instructions from the cache dispatch up to 3-instructions each cycle. One each for the fixed-point unit (FXU), the floating-point unit (FPU), and the branch processing unit (BPU). The FXU communicates with the sequencer unit for handling infrequently used instructions and certain control-intensive tasks. In addition, the FXU interfaces with the memory management unit (MMU) for cache access. The 32KB unified cache provides a 32-bit interface to the FXU, a 64-bit interface to the FPU, and a 256-bit interface to both the instruction queue and memory queue. The chip interface includes a 32-bit address bus and a 64-bit data bus. In addition, the chip supports an asynchronous serial port, the common on-chip processor (COP) bus, to support debugging and test features.

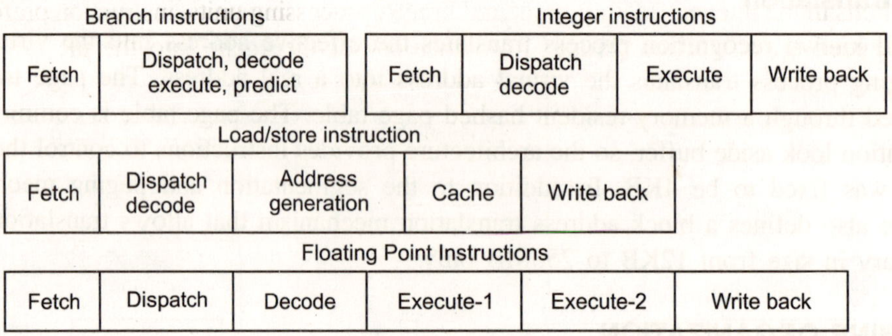

Fig. 5.3 (*b*) Pipeline description

5.4 CHIP 601

The architecture of the PowerPC defines the full instruction set available to hardware and software designers and describes the registers and other architectural features for 3-main instruction units

- a branch processor,
- a fixed-point Integer processor, and
- a floating-point processor.

A general model of the PowerPC processor design is shown in Fig. 5.4 (*a*).

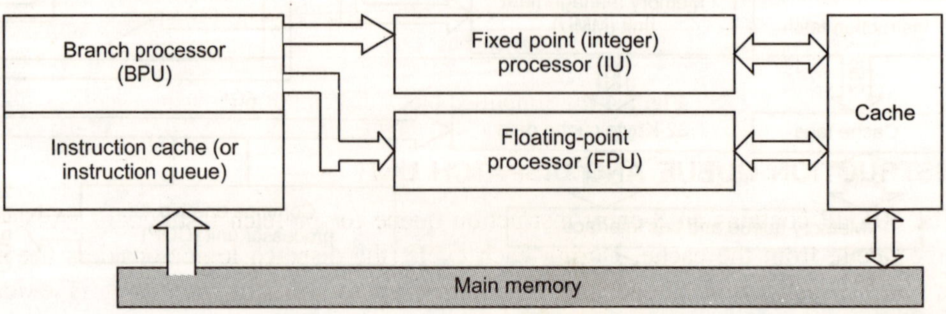

Fig. 5.4 (*a*) Logical flow of instruction and data through a PowerPC

The 601 μP contains 3-main execution units to support superscalar dispatch of instructions. They are

- Integer unit (also called IU or FXU),
- Floating-point unit (FPU), and
- Branch-processing unit.

A simplified block diagram of the 601 μP is shown in Fig. 5.4 (*b*). The branch-processing unit is actually a subsystem of the main instruction unit that is responsible for fetching and prefetching instructions. The instruction unit can store up to 8-instructions at a time in its instruction prefetch queue. Since the instruction unit can prefetch instructions concurrently with the execution of instructions in the integer, floating-point and branch processing units, instruction prefetch provides a way to speed up instruction thruput from memory or the cache into the μP.

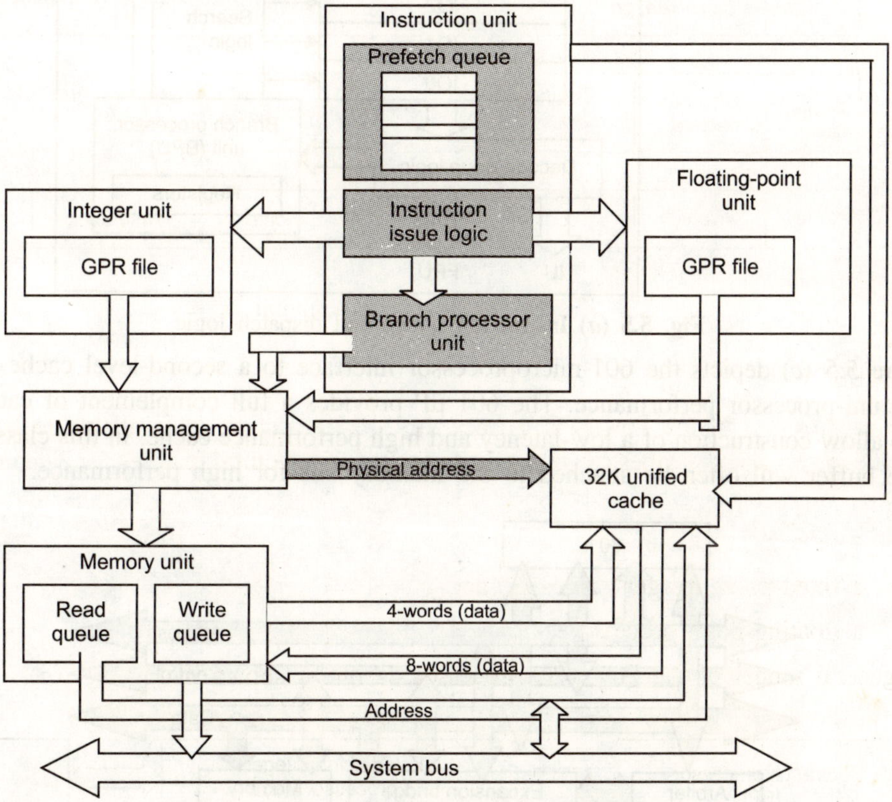

Fig. 5.4 (*b*) Simplified block diagram of 601

5.5 INSTRUCTION QUEUE AND DISPATCH UNIT

The 601 μP contains an 8-entry instruction queue for prefetch instructions. An 8-word bus feeds the queue from the cache. During each cycle, the dispatch logic considers the bottom 4-entries of the instruction queue and dispatches up to 3-instruction as it is evident from Fig. 5.4 (*b*) and Fig. 5.5 (*a*). The queue supports the full range of possible shift amounts with no instruction alignment restrictions on the program.

Figure 5.5 (*b*) is the block diagram of a 601 based single-processor system. A simple arbiter predicts whether the memory bus is owned by the 601 μP or by an expansion bridge to a system bus such as VME, NuBus, Micro channel and PCI (Peripheral Connect Interface).

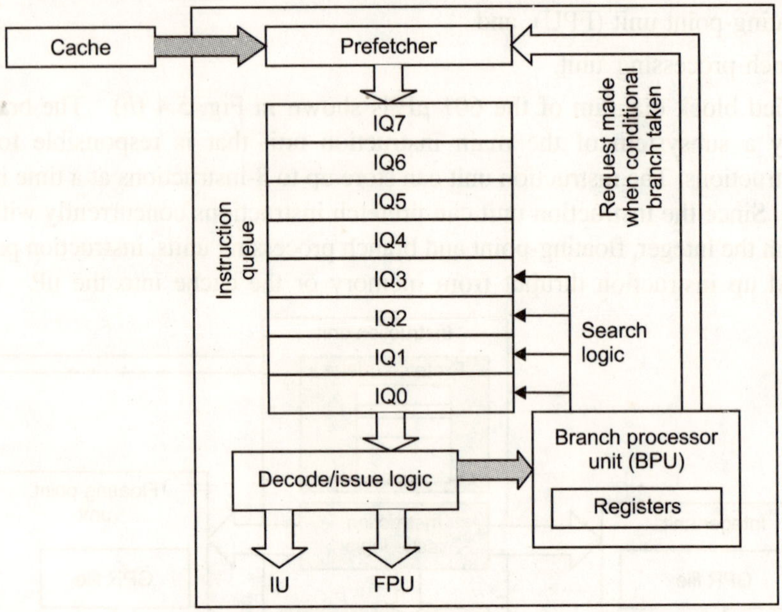

Fig. 5.5 (*a*) Instruction queue and dispatch logic

Figure 5.5 (*c*) depicts the 601-microprocessor interface to a second-level cache to provide improved uni-processor performance. The 601 μP provides a full complement of cache control signals to allow construction of a low-latency and high performance cache. In this class of system the frame buffer will often be attached to the memory bus for high performance.

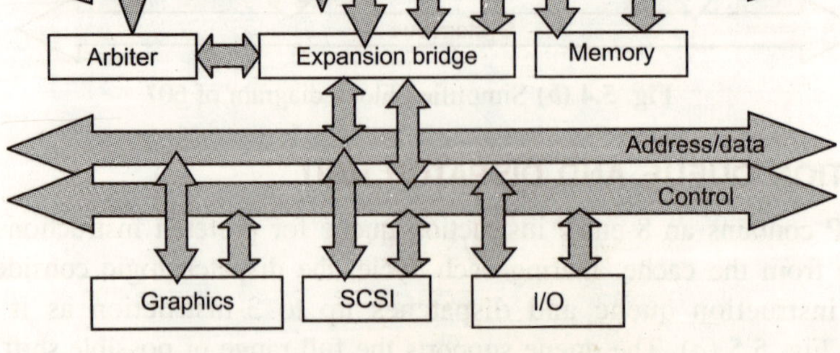

Fig. 5.5 (*b*) 601-based system with uni-processor and second level cache

Figure 5.5 (*d*) shows the arrangement of a multiprocessing system. The typical features of a 601-based multiprocessor might include a shared memory and a shared bus to facilitate hardware-enforced coherency among a number of tightly coupled processors. Also, it includes a local cache system for improved processor, system performance and cache control operations broadcast on the bus to allow processors and external hardware to control the local cache state.

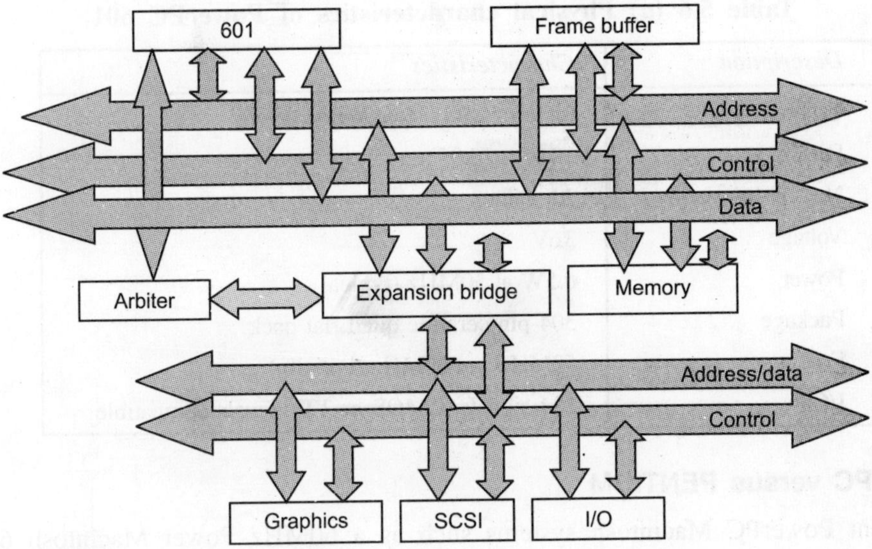

Fig. 5.5 (*c*) 601-based single processor system

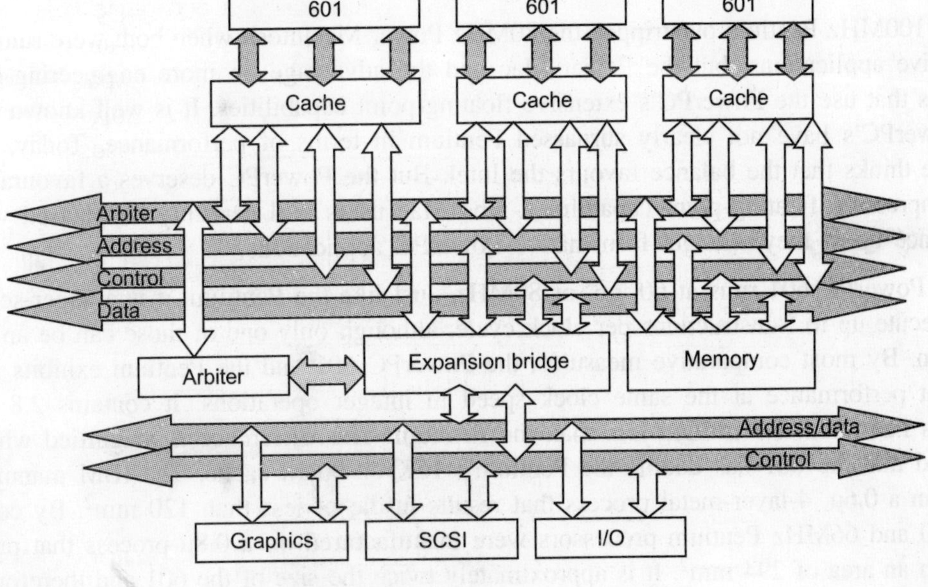

Fig. 5.5 (*d*) 601-based multiprocessing system

5.6 CHIP AND PACKAGING TECHNOLOGY

The IBM CMOS technology implemented 601 μP with 0.6μ minimum feature size and 4-layers of metal wiring. The 601 microprocessor is packaged in a 304-pin ceramic quad pack that is bonded using IBM's C4 solder ball technology. Table 5.6(*a*) summarizes other key physical characteristics of the 601 microprocessor.

Table 5.6 (*a*) Physical characteristics of PowerPC 601

Description	*Characteristics*
Technology	0.6μm CMOS; four metal levels
Die size	10.95 × 10.95 mm
No. of transistors	2.8 million
Voltage	3.6V
Power	6.5W at 50MHz (typical)
Package	304-pin ceramic quad flat pack
Frequency	50MHz and 66MHz (initially)
I/Os	184-signals, CMOS or TTL levels compatible

5.7 PowerPC versus PENTIUM

Different PowerPC Macintosh systems such as a 60MHz Power Macintosh 6100/60, a 66MHz Power Macintosh 7100/66, and an 80MHz Power Macintosh 8100/80 along with a Motorola's 68040-based Macintosh Quadra 840AV were considered for comparison. In order to gauge how well the Power Macs stand up to Intel-based systems. For the purpose, 66MHz Pentium ALR Evolution V/66 and a 100MHz Pentium system from Intel were considered.

The 100MHz Pentium outstripped the 80MHz Power Macintosh when both were running the same native applications, but the Power Mac had the advantage on more engineering-oriented operations that use the PowerPC's extensive floating-point capabilities. It is well known that the early PowerPC's have not clearly surpassed Pentium in terms of performance. Today, in fact, every one thinks that the balance favours the Intel. But the PowerPC deserves a favourable nod for its impressive floating-point capabilities. The Mac users will certainly be pleased with the performance boost they will get from native PowerPC applications.

The PowerPC 601 runs at 60, 66, or 80MHz. Just like the Pentium, it is a super-scalar μP. It can execute up to 3-instructions per clock cycle, although only one of those can be an integer instruction. By most comparative measures, the PowerPC 601 and the Pentium exhibits roughly equivalent performance at the same clock speed in integer operations. It contains 2.8 million transistors i.e., about 0.3 million less than the Pentium. This difference is magnified when it is considered that the 601 has double the Pentium's 16K on-board cache. The IBM manufactures the 601 on a 0.6μ, 4-layer-metal process that results in die of less than 120 mm^2. By contrasts, the 1st 60 and 66MHz Pentium processors were manufactured on a 0.8μ process that produced chips with an area of 294 mm^2. It is approximately twice the size of the 601 and therefore more expensive to manufacture.

Intel began shipping new Pentium µP built on its own 0.6µ process. The new 90 and 100MHz Pentium processors run faster and cooler and challenge one of the key advantages of lower power consumption of the PowerPC603. The PowerPC603 was designed to run on 3.3V and draws only 2W, making it ideal for notebooks and energy-saving desktop PCs. But the new Pentium µP runs on 3.3V and draws about 4W, compared with 13W for the 1st generation chips. The Pentium-based notebooks become available before the PowerPC 603-based notebooks.

Quantum jump

It was forecasted that Intel and the PowerPC consortium would leapfrog each other over the next several years, with the PowerPC family gaining an early edge because of the difficulty in extending the 80x86 architecture. The IBM and Motorola must have produced 100MHz-based PowerPC 601. Shortly thereafter, the PowerPC 604 will break new ground with its ability to execute up to 4-instructions. The 3-instructions are integer instructions, during each clock cycle. That will make it substantially faster than 120, 132, or 150MHz Pentiums that will likely be available around the same time.

Unlike the previous PowerPCs and any µP that Intel has announced, the PowerPC 620 is a true 64-bit processor. The Pentium, by contrast, operates on 32-bits of data at a time, although it connects to a 64-bit external bus. The PowerPC620 is expected to be informally as P6. But it is known that most software developer are today only embracing 32-bit code, it is likely to take some time before 64-bit programs that can use the PowerPC620's full power become commonplace.

Will it lead last even if PowerPC surpasses the 80×86architecture in raw performance? Not indefinitely. The balance of power cans quickly shifts. Intel will continue to develop its µPs making them look ever more RISC-like while retaining full backward compatibility.

Both Intel's competitors, AMD and Cyrix, have previewed processors that maintain 80×86 compatibility but work around many of the bottlenecks that restrict performance in today's Pentiums. Such processors suggest that today's Pentium is far from the last word in the 80×86 architecture.

The important aspects are not the clock speed of a µP, but how fast it actually runs the applications. The Intel 100MHz Pentium system surpassed the performance of the 80MHz Power Macintosh when both were running the same native applications.

The 100MHz Pentium changed the font in a large Aldus PageMaker file about a second faster than the 80MHz Power Macintosh, for instance. On the same test, a Quadra 840AV, the high end of Apple's 68K-base Macintosh line took 3s longer to make the font change than the 80MHz Power Mac.

It was observed that the Pentium was 15% to 45% faster on typical operations like opening files and redrawing screens with 66MHz Pentium system as compared to a 66MHz Power Macintosh. In other tests, the 100MHz Pentium was able to replace a text-block in a Word Perfect document in less than 4-seconds; and 80MHz Power Macintosh took nearly 13s.

5.8 INSTRUCTION FETCH UNIT

This unit coordinates instruction fetching from the cache. Several different sources in the 601 microprocessor generate instruction fetch addresses. The branch processor provides the address that results from branch instruction. The sequencer unit provides addresses associated with interrupts and other synchronizing events. The instruction fetch unit itself generates the next sequential address in the event that no branch or interrupt has occurred. An appropriate address is selected, translated, and forwarded to the cache arbitration logic for consideration to access the cache during each cycle.

The instruction queue and dispatch logic accepts and processes instruction fetched from the cache. The instruction fetcher also provides fast address translation of instruction fetch addresses through the translation shadow array (TSA). The TSA automatically keeps track of the 4-most recently used instruction address translations and provides fully associative comparison of any response to the address generation of any instruction fetch. The TSA supports both page and block oriented address translations. In the event of a miss in the TSA, the instruction fetcher arbitrates for access to the 601's primary MMU for translation, and then updates the TSA based on a least recently used replacement (LRU) algorithm.

Branch processor unit

The BPU executes branch instruction in its 2-stage pipeline. The 601 microprocessor

- dispatches,
- decodes,
- evaluates, and
- predicts the direction of branch instruction in one cycle, if necessary.

The resulting fetch can access new instructions from the cache in the next cycle. This allows the processor to quickly react to branches detected in the instruction stream and to reduce the latency of subsequent instruction. The block diagram of branch processor unit is shown in Fig. 5.8 *(a)*.

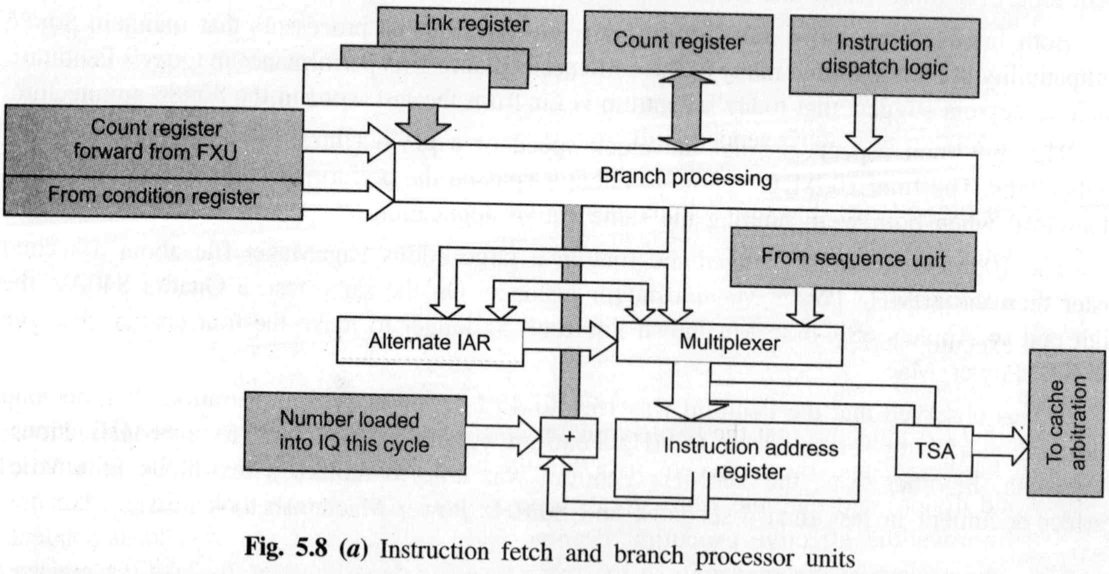

Fig. 5.8 *(a)* Instruction fetch and branch processor units

The unconditional and count register dependent branches execute during the same cycle that they are dispatched. Thus, there is no delay in the instruction dispatch stream to the FXU or the FPU and effectively execute in zero cycles. The conditional branch can either be resolved or unresolved at the time of dispatch dependent on the contents of the condition register. Each of the 8-fields of the condition register has a set of associated interlocks activated by instructions that update that field. If a conditional branch is dependent upon a non-interlocked condition register field, the branch can be resolved immediately. In this case, the condition is evaluated based on the contents of the condition register and the branch presents no delay in the instruction stream to the FXU or FPU.

The conditional branches on an interlocked condition register field remain unresolved. Under such circumstances, the 601μP employs a static branch prediction algorithm to predict the direction of the unresolved branch. The 601μP algorithm predicts that the branch will be taken if the displacement of the target address is negative and will not be taken if it is positive. As an aid for compiler directed prediction, a bit in the branch instruction opcode allows this reduction scheme to be reversed. Instruction fetched on behalf of a predicted conditional branch can be conditionally dispatched, but they do not execute until the prediction has been validated.

The condition register interlocks are eventually released and the validity of the prediction can be checked. The branch completes and all prefetched instructions are marked eligible for execution if the reduction was correct. In the event incorrect prediction, any incorrectly prefetched instructions are discarded and fetching resumes down the correct path.

The 601μP has included several features to speed up recovery after a misprediction. The 1st, a fast alternate address restores mechanism saves the alternate path address and immediately redirects the instruction fetcher. The 2nd, the FXU forwards the result of compare instruction directly to the branch processor as well as the condition register. This allows earlier resolution of the unresolved conditions. Finally, the instruction queue implements a delayed purged mechanism that retains any prefetched sequential instruction beyond the conditional branch until the prefetched predicted instructions are made available for loading into instruction queue.

5.9 FIXED-POINT UNIT

The instruction dispatcher sends instructions to the FXU. In general, the FXU serves as the master pipeline, managing the synchronization control required to achieve precise exceptions. As evident from Fig. 5.9 (*a*), the instructions execute in a 4-stage pipeline are

- fetch,
- decode,
- execute, and
- write back.

Figure 5.9 (*a*) indicates that the hardware interlocks all hazards. All architecture instructions process in hardware, and most execute in a fully pipelined manner. Some of the arithmetic (multiply and divide) and multiple word storage operations require several cycles in the execute stage. To improve the effective execution latency, the FXU forwards any register-dependent results to their respective function units at the same time the data is written back to the register file. In addition, although the differences between the POWER architecture and the PowerPC architecture at the user level are small, the 601 microprocessor also provides hardware support for all POWER architecture user mode instructions.

The FXU handles the address generation for all load and store instructions. This includes the floating-point loads and stores. The instructions also progress through the floating-point pipeline in some cases. A dependent operation following a load instruction causes a one-cycle load delay slot. The FXU sequences the execution of the load/stores multiple instructions and the load/store string instructions. The hardware handles most forms of unaligned addresses. However, in some cases, multiple accesses from the cache are necessary. The FXU features a fast execution of compare instructions and ability to forward results directly to the branch-processing unit. As discussed previously, this allows efficient handling of branches dependent on the result of the compare instruction.

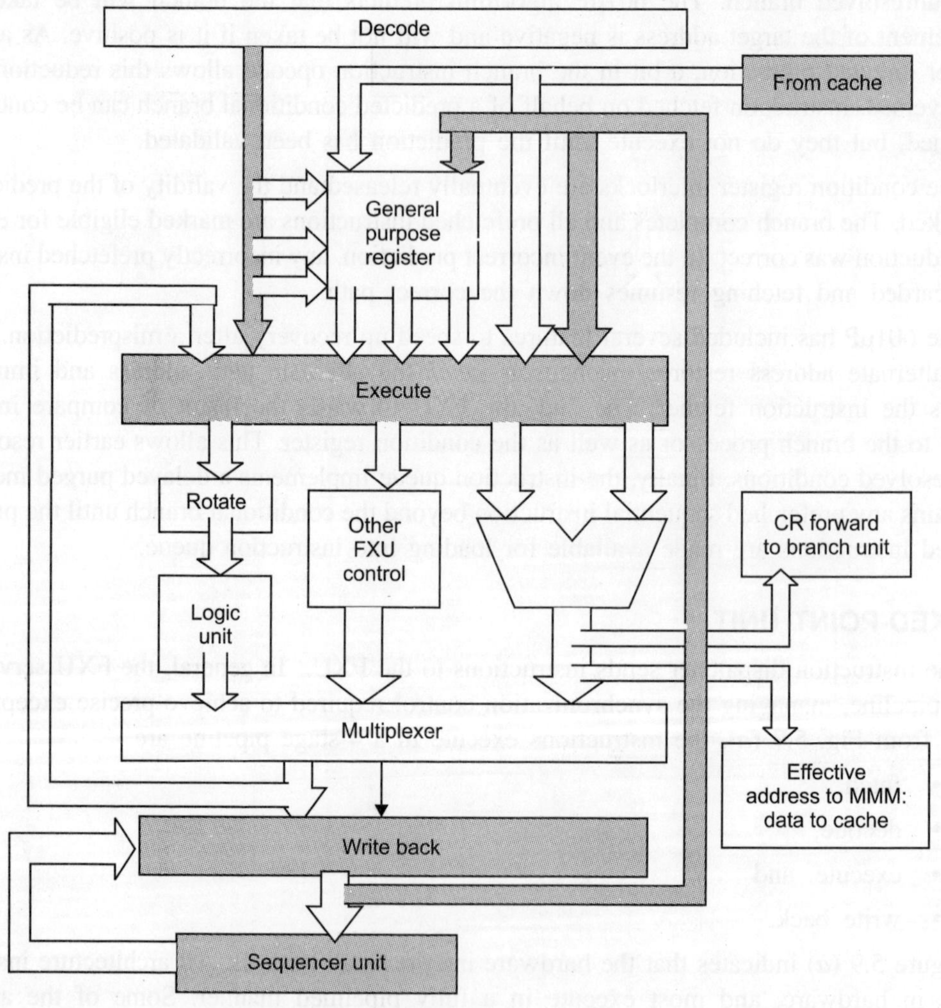

Fig. 5.9 (*a*) Fixed-point unit

5.10 FLOATING-POINT UNIT

The instruction set of the 601µP provides hardware support for all IEEE-754 floating-point data types. The PowerPC 601 User's manual states that this approach prevents the pipeline latency that occurs when an application has to perform an exception routine for unsupported floating point operation. But it is more likely true that the rich floating-point instruction set is provided to especially support the Macintosh's SANE (Standard Apple Numeric Environment) math routines. Most other desktop applications would not need the extensive floating-point hardware support provided by the PowerPC 601. But this may change as software developers add floating-point capabilities to their native PowerPC applications. The block diagram of floating-point unit is shown in Fig. 5.10 (*a*).

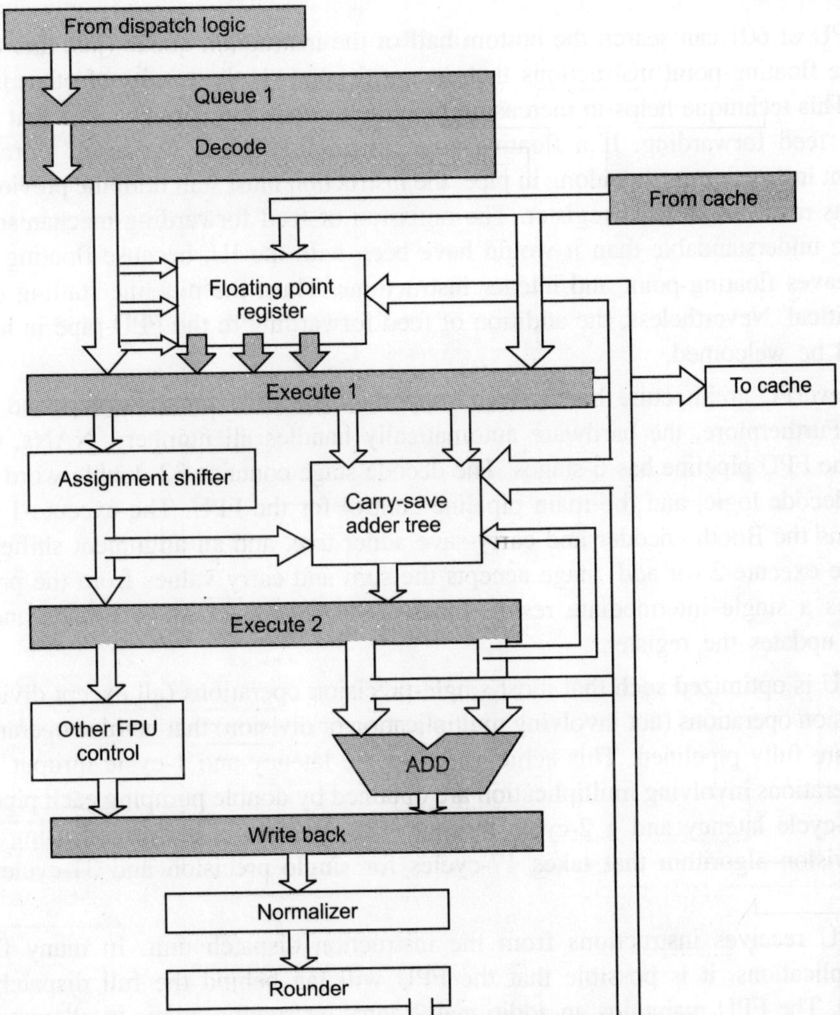

Fig. 5.10 (*a*) Floating-point unit

The FPU includes 32-registers for performing floating-point operations. As expected, each register contains 64-bits to support double-precision values. The FPU contains six pipeline stages

- fetch,
- dispatch,

- decode,
- execute-1,
- execute-2, and
- writeback.

The execute-1 and 2 stages allow 2-floating-point instructions to be executed back-to-back.

An *added bonus*

The FPU of 601 can search the bottom half of the instruction queue (just like the BPU) and then execute floating-point instructions that do not depend on the results of other instructions in the queue. This technique helps in increasing pipeline thruput. One drawback is that the FPU does not support feed forwarding. If a floating-point instruction needs the result stored by another floating-point instruction further along in pipe, the instruction must stall until the previous instruction has stored its result in an FPU register. The omission of feed forwarding mechanism in the FPU pipe is more understandable than it would have been with the IU, because floating-point routine often interleaves floating-point and integer instructions. Thus, the pipeline stalling in the FPU is often not critical. Nevertheless, the addition of feed forwarding to the FPU pipe in later PowerPC chips would be welcomed.

The PowerPC architecture has defined support of the FPU for the compound multiply-add operations. Furthermore, the hardware automatically handles all numbers, NANs, QNANs, and infinities. The FPU pipeline has 6-stages. The decode stage contains 32-double-word registers, the instruction decode logic, and the main pipeline control for the FPU. The execute-1 (or multiply) stage contains the Booth encoder and carry-save adder tree, and an alignment shifter for the 2nd operand. The execute-2 (or add) stage accepts the sum and carry values from the previous stage, and produces a single intermediate result. Finally, the write-back stage rounds and normalizes results, and updates the register.

The FPU is optimized such that most single-precision operations (all except divide) and most double precision operations (not involving multiplication or division) that involve operands of normal data types; are fully pipelined. This achieves a 4-cycle latency and 1-cycle thruput. The double-precision operations involving multiplication are obtained by double pumping each pipeline stage to achieve a 5-cycle latency and a 2-cycle thruput. The division is performed using a 2-bit non-restoring division algorithm that takes 17-cycles for single precision and 31-cycles for double precision.

The FPU receives instructions from the instruction dispatch unit. In many floating-point intensive applications, it is possible that the FPU will lag behind the full dispatch rate of the dispatch unit. The FPU maintains an additional 2-entry instruction queue to alleviate the adverse effects of this. This allows the dispatch logic to remove floating-point instructions from the primary instruction queue and expose subsequent fixed-point and branch instructions earlier.

The FPU operates independent of the FXU and can concurrently execute floating point instructions. A carefully tuned synchronization scheme allows these execution units to progress independently while maintaining precise exceptions. These pipelines cooperate in the execution of floating-point storage access instructions but are not required to proceed in lock step. For example, the FXU can process the addressing portion of a floating point store operation and pass it onto the cache unit without waiting for the FPU to produce the stored data. The cache unit will complete the store operation once the FPU provides the data.

Sequencer unit

The sequencer unit is essentially an embedded support processor that assists the core µP in handling the PowerPC architecture. It sequences the power on reset functions that include array self-testing for the cache and array initialization for all other storage cells on the chip. It maintains an on-chip real-time clock and the system decrementer function. In addition, it handles the recording and sequencing of interrupts, context-synchronization events, and errors.

The sequencer also walks through the hashed page table for address translations that misses in the TLB and it updates the storage access recording bits. Several of the less frequently used systems control registers are physically implemented in a RAM structure within the sequencer additionally to minimize the chip area overhead. As a result, the instructions that operate with these registers are passed from the FXU to the sequencer for execution.

5.11 MMU

For load and store instructions, the MMU performs virtual-to-real address translation. In addition, it acts as a backup for instruction fetch address translations that missed in the instruction fetcher's TSA. The MMU is tightly coupled to the FXU so that address translation can occur during the execute phase of the FXU pipeline. A cache access occurs in the following cycle using the translated address. Arbitration logic determines whether the FXU or the instruction fetcher has use of the MMU during a particular cycle. The instruction fetcher's use of the MMU should be infrequent since the hit rate of the TSA is high for most applications. The MMU provides support for the segment, page and block translations as indicated in Fig. 5.11 (*a*).

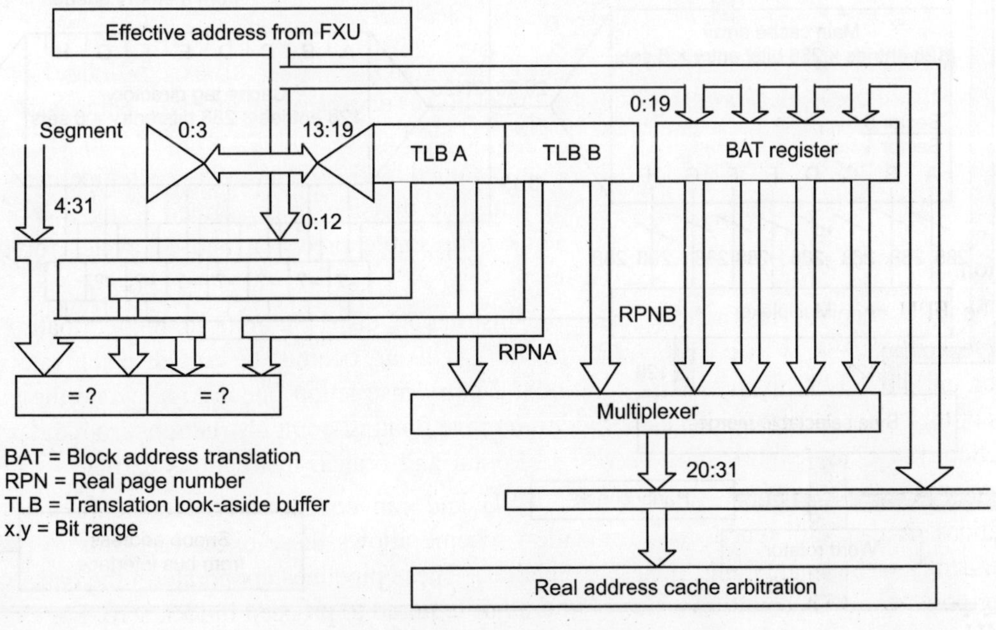

Fig. 5.11 (*a*) Memory management unit

The architecture of PowerPC specifies a hashed page table structure and a fixed-page size of 4KB. The portions of the virtual address are used to hash into the page table entries that are subsequently examined for matching virtual page numbers. A matching entry produces a real page number that can then be used to access real memory. If no match is found after 16-attempts, a page fault exception is taken. In an effort to speed up the page-based address translation, the 601 μP provides a 256-entry, 2-way set-associative TLB cache that is used for both instructions and data. Updates to the TLB and maintenance of the storage access recording (reference and change) bits are handled in hardware.

The BAT (the block address translation) mechanism allows upto 4-block mappings that can vary in size form 128KB to 8MB. This is useful for handling large sections of memory that are not subject to demand paging but still require address translation and protection. Proper use of the block translation registers can reduce TLB thrashing and increase overall processor performance.

5.12 CACHE UNIT

The 601 μP contains a 32K on-chip cache. The cache available on 601 μP is larger than cache available on any CISC μP. Also it is larger than the on-chip caches on most of the RISC μPs. The ability of Motorola and IBM to stuff a 32K cache onto an affordable and relatively small RISC processor speaks volumes about the carefully crafted design of the 601 and about the sophistication of IBM's fabrication technology. The cache organization is shown in Fig. 5.12 (*a*).

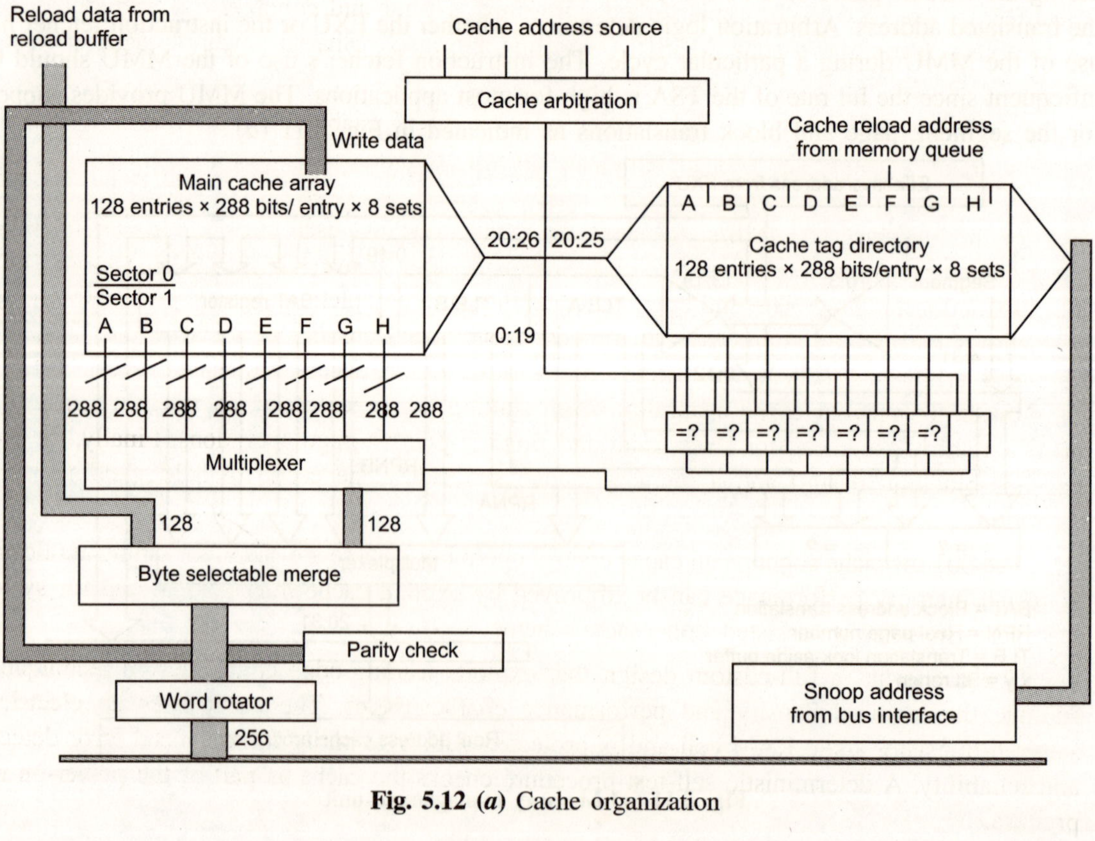

Fig. 5.12 (*a*) Cache organization

One interesting quirk is that the cache is unified. It simply means that instructions and data are stored in the same cache. The IBM pioneered split-cache concept that separates data and instructions into 2-separate caches. Split caches are a staple of the POWER RS/6000 technology. It is important to keep in mind that the main goal of the 601 design was to provide maximum performance at the lowest possible price. A unified cache is less expensive to implement than a split cache and does provide a few benefits. But a split cache provides better μP thruput than a unified cache. In fact, the PowerPC 603 chip also implements a split cache, chiefly to enhance power management support.

The cache control logic is somewhat complex and involves such concepts as set associativity, cache data and cache tag memories, translation look aside buffers and read/write protocols. These concepts of cache directly affect the way software will run on specific Macintosh and IBM PowerPC desktop systems.

The PowerPC 601 also includes a separate Memory Management Unit (MMU) and Memory Unit (MU) that translate and manage virtual and physical memory addresses and read from and write data to the on-chip cache.

The 601 μP provides a unified 32KB, 8-way set-associative cache. Each 64B holds two 32B sectors. The cache indexing and cache tag comparisons use the real (physical) address. In general, the cache makes use of a store-in (or copy back) policy to store data. However, the cache allows page and block-level control of cachability, write through, and coherency (via the MMU). An LRU cache replacement algorithm is used and coherency is maintained on a sector basis using the 4-state MESI (Modified Exclusive Shared Invalid) cache coherency protocol.

It is important to make the most effective use of the available cache bandwidth in a unified cache implementation. The 601 μP employs several techniques to achieve this. The 1st is the cache interface to the fetch unit and the memory queue is 8-word wide. At 66MHz, the data rate for this interface is 2.1GB/s. Although all 8-words are not used every cycle, this bandwidth speeds sector cast-outs (cast-outs refer to modified data moves to memory), snoops pushes and instruction fetching. The 2nd employs capability of a complete real-modify-write every cycle that allows the 601 microprocessor to process store operations in one cache cycle. The 3rd, a balanced arbitration scheme prioritizes the various cache access requests that can occur in each cycle. To prevent cache arbitration from stalling the execution units, operation queuing is provided above the cache for fixed-point and floating-point storage operations. Finally, the cache has non-blocking design and a memory queue sits below the cache to hold pending storage operation.

The 601 μP cache supports all cache control operations. These are useful for tuning applications in which memory performance can be improved by explicit cache management and for systems that perform software-assisted coherency scheme.

The cache has a full-custom design that exploits leading edge circuit design techniques to achieve the required density and performance characteristics. The cache uses an electrically stable 6-transistor static RAM cell and is protected with byte parity for enhanced error detection and reliability. A deterministic self-test procedure checks the cache as part of the power-on reset process.

Memory queue

The memory queue contains 2-entry real-address queue and a 3-entry write-address/data queue. The write queue holds up to 3-32B sectors of dirty (modified) data. The read queue, the write queue, and in some cases the cache itself, arbitrate for access to the 601 bus. The memory queue maintains full hardware-enforced coherency to the processor, cache and memory. Additionally, the memory queue supports a dynamic reload feature that permits background fetches for the other sector of a cache line after the critical sector has been successfully loaded. When enabled, this feature provides an automated cache prefetching capability for both instructions and data.

5.13 BUS INTERFACE

The BIU of the 601 translates the memory requests from the memory queue and the cache into transactions on the bus interface. It has a 32-bit address bus and a 64-bit data bus and uses an efficient bursting protocol to achieve high data thruput. The TTL or CMOS-compatible interface runs at the microprocessor clock rate or any integer multiple of the processor clock period. Figure 5.13 (*a*) is a block diagram of the BIU with the memory queue.

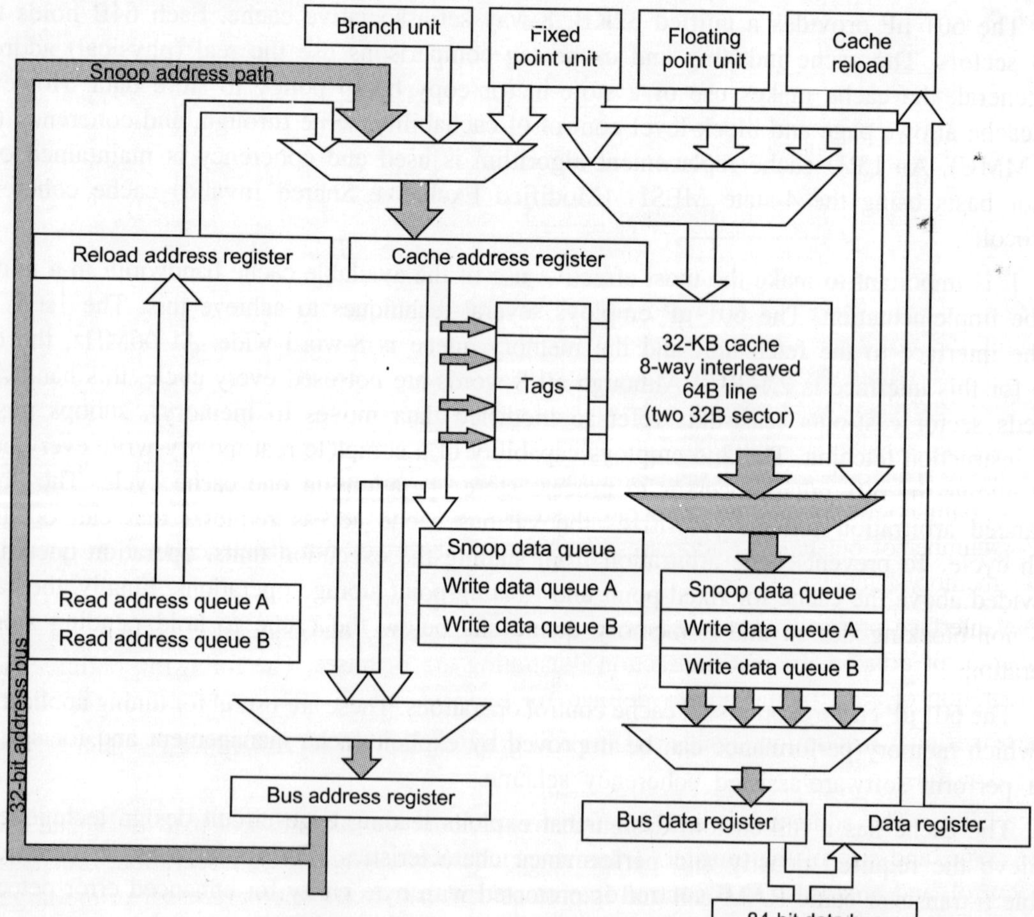

Fig. 5.13 (*a*) Memory path block diagram

The BIU employs a bursting protocol that uses one address to transfer 4-groups of data (8B each). While the theoretical limit of data bus thruput is the product of bus width and bus frequency, the realizable bandwidth is limited by factors such as interface technology, bus arbitration overhead, address transfer overhead, memory latency, and inter-processor communication. In the 601 microprocessor, consecutive burst transfers require one-cycle separation for consecutive thruput of 422MB/s at 66MHz. The protocol specifies independent operation of the address and data buses to allow the bus to approach its bandwidth limit.

A simple bus protocol employs a tenured bus in which the memory latency governs the use of the address, as shown in the 1st example in Fig. 5.13 (*b*).

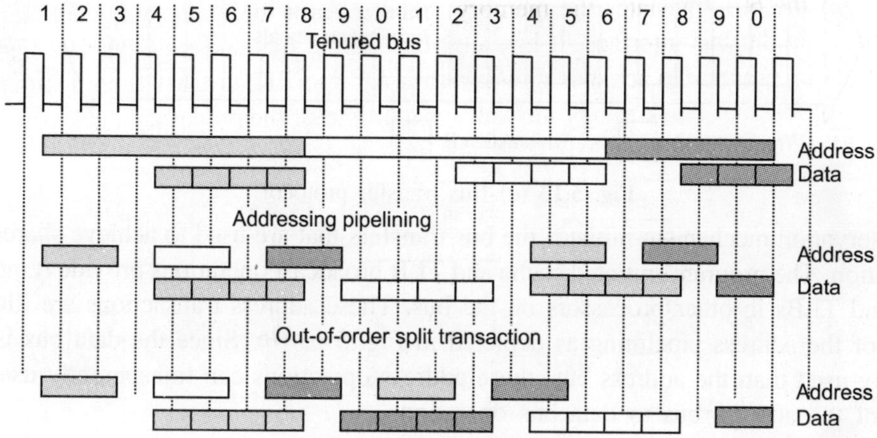

Fig. 5.13 (*b*) Bus optimization

The actual data bandwidth can significantly be degraded if the memory latency is long. The 602 address and data buses operate independently and new address can be launched before the preceding data transfer has completed, as shown in the address pipeline example, to overcome this problem. The 601 μP itself can pipeline up to 2-outstanding addresses on the bus. In multiprocessing configuration, the protocol can support out of order completion of translations and a greater number of outstanding addresses. This further improves bus efficiency in systems with variable memory latency.

The interface protocol uses 3-phases to control the address and data buses. Figure 5.13 (*c*) is a diagram of a basic 601-bus transaction illustrating the 3-phases. The following outlines the basic operation of each of the phases defined for the protocol.

Arbitration : Arbitration signals request and grant ownership separately for the address bus and the data bus.

Transfer : The address bus and the transfer attribute signals are driven to designate the location, type, and size of the transfer. Transfer include burst, single beat, reservation, broadcast cache control, and broadcast TLB control operations. Either the 601 microprocessor or the target of the transfer drives the data bus signals.

Termination : Termination signals conclude the bus tenure separately for each bus.

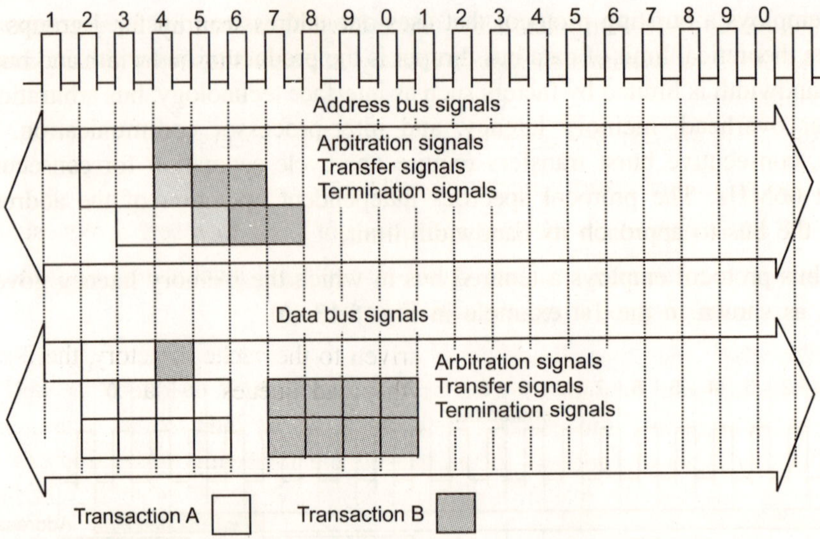

Fig. 5.13 (*c*) Bus transfer protocol

The reservation mechanism initiates the bus transfers that are used to achieve shared memory synchronization. The program-initiated cache and TLB broadcast operations provide remote control of caches and TLBs in other processors on the bus. These address transactions are illustrated in the middle of the address pipelining as depicted in Fig. 5.13 (*b*). Since the data bus is typically more heavily used than the address bus, these address operations can transparently use the spare bandwidth of the address bus to data bus operations.

The 601μP maximizes opportunities to use the bus resource by prioritization of internal request for access to the bus. As discussed earlier, the PowerPC architecture defines a weak consistency model that for most instruction streams, allows memory operations to complete in a different order than programmed. This has 2-positive effects on bus thruput. Allowing longer read operations latency to start on the bus before the older (with respect to program order) store operations, the hardware reduces the latency of the read operation. Also, placing the dirty cache sector cast-outs and the dynamic reload request lowest in priority allows these opportunistic operations to attempt to fill the gaps between higher priority bus operation.

The 601μP minimizes memory access latency through bus parking, speculative forwarding of data, and a short cache-to-memory path. Bus parking eliminates an arbitration cycle by allowing the external arbiter to grant the 601 μP the bus before the 601 makes a request. Speculative data forwarding is permitted through the definition of a data retry function that can be used to invalidate the double word transferred on the previous bus clock cycle. This is useful for systems employing secondary cache system that forward data prior to the tag comparison and for correction algorithm that protect memory ingrate. The 601 μP reflects substantial design effort to minimize latency from cache miss detection to the return of instructions or data to the cache.

5.14 BUS TIMING EXAMPLE

The system-level optimizations are critical to performance, particularly as the processor core often operates at a multiple of the bus clock frequency. Reducing memory latency at the system bus level by a bus clock cycle can potentially reduce latency as seen by the processor by several clock cycles. The 601's innovations to reduce latency are best shown by a case example using the following assumptions

- the bus is running at the same frequency as the processor;
- there are no higher priority operations queued;
- the 601 μP is parked on the bus and does not need to assert a request; and
- the 601 μP is connected to a fast, direct mapped 2nd-level cache using the speculative data forwarding capability of the 601 interface.

In the clock cycle-1, the physical address is driven to the cache directory, the 2-read queues, and the bus address register in parallel. One of the read queues is loaded as well as the bus address register. Also in cycle-1, the result of the cache directory indicates a cache miss; therefore, the bus controller can drive the address to the bus on the following cycle. Since it has already been granted the address bus, the 601 μP drives the address in clock cycle-2 as in Fig. 5.14 (*a*).

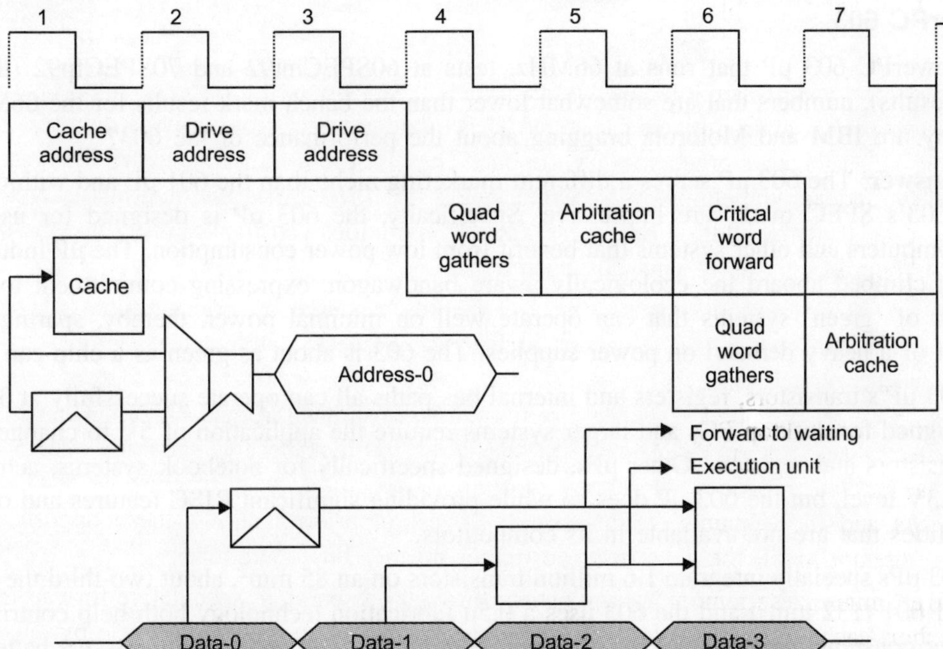

Fig. 5.14 (*a*) Cache miss timing

The memory system responds with the 1st double word of data in clock cycle-3, and the subsequent double words of data in clock cycle-4, 5, and 6. The bus interface logic gathers 2-beats (4-words) of data together before requesting access.

Deadlock prevention

In hierarchical bus environments that consist of a processor bus connected via a bridge component to an I/O bus like EISA (Extended Industry Standard Architecture), a special scenario arises. If the processor and a master on the I/O bus simultaneously request data from the other bus, one of the requests cannot continue. A processor may have its request retried so that the

I/O bus request can proceed. However, this solution may cause difficulties with devices that should see an address only once for a single request. The 601 μP overcomes this by allowing a processor request to be suspended after the address has been transferred, thus allowing the I/O bus to proceed and complete before the 601's transfer is completed.

Multiprocessing features

The 601 μP maintains coherency with a granularity of 32B (equal to a cache sector). The granularity refers to the smallest block of a memory that is acted on to maintain coherency. The 601 μP also distinguishes between global and non-global transactions. Only global transactions are monitored (snooped) by other processor. The bus supports the MESI (Modified Exclusive Shared Invalid) coherency protocol used by the cache. Each bus transaction is classified, so that the other processors maintain coherency. Other processors use this information to update the MESI state of their cache entries. Conversely, the 601 μP must snoop the bus for addresses in its own cache and generate an appropriate snoop response (shared hit, dirty hit, miss).

5.15 PowerPC 603

The PowerPC 603 μP that runs at 66MHz, tests at 60SPECint92 and 70SPECfp92 (these are IBM's results), numbers that are somewhat lower than the bench mark results for the 66MHz 601 μP. Why are IBM and Motorola bragging about the performance of the 603?

Easy answer: The 603 μP serves a different marketing niche than the 601 μP and within his niche, the 603's SPEC marks are impressive. Specifically, the 603 μP is designed for use in notebook computers and other systems that benefit from low power consumption. The μP industry has recently climbed aboard the ecologically aware bandwagon, expressing commitment to the development of 'green' systems that can operate well on minimal power, thereby, sparing the environment of a heavy demand on power supplies. The 603 is about as green as a chip can get.

The 603 μP's transistors, registers and internal bus paths all can operate successfully at 3.3V. All μPs designed for desktop PCs and larger systems require the application of 5V to change the state of transistors and registers. Other μPs, designed specifically for notebook systems, achieve this same 3.3V level, but the 603 μP does so while providing significant RISC features and other chip capabilities that are not available in its competitors.

The 603 μPs specially integrate 1.6 million transistors on an 85 mm^2, about two-third the size of the die of 601 (132 mm^2) and the 603 uses a 0.5μ fabrication technology both help contribute to low power consumption; essential for notebook systems that can not contain massive batteries or employ large heat sinks, powerful fans, and other system designs that help dissipate heat in larger, desktop systems.

The 603 μP also includes some specific design characteristics that go even further toward reducing power consumption. The 603 μP implements a split cache (separate caches for data and instructions), each cache of 8K. (The total cache size of the 603 is 16K, half the size of the 601's unified cache). The split caches help to maximize power consumption because built-in-power-consumption management technology can shut down either the instruction cache or the data cache during inactive periods.

The 603 μP actually adds functional units over and above the 3-instruction dispatch units provided by the 601 μP. Figure 5.15 (*a*) shows a block diagram of the 603 μP. The 603 incorporates a separate load/store unit (LSU) and a system register unit (SRU), both of which help to contribute to power management. The LSU manages the transfer of data between the on-chip cache and general-purpose registers. The SRU is responsible for executing instructions that use special-purpose registers (rather than the general purpose registers available in the IU) and condition registers (branch-processing instructions).

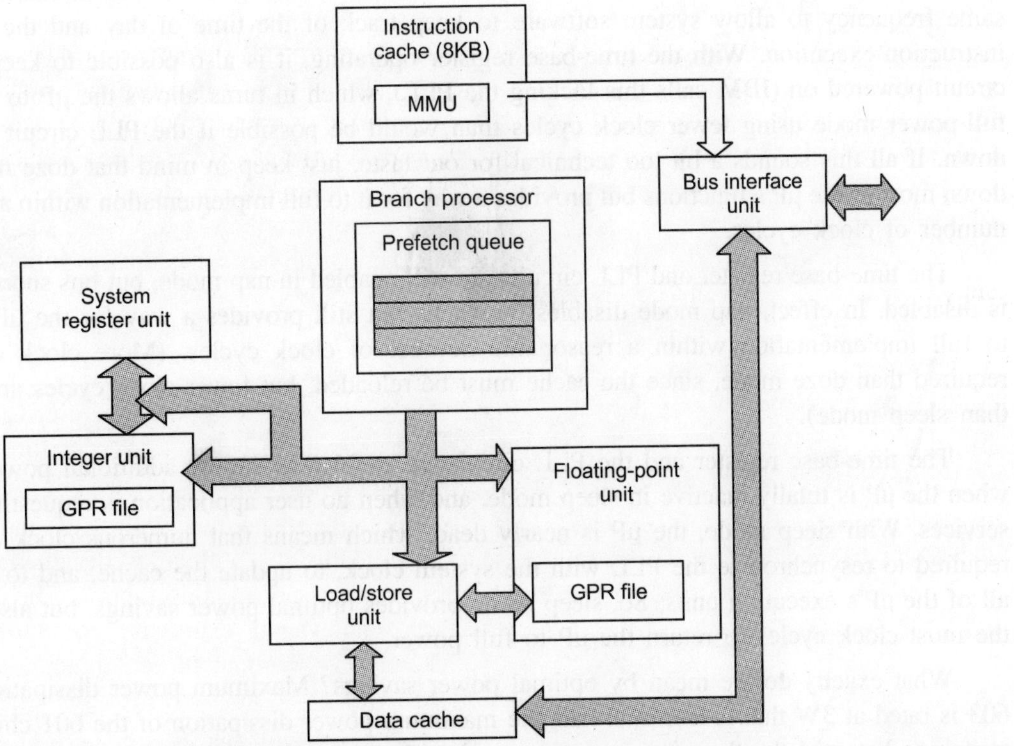

Fig. 5.15 *(a)* Block diagram of 603 chip

The additional execution units on the 603 µP are not designed to increase the number of instructions executed per clock cycle. They are included to improve power conservation. The LSU and SRU can both be disabled if they are not currently being used, providing significant power savings.

The 603 processor includes three built-in power saving modes

- doze,
- nap, and
- sleep.

The 3-modes, in general, operate under control of a phase lock loop (PLL) clock multiplier circuit. The PLL allows internal operations within the µP to run at multiples (1, 2, 3, or 4 times the speed) of the bus transfer rate. This technique is especially useful when one of the 3-power saving modes is in effect, since internal processing can zoom along even while external operations are using a slower clock speed to conserve power. In doze mode, the 603 µP shuts down all the processor's execution units and subsystems (after determining they are inactive) and allows the μP's bus snoop logic to operate. Bus snooping is a clever way of describing the ability of the μP to monitor all bus transfers to determine whether a memory access will affect the contents of the on-chip cache. In doze mode, bus snooping allows the μP to support caching of incoming data and instructions, even if the execution units of the μP itself are not busy.

The chip in doze mode continues to update the time-base register that is incremented at the same frequency to allow system software to keep track of the time of day and the timing of instruction execution. With the time-base register operating, it is also possible to keep the PLL circuit powered on (IBM calls this locking the PLL), which in turns allows the μP to recover to full-power mode using fewer clock cycles than would be possible if the PLL circuit were shut down. If all this sounds a bit too technical for our taste, just keep in mind that doze mode shuts down most of the μP's functions but provides a way for it to full implementation within a minimum number of clock cycles.

The time-base register and PLL circuits are still enabled in nap mode, but bus snooping logic is disabled. In effect, nap mode disables the cache but still provides a way for the μP to return to full implementation within a reasonable number of clock cycles. (More clock cycles are required than doze mode, since the cache must be reloaded, but fewer clock cycles are required than sleep mode).

The time-base register and the PLL circuit are enabled to supply additional power savings when the μP is totally inactive in sleep mode, and when no user application is requesting any μP services. With sleep mode, the μP is nearly dead, which means that numerous clock cycles are required to resynchronize the PLL with the system clock, to update the cache, and to power up all of the μP's execution units. So, sleep mode provides optimal power savings, but also requires the most clock cycles to return the μP to full power.

What exactly do we mean by optimal power savings? Maximum power dissipation for the 603 is rated at 3W that is less than half the maximum power dissipation of the 601 chip. But by switching between the three power-saving modes, the average power dissipation of the 603 can drop to below 2W, even as low as 1W during minimum use. This level of power saving is comparable and in many cases even better than the features of Intel's power-saving 80486 chip designed specially for notebook use. Existing Pentium designs do not even come close, and in fact would lead to full-scale meltdowns if they were implemented in notebook computers. Also it should be kept in mind that the 603 chip is available at 66MHz and 80MHz versions (in full-power mode), higher clock speeds than most 80486 notebook designed μPs can attain.

Future projection

In February 1995 at both Design Conference and the International Solid-State Circuit conference, the IBM and Motorola announced the introduction of 1st Silicon newest member of PowerPC RISC processor line, the PowerPC 602 and the PowerPC 603e. These processors have close linkage with the low-cost and low-power application PowerPC 603. The IBM and Motorola Engineers optimized the features as in Table 5.15 (*a*).

603e

It uses 0.5μ micron 4-level-metal static CMOS technology and operates with 3.3V. It packs 2.6 million transistors on 8.4 mm × 11.67 mm die. It is available both as 240-pin ceramic QFP (quad Flat Package) and a 16 × 16 BGA (Ball Grid Array). The most important enhancement is that 603e supports faster clock. It operates at 100MHz while the maximum clock frequency of 603 is 80MHz. It also supports very wide range of clock multipliers i.e., 1x, 1.5x, 2x, 2.5x, 3x, 3.5x, and 4x. This facilitates the system designer to hold own power consumption by using slower clock rate. The 603e, for an example, runs at 100MHz using 1.5 multiplier.

Table 5.15 (*a*) Comparison of PowerPCs

PowerPC	*602*	*603*	*603e*
Transistors used	1.0 million	1.6 million	2.6 million
Die size	50 mm^2 (7.07mm*7.07mm)	85.1 mm^2 (11.5mm*7.4mm)	98 mm^2
Power consumption	1.2W @66MHz	2.5W@80MHz	3W @100MHz
Pins	144-plastic QFP	240 CQFP	240-ceramicQFP
Clock frequency	66MHz	66/80MHz	100MHz
Supply voltage	3.3V	3.3V	3.3V
Address	32-bit (Multiplexed addr./data bus)	32-bit ($A_{31} \Leftrightarrow A_0$)	32-bit (Multiplexed addr./data bus)
Data	64-bit	64-bit/32-bit	64-bit
Cache size	4K, 2-way	8KB, 2-way set association	16KB, 4-way
Execution units	4	5	5
Instruction dispatched/cycle	1	2	2
Performance (Integer) (Floating Point)	40SPECint92 @66MHz N.A.	75SPECint92 @80MHz 85SPECfp92	120SPECint @100MHz 105SPECfp92

QFP ⇒ Quad Flat Package, CQFP ⇒ Ceramic Quad Flat Pack

The 2nd, important change in 603e is the large on-chip cache. It has separate 16KB 4-way set associative caches for data and instructions.

The IBM and Motorola were hard at work designing a 604 chip to outstrip the 601, chiefly by incorporating a 4-instruction per cycle design that uses 3-inter units (rather than the single integer unit available with the 601 and the 603 chips) and by fine-tuning floating-point operations. The IBM and Motorola are also developing a 620 chip that supports the full 64-bit set of instructions defined in the PowerPC architecture. The IBM is also reportedly at work on an intermediate, 615 chip that would reportedly provide full Intel 80x86 instruction set emulation on a sub-unit of the μP. Such a chip will make it possible for users to run PowerPC, DOS, and Windows applications all at near-native speeds. If the 615 chip employs sophisticated register renaming techniques (and nobody at IBM is talking right now), it would even run DOS and Windows applications at PowerPC like speeds.

Even though the 604, 615, and 620 chips are in development, they still represent speculative technology. The applications and technology opportunities will come when the PowerPC 604, 615, and 620 chips see widespread use on desktop and multiprocessor systems.

6

MOTOROLA μPs

6.1 INTRODUCTION

Motorola produced another series of popular microprocessors. The MC6800 and MC68000 are 8 and 16-bit Motorola microprocessors. It is housed in one of the following type of package

- 68-pin DIP (Dual-in-line-package),
- 68-terminal Chip Carrier,
- 68-pin Quad Pack,
- 68-pin Grid Array.

Pins $D_{15} \Leftrightarrow D_0$ are the 16-bit bi-directional data bus of MC68000. These data bus are used to transfer the data to and from memory and or I/O devices to MC68000. The $A_{23} \Leftrightarrow A_1$ are the 23-bit address bus. The lowest bit address line A_0 is obtained from decoding of the lower data strobe (LDS) and the higher data strobe (HDS) lines. Thus, MC68000 has total of 24-bit address lines. This way Motorola avoided the difficulty of multiplexed data and address bus. It included separate 16-bit data bus and 24-bit address bus. The non-multiplexed address and data bus simplifies the interfacing and possibility of the delay in the demultiplexing hardware. Figure 6.1 (*a*) shows the Pin Grid Array (PGA) packaging arrangement of the MC68000. It has several numbers of V_{CC} and ground pins. Power is, thus, distributed all around to reduce noise problems at high frequencies.

It uses low-power HMOS technology called the HCMOS. It requires only one dc power supply of +5V. It operates on several frequencies such as 6MHz, 8MHz, 10MHz, 12.5MHz, 16.67MHz, and 25MHz. It is clear from the internal organization of MC68000, in Fig. 6.1 (*b*), that it does not include the on-chip clock circuitry. Therefore, MC68000 requires a crystal controlled external clock generator/driver circuit to generate the clock. A very simple clock generator circuitry could be as shown in Fig. 6.1 (*c*). This circuit uses 2-inverters, a crystal, one resistor, and one JK-FF. The MC68000 consumes about 1.5W power.

The MC8000 series are available in different versions such as MC68000, MC68020, MC68030, and MC68040. As Motorola's MC6800 and MC68000 supports memory mapped I/O and hence it does not include any IN or OUT instruction. The MC68000 operates on 5-different data types such as

- a 1-bit,
- a 4-bit (nibble) BCD digit,
- a 8-bit (byte),
- a 16-bit (2-byte)long word.

It has 56-basic instructions in all. The Motorola also produced 32-bit RISC (reduced instruction set computer) microprocessor with simplified instruction set designated as 88000/88100.

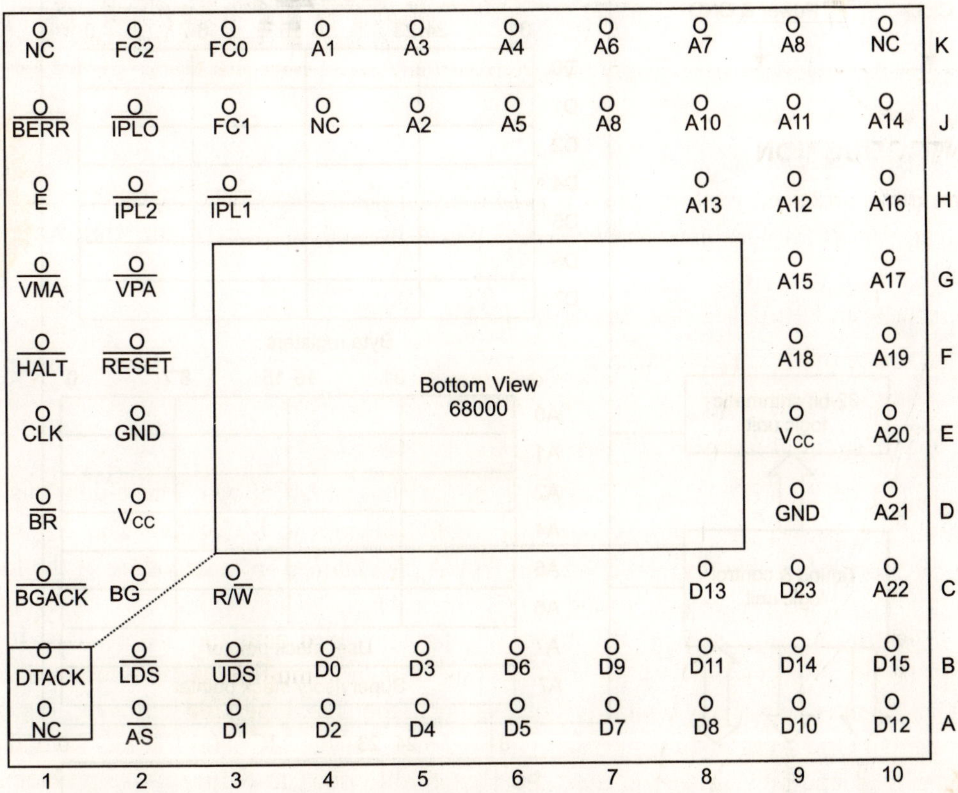

Fig. 6.1 (*a*) 68-pin PGA of MC68000

The MC68000 is known as general-purpose register microprocessor as its data register can be used as an accumulator or a scratch pad register. Even though the program counter (PC) of MC68000 is 32-bit wide, it uses only lower 24-bits to address the PC. Thus MC68000 can address up to $2^{24} = 2^{20}*2^{4}$ = 16MB memory space directly. The memory space may be extended up to 64MB using the function code lines.

6.2 REGISTERS OF MC68000

The MC68000 has a 32-bit internal architecture. It can perform most of the application on 8, 16, and 32-bits with the exception of 32-bit multiply and divide. It has 17-numbers of 32-bit registers as indicated in Figs. 6.1 (*b*) and 6.2 (*a*). Out of 17-registers, 8-numbers of them are designated as data registers $D_7 \Leftrightarrow D_0$ and 9-numbers of them are designated as address registers

$A_7 \Leftrightarrow A_0$, A7'. Again out of the 9-numbers of address registers, 2 are dedicated as the stack pointers. Depending on the mode of operation, MC68000 uses A7 or A7' as the user or supervisor stack pointer. The data such as byte (8-bit), word (16-bit), long word (32-bit), and BCD numbers (4-bit) are stored in the data registers. On the other hand, the address of the operand is stored in the address register. The organization of these registers of MC68000 is shown in Fig. 6.2 (*a*). It has 7-inerrupt levels.

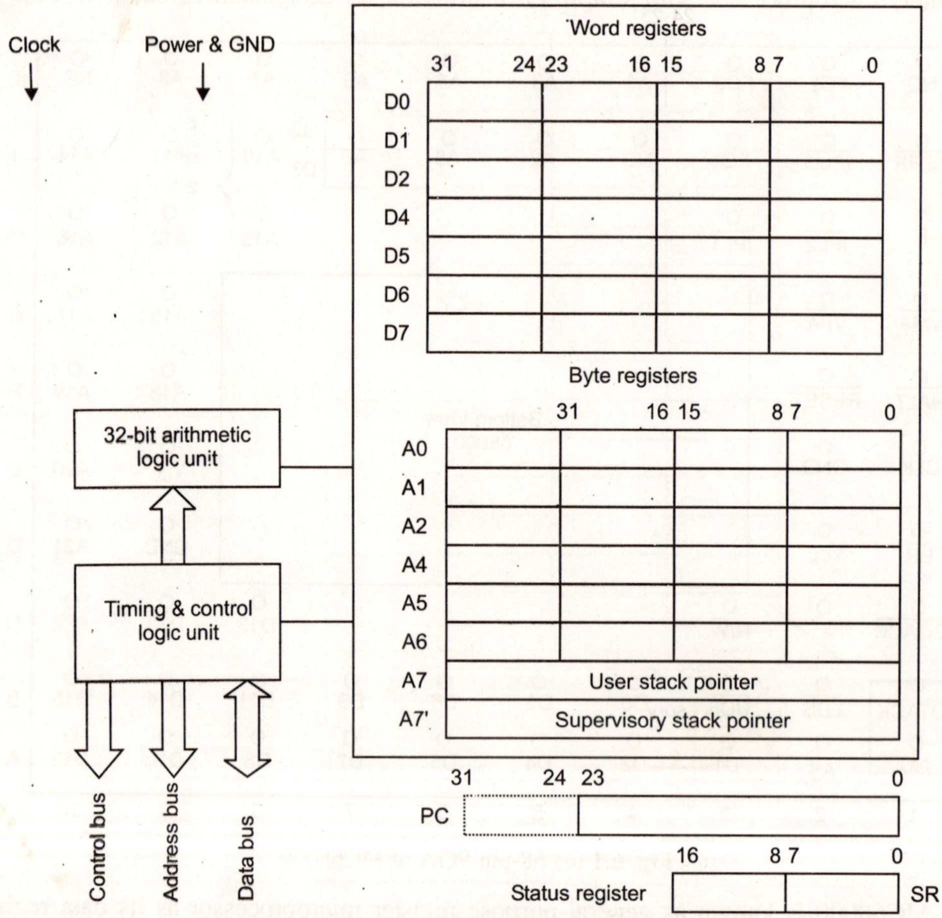

Fig. 6.1 (*b*) Internal organization of MC68000

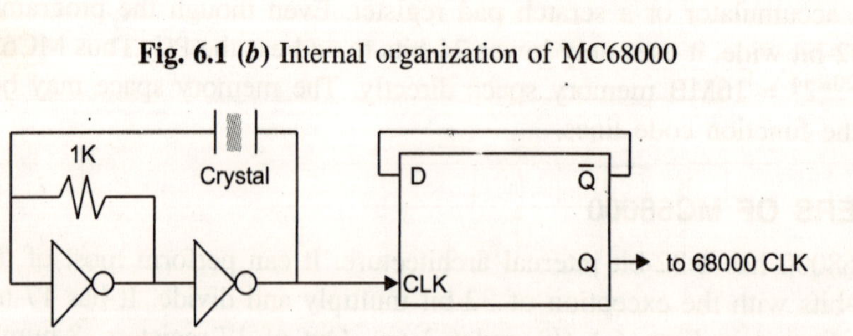

Fig. 6.1 (*c*) Clock generator

All 32-bit address registers can address either as 16-bit bus or 32-bit bus. An arrangement does exist that allows MC68000 to work as 24-bit address bus. The status register of MC68000 contains 2-bytes as indicated in Fig. 6.2 (*b*). These bytes are called

- System byte and
- User byte

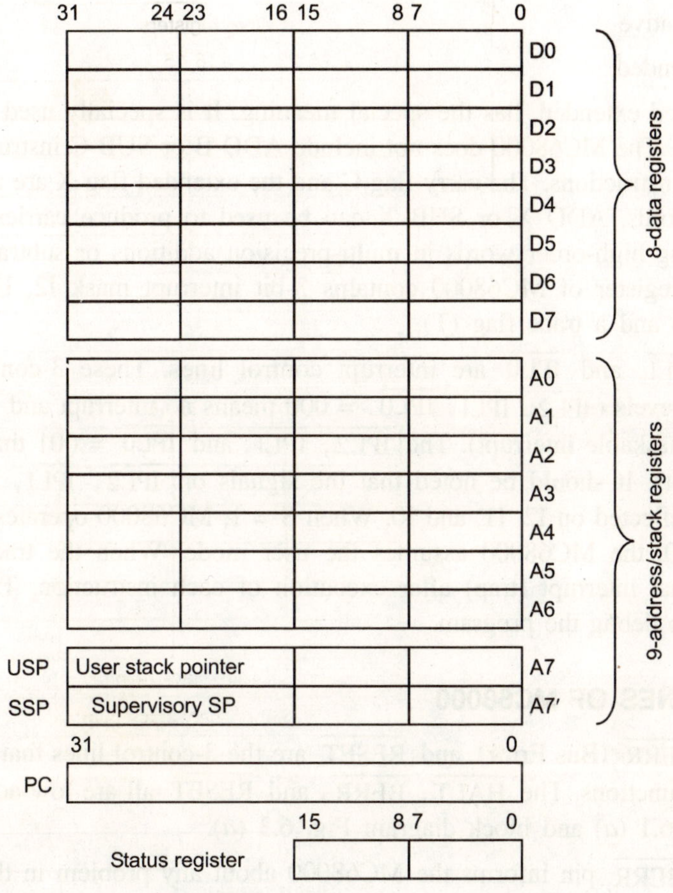

Fig. 6.2 (*a*) Register architecture of MC68000

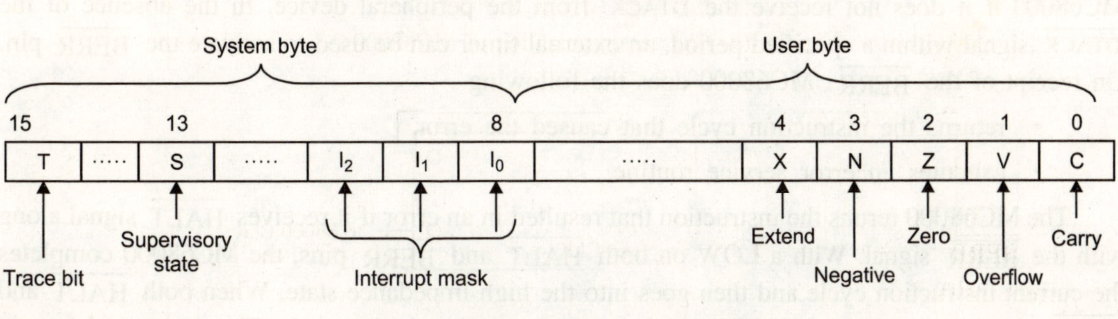

Fig. 6.2 (*b*) Status register of MC68000

The bit assignment for both bytes is shown in Fig. 6.2 (*b*). The user byte uses the conventional bits as

- C = Carry
- V = Overflow
- Z = Zero
- N = Negative
- X = Extended

The X-bit, called extended, has the special meaning. It is specially used for multi-precision arithmetic operation. The MC68000 does not include ADD B or SUB C instructions, rather it has ADD X or SUB X instructions. The carry flag C and the extended flag X are affected in identical manner. In other words, ADD X or SUB X can be used to produce carries or borrows while adding or subtracting high-order words in multi-precision additions or subtractions. The system byte of the status register of MC68000 contains 3-bit interrupt mask I2, I1, and I0. It has a supervisor flag (S) and a trace flag (T).

The $\overline{\text{IPL2}}$, $\overline{\text{IPL1}}$, and $\overline{\text{IPL0}}$ are interrupt control lines. These 3-control lines provides 7-interrupt priority levels ($\overline{\text{IPL2}}$, $\overline{\text{IPL1}}$, $\overline{\text{IPL0}}$ = 000 means no interrupt and $\overline{\text{IPL2}}$, $\overline{\text{IPL1}}$, $\overline{\text{IPL0}}$ = 111 means non-maskable interrupt). The $\overline{\text{IPL2}}$, $\overline{\text{IPL1}}$, and $\overline{\text{IPL0}}$ = 001 through 110 provides 6-maskable interrupts. It should be noted that the signals on $\overline{\text{IPL2}}$, $\overline{\text{IPL1}}$, and $\overline{\text{IPL0}}$ pins are inverted and then reflected on I2, I1, and I0. When S = 1, MC68000 operates in the supervisory mode. When S = 0, the MC68000 assumes the user mode. When the trace flag (T) = 1, it generates the internal interrupt (trap) after execution of each instruction. This provides single stepping facilities to debug the program.

6.3 CONTROL LINES OF MC68000

The $\overline{\text{HALT}}$, $\overline{\text{BERR}}$ (Bus Error), and $\overline{\text{RESET}}$ are the 3-control lines that are used to control the system related functions. The $\overline{\text{HALT}}$, $\overline{\text{BERR}}$, and $\overline{\text{RESET}}$ all are low active as indicated in PGA array of Fig. 6.1 (*a*) and block diagram Fig. 6.3 (*a*).

The input to $\overline{\text{BERR}}$ pin informs the MC68000 about any problem in the instruction cycle currently being executed. This problem may arise in the asynchronous mode of operation of MC68000 if it does not receive the $\overline{\text{DTACK}}$ from the peripheral device. In the absence of the $\overline{\text{DTACK}}$ signal within a specified period, an external timer can be used to activate the $\overline{\text{BERR}}$ pin. On receipt of the $\overline{\text{BERR}}$, MC68000 does the following

- returns the instruction cycle that caused the error,
- executes an error service routine.

The MC68000 reruns the instruction that resulted in an error if it receives $\overline{\text{HALT}}$ signal along with the $\overline{\text{BERR}}$ signal. With a LOW on both $\overline{\text{HALT}}$ and $\overline{\text{BERR}}$ pins, the MC68000 completes the current instruction cycle and then goes into the high impedance state. When both $\overline{\text{HALT}}$ and $\overline{\text{BERR}}$ pins become high, the MC68000 reruns the error instruction cycle. The cycle can be rerun repeatedly if both $\overline{\text{BERR}}$ and $\overline{\text{HALT}}$ are enabled/disabled continually.

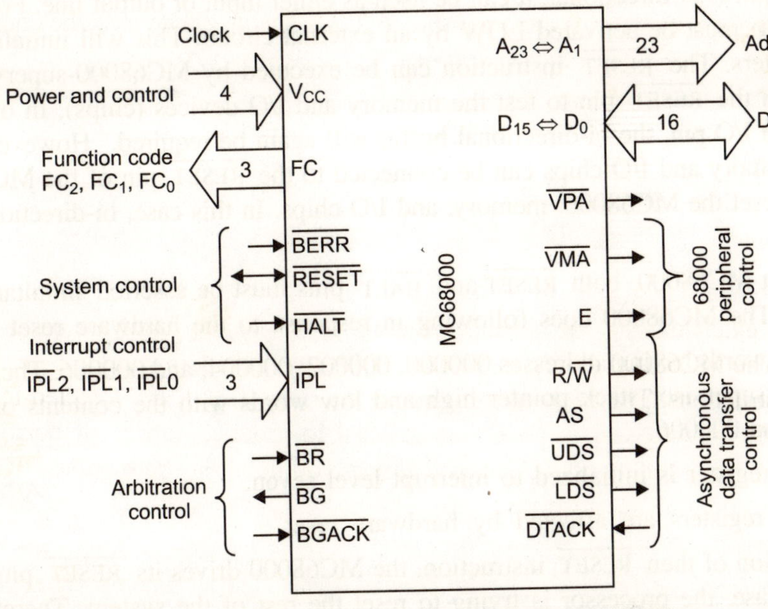

Fig. 6.3 *(a)* Bus representation of MC68000

An error service sub-routine is executed only if the $\overline{\text{BERR}}$ is received without $\overline{\text{HALT}}$. In such case, MC68000 will branch to a bus error vector address where the user writes a service routine. If two simultaneous bus errors are received via the $\overline{\text{BERR}}$ pin without $\overline{\text{HALT}}$, MC68000 automatically goes into the $\overline{\text{HALT}}$ state until it is reset.

The $\overline{\text{HALT}}$ line can also be used for performing the single stepping or providing DMA. When $\overline{\text{HALT}}$ is activated, MC68000 completes its current instruction and goes into the high impedance state until $\overline{\text{HALT}}$ returns to high state. The single stepping debugging can be achieved by enabling/disabling the $\overline{\text{HALT}}$ pin continually. However, as most of the instructions of MC68000 takes more than one instruction cycle, single stepping using $\overline{\text{HALT}}$ is usually not done. For single stepping, the trace bit of the status register is rather used. The $\overline{\text{HALT}}$ pin can be used to perform the microprocessor halt DMA. As MC68000 has separate DMA control line, the DMA using $\overline{\text{HALT}}$ pin is normally not performed. The $\overline{\text{HALT}}$ pin can also be used as an output signal. The MC68000 will assert a LOW on the $\overline{\text{HALT}}$ pin when it goes into the $\overline{\text{HALT}}$ state as a result of a catastrophic failure. The double bus error (activation of $\overline{\text{BERR}}$ twice) is an example of this type of error. When this occurs, MC68000 goes into a high impedance state until reset. The $\overline{\text{HALT}}$ line tells the peripheral device about the catastrophic failure.

The $\overline{\text{RESET}}$ line of MC68000 is also bi-directional. In order to reset MC68000, both $\overline{\text{RESET}}$ and $\overline{\text{HALT}}$ pins must be asserted at the same time. The MC68000 executes a reset service routine automatically for loading the program counter (PC) with starting address of the program. A LOW can be sent on this output line by executing the $\overline{\text{RESET}}$ instruction in the supervisory mode in order to reset external devices connected to MC68000. The execution of the $\overline{\text{RESET}}$ instruction does not affect any data, address, or status registers. Hence, the RESET instruction can be placed anywhere in the program whenever the external devices need to be reset. An external bi-directional buffer circuit is required to use the $\overline{\text{RESET}}$ pin for both input and output.

As the $\overline{\text{RESET}}$ pin is bi-directional, it can be used as either input or output line. For hardware reset, this $\overline{\text{RESET}}$ pin must be activated LOW by an external circuit. This will initialize the PC, SP, and status registers. The $\overline{\text{RESET}}$ instruction can be executed by MC68000-supervisor mode to output a LOW on the $\overline{\text{RESET}}$ pin to test the memory and I/O devices (chips). In order to use the $\overline{\text{RESET}}$ pin as an I/O pin, the bi-directional buffer will again be required. However, the reset input pins of the memory and I/O chips can be connected to the $\overline{\text{RESET}}$ pin of the MC68000 via the reset circuit to reset the MC68000, memory, and I/O chips. In this case, bi-directional buffers are not required.

In order to reset MC68000, both $\overline{\text{RESET}}$ and $\overline{\text{HALT}}$ pins must be asserted simultaneously by an external circuit. The MC68000 does following in response to the hardware reset

- It reads 4-words having addresses 000000, 000002, 000004, and 000006. The MC68000 loads the supervisor stack pointer high and low words with the contents of locations 000004 and 000006.
- Its status register is initialized to interrupt level seven.
- No other registers are affected by hardware reset.

With the execution of then $\overline{\text{RESET}}$ instruction, the MC68000 drives its $\overline{\text{RESET}}$ pin LOW for 124-clocks. In this case, the processor is trying to reset the rest of the system. Therefore, there is no effect on the external state of MC68000. All internal registers of MC68000 and the status register are unaffected by the execution of a reset instruction. All external devices connected to the $\overline{\text{RESET}}$ pin will be reset at the completion of the reset instruction.

Asserting $\overline{\text{RESET}}$ and $\overline{\text{HALT}}$ pins for 10-clock cycles will cause a processor reset, except when V_{CC} is initially applied to MC68000. In this case, an external reset must be applied for at least 100milliseconds. The reset circuit depicted in Fig. 6.3 (*b*) satisfies MC68000 reset requirements.

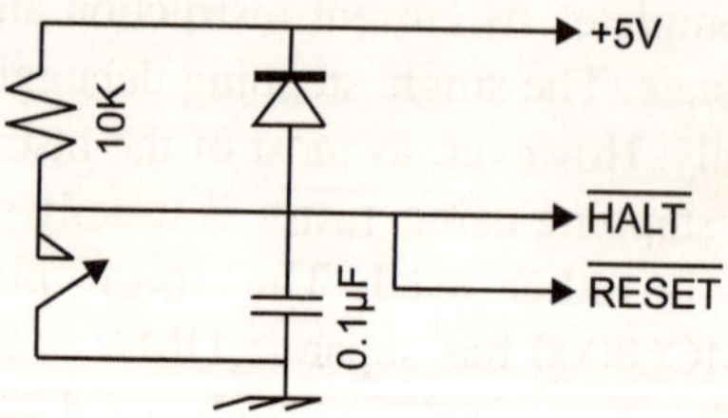

Fig. 6.3 (*b*) Reset circuit for MC68000

6.4 ADDRESSING MODES OF 68000

With 6-basic addressing modes, the MC68000 is capable of addressing in 14-different combination of modes. The 6-basic modes are

- register direct addressing,
- register indirect addressing,
- absolute addressing,
- immediate addressing,
- program counter relative addressing,
- implied addressing.

Multiprocessing

The MC68000 supports the multiprocessing in a simple and effective manner. The bus arbitration logic is in built inside the chip MC68000 for its uses in the multiprocessing environment. External hardware is not required to implement the multiprocessing environment. The 3-pins available on the chip are

- Bus request ($\overline{BR}$)
- Bus grant ($\overline{BG}$)
- Bus acknowledges ($\overline{BGACK}$)

These pins are used for the effective bus arbitration of MC68000. The MC68000 does not provide any priority of the bus request by the external processor. It works as the lowest priority device and grants request so long it is not being used by itself. An external bus request is achieved by pulling down the $\overline{BR}$ pin low. In response to the $\overline{BR}$ low, MC68000 makes $\overline{BG}$ line low after one clock duration. However, this does not mean that the bus is available for the use of external processor. In order to get satisfactory answer about the availability of the bus, certain conditions on other pins such as $\overline{AS}$ (address strobe), $\overline{DTACK}$ (data acknowledge), and $\overline{BGACK}$ must be available. When the conditions required for bus free is satisfied on all these pins, the requesting device asserts the $\overline{BGACK}$ line, informing that it has taken over the control of the bus. The MC68000 remains in the tri-stated condition until the $\overline{BGACK}$ signal is withdrawn by the requesting device.

Signal category

The MC68000 signals can be divided functionally into 5-groups, namely

- Synchronous and asynchronous controls
- System control
- DMA control
- Status control
- Interrupt control

Synchronous and asynchronous controls

The bus control of the MC68000 is asynchronous. It means that once a bus cycle is initiated, the external device must send a signal back in order to complete it. The enable (E), valid peripheral address ($\overline{VPA}$), and valid memory address ($\overline{VMA}$) are the 3-synchronous control lines of MC68000. These lines facilitate interfacing to synchronous peripherals. The synchronous operation means that the bus control is synchronized or clock uses a common system clock.

In the synchronous controls, all READ and WRITE must be synchronized with the common clock. This may create problem when interfacing the slow peripherals. This problem does not arise in asynchronous bus control.

The asynchronous operation is not dependent on the common clock. The MC68000 uses its synchronous signals (lines) to transfer its data to and from MC68000 and the peripheral device via handshaking mode. It can be interfaced to any peripheral chip regardless of the speed in its synchronous mode.

The phase-2 clock (E clock) outputs at a frequency that is 1/10th of MC68000 input clock. The $\overline{\text{VPA}}$ is an input that tells the MC68000 that a 68000 device is being addressed. Therefore, the data transfer must be synchronized with the E clock. The $\overline{\text{VMA}}$ is the processor's response to $\overline{\text{VPA}}$. The $\overline{\text{VPA}}$ signal is asserted when the memory address is valid. This also tells to the external device that the next data transfer over the data bus will be synchronized with the E clock. The $\overline{\text{VPA}}$ can be generated by decoding the address pins and strobe ($\overline{\text{AS}}$). The MC68000 asserts $\overline{\text{AS}}$ low when the address on the address bus is valid. The $\overline{\text{VMA}}$ is typically used as a chip select of the 68000 peripheral. This ensures that the 68000 peripheral is selected or deselected at proper time. The MC68000 has 5-lines to control address and data transfers asynchronously. These pins are

- Address strobe ($\overline{\text{AS}}$)
- Read/Write ($\text{R}/\overline{\text{W}}$)
- Data acknowledge ($\overline{\text{DTACK}}$)
- Upper data strobe ($\overline{\text{UDS}}$)
- Lower data strobe ($\overline{\text{LDS}}$)

The $\overline{\text{AS}}$ is used to indicate the peripheral device when the data is to be transferred. The $\overline{\text{AS}}$ is active low when MC68000 provides a valid address on the address bus. The $\text{R}/\overline{\text{W}}$ is HIGH for read and LOW for write. The $\overline{\text{DTACK}}$ is used to inform MC68000 that a transfer is to be performed. In case MC68000 wants to transfer the data asynchronously, it must activate the $\overline{\text{AS}}$ line 1st and at the same time it must issue the required address on the address line to select the peripheral device.

As the $\overline{\text{AS}}$ line tells the peripheral chip when to transfer the data, the $\overline{\text{AS}}$ line should be part of the address-decoding scheme. As soon as $\overline{\text{AS}}$ is enabled, MC68000 enters into the Wait State until it receives $\overline{\text{DTACK}}$ from the selected device. On receipt of the $\overline{\text{DTACK}}$, MC68000 knows that the peripheral device is ready for data transfer. The MC68000 in such condition uses the $\text{R}/\overline{\text{W}}$ and data lines to transfer data. The $\overline{\text{UDS}}$ and $\overline{\text{LDS}}$ signals are defined as

$\overline{\text{UDS}}$	$\overline{\text{LDS}}$	Data transfer occur via	Address
1	0	$D_7 \Leftrightarrow D_0$ pins for byte	Odd
0	1	$D_{15} \Leftrightarrow D_8$ pins for byte	Even
0	0	$D_{15} \Leftrightarrow D_0$ pins for word or long word	Even

The $\overline{\text{UDS}}$ and $\overline{\text{LDS}}$ lines are used to segment the memory into bytes instead of word. When the $\overline{\text{UDS}}$ is asserted, the contents of even address are transferred on the high-order 8-lines of the data bus, $D_{15} \Leftrightarrow D_8$. The MC68000 internally shifts these data to the low byte of the specified register. When $\overline{\text{LDS}}$ is asserted, contents of odd addresses are transferred on the low-order 8-lines of the data bus, $D_7 \Leftrightarrow D_0$. During word and long word transfer, both $\overline{UDS}$ and $\overline{LDS}$ are asserted and information is transferred on all 16-data lines, $D_{15} \Leftrightarrow D_0$. The A_0 corresponds to $\overline{UDS}$ for even address ($A_0 = 0$) and to $\overline{LDS}$ for odd address ($A_0 = 1$). The even and odd addressees are interfaced as in Fig. 6.4 (*a*).

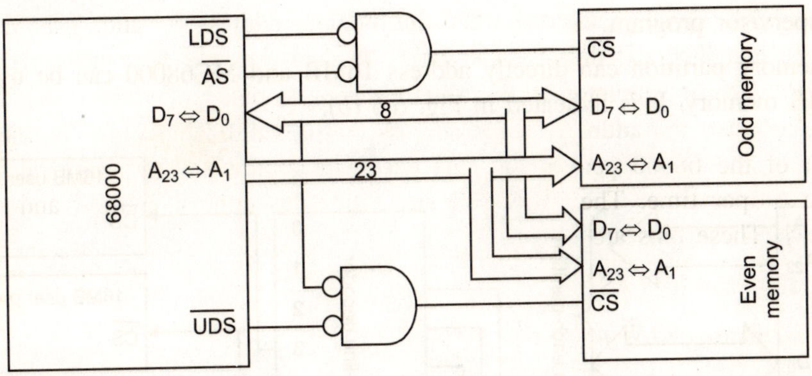

Fig. 6.4 (*a*) Even and odd memory segmentation

DMA control lines

The Bus Request ($\overline{BR}$), Bus Grant ($\overline{BG}$), and Bus Grant Acknowledge ($\overline{BGACK}$) lines are used for DMA transfers.

Status lines

The MC68000 has 3-output lines called the function code pins designated as

- FC_2
- FC_1
- FC_0

Table 6.4(*a*) indicates how these lines inform the external devices whether user data, user program, supervisor data, or supervisor program is being addressed.

Table 6.4(*a*) Status information

FC_2	FC_1	FC_0	*Operation*
0	0	0	Unassigned
0	0	1	User data
0	1	0	User program
0	1	1	Unassigned
1	0	0	Unassigned
1	0	1	Supervisor data
1	1	0	Supervisor program
1	1	1	Interrupt acknowledge

The FC_2, FC_1, and FC_0 pins are used to partition the memory into 4-functional areas such as

- user data memory
- user program memory
- supervisor data
- supervisor program.

Each memory partition can directly address 16MB and MC68000 can be used directly to address 64MB memory. It is indicated in Fig. 6.6 (*b*).

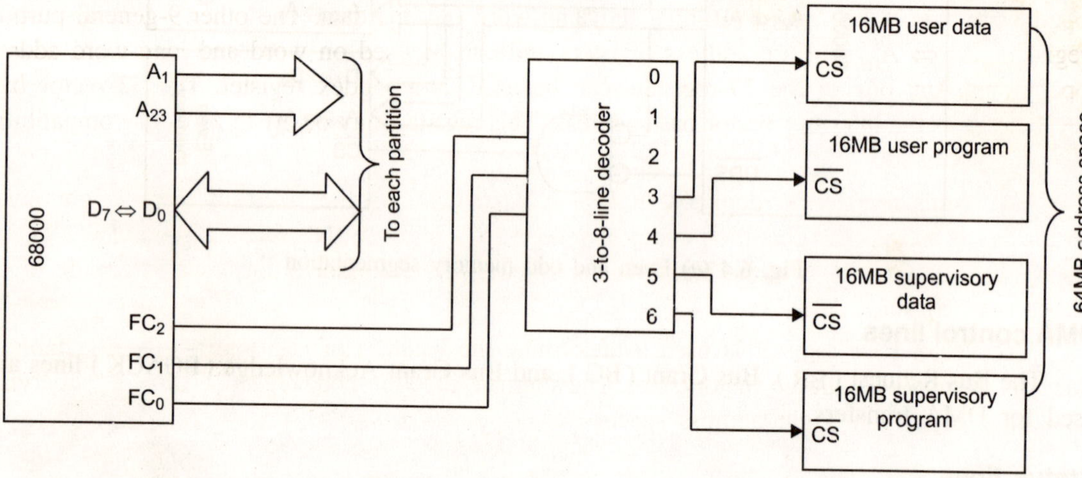

Fig. 6.4 (*b*) Partitioning of address space using FC_2, FC_1, and FC_0

6.5 MC68010

The MC68010 includes feature that improve operating system support and increases the processing speed. The Motorola's MC68010 is a 16-bit microprocessor with 32-bit registers, an expanded instruction set, and flexible addressing modes. It is identical in internal architecture of MC68000 except that this processor allows the efficient design of a virtual memory system. The MC68010 can be plugged in directly into MC68000 socket in an existing system without any additional hardware. One of the most important requirements of any new member of a processor family is that it does not require major revision of the software written for previous member of the family. For this reason the MC68010 was considered to be object code compatible with its predecessor MC68000. Since the extension provided by the MC68010 required changes in the processor interface to the operating system (i.e. exception processing, privileged instruction, and dispatching), the MC68010 was designed so that the required software modification would be confined to the operating system. It offers the following facilities to the user

- 17 number of 32-bit data and address registers,
- 16 MB direct addressing range,
- virtual memory/virtual machine support,
- 57-instruction types,

- high performance looping instructions,
- operation on 5-main data types,
- memory-mapped I/O, and
- 14-addressing modes.

Registers of MC68010

Similar to Figs. 6.1 (*b*) and 6.2 (*a*), the MC68010 also offers 17-numbers of 32-bit general-purpose registers, a 32-bit program counter, a 16-bit status register, a 32-bit vector-base register, and 2-numbers of 3-bit alternate-function code registers. $D_7 \Leftrightarrow D_0$ are 8-data registers and can operate on byte (8-bit), word (16-bit), and long-word (32-bit) data. The other 9-general purpose registers, $A_7 \Leftrightarrow A_0$, A'_7, are address registers and can be used on word and long-word address operations. Any one of the 17-registers can be used as an index register. The 32-vector base register stores the interrupt vector base address. This register may be set to zero for compatibility with the MC68000.

The status register as depicted in Fig. 6.2 (*b*) contains the interrupt mask (eight levels available) as well as the condition codes: extended (X), negative (N), zero (Z), overflow (V), and carry (C). The additional status bits indicate whether the processor is the trace (T) mode or the supervisor (S) state.

The vector base register is used to determine the location of the exception vector table in memory. It supports multiple vector tables. The alternate-function-code registers allow the supervisor to access any of the eight address spaces.

6.6 VIRTUAL MEMORY

The concept of the virtual memory came from the idea of utilizing the small expensive memory efficiently. A virtual memory system provides the user to execute the program on a very large store space of virtual address without regard to its physical existence. Figure 6.6 (*a*) provides a simple example of a system with 3-levels of hierarchy of the memory organization. The need to organize memory hierarchy is all the more acute as the clock frequencies at which processors execute continue to increase. In case of MC68000, a physical memory design that could hold the entire 16MB-address space and provide no wait-state access would involve substantial expense.

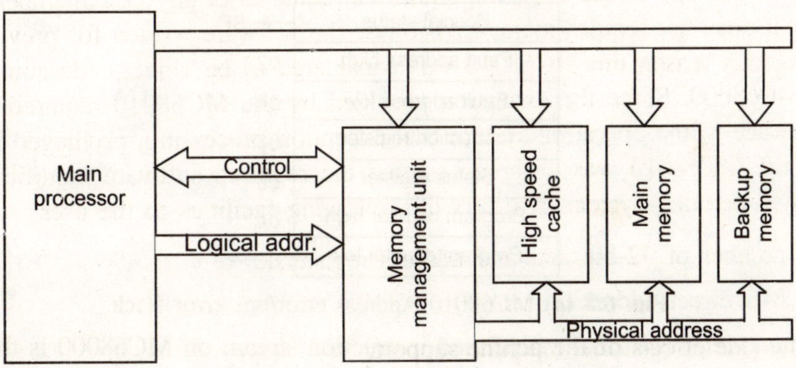

Fig. 6.6 (*a*) 3-Level memory hierarchy

The virtual and physical address spaces are divided into fixed size pages to facilitate mapping. The user need not have the knowledge about the organization of the physical memory into which his program is mapped. If an access is attempted to an address on a page that is not resident in the physical memory, a page fault occurs, interrupting the processor and initializing exception processing. The virtual memory system can correct this fault by fetching the page from an element lower in the hierarchy and substituting it for one of the pages in the physical store. While this replacement process is occurring, the processor is free to serve the other users and thus support multiprogramming more efficiently. After correcting the fault, the processor is then permitted to resume exception of the faulted program. The description is very general and the theory seems to be very memory on a processor, some of the details of implementation create several difficulties for either group designing the processor or the end user.

6.7 VIRTUAL PROCESSOR REQUIREMENT

An essential element in providing virtual memory support is the ability to recognize an access fault when it occurs. The MC68010 can recognize these faults both internally and externally. When an access fault is made to a misaligned instruction or operand, an address error is internally detected, initiating the address error-fault handling routine. Externally, the bus error ($\overline{BERR}$) pin provides the user with a method to signal that some aspect of the access has generated an error. In the context of the virtual memory, the Motorola 68451 memory management unit, or any other address translation device which can detect a fault situation, can signal a fault to the main processor via the $\overline{BERR}$ pin. The $\overline{BERR}$ is taken up as a control signal in Fig. 6.7 (*a*). Although this fault recognition capability does not need to be enhanced to support virtual memory, some of the timings associated with it were made more liberals to provide support for error detection and correction hardware.

Once the processor has recognized an access fault, the next step is to save any state information that will be needed to reconstruct the state of the machine after the fault has been corrected. The MC68000 saves only enough internal state information as indicated in Fig. 6.7 (*a*) to provide the user with an approximate indication of the state the processor when the fault occurred. This information, though providing the fault address, the function codes, and type of access, does not provide enough data to allow the internal states of the machine to be reconstructed.

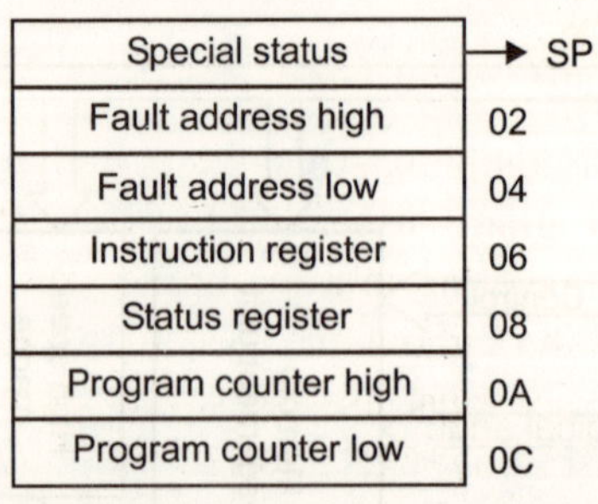

Fig. 6.7 (*a*) MC68010 Address error/bus error stack

One of the side effects of the pipelined instruction stream on MC68000 is that the program counter does not necessarily point at the instruction in which fault occurred, but rather points to the vicinity of the instruction. Furthermore, because of the pipelining, the instruction register is updated before the end of the instruction, an action that can result in the stack value of the instruction register also being misleading. In order to provide the required data, the MC68010 has to expand the size of the state that is stored on a fault from 7-words to 26-words. This stack frame is shown in Fig. 6.7 (*b*). It consists of the data stacked by MC68000 but also includes more detailed information about the access type, internal temporary registers, and various internal status bits. The stack is divided into two parts—a user-visible section in which everything that the user needs to know about the access and its correction are provided, and a non-user-visible section in which the internal status and temporaries are stored. After the internal state has been saved, the processor returns control to the operating system by providing a vector to the address error or bus error exception-handler routine. The operating system is then responsible for reconfiguring the system to either successfully complete the access or abort the process. While this activity is taking place, a different program can be dispatched to the potentially idle processor in a multi-programmed environment.

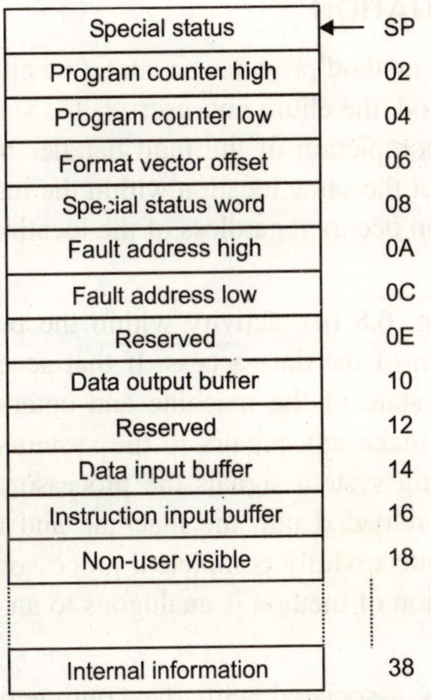

Fig. 6.7 (*b*) MC68010 address error/bus error stack

The last step is the most complex. After the operating system has made any repairs that are necessary, the state of the program suspended by the fault must be reloaded and the execution of that program resumed. If operating in a multiprocessor system, the suspended program can be dispatched to any processor throughout the system, which may not be the processor that was originally faulted. The processor must use the saved state information to reconstruct the internal state of the machine and must allow execution of the faulted program to resume. Reconstructing the internal state is composed of following steps

- reading the internal state from memory,
- and loading it into the machine,

- evaluating this state to determine what actions are needed to restore the state of the machine to its prefault condition.

In MC68000, an address error or a bus error is considered to be exception from which no recovery can be made. Hence, it has no facilities to allow the machine to rerun from these exceptions. The mechanism described here is an addition made on the MC68010.

Virtual memory implementation

There are two basic methods of implementing virtual memory on a processor

- instruction restart and
- instruction continues.

Both methods have their advantages and disadvantages. The MC68010 was implemented with less commonly used of the two; instruction continues method.

6.8 INSTRUCTION CONTINUATION

The instruction continuation method provides an attractive alternative to the instruction restart method. In the continuation method, the entire non-user-visible state of the machine is saved when an access fault is detected. On completion of the fault handler routine, the processor is allowed to resume instruction processing at the same location within the instruction at which execution was suspended by the fault. This action occurs regardless of the location within the instruction at which the access fault occurs.

As an example given in Fig. 6.8 (*a*), activity within the processor is suspended in micro-instruction B until the completion of the data access. If that access is terminated unsuccessfully, the processor saves the internal state of the machine and enters the exception handler routine. The operating system is free to make any repairs to the systems that are necessary. After these repairs are complete, the operating system signals the processor to restore the state and resume normal execution. The state is reloaded into the machine and the access that caused the fault is repeated. When the access is successfully completed, processor resumes execution with micro-instruction C. Thus the continuation of method is analogous to an interrupt operation at the micro-instruction level.

There are several problems associated with the continuation method of virtual memory support. It involves

- instructions that require execution without interruption,
- silicon resources that must be provided to support the saving and restoring of the internal state, and
- the time that is required for such saving and restoring.

In general, continuation method provides a more natural method of virtual memory support. It is less disruptive to simply suspend operations or interrupt execution at the micro-instruction level until fault has been corrected. The suspended program can then resume execution from the point at which execution was originally interrupted. Sometimes the instruction needs to be executed without interruptions. However, in this case continuation method is not suited at all to providing the required support. An example of this type of instruction is the test and set (TAS) of MC68000 instruction set. The TAS instruction provides the user with a primitive operation that can be used to implement resource protection mechanism via semaphores or other means. For the TAS instruction, the processor must provide an un-interruptible read-modify-write sequence during which the bus cannot be arbitrated away. If the fault occurs on the write portion of the access, it is necessary to reinitiate the entire read-modify-write cycle rather than simply rerun the write portion of the cycle. In this case there must be some mechanism to force the entire read-modify-write cycle to be reinitiated. The continuation that the entire internal state of the processor is saved when an access faults occurs. Because of the complexities of the MC68010, this state comprises a large number of temporary execution unit registers, sequencer state registers, and control latches. For the state to be saved, the information about it must be stored internally.

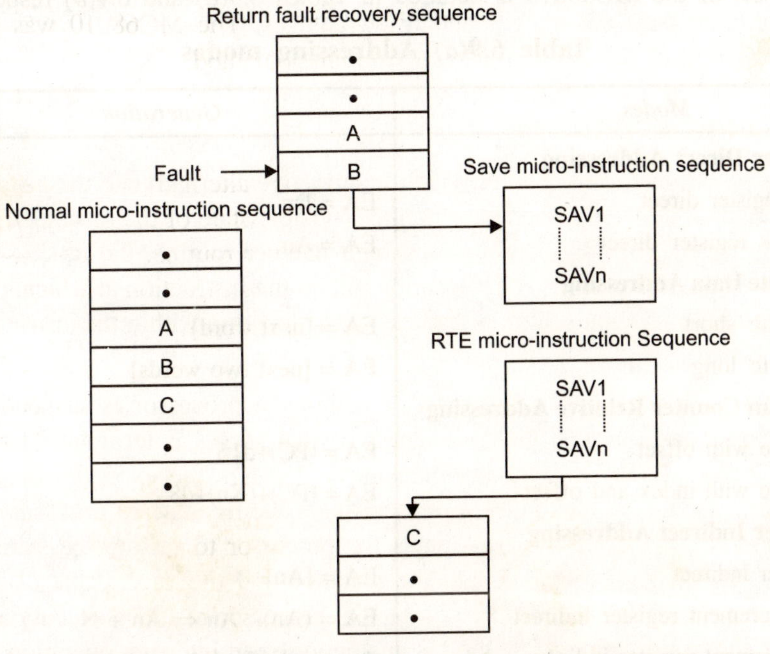

Fig. 6.8 ***(a)*** Continuation method fault recovery sequence

In addition, there must be access path from the state information, and there must be the control needed to access this information. Similarly, additional control logic is required to perform the reloading of the state. In terms of silicon area, the additional resources require a 22% increase over MC68000 in total area.

6.9 VIRTUAL OPERATION

The MC68010 provides the mechanism needed to implement a virtual machine environment in which any degree of emulation is supported. This is achieved in large part by the virtual memory mechanism. The virtual I/O, for an example, is readily achieved by defining a memory area as an I/O device, which is not physically resident. When an access is made to that address, an access fault occurs. The fault address can then be evaluated by the operating system to determine the activity that should take place. After the appropriate action has been taken, software rerun can be signaled and the return from exception (RTE) executed. Indicating to the processor that the access has been completed makes it possible to provide virtual I/O transfers. This technique can, of course, be generalized to any other type of virtual activity that the processor requests the operating system to execute.

Several microprocessors support virtual memory, each one providing different degree of such support. The MC68010 utilizing the instruction continuation method, cleanly and elegantly supports the fault detection/fault correction/program resumption process. The options available as a consequence of the use of continuation method—hardware and software rerun provide powerful support for various implementation of virtual memory. The continuation method also provides the ability to make any access virtual via the software-rerun method of return. The addressing modes and instruction set of the MC68010 is detailed in Tables 6.9(*a*) and 6.9(*b*) respectively.

Table 6.9(*a*) Addressing modes

Modes	*Generation*
Register Direct Addressing	
Data register direct	EA = Dn
Address register direct	EA = An
Absolute Data Addressing	
Absolute short	EA = [next word]
Absolute long	EA = [next two words]
Program Counter Relative Addressing	
Relative with offset	EA = (PC)+d16
Relative with index and offset	EA = (PC)+(Xn)+d8
Register Indirect Addressing	
Register indirect	EA = [An]
Post increment register indirect	EA = (An), An ⇐ An + N
Predecrement register indirect	An ⇐ An-N, EA = (An)
Register indirect with offset	EA = (An) + d16
Indexed register indirect with offset	EA = (An) + (Xn) + d
Immediate Data Addressing	Data = next word(s)
Immediate	Inherent data
Quick immediate	
Implied Addressing	EA = SR, USP, SSP, PC, VBR, SFC, DFC
Implied register	

where,

EA = Effective address
An = Address register
Dn = Data register

Xn = Address or data register used as index register

SR = Status register

PC = Program counter

() = Contents of

d8 = 8-bit offset (displacement)

d16 = 16-bit offset(displacement)

N = 1 for byte, 2 for word, and 4 for long-word if 'An' is the stack pointer and the operand size is byte,

N = 2 to keep the stack pointer on a word boundary = Replacement

Table 6.9(*b*) Instruction set summary

Mnemonics	*Description*	*Mnemonics*	*Description*
ABCD	Add decimal with extended	MULS	Signed multiply
ADD	Add	MULU	Unsigned multiply
AND	Logical AND	LSR	Logical shift right
ASL	Arithmetic shift left	NEG	Negative
ASR	Arithmetic shift right	NOP	No operation
BCC	Branch conditionality	NOT	1's complement
BCHG	Bit test and change	OR	Logical OR
BCLR	Bit test and clear	PEA	Push effective address
BRA	Branch always	RESET	Reset external device
BSET	Bit test and set	ROL	Rotate left without extended
BSR	Branch to subroutine	ROR	Rotate right without extended
BTST	Bit test	ROXL	Rotate left with extend
CHK	Check register against bounce	ROXR	Rotate right with extend
CLR	Clear operand	RTD	Return and de-allocate
CMP	Compare	RTE	Return from exception
DBCC	Decrement and branch conditionality	RTR	Return from restore
DIVS	Signed divide	RTS	Return from subroutine
DIVU	Unsigned divide	LSL	Logical shift left
EOR	Exclusive OR	SCC	Set conditional
EXG	Exchange register	STOP	Stop
EXT	Sign extended	SUB	Subtract
JMP	Jump	SWAP	Swap data register halves
JSR	Jump to subroutine	TAS	Test and set operand

(*Contd....*)

LEA	Load effective address	TRAP	Trap
LINK	Link stack	TRAPV	Trap on overflow
MOVE	Move source to destination	TST	Test
NBCD	Negate decimal with extend	UNLK	Unlink
SBCD	Subtract decimal with extend		

6.10 MC68020

The MC68020 is the 1st successful extension of Motorola's 16-bit microprocessor into its 32-bit version. It executes all user object code written for its previous member of MC68000 family. It provides a 32-bit data and address bus as well as 32-bit internal architecture. A natural and elegant implementation of MC68000 architecture, the MC68020 represents the state of the art in high performance microprocessor design. It has been designed to execute all programs written for previous members of MC68000 family at faster rate. Because of the value of the large software base available for MC68000 microprocessors, the MC68020 maintains the compatibility with the previous member of the family.

In addition to performance and compatibility, there are several system enhancements associated with the new microprocessor. It supports

- additional addressing modes,
- includes new instructions,
- extends all instructions to 32-bit operations,
- has a large register set, and
- supports more data types.

The MC68020 supports the virtual memory/virtual machine functionality of the 68010. It extends the 68010's support for modular programming, and provides a coprocessor interface for instruction set extension. Although 68000 family has 32-bit user registers, operands, and internal registers, the MC68020 adds full 32-bit data path, both internal and external. It has two 32-bit internal address path, a 32-bit execution unit, three 32-bit arithmetic unit, and an on-board instruction cache. The design goal was to provide a 4-to-6 fold performance enhancement for the MC68020 over the 8 MHz 68000. The performance improvement was realized by a combination of the higher clock frequency at which the MC68020 operates, the 32-bit data bus width, the presence of the on-chip instruction cache, and the addressing modes improvements.

The MC68020 used HCMOS technology to minimize the power consumption. It consumes about 1W to 1.5W of power. The 9.22 mm of the silicon chip of MC68020 contains 200,000 transistors. The MC68020 is packaged in a square form of the 1.345" × 1.345" pin-grid-array (PGA). Its total number of pin outs is 169 arranged in 13 × 13 matrix. Out of 169 pins only 114 pins are used. The MC68020 can operate on a number of clock frequencies such as 12.5MHz, 16.67MHz, 20MHz, 25MHz, or 33MHz. It does not include any on-chip clock generator circuitry. It supports the floating-point processor such as 68881/68882 and memory management unit (MMU) processor, the 68851. Figure 6.10 (*a*) depicts the block diagram of the MC68020.

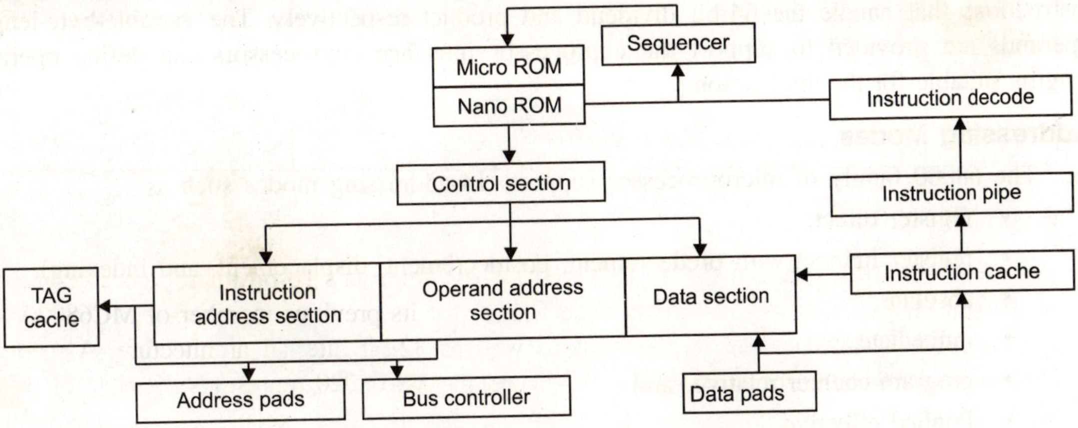

Fig. 6.10 (*a*) Block diagram of MC68020

6.11 COMPATIBILITY OF MC68020

One of the goals of the MC68020 was to maintain strictly the user object code compatibility with 68000 family. This goal was achieved easily as 68000's architecture was predominantly a 32-bit architecture. The issue of compatibility is crucial to most users, as their most significant investment is in software. The MC68020 are 68010. The changes have been restricted primarily to the exception-handling routine. Most of the difference is due to the MC68020's new stack formats and system debug features.

Instruction set enhancement

The instruction set of the MC68020 includes and extends the instruction set of the 68010. Also new instructions, new data types and new addressing modes have been to its enhancement performance.

Data types

The 68000 family of microprocessor supports data types of

- bits,
- bytes,
- words,
- long-words, and
- binary coded decimal (**BCD**).

In addition to the above types of data, the MC68020 supports 4-new data types such as

- variable-width bit fields,
- packed BCD,
- quad word (8-byte), and
- variable-byte-length operands.

The bit field instructions operate on any bit field from 1 to 32-bits in length regardless of its location in register or memory. The quad-word operands are necessary for long divide and multiply instructions that handle the 64-bit dividend and product respectively. The variable-byte-length operands are provided to support the coprocessor interface coprocessors can define operand lengths suitable for the application.

Addressing Modes

The 68000 family of microprocessor supports 14-addressing modes such as

- register direct,
- register indirect(with predecrement, postdecrement, displacement, and indexing),
- absolute,
- immediate,
- program-counter-relative, and
- implied effective.

The MC68020 provides the improved support for

- arrays, and
- lists

by expanding the indexed addressing modes to include memory indirection, 32-bit displacements, sign extension of displacements and index operands, index scaling, pre-indexing and post-indexing, and suppression of base register or index register or index value or both. These extensions are available for both data and program address space accesses. The new addressing modes are significant as because they can be used in addressing high-level-language data structures. These modes not only improve execution performance but they also make compiler-generated code much more efficient, since they are able to express more naturally the structures associated with the high-level languages.

6.12 INSTRUCTION OF MC68020

The instruction set of 68000 family required only a minor modification to make it a consistent 32-bit. Table 6.12(*a*) gives the details of the additions to 68000's instruction set that were involved in the extension for the MC68020.

Table 6.12(*a*) MC68020 instruction enhancement

Instructions	*Enhancement*
MULS, MULU, DIVS, DIVU	Operation extended to 32-bit operands
Bcc, BRA, BSR, LINK	Displacement extended to 32-bits
MOVEC	New control registers may be accessed
BKPT	Opcode substitution support

The multiply and divide instruction operate on 32-bit operands. Thus it is now possible to multiply two 32-bit operands to result into either 32, or 64-bit product. A 64/32-bit and a 32/32-bit divide operation is also available. The displacement associated with instructions was expanded from 16 to 32-bits. This is an improvement capability that allows branch and link operations to take place without restriction across the full 4-gigabyte address space. This is very significant improvement for the compiler support. The move control register (MOVEC) instruction has been expanded to accommodate the new control registers in the MC68020 and the breakpoint (BKPT) instruction has been enhanced to provide more real-time program debug support.

New instructions

Over 20-new instructions have been added as per Table 6.12(*b*) to provide new functionality to the MC68020. A large group of new instructions provide manipulation of variable-width bit field. Here, a bit field of variable length up to 32-bits, either in a data register or spanning up to 5-bytes in memory, may be cleared set, complemented, extracted, inserted, scanned, or tested. One feature of these instructions is that they make it possible, when dealing with a bit field in memory, to specify any 1- to 32-bit field regardless of its orientation in the memory. There are no restrictions as to how that field is aligned in memory. If a bit field spans five bytes, the processor can internally recognize that condition and make the appropriate adjustments.

One of the major additions to the instructions set is the group of general coprocessor instructions that support the coprocessor interface. Since it was not possible to support all of the special purpose computation needs of the users that are serviced by a general-purpose processor such as the MC68020, it was essential to develop a strategy to deal with non-general applications. To serve this need, the coprocessor interface was developed to provide a mechanism by which Motorola-defined VLSI special purpose microprocessor i.e., floating point coprocessor, the paged memory management unit, VLSI coprocessor designed by other organizations, and user defined discrete coprocessors could be used to augment the capabilities of the MC68020 without encumbering the architecture. In general purpose architecture, highly specialized instructions and data types can not be fully supported directly.

Table 6.12(*b*) New instructions of MC68020

BFCHG	Bit field change
BFCLR	Bit field clear
BFEXTS	Bit field signed extract
BFEXTU	Bit field unsigned extract
BFFF0	Bit field find first one set
BFINS	Bit field insert
BFSET	Bit field set
BFTST	Bit field test
CALLM	Call module
CAS	Compare and swap
CAS2	Compare and swap (2-operands)
CHK2	Check register against upper and lower bounds
CMP2	Compare register against upper and lower bounds
cpBcc	Coprocessor branch on coprocessor condition

(*Contd....*)

cpDBcc	Coprocessor test condition decrement and branch
cpGEN	Coprocessor general function
cpRESTORE	Coprocessor restore internal state
cpSAVE	Coprocessor save internal state
cpSETcc	Coprocessor set according to coprocessor condition
cpTRAPcc	Coprocessor trap on coprocessor condition
PACK	Pack BCD
RTM	Return from module
UNPK	Unpack BCD

The modular programming is supported by the call module (CALLM) and the return from module (RTM) instructions. These instructions provide a method of subdividing the user state into 256 access levels (levels of privilege), and then define a protocol for supporting a transition between access levels. Two different varieties of these instructions support simple module calls, which involve no access-level changes, and complex module calls, which may involve access-level changes.

The bound-checking instructions, CHK and CMP, found on earlier members of the family have been extended to allow specification of both upper and lower bounds. With CHK2 (check 2) and the CMP2 (compare 2) instructions are used to compare the upper and lower bounds in either signed or unsigned format.

The difference between the 2-instructions is that if the value is out of bounds, CHK2 causes an exception to occur in addition to setting the condition codes. The CMP2 instruction sets the condition codes according to result of the operation. The processor automatically evaluates the relationship between the 2-bounds to determine which type of comparison is appropriate.

There are several new instructions that are used to provide system support. The compare and swap (CAS) and compare and swap 2 (CAS2) instructions allow for locked-bus manipulation of byte, word, or long-word data elements to support system queue and stack functions. The pack and unpack instructions provide the ability to easily convert from packed ASCII/EBCDIC data formats to BCD format, or back, for arithmetic operations. The breakpoint instruction provides system debug assistance to the user via the breakpoint-acknowledgment mechanism. With this instruction, it is possible to breakpoint an instruction. When the processor attempts to execute the breakpoint instruction, a breakpoint-acknowledge cycle is executed and the proper op code is fetched and substituted into the instruction pipe in place of the breakpoint, if appropriate, or a breakpoint routine is executed. The PACK and UNPK instructions can be used to store BCD data, packed with 2-digits per byte and to unpack it into one BCD digit per byte.

6.13 REGISTER OF MC68020

The programmer's model of the MC68020 is identical to that of 68000 microprocessor. It is drawn in Fig. 6.13 (*a*).

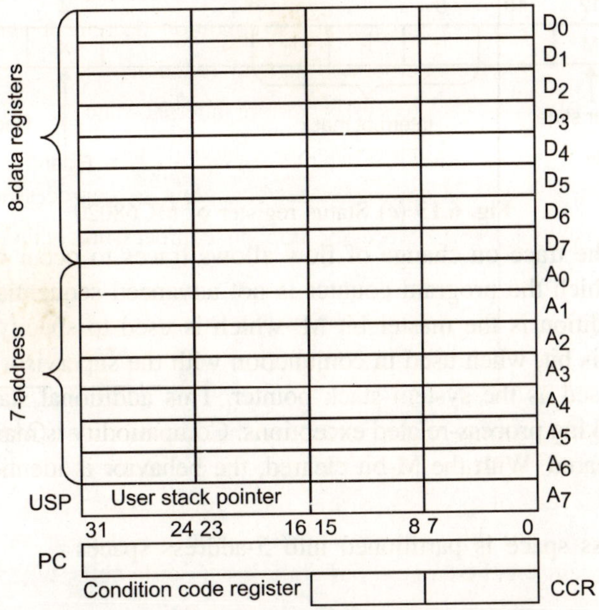

Fig. 6.13 ***(a)*** User programming model of MC68020

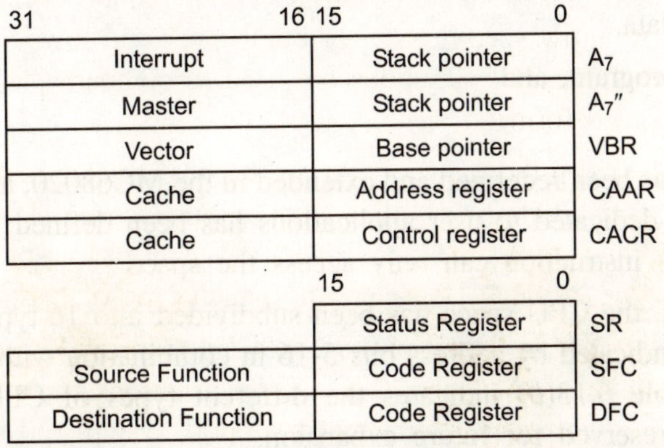

Fig. 6.13 (*b*) Supervisor programming register of MC68020

The supervisor's model as shown in Fig. 6.13 (*b*) contains some additional hardware. There are 2-registers used to control the cache and additional stack pointer for use with process-related exceptions. The cache control register (CACR) is used to control the cache and to access the status of the cache. The cache address register (CAAR) is used as a pointer into the cache for those control operations affecting only one entry in the cache. The use of the master stack pointer is taken up in detail as below.

The MC68020-status register has two newly defined bits as evident from Fig. 6.13 (*c*). An additional bit T0, when combined with the previous bit, now T1, allows for the definition of the three trace states

- no trace,
- trace on change of flow, and
- trace on instruction execution.

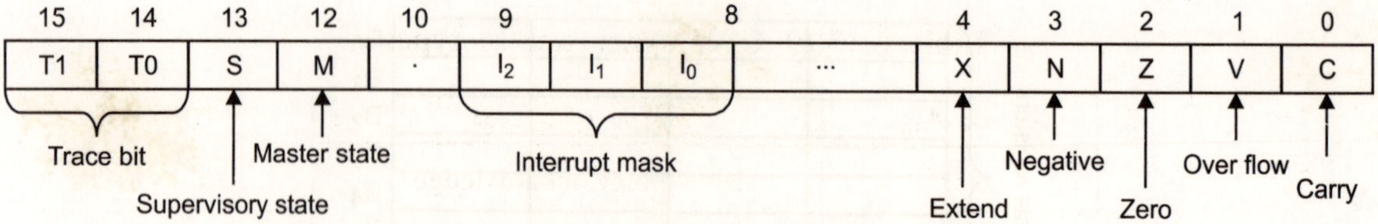

Fig. 6.13 (*c*) Status register of MC68020

The addition of the trace on change of flow allows traces to occur only after the execution of an instruction in which the program counter is not advanced sequentially i.e., a branch, JMP, or RTS. The other addition is the master bit M, which is used to specify when the processor is in the master state. This bit, when used in conjunction with the supervisor bit (s), indicates which register is to be accessed as the system stack pointer. This additional stack (if the M-bit is set) is used only in the stacking process-related exceptions. Compatibility is maintained, however, since at rest the M-bit is cleared. With the M-bit cleared, the behavior is identical to the processors in 68000 family.

The 68000-address space is partitioned into 5-address spaces

- user data,
- user program,
- supervisor data,
- supervisor program, and
- CPU space.

The partitioning has been redefined and extended in the MC68020, as shown in Table 6.13(*a*). A user-reserved space dedicated to user applications has been defined. The processor using the move space (MOVES) instruction can only access the space.

In addition to this, the CPU space has been subdivided into 16-types of accesses. The type of the CPU access is indicated by address bits 5-16 in combination with the CPU space function code (FC = 111). Table 6.13(*b*) indicates the different types of CPU accesses defined. All undefined spaces are reserved for future expansion.

Table 6.13(*a*) Function code definition

FC2	*FC1*	*FC0*	*Space*
0	0	0	Unused (reserved)
0	0	1	User data
0	1	0	User program
0	1	1	User reserved
1	0	0	Unused (reserved)
1	0	1	Supervisor data
1	1	0	Supervisor program
1	1	1	CPU space

Table 6.13(*b*) CPU space access types

$A_{19} \Leftrightarrow A_{16}$	*Type of access*
0000	Breakpoint acknowledge
0001	Memory management unit
0010	Coprocessor
0011	Reserved
0100	Reserved
0101	Reserved
0110	Reserved
0111	Reserved
1000	Reserved
1001	Reserved
1010	Reserved
1011	Reserved
1100	Reserved
1101	Reserved
1110	Reserved
1111	Interrupt acknowledge

The definition spaces are reserved for future expansion. The definition of region in the CPU space makes it possible to acknowledge breakpoints and interrupts, as well as to provide a means for communicating with coprocessors and other special devices i.e., memory management unit, without dictating memory organization for user- and supervisor-related activity.

Coprocessor interface of MC68020

The MC68020's coprocessor interface is powerful and flexible mechanism capable of extending the MC68020 basic instruction set and supporting new data types. This interface is designed to support concurrent and non-concurrent coprocessors and bus master and bus slave coprocessors without being encumbered by the specifics of an application. By providing this functionality, the interface offers a solution to the special processing problems that system designer face. With this mechanism, the system designer can use either VLSI coprocessor to solve some of the more common special-purpose processing needs (i.e., floating-point computations) or a board level design particular to some limited but very important application. The coprocessor interface is based on external bus cycles. CPU spaces reads and writes are used to transfer information between the main processor and the coprocessor. The processor interface-registers are memory-mapped into the CPU space. Sequences of writes and reads of this coprocessor control registers control coprocessor operation. Older processors in the family, which can deal with the coprocessor as a peripheral, though optimized for the MC68020, can also emulate the interface.

Coprocessor instructions (F-line instructions) from the MC68020 are passed to one of the eight possible coprocessors for evaluation. This decouples the main processor from the definition of a particular coprocessor instruction set and provides a great deal of generality. After evaluating the instruction, the coprocessor determines the service it requires from the main processor (if any) so that it can execute the instruction. These service requests are transmitted to the main processor in the form of primitives, which are then evaluated and executed by the main processor. There has been an effort to make these primitives very powerful so that there can be a minimum amount of overhead in the coprocessor interface. For an example, the main processor can perform an effective address calculation and then either pass the evaluated effective address to the coprocessor or fetch a variable-length operand and pass that to the coprocessor. This is complicated process, but by providing this service the main processor allows the coprocessor design to be much simple.

The MC68020 supports 5-types of coprocessor in instructions

- general,
- branch,
- conditional,
- save state, and
- restore.

The general instruction (cpGEN) is used for data movement and manipulation. The branch instruction (cpBcc) allows branching based on a coprocessor condition. Both 16- and 32-bit branch displacements are supported. The conditional instruction (cpScc), (cpTRAPcc), and (cpDBcc) allow evaluation of one of 64-coprocessor conditions, and they conditionally set (cpScc), trap (cpTRAP), or decrement and branch (cpDBcc). The coprocessor passes a condition for evaluation and then informs the main coprocessor of the result.

6.14 CACHE OF MC68020

Caches have been used for years in large machines to increase performance without greatly increasing system cost and complexity. Now designs are available to use cache only a fraction of the size of the main store that provides a significant decrease in the average access time to the main store. This concept has been incorporated into the MC68020. It contains 256B-on-chip instruction code cache. This small size (256B) cache reduces the number of bus cycle requirement for instruction fetches from the main memory. The reduced bus utilization by the MC68020 allows increases system performance by providing more bus bandwidth for other bus masters such as DMA devices. The MC68020 only caches instruction stream access, to avoid stale data and other data-cache-related problems. The MC68020's cache configuration is shown in Fig. 6.14 (*a*). The cache interface to the processor data paths allows complete overlap of instruction fetches with data operand accesses, and thus provides a significant increase in performance. If simultaneous instruction and data operand requests are generated by the microprocessor, a bit in the instruction cache allows concurrent fetches to take place. The instruction cache may be disabled by the cache disable pin or using the cache control register. Entries to the cache may be cleared using cache control register and cache address register.

When the enable bit of the cache control register is reset to '0', it disables the cache causing the processor to fetch data from the external memory. Setting a '1' in the enable bit of the cache control register, clears all the cache entries and causes the cache to refill with new data. When the freez bit is set, the valid data in the cache may not be replaced. With the clear entry bit set to '1', the location in the cache corresponding to the address in the cache address register is invalidated.

The cache control register is provided to allow the operating system to maintain and optimize the cache. This register is shown in Fig. 6.14 (*b*).

In order to fetch an instruction enabling the cache, the μP always checks the cache using address bits A_7 to A_2 (6-bits) to locate a 32-bit-long word in cache memory shown in Fig. 6.16 (*a*). It then compares address bits A_3 to A_8 and the function code bits FC2 with the 24-bit tag of the entry that is also stored in the cache. If the function and address bits match the tag, a cache hit occurs and the μP sets the valid bits for the cache entry. The μP selects the 32-bit instruction from the cache using the address bit A_1. When a cache miss occurs because of the tag for the instruction does not match the tag in the cache or when the valid bit is cleared, μP puts the instruction address bits A_{31} to A_8 in the tag field, sets the valid bit, then fetches one 32-bit-long word from the main memory and places it in the cache.

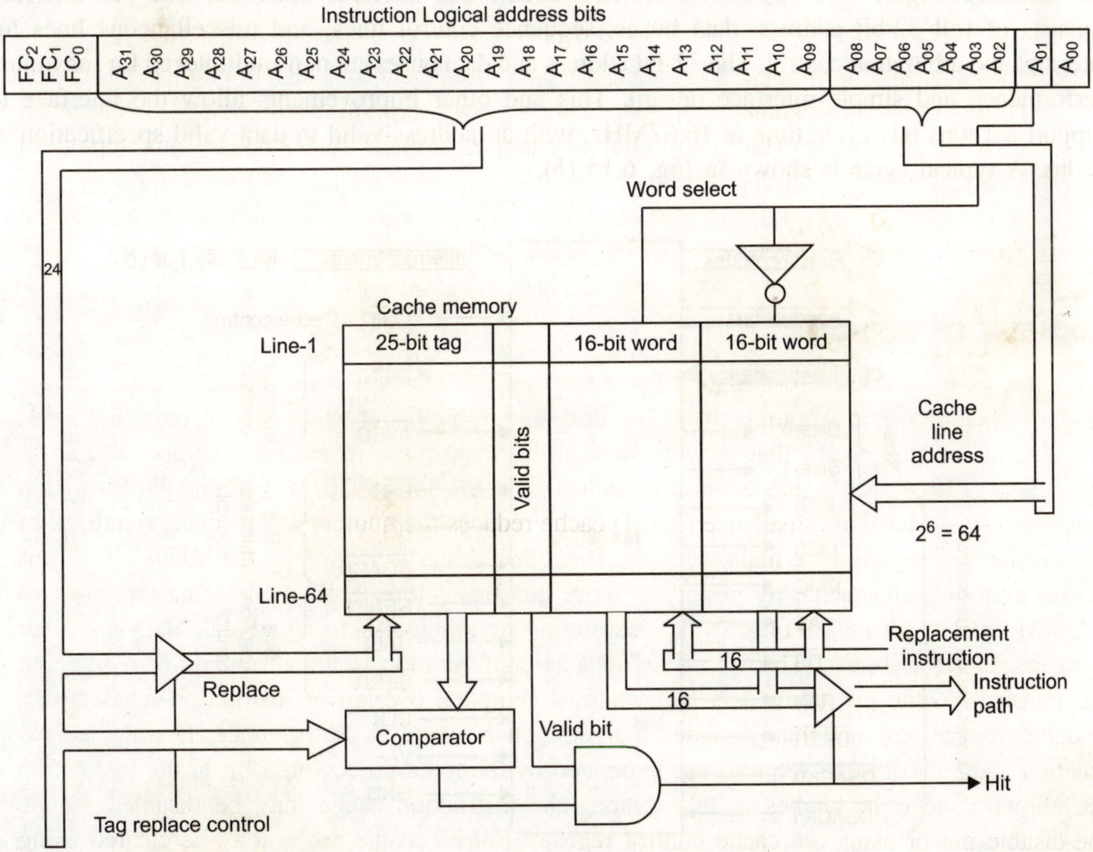

Fig. 6.14 (*a*) MC68020 on-chip cache organization

31–8	7	6	5	4	3	2	1	0
0	0	0	0	0	C	CE	P	E

C = Clear carry, CE = Clear entry, F = Freeze cache, E = Enable cache

Fig. 6.14 (*b*) MC68020 cache control register

The overall performance of the cache can be evaluated as

$$P_o = P_m + (P_h - P_m)H_r \text{ MIPS}$$

where, P_o = overall performance output

P_h = performance of the cache hit

P_m = performance of the cache miss

H_r = hit ratio

6.15 ASYNCHRONOUS BUS OF MC68020

The MC68020 maintains the fast, easy-to-use, non-multiplexed asynchronous bus interface of 68000 family. However, several significant new features have been designed into the MC68020 bus interface. Figure 6.15 (*a*) shows the MC68020's bus interface structure. The bus interface consists of full 32-bit address, data buses, sequence control lines, and miscellaneous lines for interrupt and arbitration control. The 32-bit address and data buses are non-multiplexed for maximum performance and simple interface design. This and other improvements allow the interface to support a 180ns bus cycle time at 16.67MHz, with an address-valid to data-valid specification of 120ns. A typical cycle is shown in Fig. 6.15 (*b*).

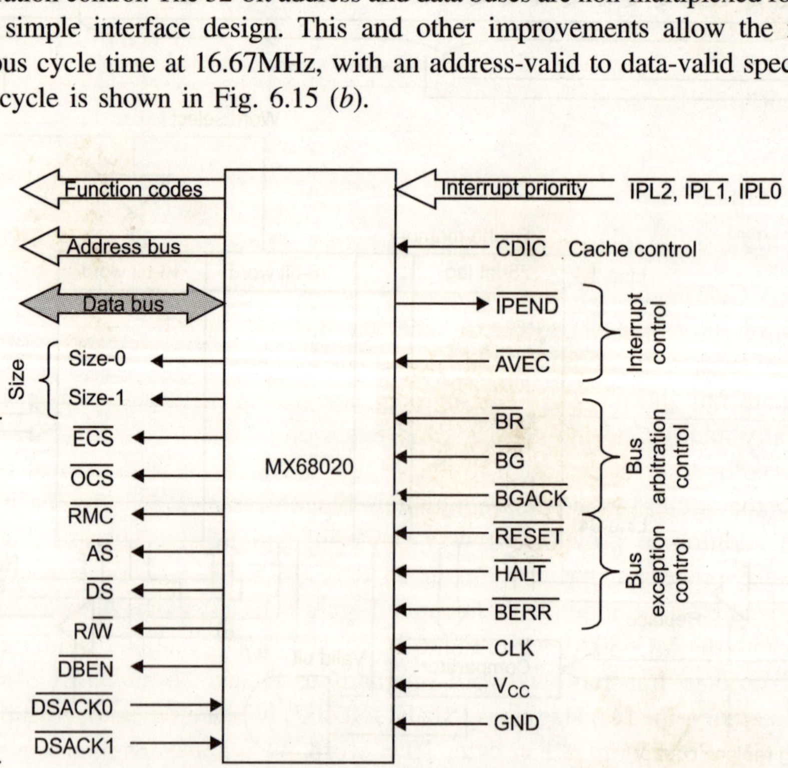

Fig. 6.15 (*a*) MC68020 bus interface

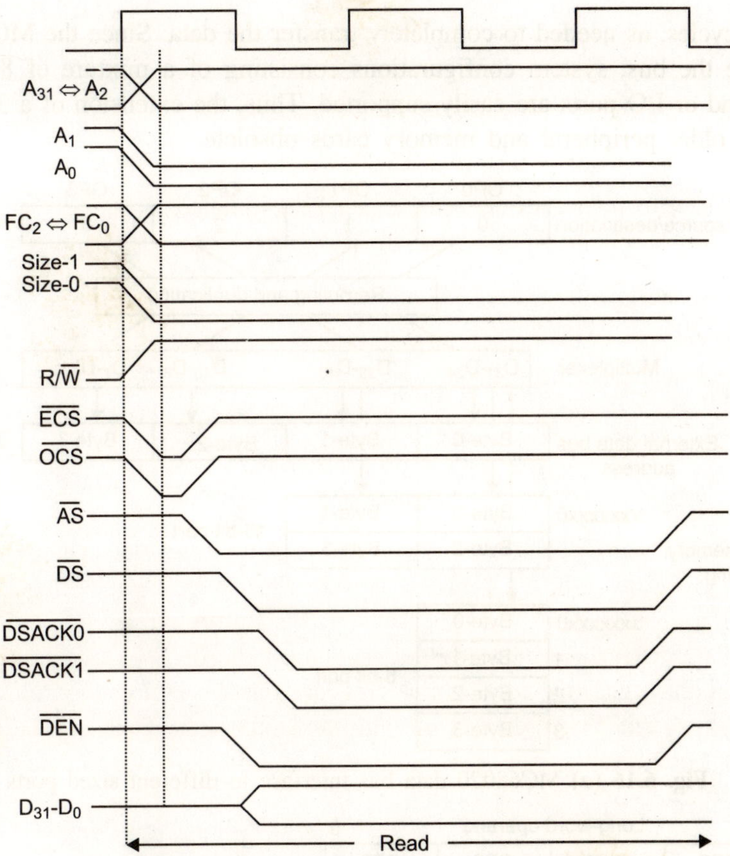

Fig. 6.15 (*b*) Typical bus cycle of MC68020

6.16 DYNAMIC BUS SIZING OF MC68020

The MC68020 bus interface supports an improved new feature known as dynamic bus sizing. This feature allows the MC68020 to 'size' the bus port on a cycle by cycle basis, and to transfer data to and from 8, 16, and 32-bit bus ports. The key elements used to implement this new concept are the data multiplexer, the SIZE outputs, and DSACKx inputs, which the MC68020 uses to dynamically interface to the various-sized buses on a bus-cycle by bus-bus-cycle basis. Dynamic bus cycle allows software design to ignore the hardware configuration of memory and I/O ports and view the address space as a contiguous block of any granularity, be it byte, word, or long-word. In addition to providing a very powerful and flexible system, it gives the user a high performance processor that can run in an 8 or 16-bit system, or in a system configured with various-sized buses to individual elements. Figure 6.16 (*a*) shows the various ports configurations possible with the MC68020, while Fig. 6.16 (*b*) shows a typical operand transfer in which dynamic sizing is used to transfer long-word operands to 8, and 16-bit buses. As the MC68020 steps through memory, the two size lines (SIZE1, SIZE2) indicate the amount of data available for each transfer (00 = long-word, 01 = byte, 10 = word, 11 = three byte).

For each example in Fig. 6.16 (*b*), the memory control responds on the DSACKx pins (DSACK0, DSACK1) to define the port size, and is used by the processor as a factor for subsequent bus cycles, as needed to completely transfer the data. Since the MC68020 is able to dynamically size the bus, system configurations consisting of a mixture of 8-, 16-, and 32-bit memory cards and or I/O ports are easily supported. Thus, the extension of a 32-bit external bus does not render older peripheral and memory cards obsolete.

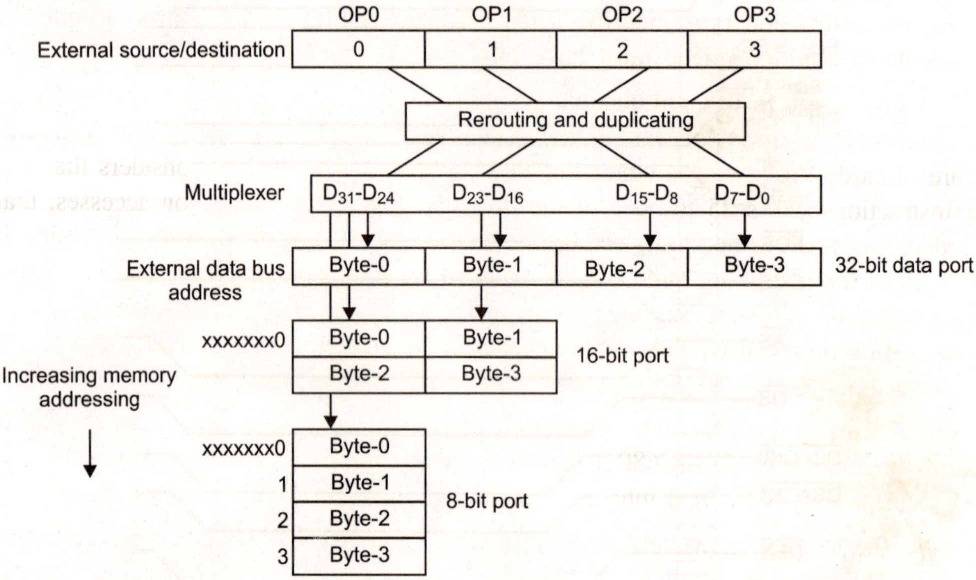

Fig. 6.16 (*a*) MC68020 data bus interface to different sized ports

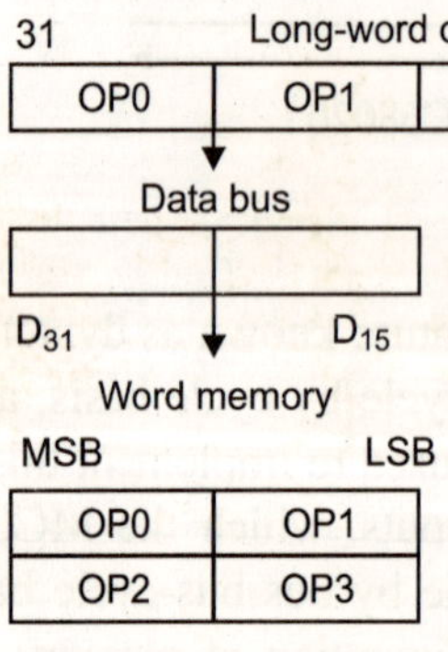

Long word transfer to word bus				Memory control	
Size1	Size0	A_1	A_0	$\overline{\text{DSACK 1}}$	$\overline{\text{DSACK 0}}$
0	0	0	0	1	0
1	0	0	1	1	0

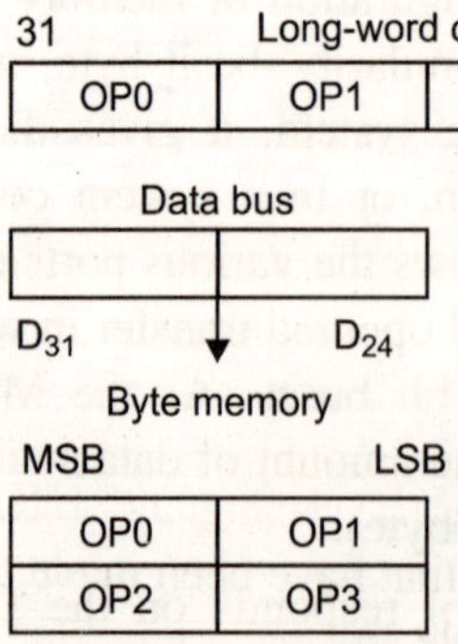

Long word transfer to byte bus				Memory control	
Size1	Size0	A_1	A_0	$\overline{\text{DSACK 1}}$	$\overline{\text{DSACK 0}}$
0	0	0	0	1	0
1	1	0	1	1	0
1	0	1	0	1	0
0	1	1	1	1	0

Fig. 6.16 (*b*) MC68020 dynamic bus sizing

Pipelining of MC68020

The micro-machine of the MC68020 is highly pipelined. The predominant pipeline mechanism is the instruction pipe, although there are several other pipelined mechanisms to provide control to various units within the micro-machine. The term instruction pipe is very often used to define an instruction queue. The distinction in this when a change of instruction flow occurs (i.e., a branch), the instruction pipe must be filled before execution of the new instruction can occur, whereas the instruction queue must have only the 1st word present.

However, as an instruction queue does not provide the performance benefits of a pipe during the execution of in-line code. Instruction queues tend to saturate the bus with instruction fetches that are discarded following a change-of-flow instruction. When one considers that a change-in-flow instruction occurs 25 to 30% of the time and that those instruction accesses, that must be discarded, can prevent necessary data accesses, the advantage of an instruction pipe is evident. The depth of the instruction pipe was increased from 2 to 3-words on the MC68020, since it was observed that on a 32-bit bus there is no performance penalty for the third element during a change-of-flow instruction. This is true because it is always possible to fetch three words in two accesses.

Parallelism is the concurrent execution of several tasks. Perhaps the most significant case of parallel operation in the MC68020 can be found in the execution unit as depicted in Fig. 6.16(*c*). The execution unit is divided into three 32-bit sections, each with a 32-bit adder

- the instruction address section, in which the instruction address are calculated and the pointers are stored;
- the operand section, in which the operand addresses are calculated, and
- data section, in which the data operation can take place.

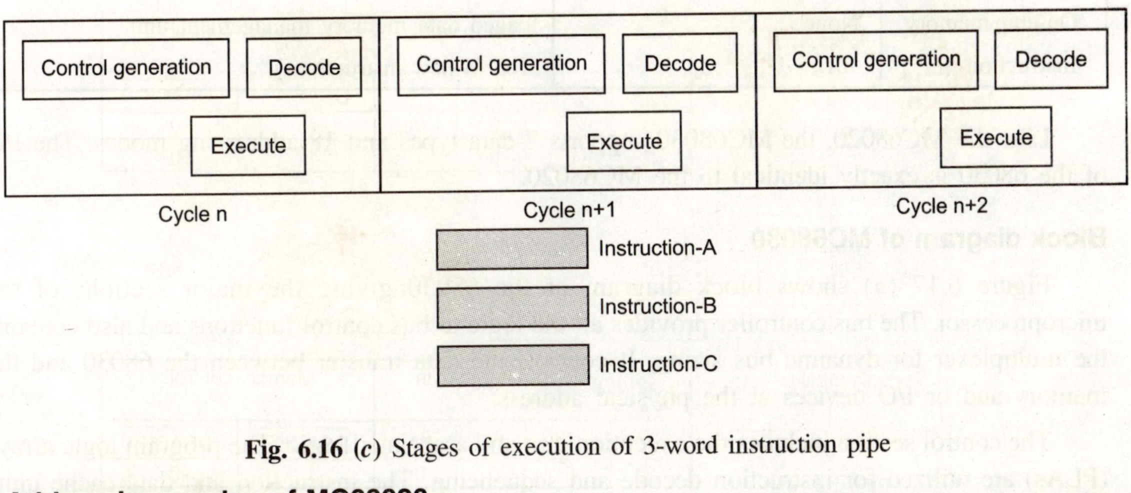

Fig. 6.16 (*c*) Stages of execution of 3-word instruction pipe

Addressing modes of MC68020

The instruction set was extended to provide further high-level-language (HLL) support. The most significant aspects of these extensions are the additions that have been made to the addressing modes. The indexed addressing mode now makes it possible to

- scale indexes,
- use memory indirection,
- have 16-bit and 32-bit displacements,
- use preindex or postindex on memory indirect, and
- suppress any element of the address calculation.

Hardware of MC68020

Special purpose hardware was designed to enhance the computational power of the basic micro-machine. This was done to provide faster execution for those instructions that require services beyond the operation performed by an ALU. A barrel shifter was added to the micro-machine to support bit field instructions; it is also used for a variety of other instructions. Special control features were added to support multiplication of two bits per micro-instruction. A significant addition was a scaling unit that, during the evaluation of an effective address, allows the scaling of index value without any delay.

6.17 MC68030

The 68030 is the Motorola's 32-bit microprocessor based on virtual memory, an extension of the MC68020 with some additional features. It also uses the HCMOS technology for lower power consumption. It can operate at frequencies such as 16.67MHz, 20MHz, and 33MHz. The differences between the 68030 and MC68020 are listed as

Characteristics	*MC68020*	*68030*
On-chip cache	256-byte instruction cache	256-byte instruction cache and 256B data cache
On-chip memory	None	paged data memory management unit
Instruction set	101	105 (4-new instructions)

Like the MC68020, the MC68030 supports 7-data types and 18-addressing modes. The I/O of the 68030 is exactly identical to the MC68020.

Block diagram of MC68030

Figure 6.17 (*a*) shows block diagram of the 68030 giving the major sections of the microprocessor. The bus controller provides all the logic to bus control functions and also contains the multiplexer for dynamic bus sizing. It controls the data transfer between the 68030 and the memory and or I/O devices at the physical address.

The control section includes the execution unit and associated logic. The program logic arrays (PLAs) are utilized for instruction decode and sequencing. The instruction and data cache units operate independently. They obtain information from the bus controller for future use. Each cache has its own address and data buses and thus permits simultaneous access. Both caches are organized as 64-long-word entries (256B) with a block size of 4-long-words. The data cache uses a write-through policy with no write allocation on cache misses.

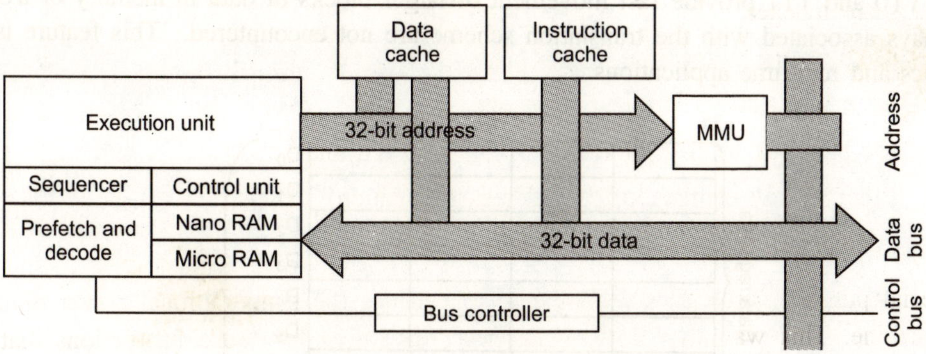

Fig. 6.17 (*a*) Block diagram of 68030

The memory management unit maps address for page sizes from 256-bytes to 32K-bytes. Mapping information stored in description resides in translation tables in memory that are automatically searched by the MC68030 on demand. Most recently used descriptors are maintained in a 22-entry fully associated cache called the Address Translation Cache (ATC) in the MMU, permitting address translation and other MC68030 functions to occur simultaneously. The MMU utilizes the ATC to translate the logic address generated by the MC68030 into a physical address.

Registers of MC68030

The register organization of the MC68030 is shown in Fig. 6.17 (*b*). The additional registers implemented on MC68030 extra to the MC68020 are meant for the MMU. All registers of MC68020 are also available on the MC68030 along with the extra cache control registers. This provides additional functions.

The additional registers on MC68030 are

- 32-bit translation control register (TC),
- 64-bit CPU root pointer (CRP),
- 64-bit supervisor root pointer (SRP),
- 32-bit transparent translation registers TT0 and TT1,
- 16-bit MMU status registers (MMUSR).

The translation control register provides several fields that control the address translation. These fields enable and disable address translation, enable and disable the use of the SRP for supervisor address space, and select or ignore the function codes in translating addresses. Other TC fields specify memory page sizes, the number of address bits used in translation, and the translation table structure.

The SRP holds a pointer to the root of the translation tree for the currently exciting MC68030 task. This tree provides the mapping information for the tasks address space. Registers TT0 and TT1 can each specify separate memory blocks as directly addressable without address translation. Logic addresses in these areas are the same as physical addresses for memory access. Therefore, registers TT0 and TT1 provide fast movement of larger blocks of data in memory or I/O space, since delays associated with the translation scheme are not encountered. This feature is useful in graphics and real-time applications.

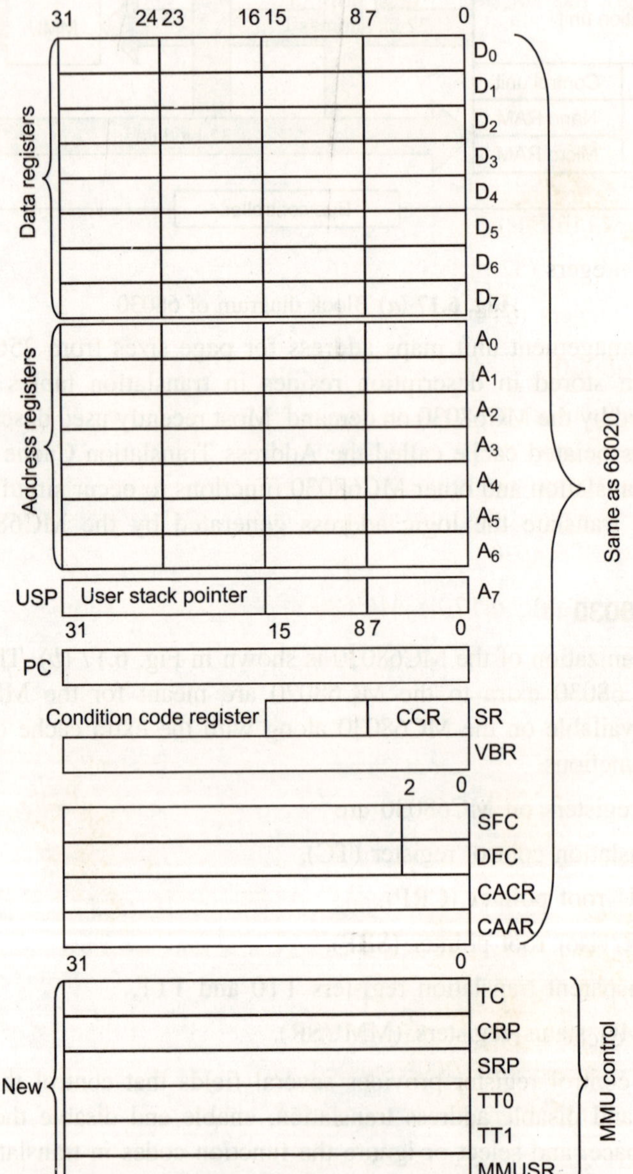

Fig. 6.17 (*b*) Programming register of MC68030 new addition to MC68020 model

ISP = Interrupt Stack Pointer, VBR = Vector Base Register, SFC and DPC = Alternate Function code Register, CACR = Cache Control Register, CAAR = Cache Address Register, MSP = Master Stack Pointer, TC = Translation Control, CRP = CPU Root Pointer, SRP = Supervisor Root Pointer, TT0 = Transparent Translation-0, TT1 = Transparent Translation-1, MMUSR = MMU Status.

The MMUSR register provides memory management status information resulting from a search of the address translation cache or the translation tree for a particular logic address.

Data types

Like the MC68020, the MC68030 also supports 7-data types as

- bits
- bit fields
- BCD digits
- Byte integers (8-bits)
- word integers (16-bits)
- long-word integers (32-bits)
- quad-word integers (64-bits).

Instructions of MC68030

The MC68030 includes all instructions of the MC68020 alongwith a subset of the MMU of the 68851 instructions. These instructions are PMOVE, PTEST, PLOAD, and PFLUSH. The MC68020 required the coprocessor 68851 interface for execution of these instructions. The MC68030 execute these instructions just as it excites the other instructions. Table 6.19(*a*) gives new instructions of the MC68030.

Table 6.17(*a*) MC68030 new instructions

Instruction	*Operand Syntax*	*Operand Size*	*Operation*
PFLUSH	(FC), #mask, [(EA)]	None	Invalidate ATC entries at effective address.
PLOAD	(FC),(EA),(R/$\overline{W}$)	None	Creates ATC entry for effective address.
PMOVE	Rn, (EA),	16,32	Register n ⇒ Destination
	(EA), Rn	16,32	Source ⇒ Register n
PTEST	(FC),(EA),#level,[R/$\overline{W}$]An	None	Information about logical address ⇒ PMMU status

Internal cache

The MC68030's cache is placed between the processor and the rest of the memory system. The effective memory access time is expressed as

$$t_{acc} = h^*t_{cache} + (1 - h)t_{external}$$

where, t_{acc} = effective system access time,

t_{cache} = cache access time,

$t_{external}$ = access time of rest memory system, and

h = hit ratio or % of time data found in the cache.

The instruction cache organization of MC68030 is shown in Fig. 6.17 (*c*). The instruction cache in MC68030 is 256B direct-memory mapped in 16-blocks. Each block contains 4-long-words. Each long-word can be accessed independently. Thus 64-entries are possible with A_1 selecting a correct word during an access.

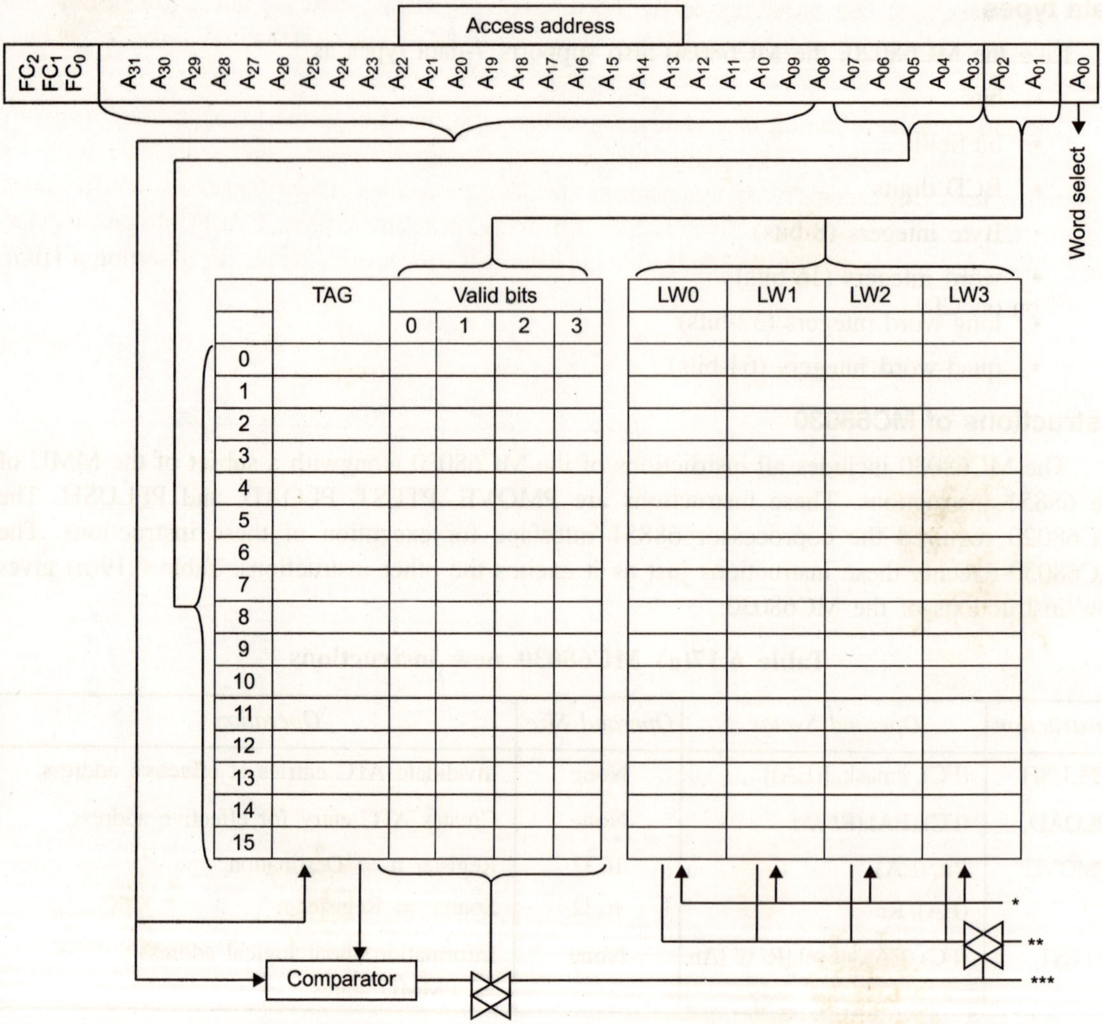

Fig. 6.17 (*c*) Instruction cache of MC68030

* = data from instruction cache bus,

** = data to instruction cache output register.

*** = cache control.

Pinouts of MC68030

Figure 6.17 (*d*) shows the functional block of MC68030. The major design goals in order of priority of the MC68030 were

- compatibility,
- performance,

- schedule, and
- integration.

The index or a line is addressed by A_4 through A_7 and A_2 and A_3 select each entry. The tag includes address lines A_8 through A_{31} along with FC_2. Since instruction must be at even address, A_0 is not selected and hence A_0 is kept zero.

An entry means a line hit has occurred and the valid bit (4-bits, one for each long-word) for the selected entry is set. If an entry miss occurs, the cache entry can be updated with the instruction read from memory. If the cache is disabled, no cache hits or updates will take place, and if the cache is enabled but frozen, hits can occur without updates. CACR is used to clear the cache entry and enable or disable the cache. The on-chip can be disabled by asserting a HIGH condition on CDIS pin.

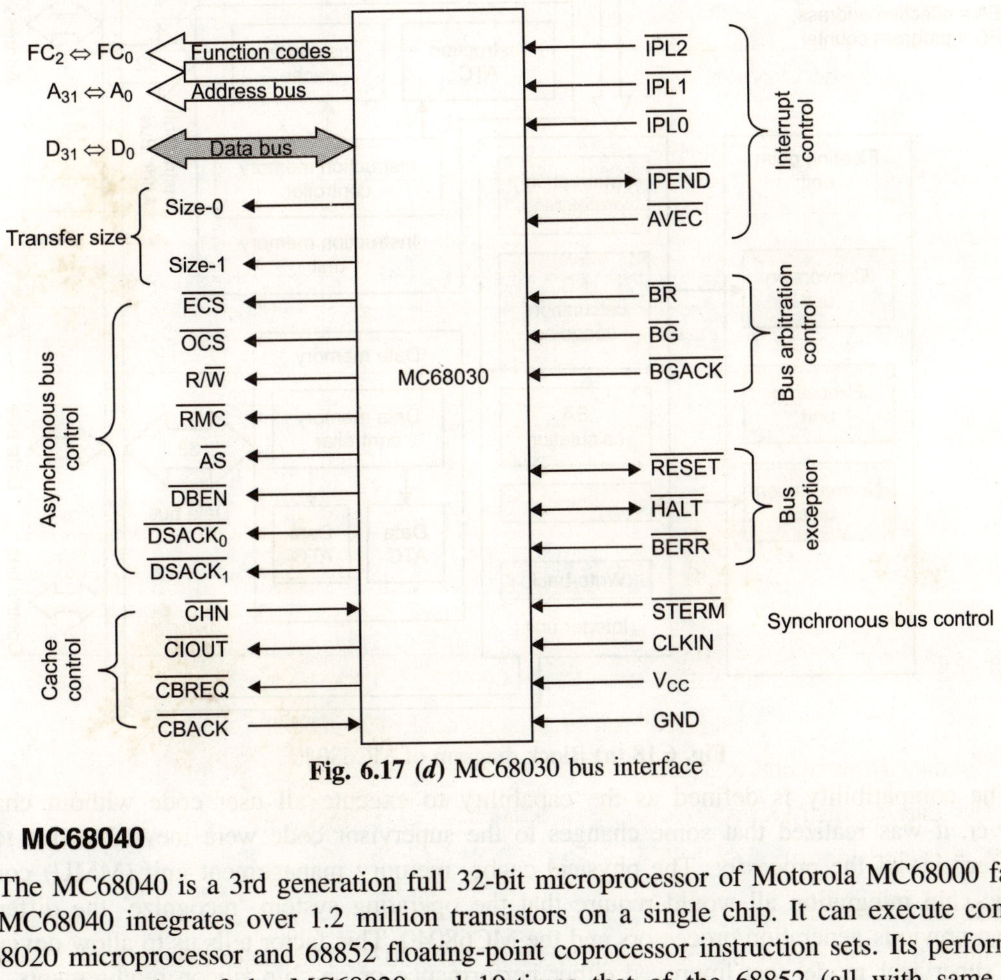

Fig. 6.17 (*d*) MC68030 bus interface

6.18 MC68040

The MC68040 is a 3rd generation full 32-bit microprocessor of Motorola MC68000 family. The MC68040 integrates over 1.2 million transistors on a single chip. It can execute complete MC68020 microprocessor and 68852 floating-point coprocessor instruction sets. Its performance level is 4-times that of the MC68020 and 10-times that of the 68852 (all with same clock frequency of 25MHz). Pipeline integer and floating-point execution units that operate concurrently with separate internal memory controllers and an autonomous bus controller contribute to this performance level. Physical caches of 4KB each for instruction and data reside on the chip.

Separate address translation caches of 64-entries a piece operate in parallel with the instruction and data cache. This arrangement provides complete memory management in a virtual demand-paged operating system. Figure 6.18 (*a*) depicts different blocks of the MC68040.

The compatibility is defined as the capability to execute all user code without change. However, it was realized that some changes to the supervisor code were inevitable due to the characteristics of the processor. The physical cache, memory management unit (MMU) control, and the chip integration all would require that the operating system 'recognize', the difference between previous generation processors and the MC68040. This factor tells us to allow deviations in the supervisor model that improved either performance or on-chip silicon requirements.

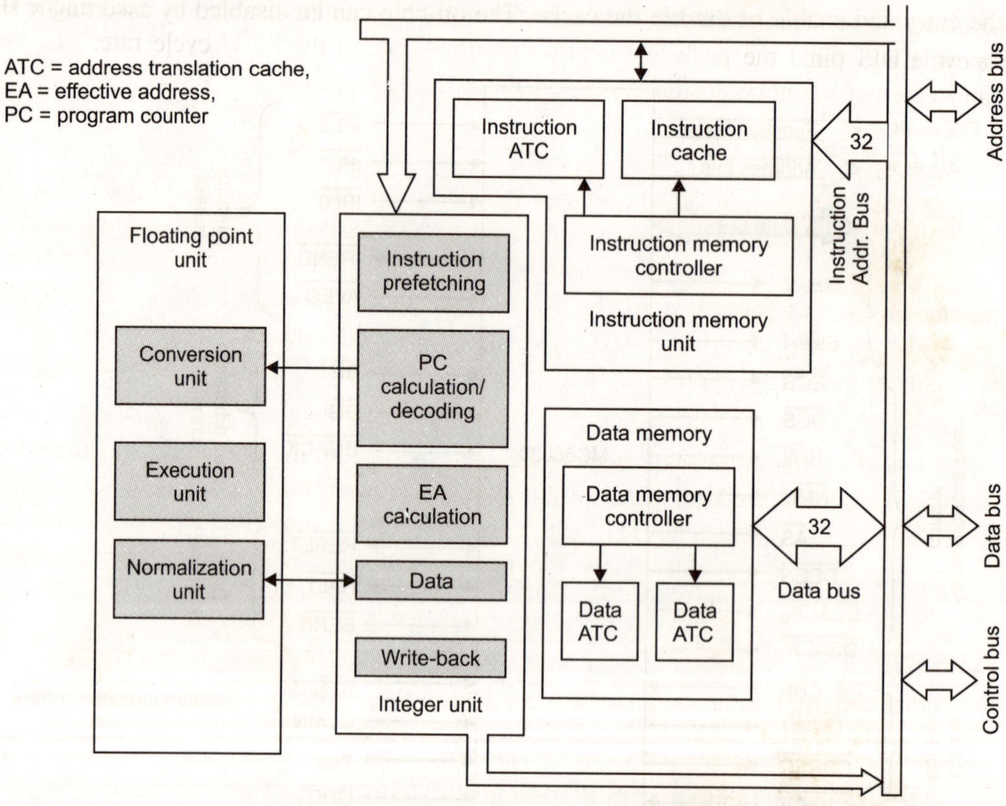

Fig. 6.18 (*a*) Block diagram of MC68040

The compatibility is defined as the capability to execute all user code without change. However, it was realized that some changes to the supervisor code were inevitable due to the characteristics of the processor. The physical cache, memory management unit (MMU) control, and the chip integration all would require that the operating system 'recognize' the difference between previous-generation processors and the MC68040. This factor tells us to allow deviations in the supervisor model that improved either performance or on-chip silicon requirements.

The integration goal was to achieve the combined effect of separate 68000 family components : integer processing, on-chip floating-point, and on-chip memory management capable of supporting a demand-paged operating system.

Philosophy

The 1st major decision was made to optimize those instructions in the 68020 that executed most frequently and to allow other instructions to be implemented conveniently. A reasonable subset of the 68020 instruction set based on trace data and user input was selected. Those instructions were examined and discovered that most of them would execute in 1-clock cycle if the IU were pipelined. A clock cycle here refers to the input clock to the chip that is also used by the system running at 25MHz. Along with this analysis, it was decided for performance reasons the overall cache architecture should continue to be based on the Harvard model (separate address and data spaces), as in the 68030. Further, the integer ALU cycle time was matched to the cache access time. This procedure allowed optimized instructions to access each cache once per clock cycle and fixed the peak instruction execution rate to the ALU cycle rate. The entire internal and external memory design rested on these decisions.

Integer unit of MC68040

The MC68040-integer unit (IU), executes the more common instructions in 1-clock cycle while maintaining the object-code compatibility with 68000 family. As discussed only supervisor mode instructions controlling the caches and MMUs differ from previous implementations. In order to increase the performance in the IU over previous generation parts, the MC68040 reduces both the numbers of ALU clock cycles per instruction and, the ALU cycle time. For a 25MHz 68020, the ALU cycles 12.5MHz, whereas the 25MHz MC68040 ALU cycle is 25MHz.

In order to reduce the number of clock cycles per instruction, the previously mentioned trace data were examined. From the knowledge of this data, dynamic instruction frequency was derived. Analysis of the instruction frequency led to the selection of the pipelined architecture to reduce load and store operations to one clock cycle. Pipelining allowed the MC68040 to optimize effective addressing modes

- register direct (Rn),
- address-register indirect, (An),
- address-register indirect with postincrement (An)+
- address-register indirect with predecrement -(An),
- address-register indirect with 16-bit offset (An,d16),
- absolute (SA address), and
- immediate #Data.

The optimized effective address (<OEA>) calculations and address fetches executes with no penalty for instructions such as

- logic,
- arithmetic, and
- data

movements. Table 5.18(*a*) compares cycle times for the most common instructions with these effective addresses in the 68020 and MC68040. Here RL stands for multiple register list, and d8 refers to an 8-bit offset.

The more complex instructions that are not in the optimized set generally execute in less than half number of 25MHz clock cycles in MC68040 than they do in a MC68020 or MC68030.

Table 6.18(*a*) Comparison of common effective address instructions in MC68020 and MC68040

Instruction	*Addressing mode*	*25MHz CLK cycle*		*Operation*
		68020	*MC68040*	
Move	Rn,Rn	2	1	Register-to-register move
Move	<OEA>,Rn	6	1	Memory-to-register move
Move	Rn,<OEA>	6	1	Register-to-memory move
Move	<OEA>,<OEA>	8	2	Memory-to-memory move
Move	(An,Rn,d8),Rn	10	3	Memory-indexed source
Move	Rn,(An,Rn,d8)	6	3	Memory-indexed destination
Move multiple	RL,<OEA>	4+2n	2+n	Move multiple register to memory
Move multiple	<OEA>,RL	6+4n	2+n	Move memory to multiple registers
Simple arithmetic	Rn,Rn	2	2	Register-to-register arithmetic
Simple arithmetic	Rn,<OEA>	6	1	Register-to-memory arithmetic
Simple arithmetic	<OEA>,Rn	8	1	Memory-to-register arithmetic
Shifts	———	4	2	Shift 1 to 31 bits
Branch taken	———	6	2	———
Branch not taken	———	4	3	———
Branch to subroutine	———	6	2	———
RET from subroutine	———	10	5	———

Integer pipeline

The MC68040 has 6-stage integer pipeline as in Fig. 6.18 (*b*). Figure 6.18 (*b*) details these stages

- instruction-prefetch (IP)
- program-counter calculation and decoding (PCD),
- effective-address calculation (EAC),
- effective-address fetching (EAF) for operands,
- data-execution (DE), and
- write back (WB).

The implementation has three independent controllers

- PC controller for the IP and PCD stages
- effective address controller for EAC, EAF, and WB stages, and
- DE count

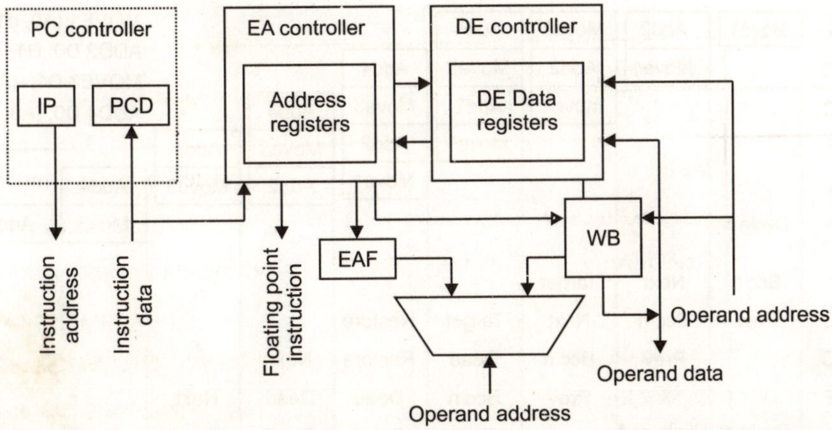

Fig. 6.18 (*b*) Integer pipeline of the MC68040

The PC controller is a finite state machine that relies on instruction-length and change-of-flow (branch, jump) decoding controls in the IP and PCD stages. The controller combines PLA decoding and microcode sequencing. Figure 6.18 (*c*) demonstrates the pipeline operation by showing a typical sequence of instructions. Each column in Fig. 6.18 (*c*) represents one 25MHz clock cycle. The notation 'dead' means that no useful activity occurs at that pipeline stage.

Branching of MC68040

The MC68040 improved branch performance relative to the MC68020 and MC68030 from 6-taken/4-not taken clock cycles to two taken/three not taken cycles (ignoring cache-miss and cache-line-alignment effects). This performance occurs despite deeper pipeline in the MC68040. Since trace data shows that 75% of all branch instructions (i.e., conditional Branch, Unconditional Branch, Branch to Subroutine, and Decrement and Branch) are taken, the branches were optimized on the MC68040. A dedicated branch adder in the PCD stage adds the branch displacement to the PC in every cycle. A dedicated branch-instruction decoder controls the processes of loading the result of the branch adder into the PC and starting a change-of-flow sequence in the stage. If the instruction is not a branch, the IU ignores the result of addition operation.

When a conditional change-of-flow sequences is started (a Conditional Branch or a Decrement Branch), the non-taken PC, the not-taken EAC (decoded instruction, and the instruction buffer) are saved in special shadow register. As shown in Fig. 6.18 (*c*), the condition codes from the instruction preceding the branch are valid at the same time the first opcode stream has been decoded.

Figure 6.18 (*c*) shows Integer Instruction flow for taken (top) and not taken (bottom) Branch Instruction. This timing allows execution to start in the EAC stage for the target in the next clock cycle. If the condition is not taken, one clock cycle is required to recover the not-taken path from the shadow buffers.

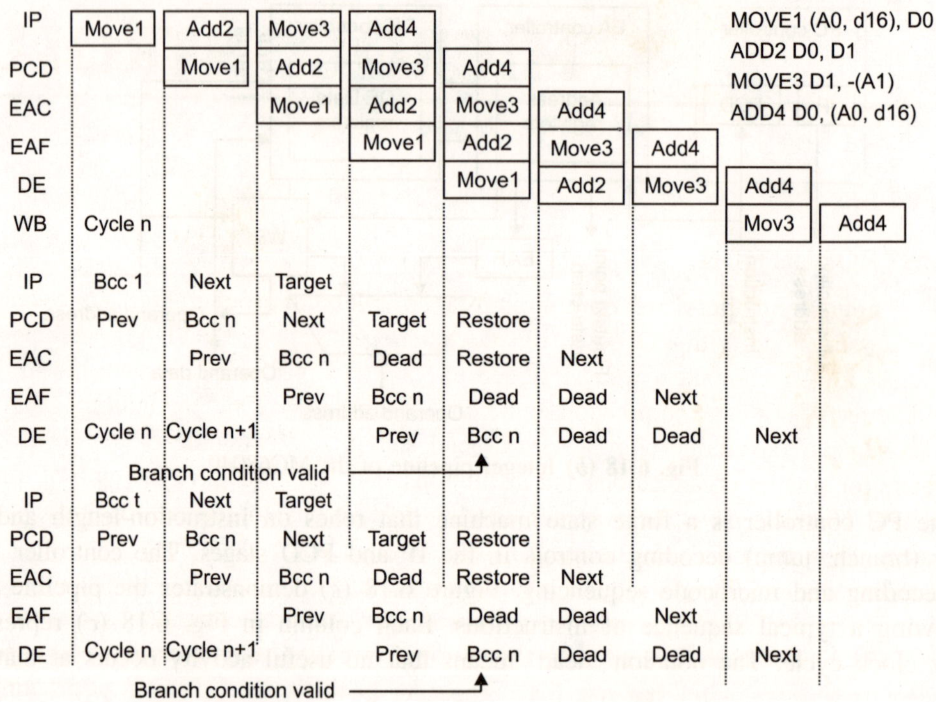

Fig. 6.18 (*c*) Integer instruction flow in the MC68040

Branch Condition Valid

Bcc.n = Branch not taken	Next = Instruction after the branch
Bcc.t = Branch taken	Prev = Instruction before the branch
Dead = Inactive pipeline stage	Target = Taken Instruction

7

MICROCONTROLLERS

7.1 INTRODUCTION

The microcontrollers are nothing but a complete microcomputer. In essence, it is a microprocessor along with I/O ports, timers/counters, memory, ADC/DAC etc. integrated on a single chip. Larger computers were designed to solve complex scientific and industrial problems and handle records of large corporations and Govt. organizations. Only big industries and institutions were able to purchase large computers. A trend started in the middle of 60s to design smaller computers for smaller organizations and institutions. This situation gave the birth of minicomputer in the late 60s that pave the way for smaller institutions, organizations, offices etc to use computers. With the rapid advancement in the semiconductor technology, it became possible to fabricate the whole CPU of a digital computer on a single chip using VLSI technology. A CPU built into a single VLSI chip is called a microprocessor. A digital computer having *microprocessor* as the CPU along with *memory* and I/O *devices* is called *microcomputer*. The prefix micro refers to its physical size, but not it's computing powers. As far as its computing power is concerned, the latest 32-bit microcomputers are as powerful as a traditional mainframe computer. Presently it is possible to construct a microcomputer having most of the features of third generation mainframe computers using just handful of ICs. The number of bits that a digital computer can process in parallel at a time is called its word length. It is a measure of the computing power of a computer. The microcomputers have word lengths of 4 to 32-bits whereas large computers have 32 to 64-bits. The 4-bit microprocessors are used for applications in domestic appliance control, calculators, video games, toys etc. A calculator is not a computer as it is not a programmable device. The user does not prepare any program for his calculations. It performs calculations using step-by-step method.

The calculator is based on a single chip microcomputer that contains 4 bit microprocessor as its CPU, semiconductor memory and I/O devices. The manufacturers have stored permanent program on-chip semiconductor memory for making calculations. The data entered by the user are stored in the memory and utilized by the processor while making calculations. The manufacturers have not made any provision for users to enter programs.

A microcontroller has been essentially designed for a dedicated application and hence it is called dedicated microcontroller. Intel introduced 8048 series of single chip microcomputers in

1976 as MCS48. It contains a 8-bit CPU, 1K/ 2K/ 4K ROM/ EPROM, 64B/ 128B/ 256B RAM, timer/ counter, parallel I/O ports, and 8-bit ADC. The 8051 has 8-bit CPU, 4/8KB ROM/EPROM, 128/256B RAM, timer/ counter, parallel I/O ports, serial I/O ports etc. It has provisions to increase its memory size using external EPROMs and RAMs. The 16-bit microcomputer MCS-96 introduced by Intel in 1983 contained 16-bit CPU, 8KB ROM, 232B RAM, timer/counter, parallel I/O ports, serial I/O ports, 10-bit ADC, high speed I/O, pulse width modulated output, watch dog timer etc. It is used in sophisticated missile guidance control to complex instrumentation system.

Integration of all basic blocks of a microcomputer system into one circuit brings at least some architectural advantages. When the device is used as a stand alone controller, it needs interfacings with its I/O peripherals. This means that the execution speed of the processing is limited only by the speed of the chip, because there is no slowdown from transferring data between memory and CPU as in multi chip design. The inclusion of data and program memories simplifies the user's hardware interface problem and system implementation.

Few examples of single chip microcomputers are Intel 8048, 8051, and 8096 series, Motorola's M6801 series, Texas Instrument's TMS1000 and Zilog's Z-80. The 16-bit microcontroller family is 8094/8095/8096/8097 and 8394/8395/8396/8397. A summary of few features of 8-bit and 16-bit single chip microcontrollers is given in Table 7.1. The 8048, 8051, 8044 series are 8-bit types whereas 8096, 8097 are 16-bit types of microcoontrollers. There has been constant improvement over one to another. Some of them have special features, such as masking capability, Boolean functions, multiplication and division instructions, serial communication processor, PLL, ADC, PWM, Watch dog timer etc. Hence, depending upon the particular requirement suitable microcontroller may be selected and can even be upgraded.

Table 7.1 8-bit and 16-bit single chip microcontrollers

Name	*ROM/EPROM*	*RAM*	*I/O port*	*External Interrupts*	*Timer*	*ADC*
8048	1KB ROM	64 B	3×8-bit	1	1	NO
8035	0	64B	3×8-bit	1	1	NO
8039	0	128B	3×8-bit	1	1	NO
8748	1KB EPROM	64B	3×8-bit	1	1	NO
8049	2KB EPROM	64B	3×8-bit	1	1	NO
8741	1KB EPROM	64B	3×8-bit	1	1	NO
8022	2KB EPROM	64B	3×8-bit	1	1	NO
8744	4KB EPROM	192B	4×8-bit	5	2	NO
8751	4KB EPROM	128B	4×8-bit	5	2	YES
8096	0	252B	5×8-bit	8	4	NO
8097	0	252B	5×8-bit	8	4	YES
8396	8KB ROM	252B	5×8-bit	8	4	NO
8397	8KB ROM	252B	5×8-bit	8	4	YES

All most all microcontrollers can be interfaced with standard peripherals such as programmable I/O controller, Universal Synchronous/Asynchronous Receiver/ Transmitter, decoder and keyboard/ display controller, timers and counters.

Despite its old age, the 8051 is one of the most popular and preferred microcontroller in use today. Many derivative microcontrollers have since been developed that are on-and compatible with–the old ones. The 8051 is a flexible microcontroller with a relatively large number of modes of operations. The operating modes of the 8051 may be changed by manipulating the values of the 8051's Special Function Registers (SFRs). SFRs are accessed as if they were normal Internal RAM. The only difference is that Internal RAM is from address 00H through 7FH ($7 \times 16^1 + 15 \times 16^0 = 127$) whereas SFR registers exist in the address range of 80H through FFH. Each SFR has an address (80H through FFH) and a name.Thus, a group of registers of 8051, addressed from 80H to FFH, are called ***special function registers*** (SFRs).

7.2 8051 (8-BIT MICROCONTROLLER)

The block diagram of the 8051 is shown in Fig. 7.1(*a*). The main features of 8051 are:

- 8-bit CPU Optimized for Control Applications
- Extensive Boolean Processing Capabilities (Single-Bit Logic)
- On-Chip Flash Program Memory upto 4 KB (89 × 51 families)
- On-Chip Data RAM 128 Byte (128 × 8 RAM)
- Bidirectional and Individually Addressable I/O Lines (4-number of 8-bit ports = 32-I/O lines)
- Two 16-bit Timer/Counters
- Full Duplex UART for serial communication
- Five Interrupts (2-External and 3-Internal)
- MUL and DIV instruction (4μs multiply and divide)
- 128K accessible external memory
- Clock frequency 12MHz
- 218-User Bit-addressable locations

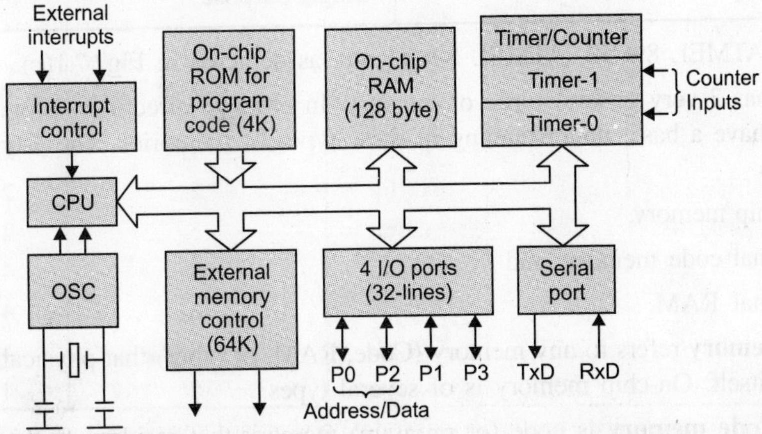

Fig. 7.1 (*a*) Block diagram of a microcontroller

The 8751 is a pin compatible EPROM version of 8051.

The microcontroller contains on-chip RAM, ROM, I/O ports etc. as shown in Fig. 7.1(*b*).

Example: Motorola's 6811, Intel's 8051, Zilog's Z8 and PIC 16X.

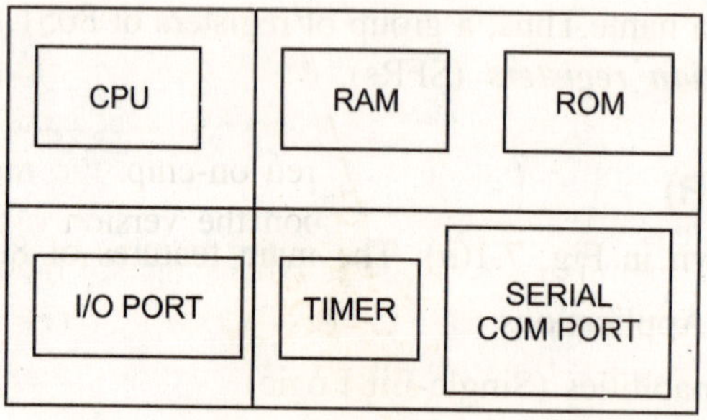

Fig. 7.1 (*b*) Internal blocks of microcontrollers

Fig. 7.1 (*c*) Top view of microcontrollers

A few important differences between microprocessor and microcontroller are:

Microprocessor	*Microcontroller*
1. CPU is stand-alone, RAM, ROM, I/O, timer are separate	CPU, RAM, ROM, I/O and timer are all on a single chip
2. Designer can decide on the amount of ROM, RAM and I/O ports	Fixed amount of on-chip RAM,ROM,I/O Ports
3. Expensive	For applications in which cost, power and space are critical
4. General-purpose	Single-purpose

Example: ATMEL 89C52, ATMEL 89C51 etc. as depicted in Fig. 7.1(*c*).

The 8051 has 3-very general types of memory. In order to effectively program the 8051 it is necessary to have a basic understanding of these types of memories. The 3-types of memory are classified as

- On-chip memory,
- External code memory, and
- External RAM.

On-chip memory refers to any memory (Code, RAM, or other) that physically exists on the microcontroller itself. On-chip memory is of several types.

External Code memory is code (or program) memory that resides off-chip. This is often in the form of an external EPROM.

External RAM is RAM memory that resides off-chip. This is often in the form of standard static or flash RAM.

Code memory

Code memory is the memory that holds the actual 8051 program that is to be executed. This memory is limited to 64K bytes and comes in many shapes and sizes: Code memory may be found ***on-chip***, either **burnt** into microcontroller as ROM or EPROM. Code may also be stored ***off-chip*** in an external ROM or more commonly in an external EPROM. Flash RAM is also another popular method of storing a program. Various combinations of these memory types may also be used. For an example, it is possible to have 4K byte of code memory on-chip, and 64K byte of code memory off-chip in an EPROM. When the program is stored on-chip, the maximum 64K byte is often reduced to 4K, 8K, 16K. This varies depending upon the version of the chip that is being used. Each version offers specific capabilities. One of the distinguishing factors from chip to chip is how much ROM/ EPROM space the chip has. However, code memory is commonly implemented as off-chip EPROM. This is essentially true in low-cost development systems and in system developed by students.

External RAM

The 8051 very well supports the *external* RAM. As the **name** suggests, external RAM is any random access memory, which is found off-chip. Since the memory is off-chip it is not flexible in terms of accessing, and is also ***slower***. For an instance, to increment an internal RAM location by 1 requires only 1 instruction and 1 instruction cycle. However, to increment 1-byte stored in external RAM requires 4-instruction and 7-instruction cycles. In **this** example, external RAM is 7-times slower.

What external RAM looses in terms of speed, it gains **in quantity**. While internal RAM is limited to 128 bytes (256 bytes in 8052), the 8051 supports external RAM up to 64K bytes.

As discussed previously, the 8051 has certain amount of *on-chip* memory. On-chip memory is of 2-types: *Internal RAM* and *Special Function Registers* (SFRs). The layout of the internal memory of the 8051 is presented in Fig. 7.1(*d*).

Internal RAM

As illustrated in Fig. 7.1(*d*), the 8051 has **banks** of 128 bytes on *internal* RAM. This internal RAM is on-chip so it is fastest RAM available and is most flexible in terms of reading, writing, and modifying its contents. Internal RAM is volatile, so on reset this memory is cleared. The lowest 32-bytes of the internal RAM are grouped into 4-banks (Bank-0, Bank-1, Bank-2, Bank-3). Each of these 4-banks **has** 8-registers (R0 to R7) and each register is of 1-byte. These registers are named as R0 through R7. Two bits of the Program Status Word (PSW) select the particular register bank in use. The default register bank is 00. If bank 1 is selected then registers 08H to 0FH will be named as R0 through R7. Similarly, if bank 3 is selected then register 18H to 1FH will be named as R0 through R7.

If we consider registers like A, B, T0 etc., then total size of the internal RAM becomes 256 bytes. These registers are for special purpose and hence they are called SFRs (Special Functions Registers). The address of these registers starts at 80H and ends at FFH. The name of these registers and its address are shown in Fig. 7.1(*f*). Thus, complete memory map is

00H to 1FH = 1FH ($1 \times 16^1 + 15 \times 16^0$ = 31-bytes) = 00 to 31 = 32-bytes $\rightarrow$ Bank Registers

20H to 2FH = 0FH = (16-bytes including 00) $\rightarrow$ Bit-addressable Register

30H to 7FH = 4FH = ($4 \times 16^1 + 15 \times 16^0$ = 79 (80-bytes including 00) $\rightarrow$ General Purpose Register

80H to FFH = 7FH = ($7 \times 16^1 + 15 \times 16^0$ = 127(128-bytes including 00) $\rightarrow$ SFRs

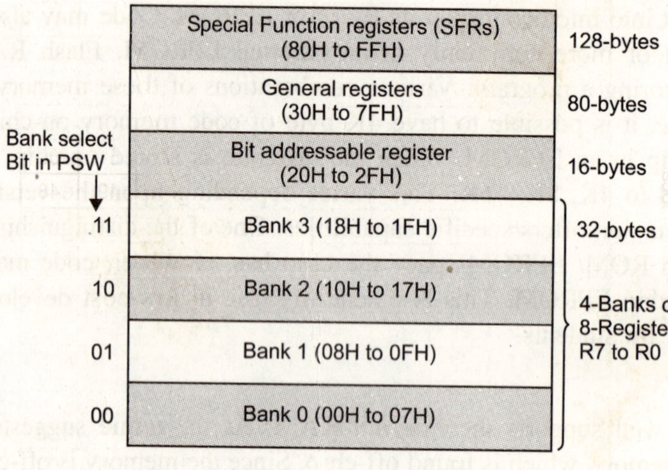

Fig. 7.1(*d*) Internal RAM of 8051

Bit Addressable Memory

The bit addressable memory is also an internal part of the RAM. The next 16-bytes above register banks form a block of bit-addressable memory space. The microcontroller instructions set includes a number of single bit instructions and these instructions can directly address 128-bits in this area. These bit address are 20H through 2FH. All registers above bit addressable space are simple register and can be used as scratch pad registers.

Example: MOV 07H,#20H

In this instruction 07H is the address of the internal RAM and 20H is the immediately available data. After execution of this instruction, register R7 in bank-0 will be filled with the data 20H.

There are two instructions of 8051 that are bit manipulated type. These are

```
CLR &lt;bit>     ;( reset specified bit 0)
SETB &lt;bit>    ;( set specified bit 1)
```

Any bit can be accessed either by their name or by their address. For an instant, the bit-5 of accumulator can be addressed as E5H or ACC.5. If we want to reset it to zero then we can do like this

```
CLR 0E5H    ; bit (0E5H)←0
```

Alternatively, we can use the bit number of the accumulator also as

```
CLR ACC.5   ; bit (ACC.5)←0
```

Registers 20H to 2FH are bit-addressable. We can address these bit directly either to set it or to reset it. For an instant, if we want to reset 3rd bit of register 25H in Fig. 7.1(*e*). The memory map of the bit-addressable area of the RAM is 2AH and we can write as

CLR 2AH

Bit-addressable locations

2F	7F	7E	7D	7C	7B	7A	79	78
2E	77	76	75	74	73	72	71	70
2D	6F	6E	6D	6C	6B	6A	69	68
2C	67	66	65	64	63	62	61	60
2B	5F	5E	5D	5C	5B	5A	59	58
2A	57	56	55	54	53	52	51	50
29	4F	4E	4D	4C	4B	4A	49	48
28	47	46	45	44	43	42	41	40
27	3F	3E	3D	3C	3B	3A	39	38
26	37	36	35	34	33	32	31	30
25	2F	2E	2D	2C	2B	2A	29	28
24	27	26	25	24	23	22	21	20
23	1F	1E	1D	1C	1B	1A	19	18
22	17	16	15	14	13	12	11	10
21	0F	0E	0D	0C	0B	0A	09	08
20	07	06	05	04	03	02	01	00

Fig. 7.1(*e*) Bit addressable area

The special function registers are **illustrated** in Fig. 7.1(*f*).

Special Function Register (SFR)

SFRs are of three types, namely,

- I/O ports SFRs
- Control SFRS, and
- Other SFRs.

The 8051 has 04 I/O ports (P0, P1, P2, P3). Each port is 8-bits wide. Thus a total of 32 I/O lines are available with these ports. Whether a given I/O line is high or low and the value read from the line is controlled by the control SFR's.

The PCON, TCON, SCON, IE, IP, and PSW are known as ***control*** SFRs. These SFRs in some way control the operation or the configuration of some aspect of the 8051. For example, **TCON** controls the timers, SCON controls the serial port.

The remaining SFRs such as SP, DPL, DPH, TL0, TL1, TH0, TH1, SBUF, ACC, B are known as auxillary SFRs in the sense that they do not directly configure the 8051 but obviously the 8051 cannot operate without them. For example, once the serial port has been configured using **SCON**, the program may read or write to the serial port using the **SBUF** register.

The SFRs (P0, PCON, P1, SCON, P2, IE, P3, IP, PSW, ACC, B) in Table 7.1(*b*) may be accessed via bit operations (i.e., using the **SETB** and **CLR** instructions). The other SFRs cannot be accessed using bit operations. As we can see that all SFRs whose addresses are divisible by 8 can be accessed with bit operations.

Address									SFR
FF									
F0	F7	F6	F5	F4	F3	F2	F1	F0	B
E0	E7	E6	E5	E4	E3	E2	E1	E0	ACC
D0	D7	D6	D5	D4	D3	D2	—	D0	PSW
B8	—	—	—	BC	BB	BA	B9	B8	IP
B0	B7	B6	B5	B4	B3	B2	B1	B0	P3
A8	AF	—	—	AC	AB	AA	A9	A8	IE
A0	A7	A6	A5	A4	A3	A2	A1	A0	P2
99	not bit addressable								SBUF
98	9F	9E	9D	9C	9B	9A	99	98	SCON
90	97	96	95	94	93	92	91	90	P1
8D	not bit addressable								TH1
8C	not bit addressable								TH0
8B	not bit addressable								TL1
8A	not bit addressable								TL0
89	not bit addressable								TMOD
88	8F	8E	8D	8C	8B	8A	89	88	TCON
87	not bit addressable								PCON
83	not bit addressable								DPH
82	not bit addressable								DPL
81	not bit addressable								SP
80	87	86	85	84	83	82	81	80	P0

Fig. 7.1(*f*) SFRs

The pin description of 8051 is shown in Fig. 7.1(*g*).

V_{CC} = +5V (supply voltage), GND = Ground.

RST = Reset input

ALE = Address Latch Enable for demultiplexing AD7 to AD0.

A15 to A8 (P2.7-P2.0): high order byte address bus for external memory.

AD7-AD0 (P0.7-P0.0): multiplexed low-byte address and data bus for external memory.

SUMMARY OF PORT PINS

The 8051 has 4-numbers of I/O ports

Port 0: pins 32-39 P0.0 to P0.7, Bidirectional bit-addressable with open collector/drain

Port 1: pins 1-8 P1.0 to P1.7, Bidirectional bit-addressable with internal pullups

Port 2: pins 21-28 P2.0 to P2.7, Bidirectional bit-addressable with internal pullups

Port 3: pins 10-17 P3.0 to P3.7, Bidirectional bit-addressable with internal pullups

Each port has 8-pins.

- Named P0.X, X = 0 ,1,...,7, P1.X, P2.X, P3.X

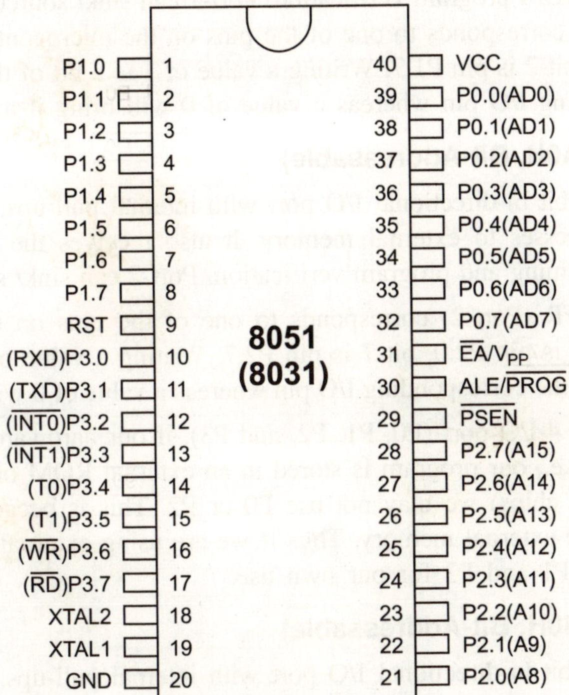

Fig. 7.1(*g*) Pinouts of the 8051

- Example: P0.0 is the bit-0 (LSB) of P0
- Example: P0.7 is the bit-7 (MSB) of P0
- These 8-bits form a byte.

P0 (Port 0, Address 80H, Bit-Addressable)

The Port-0 is **an** 8-bit open drain (of the FET) bi-directional I/O port. It works as multiplexed low order address and data bus during accesses to external memory. It also receives the instruction bytes during EPROM programming and outputs instruction bytes during program verification. (External pull-ups are required during program verification.) Port-0 can sink (and in bus operation can source) 8-LS TTL inputs.

Each bit of this SFR (Port-0) corresponds to one of the pins on the microcontroller. For example, bit-0 of Port-0 is pin P0.0, bit-7 is pin P0.7. Writing a value of 1 to a bit of this SFR will send a high level on the corresponding I/O pin whereas a value of 0 will bring it to a low level. While the 8051 has 04 I/O ports (P0, P1, P2, and P3), if our hardware uses external RAM or external code memory (i.e., our program is stored in an external ROM or EPROM chip or if we are using external RAM chips) we may not use P0 or P2. This is because the 8051 uses ports P0 and P2 to address the external memory. Thus if we are using external RAM or code memory we may only use ports P1 and P3 for our own use.

P1 (Port 1, Address 90H, Bit-Addressable)

The Port-1 is **an** 8-bit bi-directional I/O port with internal pull-ups. It receives the low order address byte during EPROM program verification. Port-1 can sink/ source 4-LS TTL inputs. Each bit of this SFR (Port-1) corresponds to one of the pins on the microcontroller. For example, bit-0 of Port-1 is pin P1.0, bit-7 is pin P1.7. Writing a value of 1 to a bit of this SFR will send a high level on the corresponding I/O pin whereas a value of 0 will bring it to a low level.

P2 (Port 2, Address A0H, Bit-Addressable)

The Port-2 is an 8-bit bi-directional I/O port with internal pull-ups. It sends the high order address byte during accesses to external memory. It also receives the high order address bits during EPROM programming and program verification. Port-2 can sink/ source 4-LS TTL inputs.

Each bit of this SFR (Port-2) corresponds to one of the pins on the microcontroller. For example, bit 0 of port-2 is pin P2.0, bit-7 is pin P2.7. Writing a value of 1 to a bit of this SFR will send a high level on the corresponding I/O pin whereas a value of 0 will bring it to a low level.

While the 8051 has 4-I/O port (P0, P1, P2, and P3), if our hardware uses external RAM or external code memory (i.e., our program is stored in an external ROM or EPROM chip or if we are using external RAM chips) we may not use P0 or P2. This is because the 8051 uses ports P0 and P2 to address the external memory. Thus if we are using external RAM or code memory we may only use ports P1 and P3 for our own use.

P3 (Port 3, Address B0H, Bit-Addressable)

The Port-3 is an 8-bit bi-directional I/O port with internal pull-ups. Port-3 can sink/source 4-LS TTL inputs. It also serves the functions of various special features of the MCS51 family as listed in Table 7.1(*a*).

Each bit of this SFR (Port-3) corresponds to one of the pins on the microcontroller. For example, bit 0 of port 3 is pin P3.0, bit-7 is pin P3.7. Writing a value of 1 to a bit of this SFR will send a high level on the corresponding I/O pin whereas a value of 0 will bring it to a low level.

Each port can be used as input or output (bi-direction). P0 has no internal pull-up resistors and does not connect to V_{CC} inside the 8051.

However, for a programmer, it is the same to program P0, P1, P2 and P3. All the ports upon RESET are configured as input ports.

Port 3 pins have dual function. Their functions are listed below (pins 10-17)

P3.0/Serial Data Receiver (RXD) (serial input port))

P3.1/Serial Data Transmission (TXD) (serial output port)

P3.2/External Interrupt-0 ($\overline{INT_0}$) (External interrupt)

P3.3/External Interrupt-1 ($\overline{INT_1}$) (External interrupt)

P3.4/Timer-0 (T0) (Timer/counter 0 external input))

P3.5/Timer-1 (T1) (Timer/counter 1 external input)

P3.6/Write Enable ($\overline{WR}$) (external data memory write strobe)

P3.7/Read Enable ($\overline{RD}$) (external data memory read strobe

Pin 18 and 19 (XTAL1 and XTAL2)

These pins are used to connect the crystal oscillator from 1MHz to 24 MHz. Ceramic Disc Capacitors of 33pF are required for stabilization as shown in Fig. 7.1(*h*).

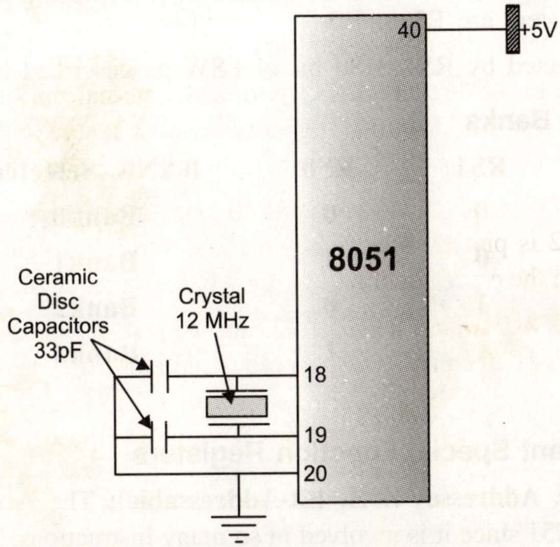

Fig. 7.1(*h*)

Port-0 pins also have dual function as shown in pin diagram (Fig. 7.1(*g*))

$\overline{\text{EA}}$ **(External Access Enable)/V_{PP} (pin 31)**

(used in Flash Burning)

$\overline{\text{EA}}$ **= 0;** microcontroller will Read Program from External ROM

$\overline{\text{EA}}$ **= 1;** microcontroller will Read Program from Internal ROM

V_{PP}:

Programming Enable Voltage = 12 V

(used during Flash Programming)

ALE (Address Latch Enable)/PROG (pin 30)

Program Pulse I/P

ALE: Used to Demultiplex Low order Address and Data line for External Memory Interfacing. They normally provide constant Pulses of Frequency 1/6 of XTAL Freq.

PROG:

Programme Pulse input during the Flash Programming.

PSEN (Programme Strobe Enable) (pin 29)

Used to Interface the external ROM. It is connected to Read Enable pin of ROM Chip

There are 4-Register Banks. The description of register banks are

Bank-0 (00H to 07H)

Bank-1 (08H to 0FH)

Bank-2 (10H to 17H)

Bank-3 (18H to 1FH)

- At a Time one Bank is selected
- Name of Registers are R7 to R0
- Banks are selected by RS1, RS0 bit of PSW as described below

Selection of Register Banks

RS1	RS0	BANK Selected
0	0	Bank0
0	1	Bank1
1	0	Bank2
1	1	Bank3

Description of Important Special Function Registers

ACC (Accumulator, Addresses E0H, Bit-Addressable): The Accumulator is one of the most-used SFRs on the 8051 since it is involved in so many instructions. The Accumulator resides as an SFR at E0H, which means the instruction **MOV A,#20H** is really the same as **MOV E0H,#20H**. However, it is a good idea to use the first method since it only requires 2-bytes whereas the second option requires 3-bytes.

ACC.7	ACC.6	ACC.5	ACC.4	ACC.3	ACC.2	ACC.1	ACC.0
bit-7	bit-6	bit-5	bit-4	bit-3	bit-2	bit-1	bit-0
E7H	E6H	E5H	E4H	E3H	E2H	E1H	E0H

- **8-bit Register**
- **Bit Addressable Register**
- **Holds the Result and the first operand**

B (B Register, Addresses F0H, Bit-Addressable)

The B register is used in 2-instructions: the *multiply* and *divide* operations. Programmers also commonly use the B register as an auxiliary register to temporarily store values.

B.7	B.6	B.5	B.4	B.3	B.2	B.1	B.0
bit-7	bit-6	bit-5	bit-4	bit-3	bit-2	bit-1	bit-0
F7H	F6H	F5H	F4H	F3H	F2H	F1H	F0H

- **8-bit Register**
- **Bit Addressable Register**
- **Specially used in MUL and DIV instructions otherwise just as a simple register**

Program Status Register

The PSW register contains information about the status of the accumulator after execution of any instruction. The PSW register has several bits that reflect the current status of the CPU. The PSW, shown in Fig. 7.1(*i*), resides in SFR space. It contains the Carry bit, the Auxiliary Carry (for BCD operations), the 2-register bank select bits, the Overflow flag, a Parity bit and 2-user definable status flags.

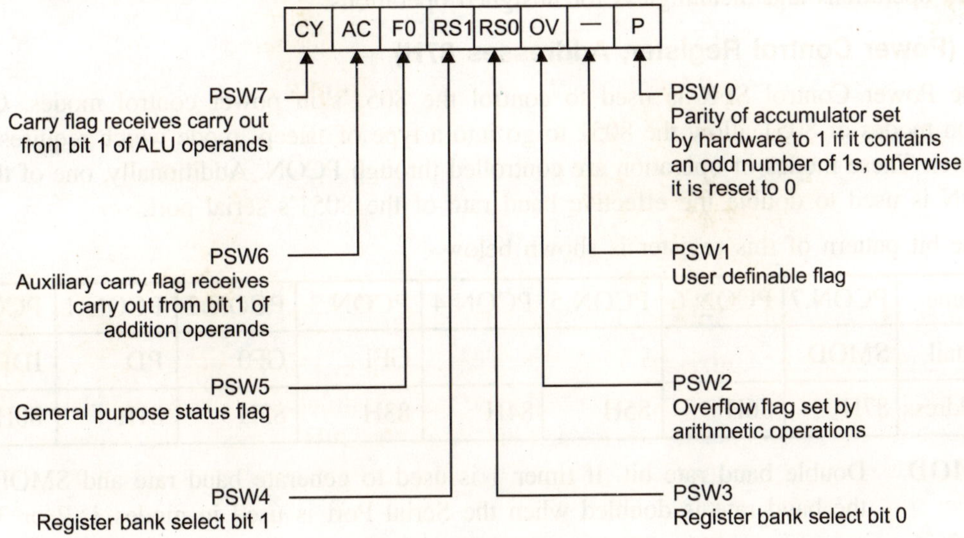

Fig. 7.1(*i*) PSW

The carry bit other than serving the functions of a carry bit in arithmetic operations, also serves as the 'accumulator' for a number of Boolean operations. The bits RS1 and RS0 are used to select one of the 4-register banks shown in Fig. 7.1(*d*). A number of instructions refer to these RAM Locations as R7 through R0. The selection of which of the 4-bank is being referred to as made on the basis of the bits RS1 and RS0 at execution time.

The lower 32-bytes are grouped into 4-banks of 8-registers. Program instruction call out these registers as R0 through R7. The 2-bits in the PSW select register bank that is in use.

The parity bit reflects the number of 1s in the accumulator: P = 1 if the accumulator contains an odd number of 1s, and P = 0 if the accumulator contains an even number of 1s. Thus the number of 1s in the accumulator plus P is always even.

Two bits in the PSW are uncommitted and may be used as general-purpose status flags.

Actually CY is used to indicate the overflow for unsigned numbers. Overflow (OV) indicates overflow for signed numbers. These bits are affected as below

CY ← C7 (Carry generated from MSB)

OV ← C7 XOR C6

This flag is set whenever the result of a signed number operation is too large, causing the high-order bit to overflow into the sign bit.

In general, carry flag is used to detect errors in unsigned arithmetic operations. The overflow flag is used to detect errors in signed arithmetic operations.

In other words, the overflow flag is set to 1 if there is carry from C6 to C7 or from C7 out, but not both.

So we can conclude that in signed number addition, OV indicates whether or not the result is valid. If OV =1, the result is erroneous, if OV = 0, the result is valid. Hence OV is valid only in signed operations and meaningless for unsigned operations.

PCON (Power Control Register, Addresses 87H)

The Power Control SFR is used to control the 8051's in power control modes. Certain operation modes of 8051 allow the 8051 to go into a type of 'sleep' mode which requires much less power. These modes of operation are controlled through PCON. Additionally, one of the bits in PCON is used to double the effective baud rate of the 8051's serial port.

The bit pattern of this register is shown below

Bit Name	PCON.7	PCON.6	PCON.5	PCON.4	PCON.3	PCON.2	PCON.1	PCON.0
Bit Detail	SMOD	-	-	-	GF1	GF0	PD	IDL
Bit Address	87H	86H	85H	84H	83H	82H	81H	80H

SMOD Double baud rate bit. If timer 1 is used to generate baud rate and SMOD = 1, the baud rate is doubled when the Serial Port is used in modes 1, 2 or 3.

GF1 General Purpose flag bit.

GF0 General Purpose flag bit.

PD Power Down bit. Setting this bit activates Power down operation in the AT89C51.

IDL Idle Mode Bit. Setting this bit activates Idle Mode operation in the AT89C51.

If 1s are written to PD and IDL at the same time. PD takes precedence.

There are two types of power control modes.

1. Idle Mode
2. Power Down Mode

Let us discuss these modes in detail.

Idle Mode

In idle mode, the CPU puts itself to sleep while all the on chip peripherals remain active. The mode is invoked by software. The content of the on-chip RAM and all the special functions registers remain unchanged during this mode. The idle mode can be terminated by any enabled interrupt or by a hardware reset; the device normally resumes program execution, from where it left off, up to two machine cycles before the internal reset algorithm takes control. On-chip hardware inhibits access to internal RAM in this event, but access to the port pin is not inhibited. To eliminate the possibility of an unexpected write to a port pin when Idle is terminated by reset, the instruction following the one that invokes idle should not be one that writes to a port pin or to external memory.

Power Down Mode

In the power-down mode, the oscillator is stopped, and the instruction that invokes power-down is the last instruction executed. The on-chip RAM and Special Function registers retain their values until the power-down mode is terminated. The only exit from power-down is a hardware reset. Reset redefines the SFRs but does not change the on-chip RAM. The reset should not be activated before V_{CC} is restored to its normal operating level and must be held active long enough to allow the oscillator to restart and stabilize.

SBUF (Serial Control, Addresses 99H)

The Serial Buffer SFR is used to send and receive data via the on-board serial port. Any value written to SBUF will be sent out the serial port's TXD pin. Likewise any value which the 8051 receives via the serial port's RXD pin will be delivered to the user program via SBUF. In other words, SBUF serves as the output port when write to and as an input port when read from.

SCON (Serial Control, Addresses 98H, Bit-Addressable)

The Serial Control SFR is used to configure the behavior of the 8051's on-board serial port. This SFR controls to the baud rate of the serial port, whether the serial port is activated to receive data, and also contains flags that are set when a byte is successfully sent or received. To use the 8051's on-board serial port, it is generally necessary to initialize the following SFR:

SCON, TCON and TMOD. This is because SCON controls the serial port. However, in most cases the program will wish to use one of the timers to establish the serial port's baud rate. In this case, it is necessary to configure timer 1 by initializing TCON & TMOD.

The bit pattern of the SCON register is shown following......

Bit name	SCON.7	SCON.6	SCON.5	SCON.4	SCON.3	SCON.2	SCON.1	SCON.0
Bit detail	SM0	SM1	SM2	REN	TB8	RB8	T1	R1
Bit Address	9FH	9EH	9DH	9CH	9BH	9AH	99H	98H

IE: Interrupt Enable Register (Bit Addressable)

The Interrupt enable SFR is used to enable and disable specific interrupts. The low 7 bits of the SFR are used to enable and disable the specific interrupts, whereas the highest bit is used to enable or disable ALL interrupts. Thus, if the high bit of IE is 0 all interrupts are disabled regardless of whether an individual interrupt is enabled by setting a lower bit.

The bit pattern of the IE register is shown as following....

Bit name	IE.7	IE.6	IE.5	IE.4	IE.3	IE.2	IE.1	IE.0
Bit detail	EA	-	-	ES	ET1	EX1	ET0	EX0
Bit address	0AFH	0AEH	0ADH	0ACH	0ABH	0AAH	0A9H	0A8H

EA IE.7 Disables all interrupts. If EA = 0, no interrupt is acknowledged. If EA = 1, each interrupt source is individually enabled or disabled by setting or clearing its enable bit.

ES IE.4 Enables or disables the serial port interrupts.

ET1 IE.3 Enables or disables the Timer1 overflow interrupt.

EX1 IE.2 Enables or Disables External Interrupt1.

ET0 IE.1 Enables or Disables the Timer 0 overflow interrupt.

IP (Interrupt Priority, Addresses B8H, Bit-addressable)

The Interrupt Priority SFR is used to specify the relative priority of each interrupt. On the 8051, an interrupt may either be of low(0) priority or high(1) priority. An interrupt may only interrupt interrupts of lower priority. For example, if we configure the 8051 so that all interrupts are of low priority except the serial interrupt, the serial interrupt will always be able to interrupt the system, even if another interrupt is currently executing. However, if a serial interrupt is executing no other interrupt will be able to interrupt the serial interrupt routine since the serial interrupt routine has the highest priority.

The bit pattern is shown as following......

Bit name	IP.7	IP.6	IP.5	IP.4	IP.3	IP.2	IP.1	IP.0
Bit detail	-	-	-	PS	PT1	PX1	PT0	PX0
Bit address	BFH	BEH	BDH	BCH	BBH	BAH	B9H	B8H

TCON Register

The timer control SFR is used to configure and modify the way in which the 8051's two timers operate. This SFR controls whether each of two timers is running or stopped and contains a flag to indicate that each timer has overflowed. Additionally some non timer related bits are located in the TCON bits-3 to 0. These bits are used to configure the way in which the external interrupts are activated and also contain the external interrupts flags which are set when an external interrupt has occurred.

TCON: timer counter control register (bit addressable)

TF1	TR1	TF0	TR0	IE1	IT1	IE0	IT0

TF1 TCON.7 Timer 1 overflow flag. Set by hardware when the timer/counter 1 overflows cleared by hardware as the processor to interrupt service routine.

TR1 TCON.6 Timer 1 run control bit set/cleared by software to turn timer/counter ON/OFF.

TF0 TCON.5 Timer 0 overflow flag set by hardware when the timer/counter 0 overflows. Cleared by hardware as the processor vectors to the service routine.

TR0 TCON.4 Timer 0 run control bit set /cleared by software to turn timer/counter 0 on/off.

IEI TCON.3 External interrupts 1 edge flag. Set by hardware when the external interrupt edge is detected. Cleared by hardware when the interrupt is processed.

ITI TCON.2 Interrupt 1 type control bit set /cleared by software to specify falling edge/ low level triggered external interrupt.

IE0	TCON.1	External interrupt 0 edge flag. Set by hardware when external interrupt edge detected. Cleared by hardware when interrupt is processed.
IT0	TCON.0	Interrupt 0 type control bit. Set/cleared by software to specify falling edge/ low level triggered external interrupt.

TMOD (Timer Mode, Addresses 89H)

The Timer Mode SFR is used to configure the mode of operation of each of the two timers. Using this SFR our program may configure each timer to be a 16-bit timer, an 8-bit autoreload timer, a 13-bit timer, or two separate timers. Additionally, we may configure the timers to only count when an external pin is activated or to count events that are indicated on an external pin.

TMOD : Timer/Counter Mode Control register (Not Bit addressable)							
Timer 1				Timer 0			
GATE	$C/\overline{T}$	M1	M0	GATE	$C/\overline{T}$	M1	M0

Gate	; when TRx (in TCON) is set and GATE =1, Timer/ Counterx runs only while the INTx pin is high (hardware control). When GATE = 0, Timer/ Counterx runs only while TRx =1 (software control).
$C/\overline{T}$ Timer or control register	; cleared for Timer operation (input from internal system, clock). Set for counter operation (input from Tx input pin)
M1	; mode selector bit
M0	; mode selector bit

TL0/TH0 (Timer 0 Low/High, Addresses 8AH/8CH)

These two SFRs, taken together, represent timer 0. Their exact behavior depends on how the timer is configured in the TMOD SFR; however, these timers always count up. What is configurable is how and when they increment in value.

TL1/TH1 (Timer 1 Low/High, Addresses 8BH/8DH)

These two SFRs, taken together, represent timer 1. Their exact behavior depends on how the timer is configured in the TMOD SFR; however, these timers always count up. What is configurable is how and when they increment in value.

Timers of 8051

The Timer of the 8051operates in following 4-modes:

M1	*M0*	*Mode*	*Description*
0	0	Mode 0	13-bit counter
0	1	Mode 1	16-bit counter
1	0	Mode 2	8-bit auto reload
1	1	Mode 3	8-bit split mode (only T0 can be programmed in Mode 3)

The details of each mode is

13-bit Timer Mode (mode 0). To maintain compatibility with it's predecessor, the 8048, the timer in mode '0' is a 13-bit timer. Generally the 13-bit timer mode is not used in new development. When the timer is in 13-bit mode, effectively, only 13 bits of the two timer bytes are being used. When timer is in 13-bit mode, TLx will count from 0 to 31. When TLx is increment from 31, it gets reset to 0 and increment THx. Thus, effectively only 13-bits of the two timer bytes are being used : Bits 0-4 of TLx and bits 0-7 of THx. This essentially indicates that the timer can contain 8192 values. If we reset a 13-bit timer to 0, it will overflow back to zero, 8192 machine cycle later.

16-bit Timer Mode (mode 1). Timer in mode '1' is a 16-bit timer. This is a very commonly used mode. It functions just like 13-bit mode except that all 16 bits are used. **16-bit timer** timer may contain up to 2^{16} = 65536 distinct values. It will overflow back to 0 after 65,536 machine cycles.

8-bit Timer Mode (mode 2). Timer in mode '2' is an 8-bit auto-reload mode.THx holds the 'reload value' and TLx is the timer itself. TLx starts counting up. When TLx reaches 255 and is subsequently incremented, instead of resetting to 0 (as in the case of modes 0 and 1) it will be reset to the value stored in THx. This mode is used specially in serial communication.

Split Timer Mode (mode 3). When Timer 0 is placed in mode 3 it behaves as two separate 8-bit timers. Timer 0 is TL0 and Timer 1 is TH0. Both timers count from 0 to 255 and overflow back to 0.All the bits that are related to Timer 1 will now be tied to TH0. While Timer 0 is in split mode, the real Timer 1 (i.e., TH1 and TL1) can be put into modes 0, 1 or 2. Normally we may not start or stop the real timer 1 since the bits that do that are now linked to TH0. The real timer 1, in this case, will be incremented every machine cycle. This mode can be used when we need to have two separate timers and, additionally, a baud rate generator. The real Timer 1 can be used as a baud rate generator and use TH0/TL0 as two separate timers.

Initializing Timer

We first decide the mode of timer. We first initialize the TMOD SFR. Since we are using timer 0 we will use lowest 4-bits of TMOD. The first two bits GATE-0 and C/T-0 are both 0 as we want the timer to be independent of the external pins. We must clear TMOD for the 16-bit timer mode-1. Thus, to initialize the timer, we must execute the instruction

```
MOV TMOD,#01H
```

Timer-0 is now in 16-bit mode. However, the timer is not running. In order to start the timer running, we must set TR0 bit. This is done like

```
SETB TR0
```

Execution of these two instructions timer-0 will immediately begin counting, being incremented once every machine cycle (every 12 crystal pulses).

Timer Reading

We may either read the actual value of the timer as 16-bit number or we may detect when timer overflows.

SP (Stack Pointer, Address 81H)

This is the stack pointer of the microcontroller. This SFR indicates where the next value to be taken from the stack will be read from in Internal RAM. If we push a value onto the stack, the value will be written to the address of SP+1. If SP holds the value 07H, a PUSH instruction will push the value onto the stack at address 08H. This SFR is modified by all instructions, which modify the stack, such as PUSH, POP, LCALL, RET, RETI, and whenever interrupts are provoked by the microcontroller.

The SP SFR, on startup, is initialized to 07H. This means the stack will start at 08H and start expanding upward in internal RAM. Since alternate register banks 1, 2, and 3 as well as the user bit variables occupy internal RAM from addresses 08H through 2FH, it is necessary to initialize SP in our program to some other value if we will be using the alternate register banks and/or bit memory. It is not a bad idea to initialize SP to 2FH as the first instruction of every one of our programs unless we are 100% sure we will not be using the register banks and bit variables.

DPL/DPH (Data Pointer Low/High, Addresses 82H/83H)

The SFRs DPL and DPH work together to represent a 16-bit value called the *Data Pointer*. The data pointer is used in operations regarding external RAM and some instructions involving code memory. Since it is an unsigned two-byte integer value, it can represent values from 0000H to FFFFH (0 through 65,535 decimal).

The DPTR is really DPH and DPL taken together as a 16-bit value. In reality, we almost always have to deal with DPTR one byte at a time. For example, to push DPTR onto the stack we must first push DPL and then DPH. We cannot simply push DPTR onto the stack. Additionally, there is an instruction to increment DPTR. When we execute this instruction, the two bytes are operated upon as a 16-bit value. However, there is no instruction that decrements DPTR. If we wish to decrement the value of DPTR, we must write our own code to do so.

8051 Instruction Set

The summary and details of each instruction is available in Appendix A and B.

Addressing Modes

There are eight addressing modes available in the 8051:

- Register
- Direct
- Indirect
- Immediate
- Relative
- Absolute
- Long
- Indexed

Register Addressing

- 8051 has 8-working registers (R0 to R7)
- Instructions using register addressing are encoded using the three least significant bits of the instruction opcode to specify a register

Example: ADD A, R7

The opcode is 00101111. The 00101 indicates the instruction and the 3-lower bits, 111, specify the register (R7).

- Some instructions are specific to a certain register, such as the accumulator, data pointer etc.

  Example: INC DPTR

  – A 1-byte instruction adding 1 to the data pointer

  Example: MUL AB

  – A 1-byte instruction multiplying unsigned values in accumulator and register B

Direct Addressing

- Direct addressing can access any on-chip memory location

  Example: ADD A, 55H

  Example: MOV P1, A

  – Transfers the content of accumulator to Port 1 (address 90H)

Indirect Addressing

One of the major problems a programmer faces is

- How is a variable identified if its address is modified while a program is running?

In 8051 solution is indirect addressing: R0 or R1 may operate as pointer registers (their content indicates an address in internal RAM where data are written or read)

- In 8051 assembly language, indirect addressing is represented by @ symbol before R0 or R1.

- Example: MOV A, @R0

  – Moves a byte of data from internal RAM at location whose address is in R0 to the accumulator.

Immediate Addressing

When the source operand is a constant rather than a variable, the constant can be incorporated into the instruction as a byte of immediate address. In assembly language, immediate operands are preceded by #. Operand may be a numeric constant, a symbolic variable or an arithmetic expression using constants, symbols and operators. Assembler computes the value and substitutes the immediate data into the instruction.

Example: MOV A,#12

Note: With one exception, all instructions using immediate addressing use 8-bit data. Exception: when initializing the data pointer, a 16-bit constant is required.

Example: MOV DPTR, #8000H

Relative Addressing

Relative addressing is used with certain jump instructions. Relative address (offset) is an 8-bit signed value (–128 to 127) which is added to the program counter to form the address of next instruction. Prior to addition, the program counter is incremented to the address following the jump (the new address is relative to the next instruction not the address of the jump instruction). This detail is of no concern to the user since the jump destinations are usually specified as labels and the assembler determines the relative offset.

Advantage of relative addressing: position independent codes.

Absolute Addressing

Absolute addressing is only used with ACALL and AJMP. The 11 least significant bits of the destination address comes from the opcode and the upper five bits are the current upper five bits in the program counter (PC).

The destination is in the same 2K (2^{11}) range of the source.

Long Addressing

Long addressing is used only with the LCALL and LJMP instructions. These 3-bytes instructions include a full 16-bit destination address as bytes-2 and 3.

- The full 64K code space is available
- The instruction is long and position dependent
  - Example: LJMP, 8AF2H : Jumps to memory location 8AF2H

Indexed Addressing

Indexed addressing uses a base register (either the program counter or data pointer) and an offset (the accumulator) in forming the effective address for a JMP or MOVC instruction

Example: MOVC A, @A+DPTR

- This instruction moves a byte of data from code memory to the accumulator. The address in code memory is found by adding the accumulator to the data pointer

8051 instructions are divided among five groups:

- Arithmetic
- Logical
- Data transfer
- Boolean variable
- Program branching

Arithmetic Instructions

ADD, ADDC, SUBB, MUL, DIV, INC, DEC, DAA:

Since we are already familiar with most of the instructions, we provide a brief overview of all the instructions:

ADD simply adds the two operands while **ADDC** adds the two operands along with the carry bit.

Since different addressing modes are available, an arithmetic instruction may be written in different ways.

When adding signed integers, OV indicates a negative number produced as the sum of two positive operands, or a positive sum from two negative operands.

Four source operand-addressing modes are allowed

- Register,
- Direct,
- Register-indirect, or
- Immediate.

Example: Suppose A_{CC} = 9CH, R0 = 40H, (R0) = 9C, R7 = 35H, select Bank 1 for registers RS1 = 0,RS0 = 1. (7F) = 45H].

ADD A,7FH (direct addressing)

ADD A,@R0 (indirect addressing)

ADD A,R7 (register addressing)

ADD A,#35H (immediate addressing)

```
             1001 1100
ADD A,7FH    0100 0101
             ---------
             1110 0001  = E1H, C = 0, AC = 1, OV = 0, P = 0, PSW = 01001000 = 48H

             10011100
ADD A,@R0    10011100
             ---------
             100111000  = 0138H, C = 1, AC = 1, OV = 1, P = 1, PSW = 11001101 = CDH

             10011100
ADD A,R7     00110101
             ---------
             1101 0001  = DlH , C = 0, AC = 1, OV = 0, P = 0, PSW = 01001000 = 48H

             10011100
ADD A,#35H   00110101
             ---------
             1101 0001  = DlH, C = 0, AC = 1, OV = 0, P = 0, PSW = 01001000 = 48H
```

Four source operand-addressing modes are allowed

- Register,
- Direct,
- Register-indirect. or
- Immediate.

All arithmetic instructions are executed in one machine cycle except INC DPTR (two cycles) and MUL AB and DIV AB (four cycles)

Example: Write a code that subtracts content of R6 from R7 and leave the result in R7.

Solution: MOV A,R7 (suppose R7 = 05H, R6 = 05H, $\overline{R6}$ + 1 = 1111 1010 + 1 = 1111 1011, select Bank0 for registers).

```
CLR C
SUBB A,R6
MOV R7,A
```

  0000 0101
  11111011
1 0000 0000   C = 1, AC = 1, P = 0, OV = 0, PSW=11000000 = C0H

- Clearing the flag is necessary because the only form of subtraction in 8051 is **SUBB** (subtract with borrow). The operation subtracts from the accumulator source byte and carry bit.
- Any memory location can be **incremented or decremented** using direct addressing without going through the accumulator.

Example: INC 7FH

– Increments the value in memory location 7FH.

- INC instruction can also work on 16-bit data pointer.

MUL AB: multiplies 8-bit unsigned values in A and B that leaves the 16-bit result in A (low byte) and B (high byte). If the product is greater than 255 (FFH), overflow flag is set.

Example: ACC = 55H, B register contains 22H, and PSW = 00H. What are the contents of these registers after execution of the MUL AB instruction?

  55
* 22

AA, (B) = 0BH, (A) = 4AH = 0100 1010B, C = 1, AC = 0, P = 1, OV = 1

AA

0B4A

Answer: ACC = 4AH, B = 0BH, P-bit in PSW is set to 1. Since the result is greater than 255, overflow flag is set.

DIV AB: divided A by B, leaving the integer result in A and remainder in B.

Note:

- For BCD arithmetic, ADD and ADDC must be followed by a **DAA** (decimal adjust) operation to ensure the result is in range for BCD.
- Note that DAA will not convert a binary number to BCD.

Example: If ACC contains BCD value of 59H then:

```
ADD A,#1
DAA
```

– First adds 1 to A, leaving 5A and then adjust the result to correct BCD value 60H.

Example: Two 4-digit BCD numbers are in internal memory at locations 40H, 41H and 42H, 43H. The most significant digits are in locations 40H and 42H. Add them and store the BCD result in locations 40H and 41H.

Solution:

```
MOV A,43H
ADD A,41H
DAA
MOV 41H,A
MOV A,42H
ADDC A,40H
DAA
MOV 40H,A
```

(40H) = BBH, (41H) = C3H, (42H) =ABH, (43H) = ABH

$$\begin{array}{r} C3\\ \underline{AB}\\ 016E\end{array}, (A) = 6EH, DAA = \begin{array}{r}6E\\ \underline{66}\\ 134\end{array}, \begin{array}{r}1\\ BB\\ \underline{AB}\\ 167\end{array}, (A) = 67H, DAA = \begin{array}{r}67\\ \underline{66}\\ 133\end{array}$$

LOGICAL INSTRUCTIONS

ANL, ORL, XRL, SWAP A, RL A, RR A

8051 logical instructions perform Boolean operations on bytes of data on a bit-by-bit basis.

ANL = logical AND

ORL = logical OR

XRL = logical XOR

Example: let us assume A = 00110101B. Instruction **ANL** A,#01010011B will leave 00010001 in accumulator

- Different modes for logical instructions: (Suppose R0=40H, ((40H)) = 9CH, R6 =33H, (55H) = 34H, select Bank1 for registers).

  ANL A,55H (0011 0100) = 0011 0100 = 34H, CY = 0, AC = 0, P = 1, OV = 0, RS1 = 0, RS0 = 1, PSW = 00001001 = 09H).

  ANL A,@R0 (1001 1100) = 0001 0100 = 14H, CY = 0, AC = 0,P = 0, OV = 0, RS1 = 0, RS0 = 1, PSW = 00001000 = 08H).

  ANL A, R6 (0011 0011) = 0011 0001 = 31H, CY = 0, AC = 0,P = 1, OV = 0, RS1 = 0, RS0 = 1, PSW = 00001001 = 09H).

  ANL A,#33H (0011 0011) = 0011 0001 = 31H, CY = 0, AC =0, P = 1, OV = 0, RS1 = 0, RS0 = 1, PSW = 00001001 = 09H).

- Logical operations can be performed on any byte in internal memory without going through accumulator

Example: XRL P1,#FFH

Eight bit in Port 1 are read, each bit exclusive ORed. The result is written back to Port 1.

- **Rotate instructions (RLA, RRA)** shift the accumulator one bit to the left or right. For a left rotation, MSB rolls into LSB position. For a right rotation, LSB rolls into MSB position.
- RLC A and RRC A perform similar operations on concatenation of carry flag and accumulator.

Example: A = 00H and C = 1 after RRC A what A and C will be?

A = 80H and C = 0

- **SWAP A:** exchanges high and low nibble within accumulator.

Example: Illustrate two ways to rotate contents of accumulator three positions to the left. Discuss each method in terms of memory and execution speed.

Suppose (A) = 0011 0101B

Solution: 1.

```
RL A        ; 0110 1010B
RL A        ; 1101 0100B
RL A        ; 1010 1001B
```

2.

```
SWAP A      ; 1001 1010B
RR A        ; 0100 1101B
```

All instruction are 1-byte and 1-cycle. So first solution uses three bytes of memory and takes three CPU cycles and second solution uses only two byte of memory and executes in two cycles.

Data Transfer Instructions

MOV <destination>, <source>: allows data to be transferred between any two internal RAM or SFR locations.

- Stack operations (pushing and popping data) are also internal data transfer instructions. The stack of 8051 is kept in the internal RAM.

Pushing increments SP before writing the data

Popping from the stack reads the data and decrements the SP

Example: Stack pointer contains 08H and A contains 55H and B contains 4AH. What internal RAM locations are altered and what are their new values after following instructions?

Solution: PUSH ACC; SP = 09H, A = 55H, (Address 09H) = 55H

PUSH B; SP = 0AH, B = 4AH, (Address 0AH) = 4AH

Address 09H will have 55H; address 0AH will have 4AH and address 81H (SP) will have 0AH.

- Instruction **XCH A,<source>** causes the accumulator and the address byte to exchange data
- Instruction **XCHD A,@Ri** causes the low-order nibbles to be exchanged.

Example: If A contains F3H, R1 contains 40H, and internal RAM address 40H contains 5BH instruction XCHD A,@R1 leaves A containing FBH and internal RAM location 40H containing 53H.

Data transfer instruction that move data between internal and external memory use indirect addressing. The address could be one byte (residing in R0 or R1) or two bytes (residing in DPTR).

- 16-bit address uses all Port 2 for high-byte and this port cannot be used for I/O
- 8-bit addresses allow access to a small external memory
- **MOVX** is used for external data transfer

Example: Read the content of external RAM locations 10F4H and 10F5H and place values in R6 and R7, respectively.

```
MOV DPTR,#10F4H   ; 16-bit data pointer loaded with 10F4H (DPH = 10H, DPL = F4H)
MOVX A,@DPTR      ; load accumulator from the location pointed by DPTR (10F4H)
MOV R6, A         ; save accumulator in register R6.
INC DPTR          ; increment DPTR (DPTR+1 = 10F5H)
```

```
MOVX A,@DPTR      ; load accumulator from the location pointed by DPTR (10F5H)
MOV R7, A         ; save accumulator in register R7.
```

Boolean Instructions

8051 contains a complete Boolean processor for single-bit operations. All bit accesses use direct addressing. Bits may be set or cleared in a single instruction

Example: **SETB** P1.7 ; set bit-7 of Port1

CLR P1.7 ; clear (reset) bit-7 of Port1.

- Carry bit in PSW is used as a single-bit accumulator for Boolean operations.
- Bit instructions that refer to carry bit as C are assembled as carry-specific instructions. Carry also has a mnemonic representation (CY), which can be used in connection with non-carry-specific instructions.

Example:

```
CLR C
CLR CY
```

Both do the same. First one is 1 byte and the second one is 2-bytes.

Example: Compute the logical AND of the input signals on bit 0 and 1 of Port 1 and output the result to bit 2 of Port 1.

```
LOOP:  MOV C,P1.0   (1 cycle)   ; load carry with bit of Port1.0
       ANL C, P1.1  (2 cycles)  ; AND content of carry with bit of P1.1 with result
                                  in carry
       MOV P1.2,C   (2 cycles)  ; load Port1.2 with content of carry
       SJMP LOOP    (2 cycles)  ; jump to LOOP
```

Branching Instructions

There are three versions of JMP instruction: SJMP, LJMP and AJMP.

- SJMP instruction specifies destination address as a relative offset. This instruction is 2 bytes and jump distance is limited to –128 to 127.
- LJMP specifies the destination address as a 16-bit constant. The destination address can be anywhere in the 64K program memory space.
- AJMP specifies the destination address as an 11-bit constant. Destination must be within 2K block of AJMP.

In all cases programmer specifies the destination address to the assembler (using label or a 16-bit constant) and the assembler puts destination address into correct format.

Subroutines and Interrupts

There are two versions of the CALL instruction: **ACALL and LCALL** using absolute and long addressing. Generic CALL may be used if the programmer does not care which way the address is coded. Either instruction pushes the content of the PC on the stack and loads PC with the address specified in the instruction. SCALL is limited to 256 bytes, ACALL is limited 2KB, and LCALL is limited to 64KB.

Note that PC will contain the address of the instruction following the CALL instruction when it gets pushed on the stack.

PC is pushed on the stack: low-byte first, high-byte second.

Example: Instruction LCALL COSINE is in code memory at address 0204H through 0206H and subroutine COSINE begins at code memory address 043AH. Assume stack pointer contains 3AH before this instruction. What internal RAM locations are altered and what their new values after this instruction will be?

Address	Content
3BH	07H
3CH	02H
81H	3CH

Subroutines should end with an RET instruction. RET pops the last two bytes off the stack and place them in PC. Jumping in or out of a subroutine any other way usually fouls up the stack and causes the program to crash.

Conditional Jumps

8051 offers a variety of conditional jump instructions

- **JZ and JNZ** test the accumulator for condition.**(jump if 0 and jump if not 0)**
- **DJNZ (decrement and jump if not zero)** is a useful instruction for building loops.

APPLICATION: To execute a loop N times, load a register with N and terminate the loop with a DJNZ to the beginning of the loop.

- **CJNE (compare and jump if not equal)** is another conditional jump instruction.

CJNE: two bytes in the operand field are taken as unsigned integers. If the first one is not equal to second; the jump occurs.

Syntax: CJNE A,#20H, [loop label]

If A is not equal to 20H; then control branches to loop label.

7.3 SAMPLE PROGRAMS

A number of programs have been written and tested on 8051 kit. These programs give a basic idea about the programming of the microcontroller and may be used as a beginning step by users in developing his-own programs.

1. Adds 2-numbers stored in the accumulator and the B-register

After completion, the result is stored in the accumulator and the flags are set/ reset according to the data conditions.

7000	74,24	MOV A, #data (24)	; Load A_{CC} with 1st data byte 24H.
7002	75,F0,42	MOV F0, #data (42)	; Load B register with 2nd data byte 42H.
7005	25,F0	ADD A, F0	; Add (A_{CC}) + (B)
7007	F8	MOV R_0, A	; Transfer result from (A_{CC}) to Bank 0, Reg 0
7008	22	RET	; Return to Command Mode.

The data given in the example is 24H and 42H. The result is

 0010 0100
+0100 0010
―――――――――
 0110 0110 = 66H stored in the Babk register R_0 and A_{CC}

Flag Status	CY	AC	F0	RS_1	RS_0	OV	–	P
	0	0	0	0	0	0	x	0 = 00H

For data 98H and 87H

 1001 1000
+1000 0111 (A_{CC}) = 1FH, (B) = 87H, (F) = 85H
―――――――――
1 0001 1111

Flag Status	CY	AC	F0	RS_1	RS_0	OV	–	P
	1	0	0	0	0	1	x	1 = 85H

2. Add 2-data bytes along with carry bit.

This program demonstrates the use of the carry flag in addition that can be set/ reset by the user.

```
7000   74,24      MOV A,#data (24) ; Load ACC with 1st data byte 24H.
7002   75,F0,42   MOV F0,#data (42); Load B register with 2nd data byte 42H.
7005   D3         SETB C           ; Set the Carry bit.
7006   35,F0      ADDC A,F0        ; Add (ACC)+(B)+(CY)
7008   F8         MOV R0, A        ; Transfer result from (ACC) to Bank 0, Register 0.
7009   22         RET              ; Return to Command Mode.
```

The data given in the example is 24H and 42H. The result is

 0010 0100
+0100 0010
―――――――――
 0110 0110 = 66H stored in the Babk register R_0

Flag Status	CY	AC	F0	RS_1	RS_0	OV	–	P
	0	0	0	0	0	0	x	0 = 00H

For data 98H and 87H

 1001 1000
+1000 0111
―――――――――
1 0001 1111 (A_{CC}) = 20H, (B) = 87H, (F) = C5H
|――→ +1
―――――――――
 0010 0000

Flag Status	CY	AC	F0	RS_1	RS_0	OV	–	P
	1	1	0	0	0	1	x	1=C5H

3. Subtract 2-data bytes.

This program demonstrates the use of the carry flag in subtraction. The SUBB instruction subtracts the indicated variable and the carry flag together from the A_{CC}, leaving the result in the A_{CC}. The SUBB sets the carry (borrow) flag if a borrow is needed for bit-7, and clears C otherwise. AC is set if a borrow is needed for bit-3, and cleared otherwise. OV is set if a borrow is needed into bit-6, but not into bit-7, or into bit-7, but not bit-6.

When subtracting signed integers OV indicates a negative number produced when a negative value is subtracted from a positive value, or a positive result when a positive number is subtracted from a negative number.

```
7000   74,24      MOV A,#data (24)   ; Load ACC with 1st data byte 24H.
7002   75,F0,E5   MOV F0,#data (E5)  ; Load B register with 2nd data byte E5H.
7005   95,F0      SUBB A,F0          ; Subtract (ACC) – (B) – (CY)
7007   35,F0      MOV R0,A           ; Transfer result from (ACC) to Bank 0, Register 0.
7008   22         RET                ; Return to Command Mode.
```

24H = 0010 0100	24H = 0010 0100	Result is negative and complement of 1110 0001 =
–42H = –0100 0010 =	$\overline{42H}$ = +1011 1101	–0001 1110 = –1EH
	1110 0001	

(A_{CC}) = 1EH and (B) = 42H.

Other data given in the (A_{CC}) = E4H and (B) = E5H. The result is

E4H = 1110 0100	E4H = 1110 0100	Result is negative and complement of 1111 1110
–E5H = –1110 0101 =	$\overline{E5H}$ = +0001 1010	= –0000 0001 = –01H
	1 1111 1110	

(A) = = (R_0) = 01H

Flag Status	CY	AC	F0	RS_1	RS_0	OV	–	P
	1	1	0	0	0	0	x	0=C0H

Note

- The result may differ by one as it depends on the value of the CY flag that may have been set by a previous program.
- If the CY flag is set by the user in the program itself using instruction SETB C before SUBB instruction, then the result will not depend upon the previous flag settings.

4. Multiply 2-data bytes

This program multiplies the number stored in A_{CC} with the number stored in B register. The LSB of the result is stored in the A_{CC} and the MSB is stored in (B). If the product is greater than 255(D) = FFH the OV flag is set; otherwise it is cleared. The carry flag is always cleared.

7000	74,E4	MOV A,#data (E4)	; Load A_{CC} with 1st data byte (E4H).
7002	75,F0,95	MOV F0,#data (95)	; Load B-Reg. with 2nd data byte (95H)
7005	A4	MUL AB	; Multiply (A_{CC}) with (B) and store result in A_{CC}
7006	F8	MOV R_0,A	; Store LSB of result in R_0.
7007	A9	MOV R_1,F0	; Store MSB of result in R_1.
7008	22		; RET

$$\begin{array}{r} E4 \\ *95 \\ \hline 474 \\ 804 \\ \hline 84B4 \end{array}$$

The result of multiplying E4H by 95H is 84B4H. The 84H is stored in register B and transferred to register R_1. B4H is stored in A_{CC} and register R_0.

Flag Status	CY	AC	F0	RS_1	RS_0	OV	–	P
	0	0	0	0	0	1	x	0 = 04H

5. Divide 2-data bytes

This program divides the data stored in A_{CC} by the data stored in register B. The A_{CC} receives the integer part of the quotient; register B receives the integer remainder. The carry and OV flags will be cleared.

Exception

If (B) = 00H, the values returned in the A_{CC} and B-register will be undefined and the OV flag will be set. The CY flag is cleared in any case.

705A	74,FB	MOV A,#data (FB)	; Load A_{CC} with 1st data byte (FBH).
705C	75,F0,12	MOV F0,#data (12)	; Load B-Reg. with 2nd data byte (12H).
705F	84	DIV AB	; Divide (A_{CC}) with (B) and store result in A_{CC}
7060	F8	MOV R_0, A	; Store LSB of result in R_0.
7061	A9	MOV R_1, F0	; Store MSB of result in R_1.
7062	22	RET	

$FBH = 16^1 * 15 + 16^0 * 11 = 251D, 12H = 16^1 * 1 + 16^0 * 2 = 18D$

$$\begin{array}{r} 18)251(13 = 0DH \\ 18 \\ \hline 71 \\ 54 \\ \hline 17 = 11H \end{array}$$

The result of dividing FBH by 12H is a quotient of 0DH that is stored in the A_{CC} and a remainder of 11H which is stored in register B. Both the results are transferred in the register banks.

Flag Status	CY	AC	F0	RS_1	RS_0	OV	–	P
	0	0	0	0	0	0	x	1 = 01H

String subtraction

Subtract 2-data bytes indicated by strings.

This program subtracts a string of 8-bit data indicated by R_1 from a string of data indicated by R_0. The number of data is indicated by R_2.

```
706A   C3                  CLR C           ; Clears the CY flag
706B   E6        SUB :     MOV A,@R0       ; Move data byte in ACC.
706C   97                  SUBB A,@ R1     ; Subtract data indicated by R1 from data in ACC.
706D   F6                  MOV @R0,A       ; Store result at R0.
706E   08                  INC R0          ; Point to next data byte.
706F   09                  INC R1          ; Point to next data byte.
7070   DA,F9               DJNZ R2,SUB     ; Decrement & jump if not zero
7072   30,D5,03            JNB OV,OV-OK    ; Check if overflow occurred on last iteration of loop.
7075   12,03,8E            LCALL DONE      ; Indication of Overflow.
7078   22        OV-OK:    RET             ; Return to Command mode.
```

The subtraction is performed for the number of data indicated by R_2 and result is stored at memory location indicated by R_0.

NOTE

The R_0 and R_1 must be set for memory location between 20H to 3FH as this is the space allocated for the user as scratch pad RAM.

Simple Logic Operation

The following set of programs test all the logic operations that may be performed on the 8051. The logic operations acceptable by the 8051 are AND, OR, Ex-OR, Complement, and Swap.

AND : Operation

(*a*) ANL A, R_n : Logical AND the contents of the Accumulator with the contents of the specified register.

7. Obtain logical AND of 2-data bytes.

```
6000   78,01   MOV R0,#01H   ; Load 1st data byte 01H in R0
6002   74,89   MOV A,#89H    ; Load 2nd data byte 89H in ACC
6004   58      ANL A, R0     ; AND contents of ACC & R0
6005   F9      MOV R1,A      ; Store result in R1
6006   22      RET           ; Return to command mode
```

By ANDing 01H = 0000 0001B with 89H = 10001001B, the result is

```
 89H = 1000 1001
*01H = *0000 0001
-----------------
       0000 0001, (A) = 01H, (B) = A8H, (F) = 01H, R0 = 01H, R1 = 01H
```

(*b*) ANL A, direct : Logical AND the contents of the accumulator with the contents of a Special Function Register (SFR).

8. Obtain logical AND of (A) and (SFR).

```
6000   74,AA      MOV A,#AA     ; Load 1st data byte AAH in ACC
6002   75,F0,F0   MOV F0,#F0    ; Load 2nd data byte F0H in B-Reg.
6005   55,F0      ANL A,F0      ; Logical AND ACC with F0.
6007   F8         MOV R0,A      ; Store result in R0.
6008   22         RET           ; Return to Command Mode
```

```
 AAH = 1010 1010
*01H = * 1111 0000
------------------
       1010 0000, (A) = A0H, (B) = F0H, (F) = 00H, R0 = A0H, R1 = 01H
```

The result A0H = 10100000B is stored in R_0.

(*c*) ANL A, @ R_i : Logical AND the contents of the accumulator with the contents stored in the memory address pointed by R_0.

9. Obtain logical AND of (A) with a byte of data indicated by the string.

```
6000   A8,40   MOV R0,@40    ;Load memory address of 1st data byte 40H in R0
6002   74,0F   MOV A,#0FH    ; Load 2nd data byte 0FH in the ACC
6004   56      ANL A,@R0     ; Logical AND these 2-data bytes
6005   F9      MOV R1,A      ; Store result in R1.
6006   22      RET           ; Return to command mode.
```

```
 7FH = 0111 1111
*01H = *0000 1111
-----------------
       0000 1111, (A) = 0FH, (B) = F0H, (F) = 00H, R0 = A0H, R1 = 0FH
```

The result is stored in R_1 = 0FH.

(*d*) ANL A, #data : Logical AND contents of the accumulator with a direct data byte.

10. Obtain logical AND of (A) with immediate data.

```
6000   74, A0   MOV A,#A0    ; Load 1st data byte A0H in ACC.
6002   54, FF   ANL A,#FF    ; Logically AND ACC with data byte FFH.
6004   F8       MOV R0,A     ; Store result.
6005   22       RET          ; Return to command mode.
```

(A) = A0H, (B) = F0H, (F) = 00H, R_0 = A0H, R_1 = 0FH

(*e*) ANL direct, A : Same operation as that in (b) with the bytes reversed. The Program is also the same with the exception that instead of the instruction ANL A, direct, the instruction ANL direct, A is used.

(*f*) ANL direct, # data : This instruction ANDs the contents of a SFR with a direct data byte. This program displays the use of the Stack Pointer through the PUSH instruction. First the SP is initialized and after the execution of the program, the result is stored in the stack.

11. Obtain logical AND of (SFR) with direct data byte.

```
6000    75,81,20   MOV SP,20H        ; Initialize the SP
6003    75,F0,74   MOV F0,#74H       ; Load 1st data byte 74H.
6006    53,65      ANL F0,#65H       ; Logical AND operation
6008    C0,F0      PUSH B            ; Store result in SP
600A    22         RET               ; Return to Command mode
```

```
  0111 0100
* 0110 0101
-----------
  0110 0100 = 64H
```

The result of Logical ANDing 74H = 0111 0100B with 65H = 0110 0101B is 64H = 0110 0100B and is stored in the stack initialized at internal memory location 20H.

(*g*) ANL C, < src = bit > : This instruction logical ANDs the contents of the carry flag with any bit of the SFR which is accessible.

12. Obtain logical AND of (CF) with any bit of SFR register.

This program sets the LSB of port-1, moves it into the carry flag, sets the MSB of the Accumulator, and ANDs these two bits. The result, which is stored in the carry flag is brought into the Accumulator with the help of the RLC A (Rotate Accumulator Left through Carry) instruction and saved in R_0.

```
6000    E4         CLR A             ; Clear the contents of ACC.
6001    D2,90      SET B P1.0        ; Set bit P1.0
6003    A2,90      MOV C,P1.0        ; Shift bit P1.0 into C
6005    D2,E7      SET B,ACC.7       ; Set MSB of ACC
6007    82,E7      ANL C,ACC.7       ; AND bit C with bit ACC.7
6009    33         RLC A             ; bring result in ACC from C
600A    F8         MOV R0, A         ; Store result in R0
600B    22         RET               ; Return to Command mode.
```

(A) = 01H, (B) = A0H, (F) = 81H, R_0 = 01H, R_1 = FEH

The result that is 01H is stored in R_0.

(*h*) ANL c / bit : Logical AND the Carry bit with the complement of another bit.

13. Obtain logical AND of carry bit with any other complemented bit.

The program is similar to program (*g*).

6000	E4	CLR A	; Clear the contents of A_{CC}.
6001	D3	SET B, C	; Set the Carry flag bit
6002	D2,E7	SET B, A_{CC}.7	; Set MSB of A_{CC}.
6004	B0,E7	ANL C,/A_{CC}.7	; AND bit C with complement of A_{CC}.7
6006	33	RLC A	; Bring result in A_{CC} from C
6007	F8	MOV R_0,A	; Store result in R_0
6008	22	RET	; Return to command mode.

(A) = 00H, (B) = A0H, (F) = 80, R_0 = 00H, R_1 = FEH.

The result which is 00H is stored in R_0.

OR-Operation

The programs for OR operation are similar to the programs for AND operation. All the different ANDing techniques displayed above are the same for ORing as well. The Hex codes for these various ORing techniques are given below for convenience. Simply replace these for their corresponding AND operations in the programs above and the results will change for OR-operation.

(*a*) ORL A, R_n: Logical AND the contents of the Accumulator with the contents of the specified register.

14. Obtain logical OR with (A) with contents of any specific register.

6000	78, 01	MOV R_0,#01H	; Load 1st data byte 01H in R_0
6002	79, 89	MOV A,#89H	; Load 2nd data byte in A_{CC}
6004	58	ORL A,R_0	; AND contents of A_{CC} and R_0
6005	F9	MOV R_1,A	; Store result in R_1
6006	22	RET	; Return to command mode

By ORing 01H = 0000 0001H with 89H = 1000 1001B, the result is

$$\begin{array}{r} 89H = 1000\ 1001 \\ +\ 01H = \underline{0000\ 0001} \\ 1000\ 1001 = 89H = R_0 \end{array}$$

(*iii*) XOR operation: This operation can be performed the same way as the AND and the OR operations.

(*a*) XRL A, R_n: Logical XOR the contents of the accumulator with the contents of the specified register.

15. Obtain logical XOR of (A) with contents of any specific register.

6000	78,01	MOV R_0,#01H	; Load 1st data byte in R_0
6002	74,89	MOV A,#89H	; Load 2nd data byte in A_{CC}
6004	68	XRL A,R_0	; XOR the contents of A_{CC} and R_0
6005	F9	MOV R_1,A	; Store result in R_1
6006	22	RET	; Return to command mode

By ORing 01H = 0000 0001B with 89H = 1000 1001B, the result is

$$
\begin{array}{r}
89H = 1000\ 1001 \\
\oplus\ 01H = \underline{0000\ \ 0001} \\
1000\ 1000 = 88H = (R_1)
\end{array}
$$

Complement

CPL A: This instruction complements the contents of the accumulator.

16. Complement the (A).

This program stores data FFH in the accumulator and then complements it so that the data is changed to 00H.

```
6000    74,F0    MOV A,F0H     ; Load ACC with the data
6002    F4       CPL A         ; Complement data in ACC
6003    F8       MOV R0,A      ; Store result in R0
6004    22       RET           ; Return to command mode
```

The result i.e., 0FH is stored in R_0.

(*b*) CPL bit addr: This instruction complements any accessible bit in the SFRs.

17. Complement any accessible bit of the SFRs.

```
6000    C2       CLR B, P1.1   ; Clear bit P1.1
6001    B2, 91   CPL P1.1      ; Complement bit P1.1
6003    C3       CLR C         ; Clear carry flag bit
6004    A2,91    MOV C, P1.1   ; Load P1.1 bit in C
6006    33       RLC A         ; Move result in ACC
6007    F8       MOV R0,A      ; Store result in R0
6008    22       RET           ; Return to command mode
```

The result that is 01H is stored in R_0.

(*a*) SWAP A

18. SWAP 4-MSBs with 4-LSBs of (A).

This program swaps the nibbles of the accumulator meaning that the 4-MSBs are interchanged with 4-LSBs in the accumulator.

```
6000    74,F0    MOV A, F0H    ; Load F0H in ACC
6002    C4       SWAP A        ; Swap nibbles in ACC
6003    F8       MOV R0,A      ; Store result in R0
6004    22       RET           ; Return to command mode
```

The result which is 0FH is stored in R_0.

19. Compare 3-B and arrange in ascending order.

```
6000    78,data  MOV R0,#data  : Load 1st data in R0
6002    79,data  MOV R1,#data  ; Load 2nd data in R1
6004    F8       MOV A, R0     ; Shift (R0) to ACC
6005    99       SUBB A, R1    ; Subtract (R1) to ACC
```

6006	50,04	JNC END	; Jump to END if no carry
6008	F8	MOV A, R_0	; Shift (R_0) to A_{CC}
6009	FA	MOV R_2,A	; Shift (A_{CC}) to R_2
600A	22	RET	; Return to command mode.

Note

If the 1st number is less than the 2nd number then it is shifted into R_2 so that the numbers in descending order start from R_1. If the 1st number is greater than the 2nd number then the greater one remains in R_0 and the lesser one remains in R_1 thereby again making the arrangement in descending order. Place 01H in R_0 and 06H in R_1. After the execution of the program, the 01H data is shifted into R_2. Now place 06H in R_0 and 01H in R_1. Now after the execution of the program, there is no change as the numbers are already arranged in descending order.

The opcodes of 8031/8051 are arranged Appendix-(A) and B.

The details of all registers of 8051 and its conditions after reset, interrupt conditions and Data and program memory are described in short below.

Summary of all registers inside the 8051

Register	*Bit*	*Primary function*	*Bit addressable*	*Condition after RESET*
A	8	Math, data manipulation etc.	Y	00
B	8	Math	Y	00
PC	16	Address program bytes	N	0000
DPTR	16	Address code and external data	N	0000
SP	8	Address internal RAM stack data	N	07
PSW	8	Processor status word	Y	00
P0-P3	8	Store I/O port data	Y	FF
TH0/TL0	8/8	Timer/counter 0	N	00/00
TH1/TL1	8/8	Timer/counter 1	N	00/00
TCON	8	Timer/counter control	Y	00
TMOD	8	Timer/counter mode control	N	00
SBUF	8	Serial port data	N	XX
SCON	8	Serial port control	Y	00
PCON	8	Power control	N	XXX00000b
IE	8	Interrupt enable control	Y	XXX00000b
IP	8	Interrupt priority control	Y	XXX00000b

Input conditions		*Data and program memory*	
Interrupt	HEX destination	Internal	Function
IE0	0003	RAM (128B)	R7-R0 register, data storage, stack
TF0	000B	ROM (4KB)	Program storage
IE1	0013		
TF1	001B	External	Function
SERIAL	0023	RAM (64KB)	Data storage
		ROM (64KB)	Program storage

7.4 APPLICATION AREAS

1. Interfacing of LED

The circuit is shown in the Fig. 7.4 (*a*). Here we are interfacing the LED with Port 1 to make it ON. The LED's Anode is connected with resistor of 220 ohms and Cathode is connected to the P1.0 of 8051 microcontroller. Crystal osc and ceramic disc capacitors are connected as usually. Pin 40 and Pin 31 are given +5V supply.

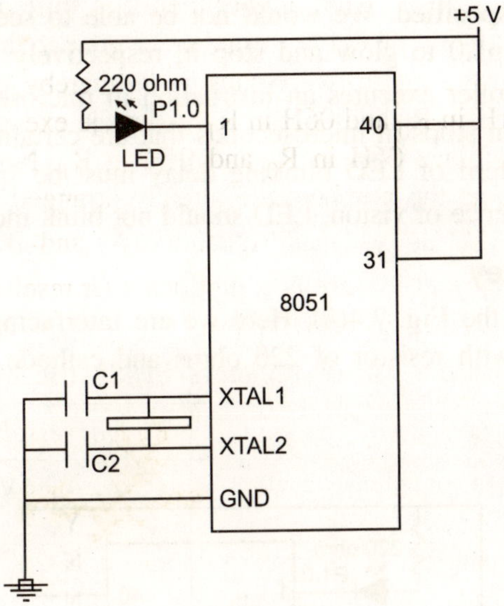

Fig. 7.4 (*a*) Interfacing of LED

To glow LED

$mod51 is used as the header file else our program may result in error on assembling.

```
1.      $MOD51              ;
        ORG 0000H           ; initialization of memory address
        CLR P1.0            ; clear Port P1.0
here:   SJMP here           ; jump to label here
        END                 ; end of program
```

To blink LED

```
        $MOD51
        ORG 0000H           ; initialization of memory address
here:   CLR P1.0            ; clear Port P1.0
        ACALL delay         ; absolute call for delay subroutine
        SETB P1.0           ; set bit 0 of Port P1
        ACALL delay         ; call delay subroutine
        SJMP here           ; jump to label here
                            ; delay subroutine
delay:  MOV R0,#FFH         ; load R0 with FFH
again:  MOV R1,#FEH         ; load R1 with FEH
```

```
       NOP               ; no operation
       NOP               ; no operation
loop:  DJNZ R1,loop      ; decrement R1 and jump if not zero to loop
       DJNZ R0,again     ; decrement R0 and jump if not zero to again
       RET               ; return to command mode
       END               ; end program
```

Note: If no delay was specified. We would not be able to see LED blinking even though we cleared and set the port p1.0 to glow and stop it, respectively.

Reason: The microcontroller executes an instruction in microseconds. So without the delay specified LED glows and then stops in microseconds that are certainly not visible to naked eyes. Hence to clearly see the pattern of LED blinking delay must be specified.

Scientifically, for persistence of vision, LED should not blink more than 16 times per second.

2. Interfacing of Single Key

The circuit is shown in the Fig. 7.4(*b*). Here we are interfacing the LED with Port 1. The LED's Anode is connected with resistor of 220 ohms and cathode is connected to the P1.0 of 8051 microcontroller.

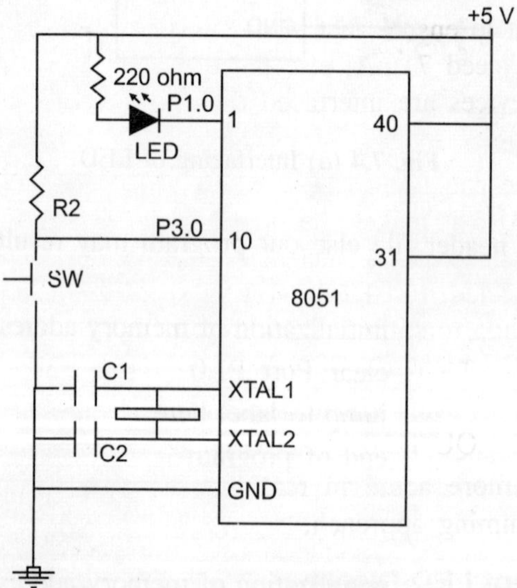

Fig. 7.4(*b*) Interfacing of single key

Push button key is connected with P3.0 of microcontroller. When key is open then P3.0 will have HIGH status, and when key is closed then P3.0 will have LOW status. By continuously monitoring the status of P3.0 we can check the status of the key. When the key is pressed then the status of LED will be complemented.

Crystal osc and ceramic disc capacitors are connected as usually. Pin 40 and Pin 31 are given +5V supply.

```
1.       $MOD51
         ORG 0000H
here :   JNB P3.0,next          ; jump if not bit P3.0 to next
         SETB P1.0              ; set Port P1.0
         SJMP here              ; jump to here
next :   CLR P1.0               ; clear Port P1.0
         SJMP here              ; jump to here
         END                    ; end program
2.       $MOD51
         ORG   0000H            ; initialization of memory address
         LED EQU P1.0           ; LED equivalent to Port P1.0
         SW EQU P3.0            ; SW equivalent to Port P3.0
BEG :    MOV C,P3.0             ; load carry with P3.0
         MOV P1.0,C             ; load Port P1.0 with carry bit
         SJMP BEG               ; jump to label BEG
         END                    ; end of program
```

It is to be noted that microcontroller can source the current in microamperes but can sink the current in milli amperes. Generally, at least 10mA current must pass through LED to glow it. So any device must be interfaced at negative logic.

Also care must be taken to ensure that at any instant of time, the current flowing through microcontroller should not exceed 71mA; else the electrical behavior of microcontroller would become unpredictable. So devices are interfaced taking these two points into consideration.

Use of equate statement: =

Now instead of P1.0 we can use LED word in programming. This makes program more readable and user friendly. Also it makes the program flexible. Suppose we wrote the program by assuming that particular LED was connected at P1.0 but actually the hardware was made by some other fellow who connected it at P2.0. Now, it is always better to change software rather than modify hardware according to software. So to do the job everywhere in program we would have to replace P1.0 by P2.0 which may be very time consuming.

We can solve this problem using equate statement. We just need to change one statement i.e., LED EQU P1.0 to LED EQU P2.0 in above program. This saves lot of time and energy.

The problem becomes more acute in real life industrial projects. Hence use of equate statement is a sound programming approach.

3. Interfacing of 7 Segment LED Display

The circuit is shown in the Fig. 7.4(*c*). Here we are interfacing the 7 segments LED display with Port 1. The LED display is given Vcc from pin no. 3 and 8, which are anode pins. Rest of circuit is same as previous circuits.

Crystal osc and ceramic disc capacitors are connected as usually. Pin 40 and Pin 31 are given +5V supply.

The numbers will be displayed according to the binary patterns applied on the P1. These patterns are decided according to the logic that when 0 is applied to the pin, then its corresponding LED segment will glow.

Program below displays 0 to 9 continuously. C0H refers to display 0; F9H refers to displaying 1 etc.(See Fig. 7.4(*c*))

```
        $MOD51
        ORG 0000H            ; Initializing the program address
BEG:    MOV P1,#0C0H         ; Load Port P1 with C0H (display 0)
        ACALL DELAY          ; Calling the subroutine delay
        MOV P1#0F9H          ; Load Port P1 with F9H (display 1)
        ACALL DELAY          ; Calling the subroutine delay
        MOV P1#0A4H          ; Load Port P1 with A4H (display 2)
        ACALL DELAY          ; Calling the subroutine delay
        MOV P1#0B0H          ; Load Port P1 with B0H (display 3)
        ACALL DELAY          ; Calling the subroutine delay
        MOV P1#99H           ; Load Port P1 with 99H (display 4)
        ACALL DELAY          ; Calling the subroutine delay
        MOV P1#92H           ; Load Port P1 with 92H (display 5)
        ACALL DELAY          ; Calling the subroutine delay
        MOV P1#82H           ; Load Port P1 with 82H (display 6)
        ACALL DELAY          ; Calling the subroutine delay
        MOV P1#0F8H          ; Load Port P1 with F8H (display 7)
        ACALL DELAY          ; Calling the subroutine delay
        MOV P1#80H           ; Load Port P1 with 80H (display 8)
        ACALL DELAY          ; Calling the subroutine delay
        MOV P1#90H           ; Load Port P1 with 90H (display 9)
        ACALL DELAY          ; Calling the subroutine delay
        SJMP BEG             ; Jump to label BEG
DELAY : MOV R1,#0AH          ; load R1 with 0AH
LOOP1 : MOV R0,#32H          ; load R0 with 32H
LOOP :  MOV TCON,#00H        ; Load TCON register with 00H
        MOV TMOD,#01H        ; Load TMOD with 01H (mode 1)
        MOV TH0,#0B1H        ; Load TH0 with B1H
        MOV TL0,#0EFH        ; load TL0 with EFH
        CLR TF0              ; CLR timer overflow for count 0
        SETB TR0             ; set timer run for counter 0
Wait :  JNB TF0,wait         ; jump if not bit TF0 to wait
        CLR TR0              ; CLR timer run for count 0
        DJNZ R0,LOOP         ; decrement R0 and jump if not zero to loop
        DJNZ R1,LOOP1        ; decrement R1 if not zero to loop1
        RET                  ; return to control mode
        END                  ; end program
```

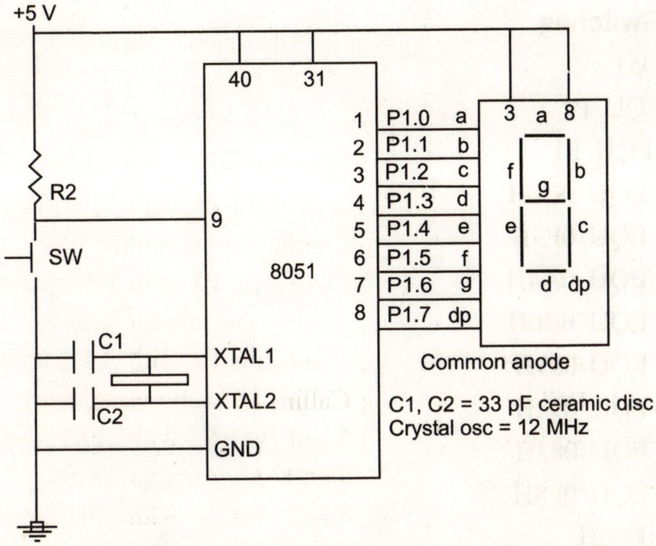

Fig. 7.4(*c*) Interfacing of 7-segment display

4. Interfacing of DIP Switch

The circuit is shown in the Fig. 7.4(d) Here we are interfacing the 7 segments, LED display with Port 1. The LED Display is given Vcc from pin no. 3 and 8, which are anode pins. For keys we are using push button keys which are connected with port 2. Here we are using resistors of 220 ohms with each key. When user will press any key then that particular pin becomes LOW and we will identify it. If not any key is pressed then P2 will read as FFH otherwise value will be other than FFH. We are writing the software in such a way that until any key is pressed because of previous action control will not move ahead. When it is confirmed that all keys are open then scanning of keys will start.

The content of the P2 is copied to the accumulator, after that each bit of the accumulator is checked in a sequence starting from bit 0 to bit 7.

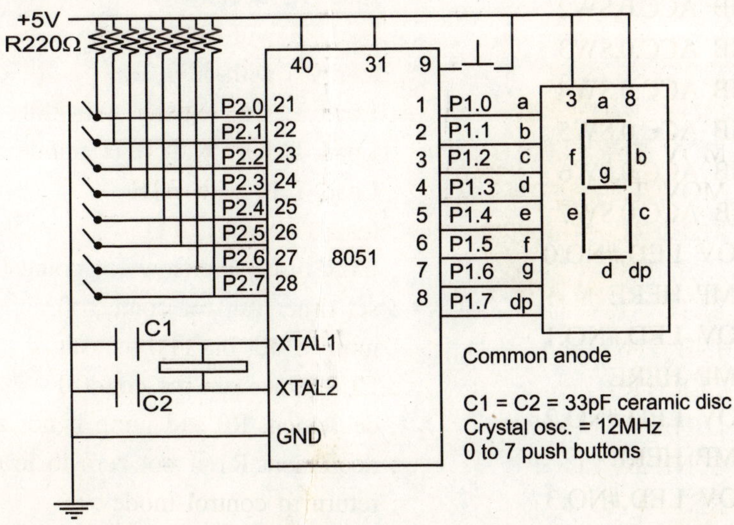

Fig. 7.4(*d*) Interfacing DIP switches

Program for DIP Switches

```
        $MOD51
        DIP EQU P2
        LED EQU P1
        NO.0 EQU 0C0H
        NO.1 EQU 0F9H
        NO.2 EQU 0A4H
        NO.3 EQU 0B0H
        NO.4 EQU 099H
        NO.5 EQU 092H
        NO.6 EQU 082H
        NO.7 EQU 0F8H
        ORG 0000H
        MOV DIP,#0FFH           ; I/P PORT CONFIGURATION
HERE:   MOV A,DIP               ; Move content of P2 to Acc.
        CJNE,#0FFH,NEXT         ; compare acc. With OFFHand jump to next if not equal
        SJMP HERE
NEXT:   ACALL DELAY_300ms       ; debouncing effect (checking for bouncing)
        MOV A,DIP
        CJNE,#0FFH,NEXT1
        SJMP HERE
NEXT1:  JNB ACC.0,SW0
        JNB ACC.0,SW1
        JNB ACC.0,SW2
        JNB ACC.0,SW3
        JNB ACC.0,SW4
        JNB ACC.0,SW5
        JNB ACC.0,SW6
        JNB ACC.0,SW7
SW0:    MOV LED,#NO.0
        SJMP HERE
SW1:    MOV LED,#NO.1
        SJMP HERE
SW2:    MOV LED,#NO.2
        SJMP HERE
SW3:    MOV LED,#NO.3
        SJMP HERE
```

```
SW4:    MOV  LED,#NO.4
        SJMP HERE
SW5:    MOV  LED,#NO.5
        SJMP HERE
SW6:    MOV  LED,#NO.6
        SJMP HERE
SW7:    MOV  LED,#NO.7
        SJMP HERE
DELAY_300ms:
        MOV  RO,#15D
LOOP:   MOV  TCON,#00H
        MOV  TMOD,#01H
        MOV  TH0,#0B1H
        MOV  TL0,#0EFH
        CLR  TF0
        SETB TR0
WAIT:   JNB  TF0, WAIT
        CLR  TR0
        DJNZ R0, LOOP
        RET
```

5. Intelligent LCD Display

FIgs. 7.4(*e*) and (*f*) are block diagram and detail circuit representation of LCD interfacing with microcontoller. We know that the LCD is itself an embedded system having RAM, ROM, registers, microcontrollers, etc., as depicted in Fig. 7.4(*e*). If we select single line display, RAM locations of 80 and 87 and C0 to C7 represent single line and one character can be written in each of these locations. If double line display is used then 80 to 8F and C0 to CF represent double line.

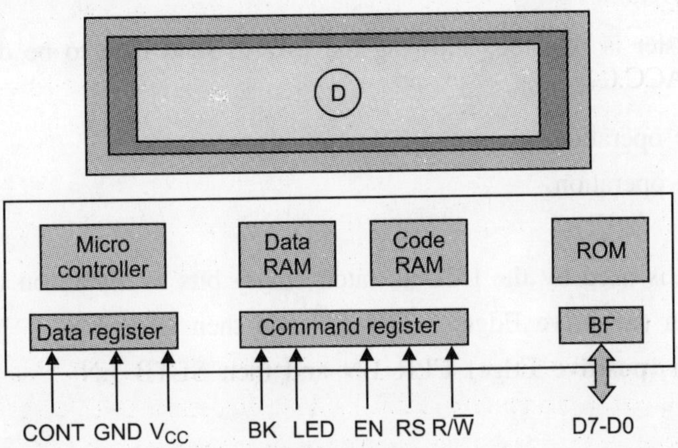

Fig. 7.4(*e*) Block diagram of LCD

The intelligent displays are of two types:

1. Text Display
2. Graphics Display

Advantages of LCD

- The ability to display the numbers, characters and graphics which is not possible in LEDs, since they can display numbers and few characters.
- Incorporation of a refreshing controller into the LCD, thereby relieving the CPU of the task of refreshing the LCD. In contrast, the LED must be refreshed by the CPU (or in same other way) to keep displaying the data.

1	2	3	4	5	6	7	8	9	10	11	12	13	14	15	16
GND	**+V**	**C**	**RS**	**RW**	**EN**	**D0**	**D1**	**D2**	**D3**	**D4**	**D5**	**D6**	**D7**	**LED+**	**LED-**

DESCRIPTION

Microcontroller: It is the brain of LCD display.

Data RAM: This RAM is storing the ASCII values of corresponding characters

Code RAM: This RAM store the binary pattern according to the character.

Data Register: This register work as buffer for data lines and the internal buses of LCD. The ASCII values of characters will be given to the data register.

BF (Busy Flag): It indicates the internal working of the LCD.

If BF=0 (LCD is idle we can proceed for next operation)

If BF=1 (LCD is busy we can't proceed for next operation)

RS (Register Select):

The RS pin is used to select DR or CR.

RS=0: CR register is selected, allowing the user to send a command such as clear display, curser at home etc.

RS=1: DR resister is selected, allowing the user to send data to be display on the LCD.

Read/Write:

= 0, Write operation.

= 1, Read operation.

EN (Enable):

The enable pin is used by the LCD to latch binary bits available on its data pins.

Write operation (**negative** Edge) SETB EN and then CLR EN

Read operation (**positive** Edge) CLR EN and then SETB EN

D7-D0:

This is 8-bit data pins. D7-D0

BK-LED:

These pins are used to give the supply to the backlight of the LCD display.

Action	*RS*	*RW*	*D7*	*D6*	*D5*	*D4*	*D3*	*D2*	*D1*	*D0*	*Description*
Clear display	0	0	0	0	0	0	0	0	0	1	Clear complete display, sets RAM address counter at 0
Return home	0	0	0	0	0	0	0	0	1	X	Home cursor, shift shifted string to original position, do not clear RAM contents
Entry mode set	0	0	0	0	0	0	0	1	I/D	S	Sets cursor movement direction and specifies shift of display. These operations are performed during data write and read S = 0, cursor S = 1, screen I/D = 1, increment (right shift) I/D = 0, decrement (left shift)
Display ON/ OFF control	0	0	0	0	0	0	1	D	C	B	Set ON/OFF of entire display (D), cursor ON/OFF (C), and blink of cursor position character (B) D = 1, display ON = 0, display OFF C = 1, cursor ON = 0, cursor OFFB = 1, Blinking ON = 0, Blinking OFF
Cursor or display shift	0	0	0	0	0	1	S/C	R/L	-	-	Moves cursor and shift display without changing DD RAM contents
Function set	0	0	0	0	1	DL	N	F	-	-	Sets interface data length (DL), No. of display lines (L) and character font (F), DL = 1, 8-bits/character, DL = 0, 4-bits/ character, N=1, 2 line display N=0, 1-line display, F = 1 Big font(10*5) F=0, small font (7*5)
Set CG RAM Address	0	0	0	1	AGC						Sets code RAM address
Set DD RAM	0	0	1	A6	A5	A4	A3	A2	A1	A0	Sets data RAM address
Read Busy Flag & Address	0	1	BF	A6	A5	A4	A3	A2	A1	A0	Reads busy flag (BF) indicating internal operation is being performed and read address counter content.

Program for LCD Display

```
$MOD51
ORG 0000H
```

```
         MOV A,#38H        ; initialize LCD 2 lines, 8-bits/character, small fonts
         ACALL command     ; call command subroutine
         ACALL delay       ; call delay subroutine
         MOV A,#0EH        ; lcd (display) on,cursor on, blinking OFF.
         ACALL command     ; call command subroutine
         ACALL delay       ; call delay subroutine
         MOV A,#01H        ; CLR LCD
         ACALL command     ; call command subroutine
         ACALL delay       ; call delay subroutine
         MOV A,#06H        ; shift cursor right
         ACALL command     ; call command subroutine
         ACALL delay       ; call delay subroutine
         MOV A,#80H        ; initialize the starting position of display (load A with 80H)
         ACALL command     ; call command subroutine
         ACALL delay       ; call delay subroutine
         MOV A,#'m'        ; load accumulator with character 'm'
         ACALL display     ; call display subroutine
         ACALL delay       ; call delay subroutine
         MOV A,#'m'        ; load accumulator with character 'm'
         ACALL display     ; call display subroutine
         ACALL delay       ; call delay subroutine
         MOV A,#'m'        ; load accumulator with character 'm'
         ACALL display     ; call display subroutine
         ACALL delay       ; call delay subroutine
         MOV A,#'e'        ; load accumulator with character 'e'
         ACALL display     ; call display subroutine
         ACALL delay       ; call delay subroutine
         MOV A,#'c'        ; load accumulator with character 'c'
         ACALL display     ; call display subroutine
         ACALL delay       ; call delay subroutine
here:    SJMP here         ; jump to here
command:                   ; command subroutine
         MOV P1,A          ; load Port P1 with contents of accumulator
         CLR P3.5          ; rs = 0 for command register
         CLR P3.3          ; r/w = 0 to write
         SETB P3.4         ; }
         CLR P3.4          ; }  high to low transition
         RET               ; return to control mode
display:                   ; display subroutine
```

```
        MOV  P1,A          ; load Port P1 with contents of accumulator
        SETB P3.5          ;    rs = 1 for data register
        CLR  P3.3          ;    r/w = 0
        SETB P3.4          ; } high to low transistion for write command
        CLR  P3.4          ; }
        RET                ; return to subroutine
delay:                     ; delay subroutine
        MOV  R3,#50        ; load R3 with 50D
here2:                     ; here2 label
        MOV  R4,#255       ; load R4 with 255D
here1:                     ; here1 label
        DJNZ R4,here1      , decrement R4 and jump if not 0 to here1.
        DJNZ R3,here2      , decrement R3 and jump if not 0 to here2.
        RET                ; return to command mode
        END                ; end program
```

80	81	82	83	84	85	86	87	C0	C1	C2	C3	C4	C5	C6	C7

Single line display

80	81	82	83	84	85	86	87	88	89	8A	8B	8C	8D	8E	8F
C0	C1	C2	C3	C4	C5	C6	C7	C8	C9	CA	CB	CC	CD	CE	CF

Double line display

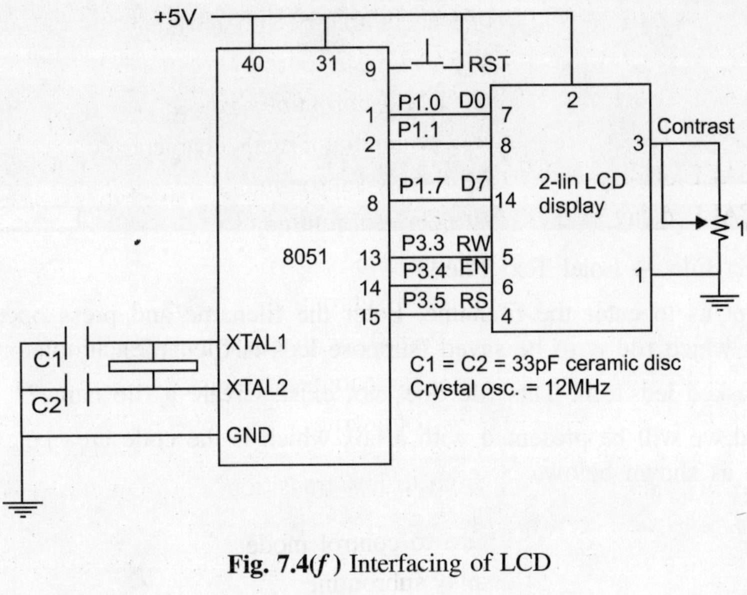

Fig. 7.4(*f*) Interfacing of LCD

Working on Topview Simulator

STEP 1: Install Topview Simulator.

STEP2: When we open up the application; we are presented with the following window.

From the pop down menu select suitable device; we will use 89c51 or 89c52. Enter the value of operating frequency; usually 12MHz or 12000000 Hz. Press OK.

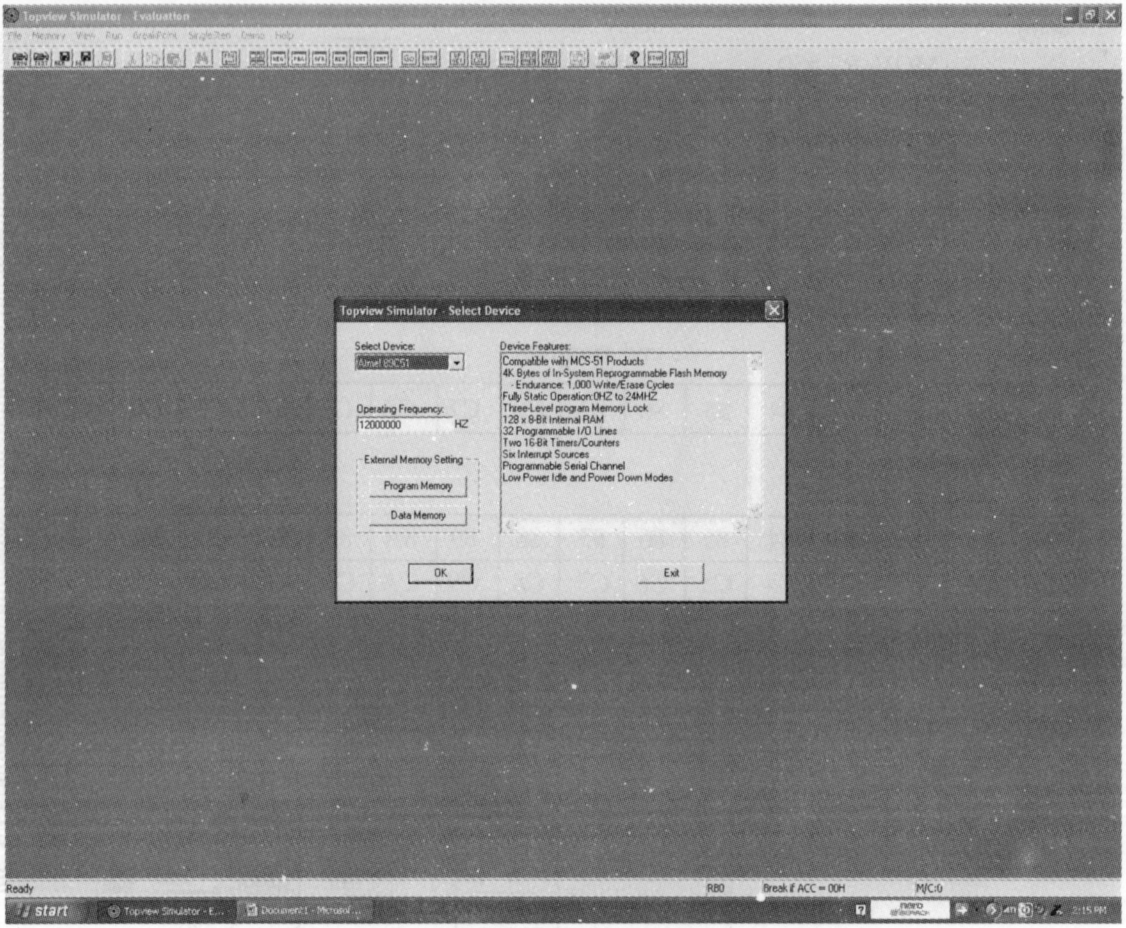

STEP3: Select File → Load Text File.

It will prompt us to enter the filename. Enter the filename and press open after opening required folder in which file is to be saved (suppose leds is filename).

We will be asked leds.asm. This file does not exist. Create a file name?

Press yes and we will be presented with a box which is the code area i.e., where we have to enter the code as shown below.

STEP4: After entering the code; press RUN ASM on task bar which runs the assembler and generates .lst file and .hex file. It will also display whether there are errors or no errors. In case of errors; correct the code and press RUN ASM again.

If no errors present; "file loaded successfully message displayed". Press OK.

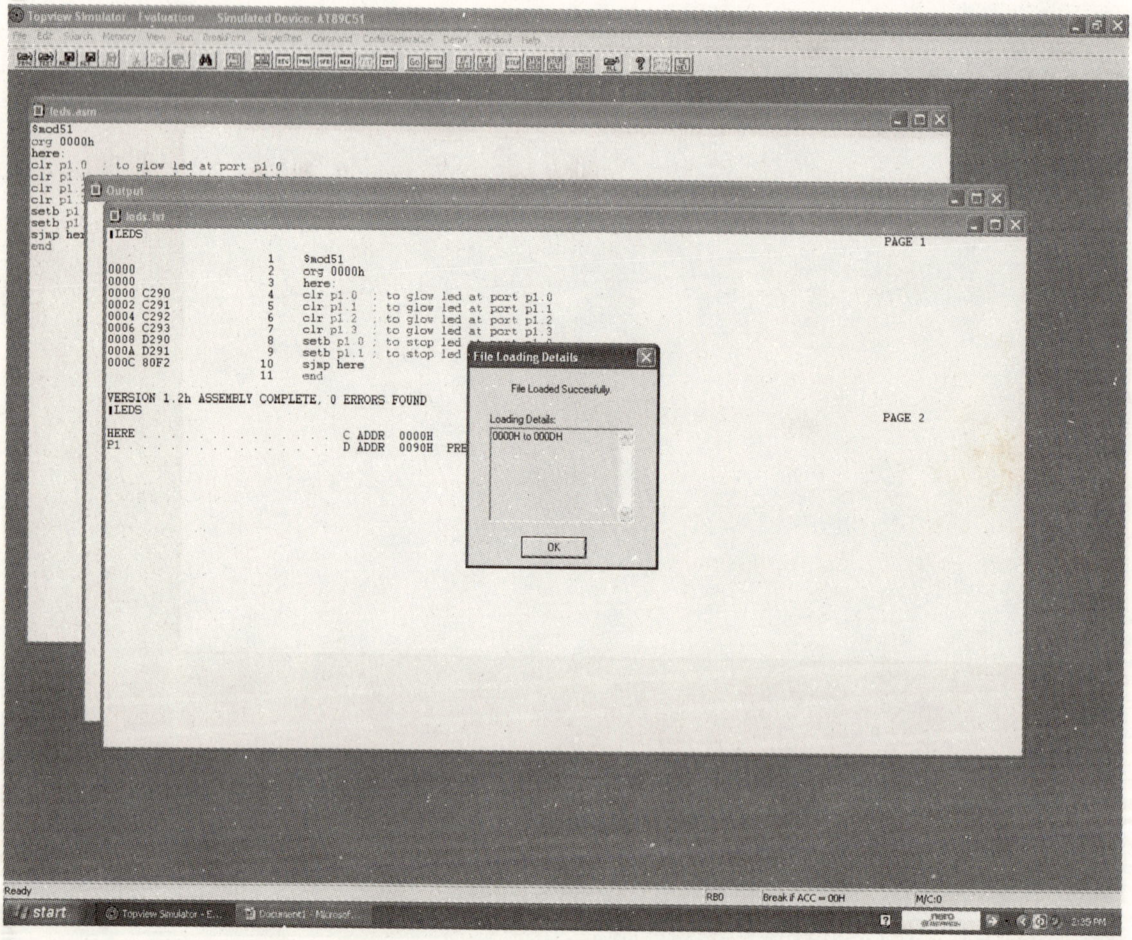

STEP 5: Go to File → External Modules Setting. Select required module out of LED, LCD, KEYBOARD, IIC BUS etc…

Here we select LED.

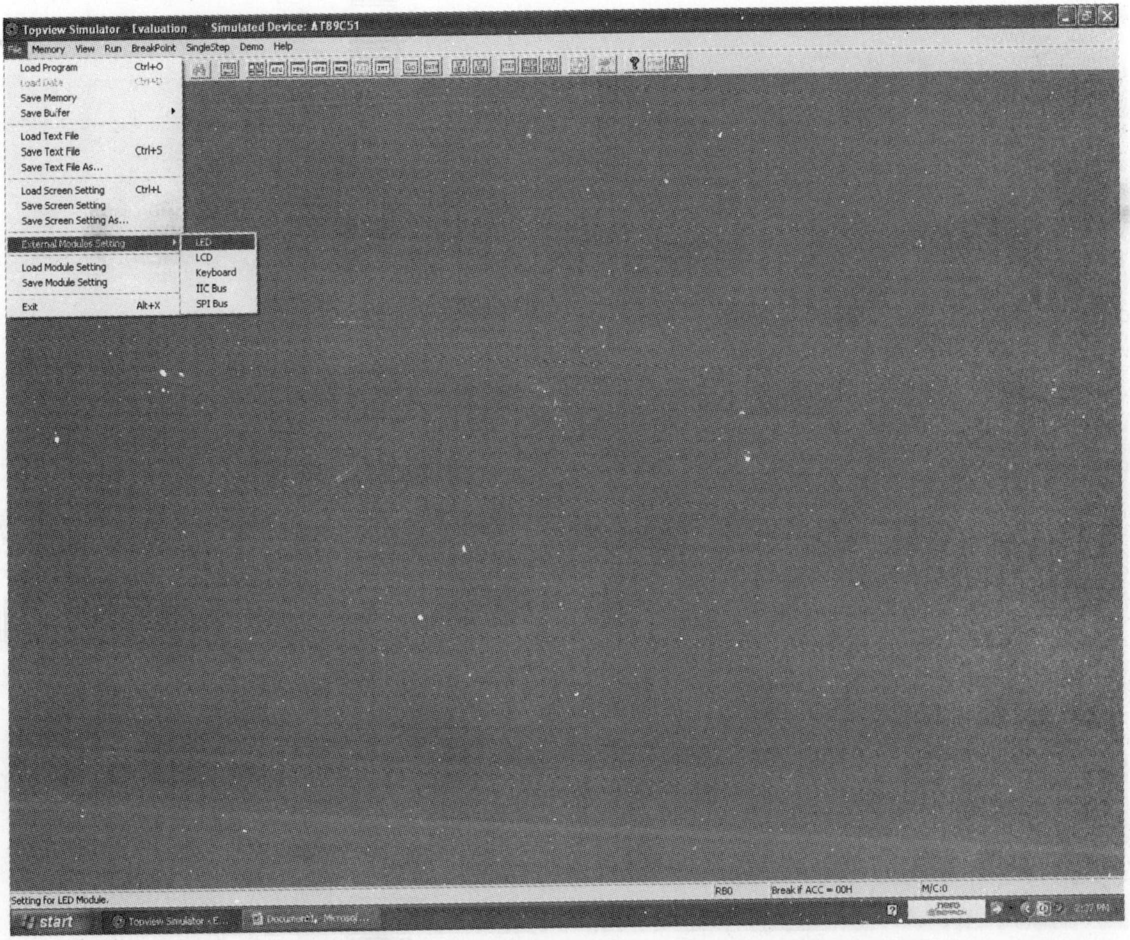

STEP 6: After selecting LED we will be presented with following window.

Check the boxes corresponding to the ports to which LEDs are to be connected as shown. We can also select LED colour red or green. Since no programming of seven segments display is there; check the box for no seven segment display.

Now press OK.

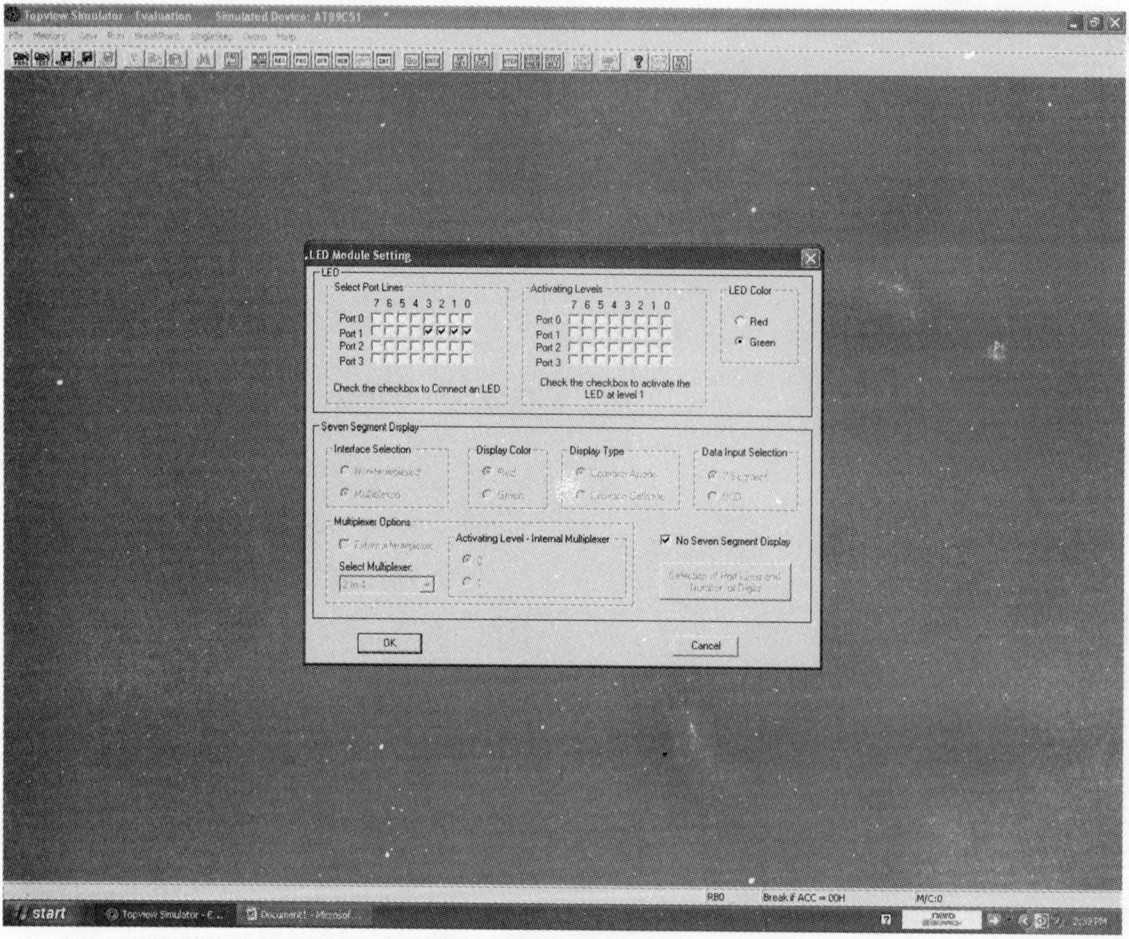

STEP 7: Go to VIEW → EXTERNAL MODULES → LED

STEP 8: Press '<u>reset</u>'and then '<u>go</u>'on task bar to see the output as shown.

NOTE: IN CASE "PC EXCEEDS PROGRAM MEMORY CAPACITY" Reported; press reset and go few times to see output.

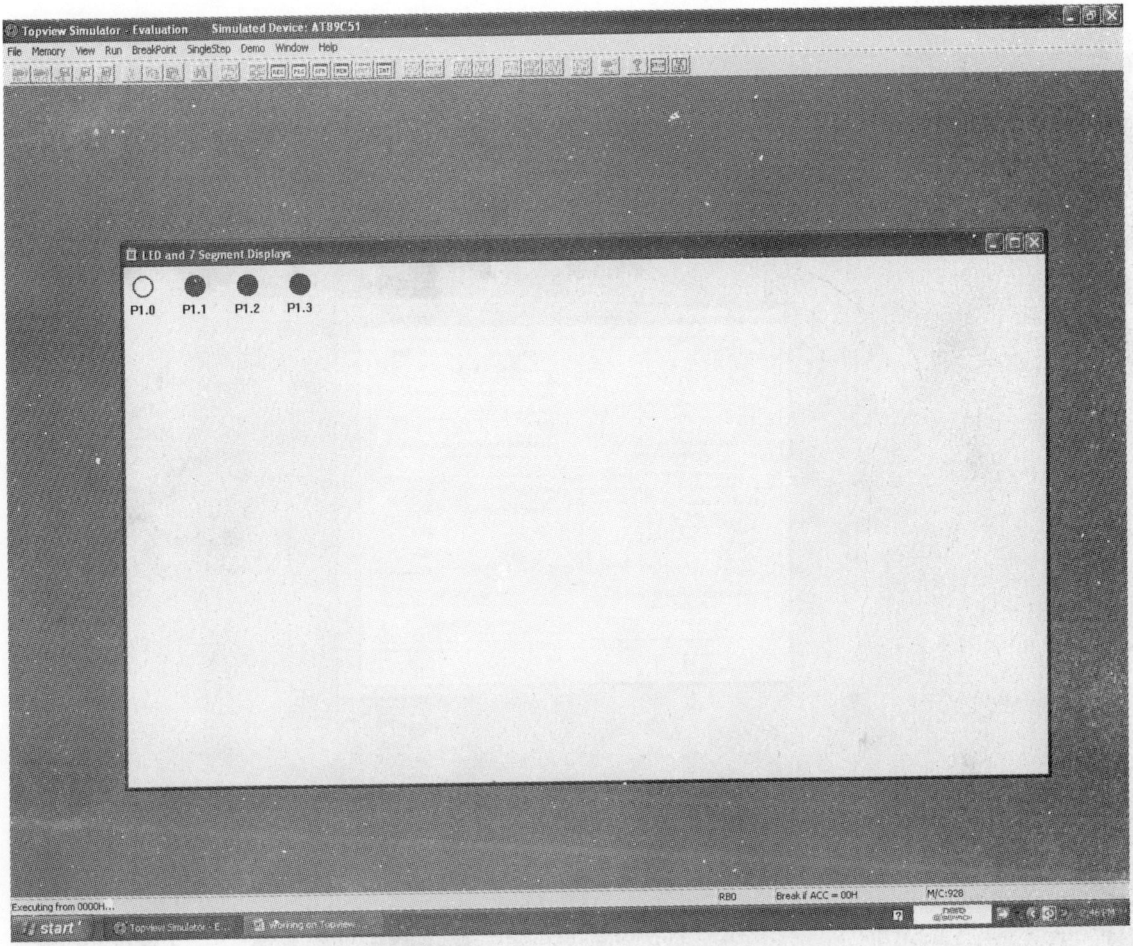

Hence; the working on topview simulator is very easy and a little practice will make we an expert.

Working on other modules is similar except for external modules setting.

Example of external modules setting **: matrix keyboard**.

1.Select FILE → EXTERNAL MODULES SETTING → KEYBOARD. We will be presented with the following window

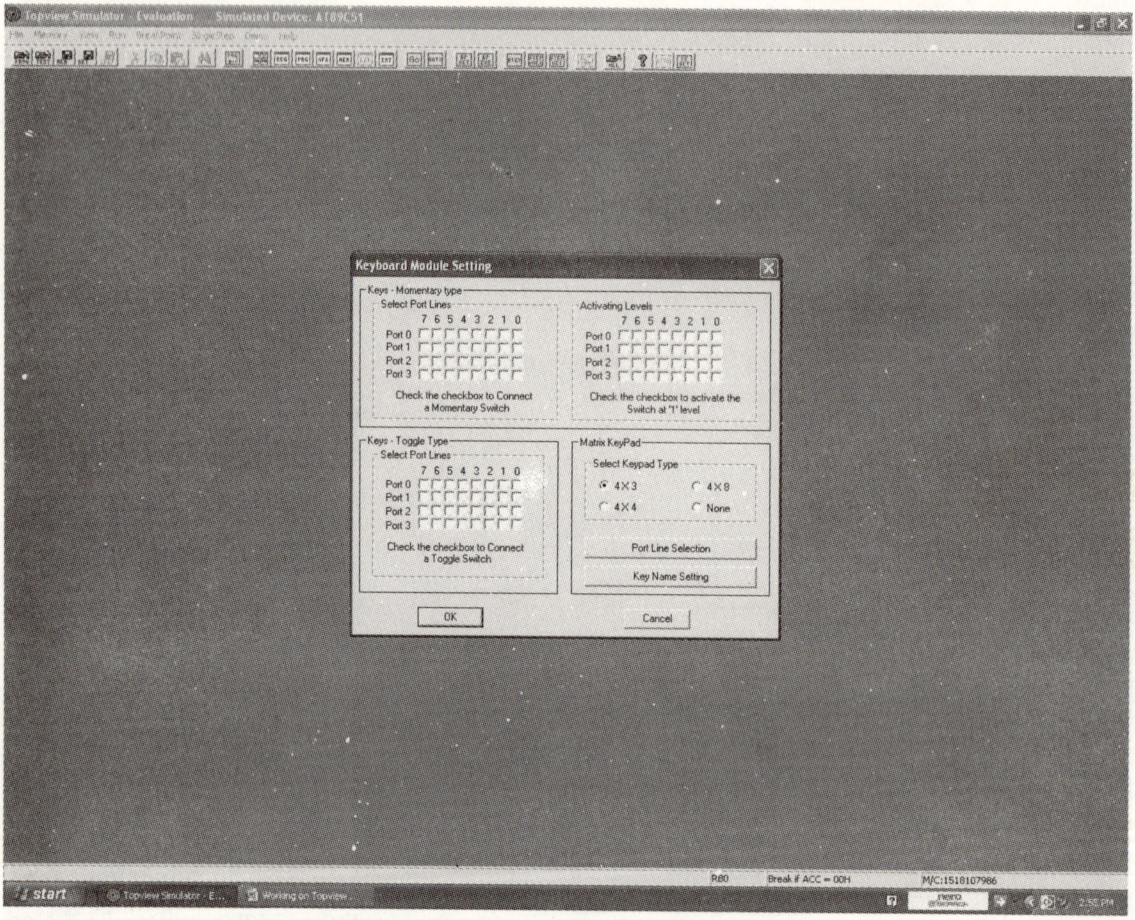

Now select keypad type like 4 × 3 or 4 × 4 etc..

There are two more settings: port line selection and key name setting. Just click on them; interface to right ports and give proper key names etc... (In accordance with our programming requirements).

Both the windows are shown below.

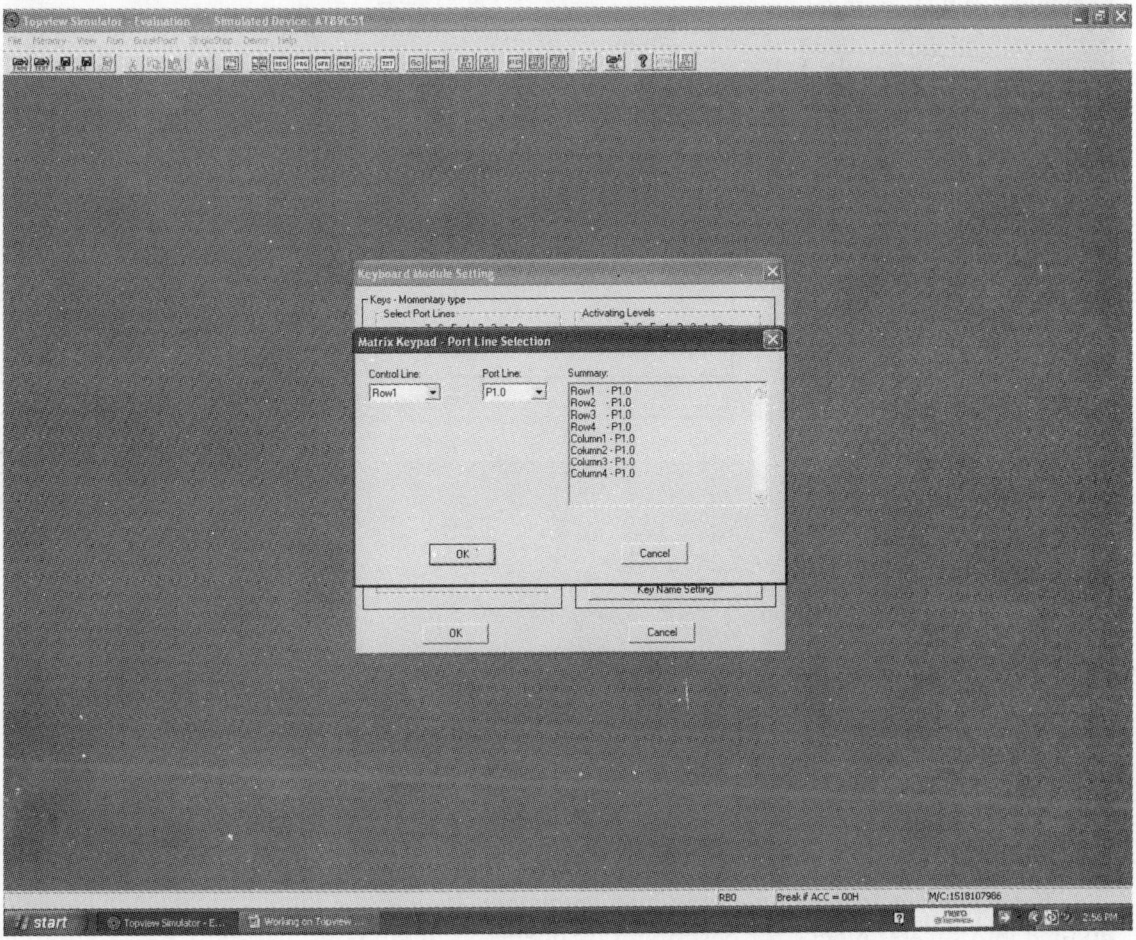
Topview Simulator - Evaluation
Simulated Device: AT89C51
Keyboard Module Setting
Keys - Momentary type
Select Port Lines
Activating Levels
Matrix Keypad - Port Line Selection
Control Line:
Row1
Port Line:
P1.0
Summary:
Row1 - P1.0
Row2 - P1.0
Row3 - P1.0
Row4 - P1.0
Column1 - P1.0
Column2 - P1.0
Column3 - P1.0
Column4 - P1.0
OK
Cancel
Key Name Setting
OK
Cancel
RB0
Break if ACC = 00H
M/C:1518107986
start
2:56 PM

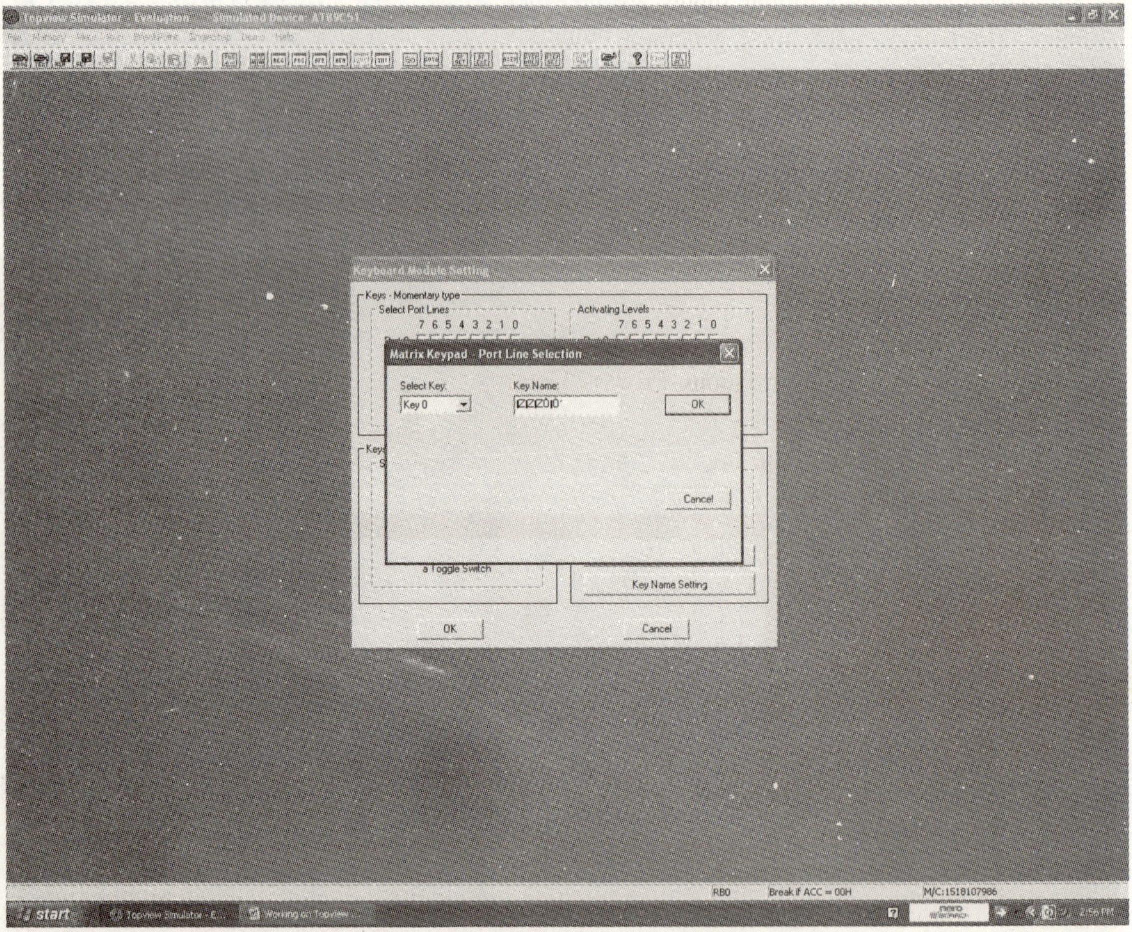

Similarly other modules can be set and then their output can be verified on simulator.

7.5 8096/97 (16-BIT MICROCONTROLLERS)

The 8096 is a 16-bit single chip microcomputer. Its internal circuit arrangements can be divided into several blocks for describing its operation. It contains a *16-bit CPU*, a *programmable high speed I/O unit*, an *ADC*, a *serial port*, and a *pulse width modulated output* for digital to analog conversion. In addition to these functional units, there are some sections that support overall operation of the chip such as the clock generator and the clock-bias generator. The CPU and the programmable I/O make the 8096 very different from any other microcontroller.

The family members of 16-bit microcontroller 8096 are 8094/8095/8096/8097/8394/8395/8396/8397. All of these members have been designed for high-speed high performance control functions. Its CPU supports bit, byte, and word operations. The 32-bit double-word are supported for a subset of the instruction set. With a 12MHz input frequency, the 8096 can do a 16-bit addition in 1μs and a 16 × 16 multiplication or 32/16-bit divide in 6.5μs. The average instruction execution time is 1 to 2μs in typical applications.

The 4-high-speed trigger inputs are provided to record the times at which external events occur. The 6-high-speed pulse generator outputs are provided to trigger external events at preset times. The high-speed output unit can simultaneously perform timer functions. Up to 4-such 16-bit Software Timers can be in operation at once.

An on-chip ADC converts up to 4 (in the 48-pin version) or 8 (in the 68-pin version) analog input channels to 10-bit digital values. This feature is only available on the 8095, 8395, 8097, and 8397. Some important applications of the 8096 microcontroller are

Industrial

- Motor control
- Robotics
- Discrete and Continuous Process Control
- Numeric Control
- Intelligent Transducers

Instrumentation

- Medical Instrumentation
- Liquid and Gas Chromatographs
- Oscilloscopes

Consumer

- Video Recorder
- Laser Disk Drive
- High-end Video Games

Guidance & Control

- Missile Control
- Torpedo Guidance Control
- Intelligent Annunciation
- Aerospace Guidance Systems

Data Processing

- Plotter
- Colour and B & W Copier
- Winchester Disk drive
- Tape Drive
- Impact and Non-Impact Printers

Telecommunications

- Modem
- Intelligent Line Card Control

Automotive

- Ignition Control
- Transmission Control

Anti Skid Braking

Emission Control

The internal architecture of the 8096 can broadly be partition into three sections;

- CPU,
- Peripheral,
- Bus control.

The CPU contains memory, ALU (RALU) and microcode control unit. The memory of the 8096 is organized into two sections as;

- 256 byte RAM,
- 232 byte SFRs.

The second section of the 8096 contains on chip peripherals, on chip signal processing and conversion circuits. The on chip ADC along with sample and hold circuits and a multiplexer offers 8-bit or 16-bit conversion facilities. The ADC input port is multiplexed with 8-bit I/O Port-0. Three PWM outputs are available from On chip PWM modulator. Each of these outputs provides DAC using external 8-bit PWM registers. Two numbers of 16-bit timers are available to support timing and counting applications.

Register Section

The On chip 256 byte RAM is used as 8-bit scratchpad registers. In addition, it has 232 bytes of RAM that is register file circuit that can be accessed as 8-bit, 16-bit or 32-bit data memory. The RALU contains special function register banks that have 80H (128 byte) 8-bit registers called SFRs.

Addressing Modes

Register Direct Mode

The register direct mode is used to directly access a register from the 256 byte on-board register file. Thus, the register file address should be allotted with symbolic register names of the 80x96 family.

Example	ADD AX, BX,CX;	;AX ← BX+CX
	MUL AX,CX	;AX ← AX.CX

Indirect Mode

Address of the operand in the memory is available in 16-bit register of the register file.

Example	ADD AL,BL,[CX]	;AL ← BL+Byte [CX]
	ADD AX,[CX]	;AX ← AX+word [CX]
	INCB CL	;CL ← CL+1

Indirect with Auto Increment Mode

The indirect mode is used to access the operand by placing its address in a WORD variable register file. The calculated address must conform to the alignment rules of the operand type. The 16-bit address is automatically incremented after the data access

```
ADD AX,[CX]     ; AX ← AX+word[CX]
                ; CX ← CX+2
ADD AL,[BX]     ; AL ← AL+byte[BX]
                ; CX ← BX+1
```

Immediate Mode

Data is immediately available after the opcode.

Example ADD A,#05H ; A ← A+05

Short Indexed Mode

The 16-bit value in the register file serves as an index and 8-bit signed extension serves as displacement to/ from address of the operand

Example

```
ADD AX,15[BX]   ; AX ← AX+word[BX+15]
LD AX,5[CX]     ; AX ← word [CX+5]
```

Long Indexed Mode

Here 16-bit signed displacement is used

Example

```
AND AX,CX,1234[BX]    ; AX ← CX AND[1234+BX]
ADD AL, BL,1234[CX]   ; AL ← BL+byte [1234+CX]
```

The 8096A has following bocks

- serial port,
- a watchdog timer (watches the 8-states to trigger internally the timer-1 and timer-2 is synchronized to external occurrences), and
- a pulse-width modulated output signals.

The 8096 are available with and without ADC and with and without on-chip ROM. The 8097 is the ROMless version of 8096 with on-chip ADC. Table 7.5(*a*) details some of the features of the 8096.

Table 7.5(*a*) Features and benefits

Features	*Benefits*
16-bit CPU	Efficient machine with higher throughput
8KB ROM	More program space
232B RAM	Large on-board register file
Hardware MUL/ DIV	Better math capability 16 by 16 multiply or 32 by 16 divide in 6.5μs @12MHz
6-Addressing Modes	Greater flexibility in programming/data
High Speed I/O Lines 4-dedicated I/O 4-progammable	Can measure and generate pulse with high resolution (2μs @ 12MHz)

10-bit A/D Converter	Reads the external analog inputs
Full Duplex Serial Port	Provides asynchronous serial to the other processors
Up to 40 I/O Ports	Provides TTL compatible digital data I/O
Programmable 8-source Priority Interrupt System	Respond to asynchronous events
PWM Output	Provides programmable pulse train with variable duty cycle. Also used to generate analog output.
Watchdog Timer	Ability to recover from software malfunction or hardware upset

The 8096 replaces multiple chip solution by integrating a 16-bit CPU with all the I/O features and interface resources on a single piece of silicon that represents highest level of integration in context with the microcontrollers.

Pin description of 8096

ANGND: Reference ground for the ADC . It should be held at nominally the same potential as V_{SS} = 0V.

V_{CC} : Main supply voltage = +5V.

V_{SS} : Digital circuit ground (0V). There are 2 numbers of V_{SS} pins (Pins 10 and 42), both must be tied to ground.

V_{PD} : RAM standby power supply = +5V. This voltage must be present during normal operation.

V_{BB} : Substrate voltage from the on-chip back-bias generator. This pin should be connected to the ANGND through a 0.01μF capacitor. The capacitor is not required if ADC is not used.

V_{REF} : Reference voltage and power supply for the analog portion of the ADC is nominally at +5V.

XTAL1 : Input of the oscillator inverter and input to the internal generator.

XTAL2 : Output of the oscillator inverter.

CLKOUT : Output of the internal clock generator. The frequency of CLKOUT is 1/3 the oscillator frequency. It has 33% duty cycle. The CLKOUT can drive one TTL input.

$\overline{\text{RESET}}$: Reset input to the chip and also output to other chip. Input low for at least 2 state time resets the chip. $\overline{\text{RESET}}$ has strong internal pullup.

$\overline{\text{TEST}}$: Low input enables a factory test mode. The user should tie this pin to V_{CC} for normal operation.

NMI : A low to high transition a vector to external memory location 0000H. Reserved for the Intel development system.

INST : Output high while the address is valid during an external read that indicates the read is an instructional fetch.

$\overline{\text{EA}}$: Input for memory select (external access). $\overline{\text{EA}}$ = 1 causes memory access to location

2000H though 3FFFH to be directed to on-chip ROM. $\overline{EA}$ = 0 causes access to these locations to be directed to off chip memory. $\overline{EA}$ has an internal pulldown, so it goes to 0 unless driven to 1.

ALE : Address latch enable output. It is activated only during external memory access. It is used to latch up the address from the multiplexed address data bus.

$\overline{RD}$: Read signal outputs to external memory. It is activated during external memory reads.

$\overline{WR}$: Write signal outputs to external memory. $\overline{WR}$ is activated during external memory writes.

$\overline{BHE}$: Bus high enable signal outputs to external memory. $\overline{BHE}$ = 0/1 selects/deselects the bank of memory that is connected to the high byte of the data bus.

READY: The READY input is used to lengthen external memory bus cycle up to the time specified in the data sheet.

HSI : High impedance inputs to HSI units. Four HSI pins are available HSI.0, HSI.1, HSI.2, and HSI.3. Two of them HSI.2 and HSI.3 are shared with HSO unit.

HSO : Outputs from HSO unit. Six HSO pins are available; HSO.0, HSO.1 , HSO.2, HSO.3, HSO.4, and HSO.5. Two of them HSO.4 and HSO.5 are shared with HSI unit. All HSO pins are capable of driving one TTL input.

Port 0 : (Analog channel) Only high impedance input port. This port can be used as digital inputs and/or analog inputs to the on-chip ADC.

Port 1 : Quasi-bidirectional I/O port. All pins of P1 are capable of driving one LSTTL input.

Port 2 : Multi-functional port. Six of its pins are shared with other functions as in Table 7.5(*b*).

Table 7.5(*b*) Functions of pins

Port	*Function*	*Alternative function*	*Port*	*Function*	*Alternative function*
P2.0	Output	TXD (serial port transmit)	P2.4	Input	T2RST(timer2 reset)
P2.1	Input	RXD (serial port receive)	P2.5	output	PWM (modulator)
P2.2	Input	EXTINT (external interrupt)	P2.6	Quasi-bidirectional	
P2.3	Input	T2CLK (timer2 input)	P2.7	Quasi-bidirectional	

Port 3 and 4 : 8 bit bi-directional I/O ports. These pins are shared the multiplexed address/data bus when accessing external memory, with the port 3 pins accessing the low byte and port 4 accessing the high byte. They are open drain except when being used as system bus pins.

Its internal circuit arrangements can be divided into several blocks for describing its operation as in Fig. 7.5 (*a*). Its Pinouts is shown in Fig. 7.5 (*b*).

The block diagram of the 8096 contains

- a CPU,

- a programmable high speed I/O unit,
- an ADC,
- a serial port, and
- a pulse width modulated output for digital to analog conversion.

The 8096 single chip microcontrollers is designed for use in sophisticated real-time demanding applications such as industrial control, instrumentation and intelligent computer peripherals. The wide base applications cut across all industry segments have already been enumerated in applications of microcontrollers. With 16-bit CPU horsepower, high-speed math processing and high-speed I/O, the 8096 is ideal for complex motor control and axis control systems. Example includes 3-phase, large horsepower AC motors and robotics. With 10-bit ADC option, the device finds usage in data acquisition systems and closed-loop analog controllers.

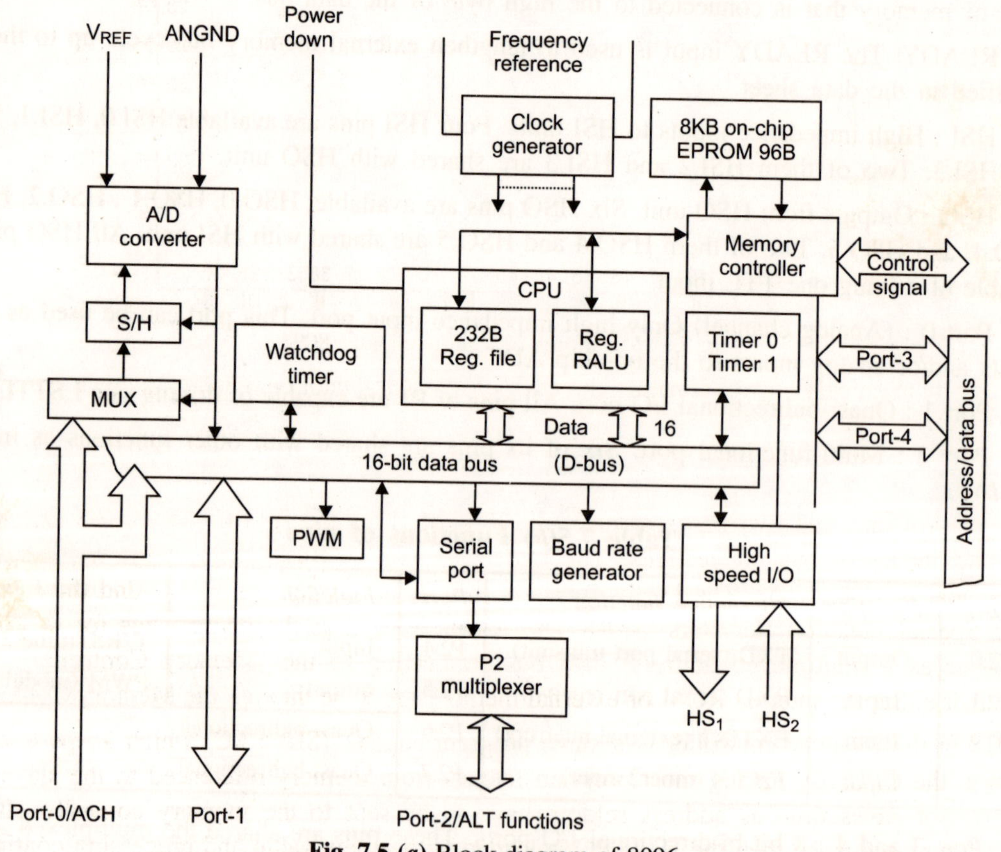

Fig. 7.5 (*a*) Block diagram of 8096

CPU

The major component of the CPU of 8096 has register file and the RALU (Register Arithmetic Unit). Communication with the outside word is done either through the special function registers (SFRs) or the memory controller. The RALU does not use an accumulator. It operates directly on the 256B-register space made up of the register file and SFRs. Efficient I/O operations are possible by directly controlling the I/O through the SFRs. The main benefits of this structure

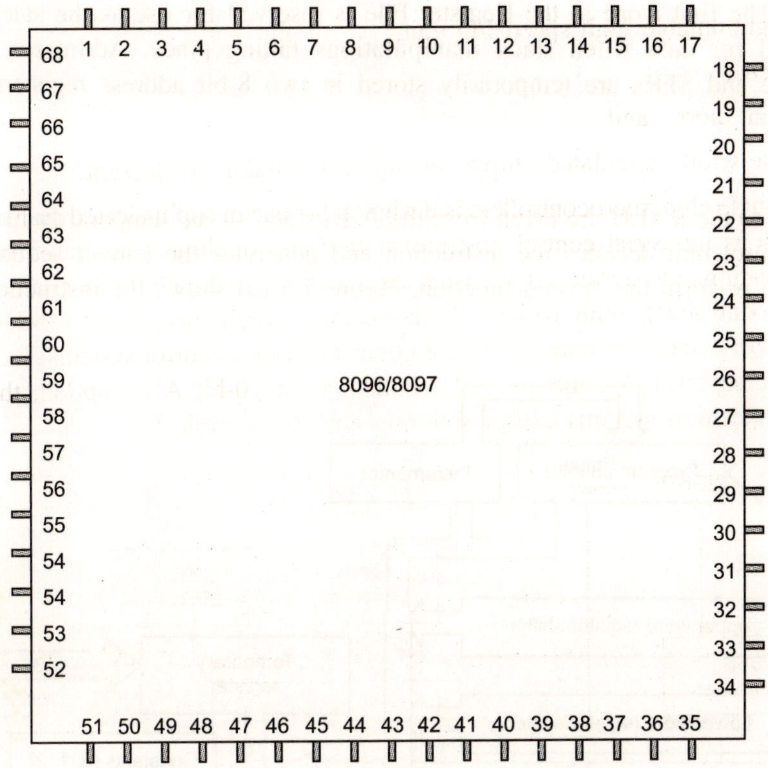

Fig. 7.5 (*b*) Pinouts of microcomputer 8096

are the ability to quickly change contents, the absence of accumulator bottleneck, and fast throughput and I/O times.

Buses

A control unit and two buses connect the Register File and RALU. Figure 7.5 (*c*) shows CPU with its major bus connections. The two buses are 'A-Bus' that is 8-bit wide and the 'D-Bus' that is 16-bit wide. The D-Bus transfers data only between the RALU and the Register File or Special Function Registers (SFR). The A-Bus is used as the address bus for the above transfer or as a multiplexed address/ data bus connecting to the 'Memory Controller'. Any accesses of either the internal ROM or external memory are done through the Memory Controller.

Within the memory controller is a slave program counter (Slave PC) which keeps track of the PC in the CPU. By having most program fetches from memory referenced to the slave PC, the processor saves time as address seldom have to be sent to the memory controller. If the address jumps sequence then the slave PC is loaded with a new value and processing continues. Data fetches from memory are done through the memory controller, but the slave is bypassed for this operation.

Register file

The Register File contains 232B of RAM that can be accessed as bytes, words, or double words. Since each of these locations can be used by the RALU, there are essentially 232 'accumulators'. The first word in the Register File is reserved for use as the stack-pointer so it can not be used for data when stack manipulations taking place. Addresses for accessing the Register File and SFRs are temporarily stored in two 8-bit address registers by the CPU hardware.

RALU

Instructions to the RALU are taken from the A-Bus and stored temporarily in the instruction register. The control unit decodes the instruction and generates the correct sequence of signals for the RALU to perform the desired function. Figure 7.5 (*a*) shows the instruction register and the control unit.

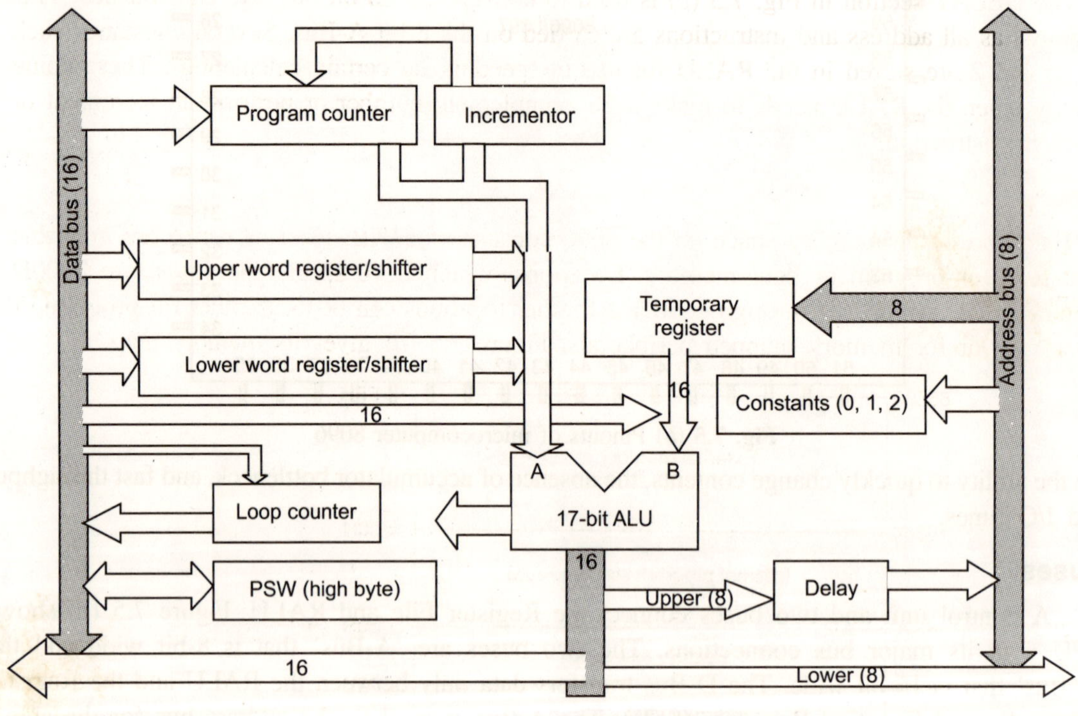

Fig. 7.5 (*c*) RALU block diagram

The 8096 performs most of the calculations in the RALU. The block diagram arrangement of RALU is shown in Fig. 7.5 (*c*). The RALU contains

- a 17-bit ALU,
- the Program Status Word (PSW),
- the Program Counter (PC),
- a loop counter, and
- 3-temporary registers.

All registers are 16-bit or 17-bit (16+sign extension) wide. Some of the registers have the ability to perform simple operations to off-load the ALU. A separate incrementor is used for the PC; however, jumps must be handled through the ALU. Two of the temporary registers have their own shift logic. These registers are used for the operations that require logical shift, including Normalize, Multiply, and Divide. The 'Lower Word' register is used only when double word quantities are being shifted, the 'Upper Word' register is used whenever a shift is performed or as a temporary register for many instructions. Repetitive shifts are counted by the 5-bit 'Loop Counter'.

A temporary register is used to store the second operand of two operand instructions. This includes the multiplier during multiplication and the divisor during the multiplication and divisions. In order to perform subtractions, the output of this register can be complemented before being placed into the 'B' input of the ALU.

The DELAY section in Fig. 7.5 (*c*) is used to convert the 16-bit bus into an 8-bit bus. This is required as all address and instructions are carried on the 8-bit A-Bus. Several constants, such as 0, 1, and 2 are stored in the RALU for use in speeding up certain calculations. These come in handy when the RALU needs to make a 2's complement number or perform an increment or decrement instruction.

Memory

The addressable memory space on the 8096 consists of 64KB, most of which are available to the user for program or data memory. Locations, which have special purposes, are 0000H through 00FFH and 1FFFH through 2010H. All other locations can be used either for program or data storage or for memory mapped peripherals. Figure 7.5 (*d*) gives its memory map.

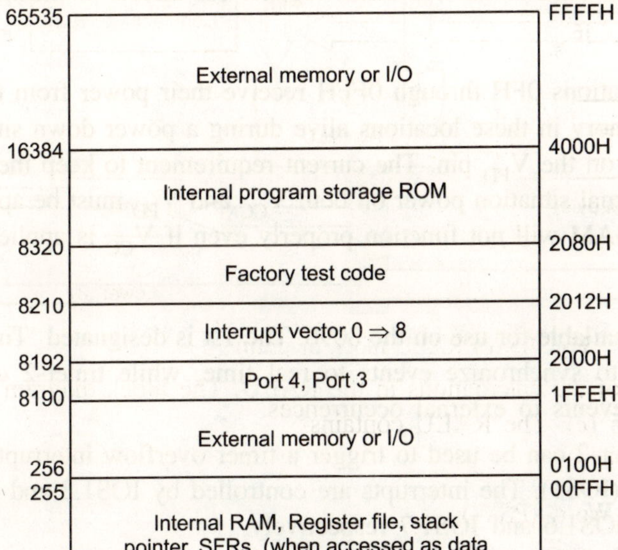

Fig. 7.5 (*d*) Memory map of 8096

The internal register locations on the 8096 are divided into 2-groups

- a register file
- a set of Special Function Registers (SFRs).

Locations 00H through 0FFH contain Register File and Special Function Registers (SFRs). No code can be executed from this internal RAM section. If an attempt to execute instructions from location 00H through 0FFH is made, the instruction will be fetched from *external memory*. This section is reserved for the development. Execution of the non-maskable interrupt (NMI) will force a call to external location 0000H, therefore, the NMI and TRAP interrupt are also reserved for Intel development tools.

Reserved memory

Locations 1FFEH and 1FFFH are reserved for Port-3 and Port-4 respectively. Locations 2000H through 2011H stores 9-vector interrupts. The 9th vector is used by Intel for its system development. Locations 2012H through 2077H are reserved for Intel factory test.

The RALU can operate on any of the 256 internal register locations. Locations 00H through 17H are used to access the SFRs. Locations 18H and 19H contain the Stack Pointer. There are no restrictions on the use of the remaining 230 locations except that the code cannot be executed from them.

All I/O on the 8096 is controlled through the SFRs. Many of these registers serve two functions

- if they are read from and
- if they are written to.

These registers cannot be used as base or index register for directs or indexed instructions.

Power down

The upper 16-RAM locations 0FH through 0FFH receive their power from the V_{PD} pin. If it is desired to keep the memory in these locations alive during a power down situation, all that is needed is to keep voltage on the V_{PD} pin. The current requirement to keep the RAM alive is approximately 1mA. For normal situation power on both V_{CC} and V_{PD} must be applied. If power on V_{PD} is not applied, the RAM will not function properly even if V_{CC} is applied.

Timers

Two 16-bit timers are available for use on the 8096. The 1st is designated 'Timer-1', the 2nd 'Timer-2'. Timer-1 is used to synchronize events to real time, while timer-2 can be clocked externally and synchronize events to external occurrences.

Both Timers-1 and Timer-2 can be used to trigger a timer overflow interrupt and set a flag in the I/O Status Register-1 (IOS1). The interrupts are controlled by IOS1.2 and IOS1.3 respectively. The flags are set in IOS1.6 and IOS1.7 respectively.

High speed inputs

The high-speed input unit (HSI) can be used to record the time at which an event occurs w.r.t. timer-1. There are 4-lines (HSI.0 through HSI.3) which can be used in this mode and up to a total of 8-events can be recorded. HSI.2 and HSI.3 share pins with HSO.4 and HSO.5. The I/O control registers (IOC0 and IOC1) are used to determine the functions of these pins.

High speed outputs

The high-speed output unit (HSO) is used to trigger events at specific times with minimal CPU overhead. These events include, starting an ADC, resetting timer-2, setting 4-software flags, generating interruots, and switching up to 6-output lines. Up to 8-events can be pending at any time.

Analog inputs

The ADC on the 8096 provides a 10-bit result on one of 8-input channels. Conversion is done using successive approximation with a result equal to the ratio of the input voltage divided by the analog supply voltage. If the ratio is 1.0, then the result will be all ones. The ADC is available on the 8096, 8397, 9095, and 8395 members of the MCS-96 family. The input voltage must be in the range of 0 to V_{REF}, the analog voltage and supply voltage. For proper operation, V_{REF} (the reference voltage and the analog power supply) must be held nominally at 5V. The ADC result is calculated from the formula

$$1024 \frac{\text{Input voltage} - \text{ANGND}}{V_{REF} - \text{ANGND}}$$

The ANGND must be tied to V_{SS} (digital ground) for proper operation of the 8096.

Serial port

The serial port is compatible with the MCDS-51 serial port. It is full duplex, meaning it can transmit and receive simultaneously. It is also receive-buffered, meaning it can commence reception of a second byte before a previously received byte has been read from the receive register. The serial port registers are both accessed at location 07H. A write to this location accesses the transmit register, and a read accesses a physically separate receive register.

The serial port can operate in 4-modes. Selection of these modes is done through the serial port control register at location 11H.

I/O Ports 0, 1, 2, 3, and 4

The 8096 has five 8-bit I/O ports. Some of these ports are input only, some output only some bi-directional and some have alternate functions. Input ports connect to the internal bus through an input buffer. Output ports connect through an output buffer to an internal register that holds the output bits. Bi-directional ports consist of an internal register, an output buffer, and an input buffer.

When an instruction accesses a bi-directional port as a source register, the question often arises as to whether the value that is brought into the CPU comes from the internal port register or from the pins through input buffer. In the 8096, the value always comes from the port pins, never from the internal register.

Watchdog timer

This feature is provided as a means of graceful recovery from a software upset. If the software fails to reset the watchdog at least every 64K-state times, a hardware reset will be initiated.

The software can be designed so that the watchdog times out if the program does not progress properly. The watchdog will also time-out if the software error was due to ESD

(Electrostatic Discharge) or other hardware related problems. This prevents the controller from having a malfunctioning for longer than 16ms if a 12MHz oscillator is used.

Packaging

The 8096 is available in 48-pin, 64-pin and 68-pin packages, with and without ADC and with and without on-chip ROM or EPROM. Table 7.5(*c*) gives packaging details of the 8096.

Table 7.5(*c*) Pin definition of the 8096 microcontroller

Name	*PGA 68-pin*	*DIP 48-pin*	*Name*	*PGA 68-pin*	*DIP 48-pin*	*Name*	*PGA 68-pin*	*DIP 48-pin*
ACH0/P0.0	4	-	P0.1/ACH1	5	-	P3.5/AD5	23	27
ACH1/P0.1	5	-	P0.2/ACH2	3	-	P3.6/AD6	24	26
ACH2/P0.2	3	-	P0.3/ACH3	6	-	P3.7/AD7	25	25
ACH3/P0.3	6	-	P0.4/ACH4	67	43	P4.0/AD8	26	24
ACH4/P0.4	67	43	P0.5/ACH5	68	42	P4.1/AD9	27	23
ACH5/P0.5	68	42	P0.6/ACH6	2	40	P4.2/AD10	28	22
ACH6/P0.6	2	40	P0.7/ACH7	1	41	P4.3/AD11	29	21
ACH7/P0.7	1	41	P1.0	59	-	P4.4/AD12	30	20
ALE/ ADV	16	34	P1.1	58	-	P4.5/AD13	31	19
ANGND	66	44	P1.2	57	-	P4.6/AD14	32	18
$\overline{\text{BHE}}$ / $\overline{\text{WRH}}$	37	15	P1.3	56	-	P4.7/AD15	33	17
BUS WIDTH	14	-						
CLKOUT	13	-	P1.4	55	-	$\overline{\text{RD}}$	17	33
$\overline{\text{EA}}$	8	39	P1.5	48	-	READY	35	16
EXTINT/P2.2	63	47	P1.6	47	-	$\overline{\text{RESET}}$	62	48
HSI.0	54	3	P1.7	46	-	RXD/P2.1	61	1
HSI.1	53	4	P2.0/TXD	60	2	$\overline{\text{PROG}}$ / $P_{2.2}$	63	47
HSI.2/HSO.4	52	5	P2.1/RXD	61	1	TXD/P2.0	60	2
HSI.3/HSO.5	51	6	P2.2/EXTINT	63	47	T2CLK/P2.3	34	-
HSO.0	50	7	P2.3/T2CLK	34	-	T2RST/P2.4	36	-
HSO.1	49	8	P2.4/T2RST	36	-	V_{PP}	41	12
HSO.2	44	9	P2.5/PWM	39	13	V_{CC}	9	38
HSO.3	43	10	P2.6	45	-	V_{PD}	64	46
HSO.4/HSI.2	52	5	P2.7	40	-	V_{REF}	65	45
HSO.5/HSI.3	51	6	P3.0/AD0	18	32	V_{SS}	10	11
INST	15	-	P3.1/AD1	19	31	V_{SS}	42	37
NMI	7	-	P3.2/AD2	20	30	$\overline{\text{WR}}$	38	14
PWM/P2.5	39	13	P3.3/AD3	21	29	XTAL1	11	36
P0.0/ACH0	4	-	P3.4/AD4	22	28	XTAL2	12	35

	Option	68-pins	48-pins
Digital I/O	ROMless	8096	8094
	ROM	8396	8394
Analog and Digital I/O	ROMless	8097	8095
	ROM	8397	8395

Table 7.5(*c*) summarizes the Opcodes of the 8096.

Table 7.5(*d*) Opcode and state time of 8096 microcontroller

Mnemonics	*DIRECT*				*IMMEDIATE*			*INDIRECT*✪ *NORMAL*			*INDIRECT*✪ *AUTO-INC*		*INDEXED*✪ *SHORT*			*INDEXED*✪ *LONG*	
	Operand	Opcode	Bytes	State Times	Opcode	Bytes	State Times	Opcode	Bytes	State Times	Bytes	State Times	Opcode	Bytes	State Times	Bytes	State Times
ARITHMETIC INSTRUCTIONS																	
ADD	2	64	3	4	65	4	5	66	3	6/11	3	7/12	67	4	6/11	5	7/12
ADD	3	44	4	5	45	5	6	46	4	7/12	4	8/13	47	5	7/12	6	8/13
ADDB	2	74	3	4	75	3	4	76	3	6/11	3	7/12	77	4	6/11	5	7/12
ADDB	3	54	4	5	55	4	5	56	4	7/12	4	8/13	57	5	7/12	6	8/13
ADDC	2	A4	3	4	A5	4	5	A6	3	6/11	3	7/12	A7	4	6/11	5	7/12
ADDCB	2	B4	3	4	B5	3	4	B6	3	6/11	3	7/12	B7	4	6/11	5	7/12
SUB	2	68	3	4	69	4	5	6A	3	6/11	3	7/12	6B	4	6/11	5	7/12
SUB	3	48	4	5	49	5	6	4A	4	7/12	4	8/13	4B	5	7/12	6	8/13
SUBB	2	78	3	4	79	3	4	7A	3	6/11	3	7/12	7B	4	6/11	5	7/12
SUBB	3	58	4	5	59	4	5	5A	4	7/12	4	8/13	5B	5	7/12	6	8/13
SUBC	2	A8	3	4	A9	4	5	AA	3	6/11	3	7/12	AB	4	6/11	5	7/12
SUBCB	2	B8	3	4	B9	3	4	BA	3	6/11	3	7/12	BB	4	6/11	5	7/12
CMP	2	88	3	4	89	4	5	8A	3	6/11	3	7/12	8B	4	6/11	5	7/12
CMPB	2	98	3	4	99	3	4	9A	3	6/11	3	7/12	9B	4	6/11	5	7/12
MULU	2	6C	3	25	6D	4	25	6E	3	27/32	3	28/33	6F	4	27/32	5	28/33
MULU	3	4C	4	26	4D	5	27	4E	4	28/33	4	29/34	4F	5	28/33	6	29/34
MULUB	2	7C	3	17	7D	3	17	7E	3	19/24	3	20/25	7F	4	19/24	5	20/25
MULUB	3	5C	4	18	5D	4	18	5E	4	20/25	4	21/26	5F	5	20/25	6	21/26
MUL	2	②	4	29	②	5	30	②	4	31/36	4	32/37	②	5	31/36	6	32/37
MUL	3	②	5	30	②	6	31	②	5	32/37	5	33/38	②	6	32/37	7	33/38

MULB	2	②	4	21	②	4	21	②	4	23/28	4	24/29	②	5	23/28	6	24/29
MULB	3	②	5	22	②	5	22	②	5	24/29	5	25/30	②	6	24/29	7	25/30
DIVU	2	8C	3	25	8D	4	26	8E	3	28/32	3	29/33	8F	4	28/32	5	29/33
DIVUB	2	9C	3	17	9D	3	17	9E	3	20/24	3	21/25	9F	4	20/24	5	21/25
DIV	2	②	4	29	②	5	30	②	4	32/36	4	33/37	②	5	32/36	6	33/37
DIVB	2	②	4	21	②	4	21	②	4	24/28	4	25/29	②	5	24/28	6	25/29
LOGICAL INSTRUCTIONS																	
AND	2	60	3	4	61	4	5	62	3	6/11	3	7/12	63	4	6/11	5	7/12
AND	3	40	4	5	41	5	6	42	4	7/12	4	8/13	43	5	7/12	6	8/13
ANDB	2	70	3	4	71	3	4	72	3	6/11	3	7/12	73	4	6/11	5	7/12
ANDB	3	50	4	5	51	4	5	52	4	7/12	4	8/13	53	5	7/12	6	8/13
OR	2	80	3	4	81	4	5	82	3	6/11	3	7/12	83	4	6/11	5	7/12
ORB	2	90	3	4	91	3	4	92	3	6/11	3	7/12	93	4	6/11	5	7/12
XOR	2	84	3	4	85	4	5	86	3	6/11	3	7/12	87	4	6/11	5	7/12
XORB	2	94	3	4	95	3	4	96	3	6/11	3	7/12	97	4	6/11	5	7/12
DATA TRANSFER INSTRUCTIONS																	
LD	2	A0	3	4	A1	4	5	A2	3	6/11	3	7/12	A3	4	6/11	5	7/12
LDB	2	B0	3	4	B1	3	4	B2	3	6/11	3	7/12	B3	4	6/11	5	7/12
ST	2	C0	3	4	-	-	-	C2	3	7/11	3	8/12	C3	4	7/11	5	8/12
STB	2	C4	3	4	-	-	-	C6	3	7/11	3	8/12	C7	4	7/11	5	8/12
LDBSE	2	BC	3	4	BD	3	4	BE	3	6/11	3	7/12	BF	4	6/11	5	7/12
LDBZE	2	AC	3	4	AD	3	4	AE	3	6/11	3	7/12	AF	4	6/11	5	7/12
STACK OPERATION (Internal Stack)																	
PUSH	1	C8	2	8	C9	3	8	CA	2	11/15	2	12/16	CB	3	11/15	4	12/20
POP	1	CC	2	12	-	-	-	CE	2	14/18	2	14/18	CF	3	14/18	4	14/20
PUSHF	0	F2	1	8													
POPF	0	F3	1	9													
STACK OPERATION (External Stack)																	
PUSH	1	C8	2	12	C9	3	12	CA	2	15/19	2	16/20	CB	3	15/19	4	16/20
POP	1	CC	2	14	-	-	-	CE	2	16/20	2	16/20	CF	3	16/20	4	16/20
PUSHF	0	F2	1	12													
POPE	0	F3	1	13													

JUMPS AND CALLS

Mnemonics	*Opcode*	*Bytes*	*States*	*Mnemonics*	*Opcode*	*Bytes*	*States*
LJMP	E7	3	8	LCALL	EF	3	13/16[5]
SJMP	20-27[4]	2	8	SCALL	28-2F[4]	2	13/16[5]
INDJMP	E3	2	8	RET	F0	1	12/16[5]
				TRAP[3]	F7	1	21/24

CONDITIONAL JUMPS							
All conditional jumps are 2-bytes instructions. It require 8-states for jumps taken and 4 if not taken.							
Mnemonics	*Opcode*	*Mnemonics*	*Opcode*	*Mnemonics*	*Opcode*	*Mnemonics*	*Opcode*
JC	DB	JE	DF	JGE	D6	JGT	D2
JNC	D3	JNE	D7	JLT	DE	JLE	DA
JH	D9	JV	DD	JVT	DC	JST	D8
JNH	D1	JNV	D5	JNVT	D4	JNST	D0

JUMPS ON BIT CLEAR OR BIT SET								
These instructions are 3-byte instructions. It require 9-state time if jumps taken and 5 if not taken								
	BIT NUMBER							
Mnemonics	0	1	2	3	4	5	6	7
JBC	30	31	32	33	34	35	36	37
JBS	38	39	3A	3B	3C	3D	3E	3F

LOOP CONTROL			
Mnemonics	*Opcode*	*Bytes*	*STATE TIMES*
DJNZ	E0	3	5/STATE TIMES (NOT TAKEN/TAKEN)[8]

SINGLE REGISTER INSTRUCTIONS							
Mnemonics	*Opcode*	*Bytes*	*States*	*Mnemonics*	*Opcode*	*Bytes*	*States*
DEC	05	2	4	EXT	06	2	4
DECB	15	2	4	EXTB	16	2	4
NEG	03	2	4	NOT	02	2	4
NEGB	13	2	4	NOTB	12	2	4
INC	07	2	4	CLR	01	2	4
INCB	17	2	4	CLRB	11	2	4

SHIFT INSTRUCTIONS									
Mnemonics	*Word*		*Mnemonics*	*Byte*		*Mnemonics*	*DBL WD*		*State Times*
	OP	*B*		*OP*	*B*		*OP*	*B*	
SHL	09	3	SHLB	19	3	SHLL	0D	3	7+1 Per Shift⑦
SHR	08	3	SHRB	18	3	SHRL	OC	3	7+1 Per Shift⑦
SHRA	0A	3	SHRAB	1A	3	SHRAL	OE	3	7+1 Per Shift⑦

SPECIAL CONTROL INSTRUCTIONS							
Mnemonics	*Opcode*	*Bytes*	*States*	*Mnemonics*	*Opcode*	*Bytes*	*States*
SETC	F9	1	4	DI	FA	1	4
CLRC	F8	1	4	EI	FB	1	4
CLRVT	FC	1	4	NOP	FD	1	4
RST[6]	FF	1	166	SKIP	00	2	4

NORMALIZE			
Mnemonics	*Opcode*	*Bytes*	*States Times*
NORMAL	0F	3	11+1 Per Shift

⑥The instruction takes 2-state to pull RST low, then holds it low for 2-states to initiate a reset. The reset takes 12-states, at which time the program restarts at location 2080H

②The opcode for signed multiply and divide is the opcodes for the unsigned functions with 'FE' appended as a prefix.

④The least significant 3-bits of the opcode are concatenated with following 8-bits to form an 11-bits. 2's complement, offset for the relative call or jump.

⑤State times for stack located internal/external

Conditions of registers after power ON or RESET							
Register	*Contents*	*Register*	*Contents*	*Register*	*Contents*	*Register*	*Contents*
PC	0000H	P0-P3	FFH	TH0	00H	RCAP2H	00H
A_{CC}	00H	IP	xx000000H	TL0	00H	RCAP2L	00H
B	00H	IE	0x000000H	TH1	00H	SCON	00H
PSW	00H	TMOD	00H	TL1	00H	SBUF	Not known
SP	07H	TCON	00H	TH2	00H	PCON	0xxx0000
DPTR	0000H	T2CON	00H	TL2	00H		

REVIEW QUESTIONS AND ANSWERS

Q.1. What is the difference between a microprocessors and microcontrollers?

Ans. The microprocessor is one unit of different blocks of the microcontroller.

Q.2. Why microcontroller was developed?

Ans. The trend to develop small microcomputers that costs less forced the manufacturers to develop the microcontrollers.

Q.3. What are applications of microcontrollers?

Ans. The applications of microcontrollers extends from home appliances such as audio and video systems, intelligent cooker, washing machine, video games, toys etc to scientific equipments such as calculators etc.

Q.4. What is the definition of a microcontroller?

Ans. A microcontroller is a single chip microcomputer containing all essential elements of a microcomputer such as a µP, ROM/EPROM, RAM, I/O ports, timer, counter, decoder, ADC etc.

Q.5. What is a dedicated microcontroller?

Ans. A dedicated microcontroller is developed for a specific work. It can include the special functions such as masking capabilities, Boolean functions, multiplication/ division instructions, serial/ parallel ports, PLL, ADC, PWM, Watch dog timer etc.

Q.6. What are sizes of different elements of 8748 microcontroller?

Ans. It contains 8-bit µP, external interrupts along with ROM/ EPROM = 1/2/4KB, RAM = 64/128/ 256B, Timer = 1/2 Nos, I/O ports = 1/2/4 Nos.

Q.7. How does MCS-8051 differ from MCS8784?

Ans. The MCS8051 has higher capacity of ROM/ EPROM (4/8KB), RAM (128/256B), and a serial port in addition to parallel ports.

Q.8. What does MCS8096 contains?

Ans. It contains μP = 16-bit, ROM = 8KB, RAM = 232B, 10-bit ADC, Serial and parallel I/O ports, PWM, Watch dog timer.

Q.9. What are essential sections of a microcontroller?

Ans. The essential sections of a microcontroller are

- Arithmetic section,
- Program memory,
- Data memory,
- I/O ports,
- Program counter and stack counter
- PSW,
- Timer/counter,
- Program/verification
- etc.

Q.10. What does the arithmetic section of the microcontroller contains?

Ans. The arithmetic section contains the following basic blocks

- Instruction decoder,
- ALU
- Accumulator
- Flags

Q.11. What is the size of the program in 8748?

Ans. Program memory = 1KB × 8 ($2^{10} - 1$ = 1023D = 3FFH *i.e.*, 000H to 3FFH).

Q.12. How EPROM of the 8751 microcontroller is programmed?

Ans. Figure Prob. 12 is used to program the EPROM of the 8751 microcontroller. Only RST pin requires 2.5V for its high logic. The $\overline{EA}/V_{PP}$ is held normally high and pulsed to +21V. While $\overline{EA}/V_{PP}$ is held high, the $ALE/\overline{PROG}$ which is normally being held high, is pulsed at low for 50ms.

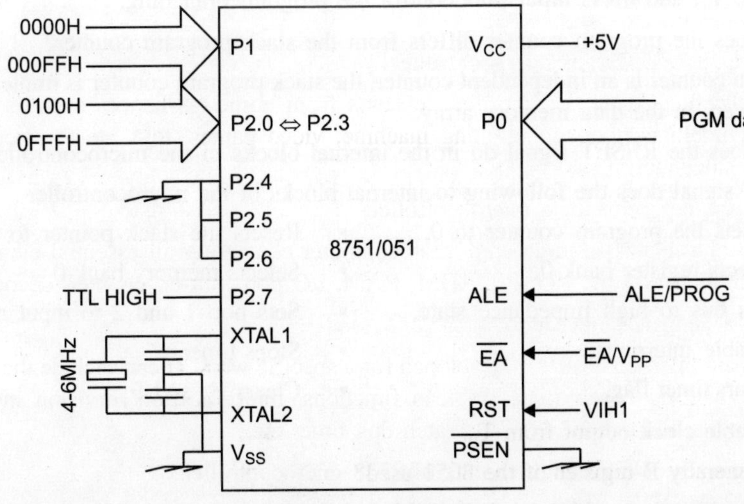

Fig. Prob. 12 Programming of the EPROM of the 8751 microcontroller

Q.13. How program verification is achieved in 8751 microcontroller?

Ans. The program verification can be done if program security bit has not been set up. The set up is the same as for the programming of the EPROM, except that the Pin2.7 is held at TTL low. The address of the program memory location to be read is applied to Port-1 and Pins P2.0⇔P2.3. The other port-2 pins and $\overline{\text{PSEN}}$ are held low. The ALE, $\overline{\text{EA}}$, and RST are held TTL high. The contents of the addressed location will come out on Port-0. The external pullups are required at Port-0 for this purpose. Figure Prob. 13 shows program verification scheme.

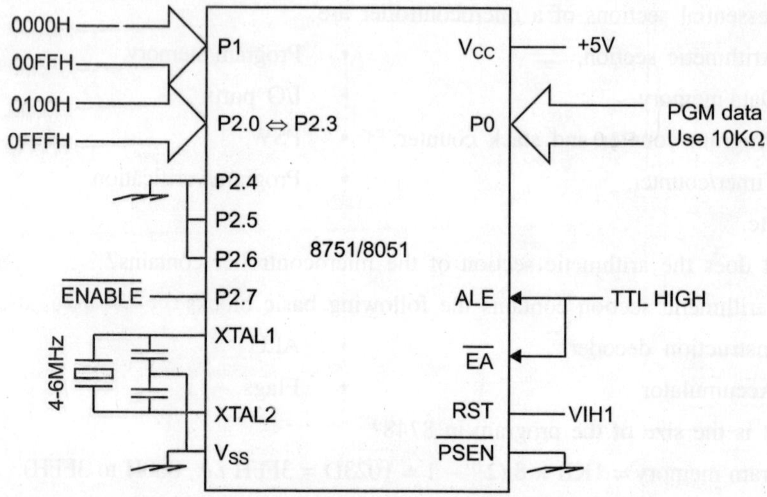

Fig. Prob. 13 Program verification of EPROM of the 8751 microcontroller

Q.14. How much external memory can be connected to the 8748 microcontroller?

Ans. 12-bit program counter will address = 2^{12} = 4KB, but 1KB × 8 (000H to 3FFH) is internally addressed, and hence the external memory address = 4FFH to FFFH.

Q.15. Which pins control the program branching?

Ans. Pins T_0, T_1, and $\overline{\text{INT}}$ input pins control the program branching.

Q.16. How does the program counter differs from the stack program counter?

Ans. Program counter is an independent counter, the stack program counter is implemented using pair of registers in the data memory array.

Q.17. What does the RESET signal do in the internal blocks of the microcontroller?

Ans. RESET signal does the following to internal blocks of the microcontroller

- Resets the program counter to 0,
- Resets the stack pointer to 0,
- Selects register bank-0,
- Selects memory bank-0
- Sets bus to high impedance state,
- Sets port-1 and 2 to input mode,
- Disable interrupts,
- Stops timer,
- Clears timer flag,
- Clears F_0 and F_1,
- Disable clock output from T_0.

Q.18. How generally B-register of the 8051 used?

Ans. It is generally for multiply and divide functions. It is used as a scratch pad register for other instructions.

Q.19. What is the function of the RS1 and RS0 bits of the PSW of the 8051 microcontroller?

Ans. It is used to select any one of the 4-register banks (bank-0, bank-1, bank-2, bank-3).

Q. 20. What is program memory security in 8751? How it is done?

Ans. The 8751 has a security bit. Once this security bit is programmed, it denies the electrical access by any external means to the on-chip memory. Figure Prob. 20 illustrates the set for it. It is similar to the set up for programming the EPROM, except that the pin P2.6 is held at TTL high. Port-0, Port-1, and Pins P2.0⇔P2.3 of Port-2 may be in any state.

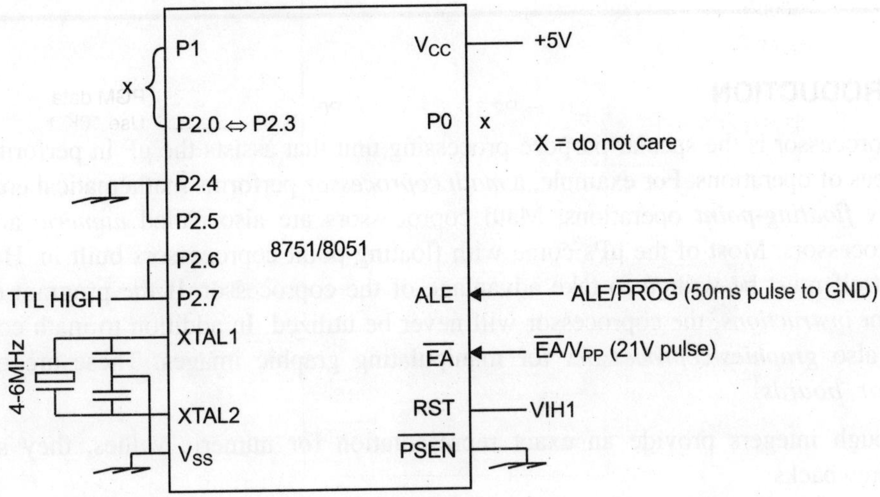

Fig. Prob. 20 Programming the security bit in the 8751 microcontroller

8

COPROCESSOR

8.1 INTRODUCTION

A coprocessor is the special-purpose-processing unit that assists the μP in performing certain special types of operations. For example, a *math coprocessor* performs mathematical computations, particularly ***floating-point*** operations. Math coprocessors are also called *numeric* and *floating-point* coprocessors. Most of the μPs come with floating-point coprocessors built in. However, the program itself must be written to take advantage of the coprocessor. If the program contains no coprocessor *instructions*, the coprocessor will never be utilized. In addition to math coprocessors, there are also *graphics-coprocessors* for manipulating graphic images. These are often called *accelerator boards*.

Although integers provide an exact representation for numeric values, they suffer from 2-major drawbacks

- the inability to represent fractional values and
- a limited dynamic range.

Floating-point arithmetic solves these 2-problems at the expense of accuracy and on some processors, speed. Most programmers are aware of the speed loss associated with floating-point arithmetic; however, they are blithely unaware of the problems with accuracy.

For many applications, the benefits of floating-point outweigh the disadvantages. However, to properly use floating-point arithmetic in any program, we must learn how floating-point arithmetic operates. Intel, understanding the importance of floating-point arithmetic in modern programs, provided support for floating-point arithmetic in the earliest designs of the 8086 - the 80x87 FPU (floating-point unit or math coprocessor). However, on processors earlier than the 80486 (or on the 80486SX), the floating-point processor was an optional device (as if this device was not present) and we were forced to simulate it in software.

All numbers in fixed-point are represented as integers or fraction. Signed integer or BCD numbers are referred to as fixed-point numbers as they do not contain any information about the position of decimal point. In fixed-point representation, the decimal point is assumed either at the extreme left or extreme right of the number. If the decimal point is assumed at the extreme left, all numbers are positive or negative fraction. Similarly, if the decimal point is chosen at the extreme right, all numbers are either positive or negative integers.

Example

$$120.364 = 0.120364\times10^{3} = \text{Fraction representation}$$
$$= 120364\times10^{-3} = \text{Integer representation}$$

Fractions of large decimal places are always used in scientific calculations. Hence, some system should keep track of the decimal place to give the correct result. This system of representation is called floating-point representation. A number having *both integer* as well as *fractional parts* is also called *real* or ***floating-point*** number. A real number may either be positive or negative.

Example

Numbers such as 125.525, 0.236, –125.25, –0.2525 etc are positive and negative real numbers. The example of positive and negative binary numbers are 110.101, 0.101, –101.001, –0.001 etc. The binary floating-point number is represented as a bit string characterized by 3-components

- a sign,
- a signed integer exponent and
- a significand.

The **sign bit** is represented by '*s*', identifies positive when *s* = 0 and negative when *s* = 1. The **exponent** denoted by '*e*', signifies the powers to which 2 is raised in determining the magnitude of the represented number. The **significand**, denoted by 1.F, means an explicit or implicit leading bit/ to the left of its binary point and a fraction field F to the right of the binary point. This fraction communicates the precision of the number. The more significand bits the fraction contains, the better the precision in representing a number. Similarly, the more bits in the exponent, the wider the dynamic range of numbers being represented. Therefore, the exponent sets off the range, while the fraction notes the exact value of the number within that range.

A number having both integer as well as fractional parts can be represented in many ways.

Example

$$125.525 = 12.5525\times10^{1} = 1.25525\times10^{2} = .125525\times10^{3} \;\text{etc}\; = m\times r^{e}$$
$$\uparrow \qquad \uparrow \qquad \uparrow \qquad \uparrow$$

Here, m ⇒ mantissa, r ⇒ radix and e ⇒ exponent

As in the above example, the decimal point indicated by arrows in mantissa (*m*) *floats around by adjusting the magnitude of the exponent* (*e*)*, this notation is called floating-point representation.*

Since the scientific applications require very large integer as well as very large decimal places of the fractional numbers, the integer and the fractional parts may be considered separately for arithmetic operation and carry and borrow may be accounted after arithmetic operations are performed. Scientists have represented the number as multiplication factor of fractional part in some powers of 10. The fractional part is called ***mantissa*** and the **multiplication factor** of the fraction that is some powers of 10 is called ***exponent***.

Example

$$125.525 = 0.125525\times10^3 = \underbrace{\overset{\text{Sign}}{0}\quad 125525}_{\text{mantissa}}\ \underbrace{\overset{\text{Sign}}{1}\quad 03}_{\text{exponent}}$$

A zero in the leftmost position of the mantissa indicates positive (+) sign. Though most of computers use floating point numbers, there are computers which use integer number system.

In integer number system, the number $125.525 = 125525\times10^{-3} = \underbrace{\overset{\text{Sign}}{1}\quad .125525}_{\text{mantissa}}\ \underbrace{\overset{\text{Sign}}{0}\quad 03}_{\text{exponent}}$

Negative fixed-point number system $-125.525 = -0.125525\times10^3 = \underbrace{\overset{\text{Sign}}{0}\quad 125525}_{\text{mantissa}}\ \underbrace{\overset{\text{Sign}}{1}\quad 03}_{\text{exponent}}$

The fractional negative number $-0.5253 = -5253\times10^{-4} = \underbrace{\overset{\text{Sign}}{1}\quad .5253}_{\text{mantissa}}\ \underbrace{\overset{\text{Sign}}{1}\quad 04}_{\text{exponent}}$

The floating-point binary is also represented in the same manner.

Example

$$1001.11101 = 0.100111101\times2^4 = \underbrace{\overset{\text{Sign}}{0}\quad .100111101}_{\text{mantissa}}\ \underbrace{\overset{\text{Sign}}{0}\quad 000100}_{\text{exponent}}$$

The mantissa occupies 9-bits (including the sign bit) and the exponent 7-bits (including sign bit). The decimal point is not physically placed in the register, but assumed to be present.

Floating-point number

The operation of removing all leading zeros by shifting the significand left (until the leading significand bit becomes 1) and decrementing the exponent accordingly, is called normalization. In other words, if the most significant bit of the mantissa is nonzero, the floating-point number is called normalized. Any non-zero real number can be represented as normalized floating-point number form as

$$(-1)^s 2^e \left(\ell.\text{F}\right)$$

Example

0.000523×10^{-3} and 0.0125525×10^4 are not normalized numbers.

0.523×10^{-6} and 0.125525×10^3 are corresponding normalized numbers.

Similarly, the binary numbers 0.1101×2^6 and 0.10011×2^{-7} are normalized number whereas 0.0010110×2^8 and 0.0011101×2^{-5} are non-normalized numbers.

Unnormalized number

A number whose leading significand bit is a 0 is called an unnormalized number. For add/subtract operation, the unnormalizing is done if the exponents of the 2-operands X and Y are not equal. In such circumstances, the binary points of X and Y operands are shifted by unnormalizing the operands with the smaller exponent (i.e., shifting its signficand to the right and incrementing its exponent) until the 2-exponents become equal. In general, this right shift is done up to '*n*' times where *n* is the number of bits in the significand. If the right shift requirement is more than '*n*', then 2-operands are considered to be incommensurate and the result takes the value of operand with the larger exponent.

Denormalized number

The result of a floating-point is delivered to a **destination** that may either be explicitly designed by the user or implicitly supplied by the system. The destination format shall be at least as wide as the operand's format. The result also may be denormalized. To denormalize a binary floating-point result, its significand is shifted to the right while its exponent is decremented until it reaches that of the smallest normalized number representable in the destination. Denormalization on the result is performed before rounding.

Stored number range

The range of stored number depends on the number of bits present in the mantissa as well as in the exponent. Out of 16-bits, 9-bits are used for storing the mantissa including the sign bit and the 7-bits for storing the exponent including sign bit. The maximum value of the number that can be stored is shown in Fig. 8.1(*a*). The maximum value can be expressed as

0.11111111E0111111.

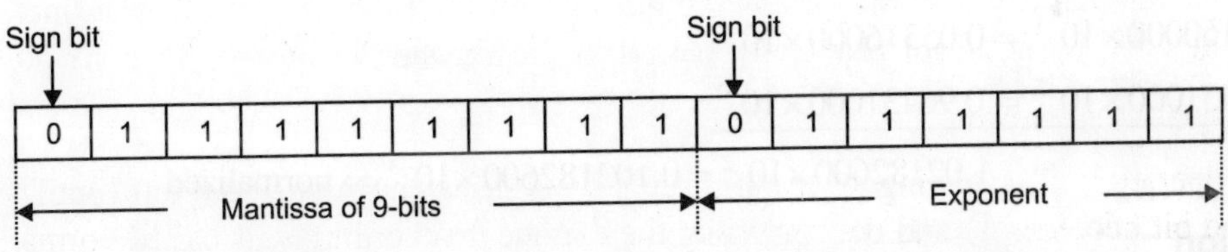

Fig. 8.1(*a*) 16-bit word maximum value of stored number

Here, E stands for the exponent. A zero (0) on the right side of E indicates that the exponent is positive.

$$\text{The maximum value} = (1-2^{-8})(2^{2^6-1}) = (1-2^{-8})(2^{63}) = (2^{63}-2^{55}) \cong 2^{63} = 10^{19}$$

The minimum value of the number that can be stored is shown in Fig. 8.1(*b*). The minimum value of the number = 0.10000000E1111111. Bit 1 on right side of E indicates that the exponent is negative.

$$\text{The minimum value of the number} = (2^{-1})\{2^{-(2^6-1)}\} = (2^{-64}) = 10^{-19}$$

Thus, the range of number for 16-bit word = 2^{-64} to 2^{63} or 10^{-19} to 10^{19}.

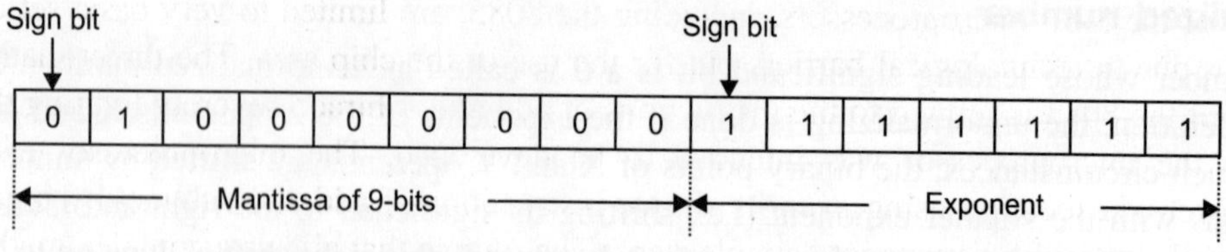

Fig. 8.1(*b*) 16-bit word minimum value of number

Similarly, in 32-bit word, 24-bit will be used for the mantissa and 8-bits for the exponent.

$$\text{The maximum value} = (1-2^{-23})(2^{2^7-1}) = (1-2^{-23})(2^{128-1}) = (1-2^{-23})(2^{127})$$

$$= (2^{127} - 2^{127}\times 2^{-23}) = 2^{127} = 10^{38}$$

$$\text{The minimum value} = (2^{-1})\{2^{-(2^7-1)}\} = (2^{-1})\{2^{-127}\} = 2^{-128} = 10^{-38}$$

Thus, the range of number for 32-bit word $= 2^{-128}$ to 2^{127} or 10^{-38} to 10^{38}.

Floating-point arithmetic

The *floating-point arithmetic* operations are ***complicated*** w.r.t. fixed-point arithmetic. The *circuitries* are also *complicated* and the *execution time is more*. The exponents of 2-numbers must be made equal for addition. This process of making the exponents equal is known as scaling or alignment of numbers.

Addition

Add 0.43560000×10^{-4}, 0.53160000×10^{-3}, and 0.96431000×10^{-2} rounded to 6.

$0.43560000\times10^{-4} = 0.00435600\times10^{-2}$

$0.53160000\times10^{-3} = 0.05316000\times10^{-2}$

$0.96431000\times10^{-2} = 0.96431000\times10^{-2}$

$1.02182600\times10^{-2} = 0.102182600\times10^{-3} \Rightarrow$ normalized

Subtraction

Subtract 0.16436100×10^{3} from 0.46373900×10^{6}

$0.46373900\times10^{6} = 0.46373900\times10^{6}$

$0.16436100\times10^{3} = 0.00016436\times10^{6}$

0.46357464×10^{6}

Multiplication

Multiply 0.23410000×10^{8} and 0.1307000×10^{-6}

0.23410000×10^{8}

$\times 0.1307000\times10^{-6}$

0.03059687×10^{-2}

Most of the 8-bit microprocessors, including the 8085, are limited to very basic set of simple instructions due to technological barriers put for the use of the chip area. The direct mathematical capability of the 8085 is limited to the instructions of add and subtract, severely limiting the power for which the microprocessor was intended to be developed. The microprocessor exhibits its greater power of execution using even its simple instructions of add and subtract by breaking the complex mathematical functions into simple steps. The routine that allows a complex mathematical function to be broken into simple steps for calculation is called **algorithm**. In order to create an algorithm, the function must be defined in a set of simple steps of the type that the particular microprocessor can execute. For an instance, the mathematical function 'Sinx' is expressed as

$$\text{Sin}x = x - \frac{x^3}{3!} + \frac{x^5}{5!} - \frac{x^7}{7!} + \frac{x^9}{9!} - ....$$

Since, the function 'Sin*x*' is expressed as an infinite series, the emulation must be carried out indefinitely to get the exact result of the Sinx for any value of the angle. The indefinite calculation can not be carried out, no matter what the speed of the microprocessor. The series has to be truncated at some point that represent the close approximation of the exact value. Thus, the microprocessor may approximate the infinite series by performing a large number of multiplication and division operations that in turn are algorithms consisting of the simple add and subtract functions. Thus, the 8-bit microprocessors do neither include hardware nor firmware for performing scientific computations such as floating-point arithmetic, matrix manipulations, graphical data processing etc. Therefore, users have to write separate programs compatible with the 8-bit microprocessors. However, for high-speed calculation of complicated mathematical function under program emulation can not approach the real time on 8-bit microprocessors. Thus, we know that many real word problems can not be solved with integer arithmetic and 4-basic math functions such as add, subtract, multiply and divide provided by even 16-bit microprocessor (8086). In some application, execution of these types of functions become excessive. Hence, a dedicated processor must be used for such complex mathematical function called **co-processor**.

Thus, we assume that for high-speed computation of complex mathematical functions, a chip is built which works as the companion of a particular microprocessor. Only the companion (co-processor) processor must interpret each encoded instruction of the host processor. After receiving any encoded instruction from the host processor, the coprocessor start working independently. This processing can also be assumed as parallel processing for 2-processors : the host and the companion. The host processor demands the value of a complex mathematical function and proceeds on the same routine assuming that the companion (co-processor) will provide the value in the real time. The high-end microprocessors such as 80386, 80486, Pentium, 68020 etc. have incorporated the coprocessors on the chip itself.

The coprocessor is capable of communicating with the microprocessor through protocols called coprocessor interface. As it performs the complex mathematical functions, the coprocessor integrates additional special purpose instructions, registers and operand on the chip.

8.2 IEEE FLOATING-POINT FORMATS

When Intel planned to introduce a floating-point coprocessor for their new 8086 microprocessor, they were smart enough to realize that the electrical engineers and solid-state physicists who design chips were, perhaps, not the best people to do the necessary numerical analysis to pick the best possible binary representation for a floating-point format. So Intel went out and hired the best numerical analyst they could find to design a floating-point format for their 8087 FPU. That person then hired 2-other experts in the field and the 3-of them (Kahn, Coonan and Stone) designed Intel's floating-point format. They did such a good job designing the KCS floating-point standard that the IEEE organization adopted this format for the IEEE floating-point format.

To handle a wide range of performance and accuracy requirements, Intel actually introduced 3-floating-point formats

- single precision,
- double precision and
- extended precision.

The single and double precision formats corresponded to C's float and double types or FORTRAN's real and double precision types. Intel intended to use extended precision for long chains of computations. Extended precision contains 16 extra bits that the calculations could use for guard bits before rounding down to a double precision value when storing the result.

The single precision format uses a 1's complement 24-bit mantissa and an 8-bit excess-128 exponent. The mantissa usually represents a value between 1.0 to just under 2.0. The H.O. bit of the mantissa is always assumed to be 1 and represents a value just to the left of the binary point. The remaining 23 mantissa bits appear to the right of the binary point. Therefore, the mantissa represents the value

1.nnnnnnn nnnnnnnn nnnnnnnn

The 'nnnnnn.....' characters represent the 23-bits of the mantissa. Keep in mind that we are working with binary numbers here. Therefore, each position to the right of the binary point represents a value (zero or one) times a successive negative power of 2. The implied 1-bit is always multiplied by 2^0, which is 1. This is why the mantissa is always ≥ 1. Even if the other mantissa bits are all 0, the implied 1-bit always gives us the value 1. Of course, even if we had an almost infinite number of 1-bit after the binary point, they still would not add up to 2. This is why the mantissa can represent values in the range of one to just under 2.

Although there are an infinite number of values between 1 and 2, we can only represent 8 million of them because we use 23-bit mantissa (the 24th bit is always one). This is the *reason* for *inaccuracy in floating-point arithmetic* → we are limited to 23-bits of precision in computations involving single precision floating-point values.

The mantissa uses a 1's complement format rather than 2's complement. This means that the 24-bit value of the mantissa is simply an unsigned binary number and the sign bit determines whether that value is positive or negative. The 1's complement numbers have the unusual property that there are 2 representations for zero (with the sign bit set or clear). Generally, this is important only to the person designing the floating-point software or hardware system. We will assume that the value zero always has the sign bit clear.

In order to represent values outside the range 1.0 to just under 2.0, the exponent portion of the floating-point format comes into play. The floating-point format raises 2 to the power specified by the exponent and then multiplies the mantissa by this value. The exponent is 8-bits and is stored in an excess-127 format. In excess-127 format, the exponent 20 is represented by the value 127 (7th). Therefore, to convert an exponent to excess-127 format simply adds 127 to the exponent value. The use of excess-127 format makes it easier to compare floating-point values. The single precision floating-point format takes the form shown in Fig. 8.2(*a*).

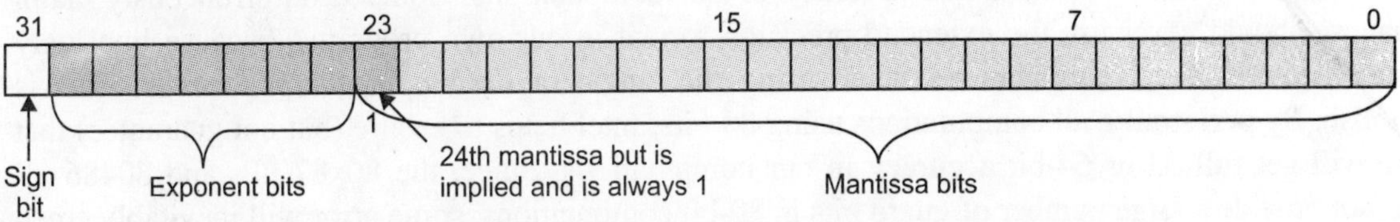

Fig. 8.2(*a*) Single precision floating-point format

With a 24-bit mantissa, we will get approximately 6-1/2 digits of precision (one half digit of precision means that the first 6-digits can all be in the range 0.9 but the seventh digit can only be in the range 0.x where x< 9 and is generally close to 5). With an 8-bit excess-128 exponent, the dynamic range of single precision floating-point numbers is approximately $2^{\pm128}$ or about $10^{\pm138}$.

Although single precision floating-point numbers are perfectly suitable for many applications, the dynamic range is somewhat small for many scientific applications and the very limited precision is unsuitable for many financial, scientific and other applications. Furthermore, in long chains of computations, the limited precision of the single precision format may introduce serious error.

The double precision formats overcome the problems of single precision floating-point. Using twice the space, the double precision format has an 11-bit excess-1023 exponent and a 53-bit mantissa (with an implied H.O. bit of one) plus a sign bit. This provides a dynamic range of about 10±308 and 14-1/2 digits of precision, sufficient for most applications. Double precision floating-point values take the form shown in Fig. 8.2(*b*).

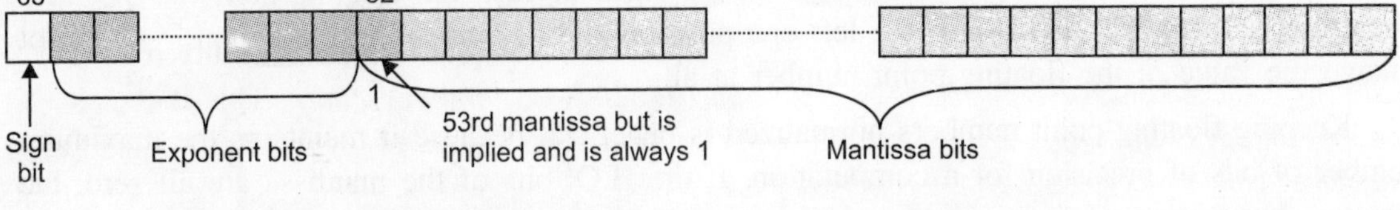

Fig. 8.2(*b*) Double precision floating-point format

In order to ensure accuracy during long chains of computations involving double precision floating-point numbers, Intel designed the extended precision format. The extended precision format uses 80-bits. Twelve of the additional 16-bits are appended to the mantissa; 4 of the additional bits are appended to the end of the exponent. Unlike the single and double precision values, the extended precision format does not have an implied H.O. bit which is always one. Therefore, the extended precision format provides a 64-bit mantissa, a 15-bit excess-16383 exponent and a 1-bit sign. The format for the extended precision floating-point value is shown in Fig. 8.2(*c*).

On the 80x87 FPUs and the 80486 μP, all computations are done using the extended precision form. Whenever we load a single or double precision value, the FPU automatically converts it to an extended precision value. Likewise, when we store a single or double precision value to memory, the FPU automatically rounds the value down to the appropriate size before storing it. By always working with the extended precision format, Intel guarantees a large number of guard bits that are present to ensure the accuracy of our computations. Some texts erroneously claim that we should never use the extended precision format in our own programs, because Intel only guarantees accurate computations when using the single or double precision formats. This is foolish. By performing all computations using 80-bits, Intel helps to ensure (but not guarantee) that we will get full 32 or 64-bit accuracy in our computations. Since, the 80x87 μPs and 80486 μP do not provide a large number of guard bits in 80-bit computations, some error will inevitably creep into the L.O. bits of an extended precision computation. However, if our computation is correct to 64-bits, the 80-bit computation will always provide at least 64 accurate bits. Most of the time we will get even more. While we cannot assume that we get an accurate 80-bit computation, we can usually do better than 64 when using the extended precision format.

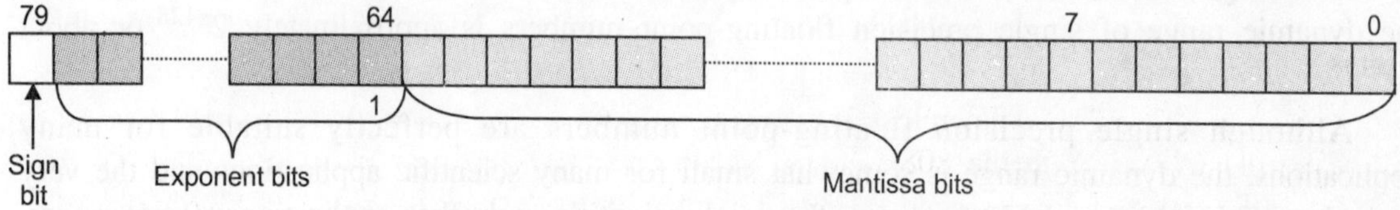

Fig. 8.2(*c*) Double precision floating-point format

In order to maintain maximum precision during computation, most computations use normalized values. A normalized floating-point value is 1 that has a H.O. mantissa bit equal to 1. Shifting the mantissa bits to the left and decrementing the exponent by 1 until a one appears in the H.O. bit of the mantissa can normalize almost any non-normalized value. Remember that the exponent is a binary exponent. Each time we increment the exponent, we multiply the floating-point value by 2. Likewise, whenever we decrement the exponent, we divide the floating-point value by 2. By the same token, shifting the mantissa to the left one-bit position multiplies the floating-point value by 2; likewise, shifting the mantissa to the right divides the floating-point value by 2. Therefore, shifting the mantissa to the left one position and decrementing the exponent does not change the value of the floating-point number at all.

Keeping floating-point numbers normalized is beneficial because it maintains the maximum number of bits of precision for a computation. If the H.O. bits of the mantissa are all zero, the mantissa has that many fewer bits of precision available for computation. Therefore, a floating-point computation will be more accurate if it involves only normalized values.

There are 2-important cases where a floating-point number cannot be normalized. The value 0.0 is a special case. Obviously it cannot be normalized because the floating-point representation for zero has no one bits in the mantissa. This, however, is not a problem since we can exactly represent the value zero with only a single bit.

The 2nd case is when we have some H.O. bits in the mantissa which are zero but the biased exponent is also zero (and we cannot decrement it to normalize the mantissa). Rather than disallow certain small values, whose H.O. mantissa bits and biased exponent are zero (the most negative exponent possible), the IEEE standard allows special denormalized values to represent these smaller values. Although the use of denormalized values allows IEEE floating-point computations to produce better results than if underflow would have occurred, keeping in mind that denormalized values offer less bits of precision and are inherently less accurate.

Since the 80x87μPs and 80486 μP always convert single and double precision values to extended precision, extended precision arithmetic is actually faster than single or double precision. Therefore, the expected performance benefit of using the smaller format is not present on these chips. However, when designing the Pentium/586 μPs, Intel redesigned built-in floating-point unit to better compete with RISC chips. Most RISC chips support a native 64-bit double precision format that is faster than Intel's extended precision format. Therefore, Intel provided native 64-bit operations on the Pentium to compete against the RISC chips. Therefore, the double precision format is the fastest on the Pentium and later chips.

8.3 8087

The semiconductor technology was not to the point where Intel could put floating-point instructions directly on the 8086 μP when it appeared in the late 1970's. Therefore, they devised a scheme whereby they could use a second chip to perform the floating-point calculations—the floating-point unit (or FPU). They released their original floating-point chip, the 8087, in 1980. This particular FPU worked with the 8086, 8088, 80186 and 80188 μPs. When Intel introduced the 80286 μP, they released a redesigned 80287 FPU chip to accompany it. Although the 80287 was compatible with the 80386 μP, Intel designed a better FPU, the 80387, for use in 80386 systems. The 80486 μP was the first Intel μP to include an on-chip floating-point unit. Shortly after the release of the 80486, Intel introduced the 80486SX μP that was an 80486 without the built-in FPU. To get floating-point capabilities on this chip, the Intel had to add an 80487 chip, although the 80487 was really nothing more than a full-blown 80486 that took over for the SX chip in the system. Intel's Pentium/586 chips provide a high-performance floating-point unit directly on the μP. There is no floating-point coprocessor available for the Pentium chip.

Collectively, we will refer to all these chips as the 80x87 FPU. Given the obsolescence of the 8086, 80286, 8087 and 80287 chips, this text will concentrate on the 80387 and later chips. There are some differences between the 80387/80486/Pentium floating-point units and earlier FPUs. If we need to write code that will execute on those earlier machines, we should consult appropriate Intel documentation for those devices.

Thus, the **8087 numeric processor extension** or 'Math Coprocessor' was developed by Intel to extend the capabilities of the 8088/8086 family of processor to include floating-point arithmetic. The 8087 is more than a true microprocessor and capable of executing special instructions in parallel with the host microprocessor.

Hardware

Physically, the original 8087 chip was designed to be installed in parallel with the 8086 bus. As such, it could monitor bus activity and intercept instruction codes intended for floating-point operations. These 8-special codes are found in the 8086 instruction set as ESC #, covering hexadecimal opcodes $D_F \Leftrightarrow D_8$. In other words, any instruction beginning with an opcode in the range $D_F \Leftrightarrow D_8$ cannot be executed by the 8086 but (provided an 8087 is installed) the math coprocessor will handle the instruction automatically. Note that this implies *there is no immediate improvement* gained by installing an 8087; only when its specific instructions are incorporated into a program will it ever come into play. Also that certain programs and packages test for the presence of an 8087 and invoke different routines accordingly; these programs create false impression that the 8087 magically speeds up general purpose computing. The pinouts and the block diagrams of the 8087 is shown in Figs. 8.3(*a*), (*b*) and (*c*). A very simple circuit of interfacing between the 8086 and 8087 is shown in Fig. 8.3(*d*).

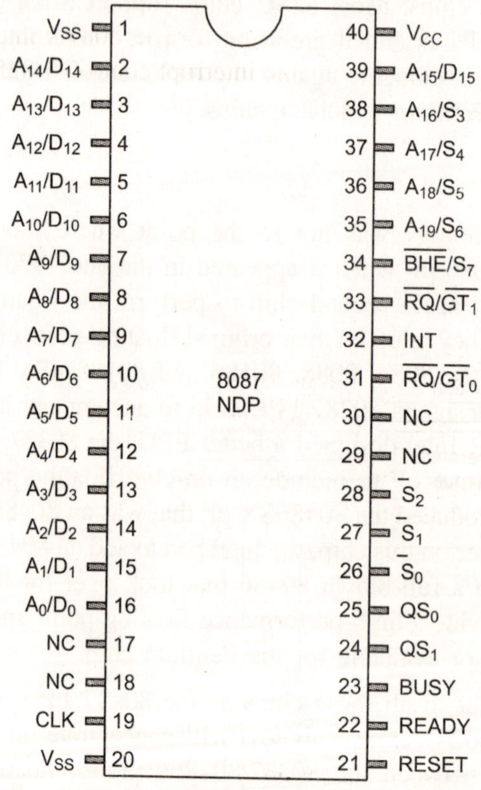

Pin	Signal	Pin	Signal
1	V_{SS}	40	V_{CC}
2	A_{14}/D_{14}	39	A_{15}/D_{15}
3	A_{13}/D_{13}	38	A_{16}/S_3
4	A_{12}/D_{12}	37	A_{17}/S_4
5	A_{11}/D_{11}	36	A_{18}/S_5
6	A_{10}/D_{10}	35	A_{19}/S_6
7	A_9/D_9	34	$\overline{BHE}/S_7$
8	A_8/D_8	33	$\overline{RQ}/\overline{GT_1}$
9	A_7/D_7	32	INT
10	A_6/D_6	31	$\overline{RQ}/\overline{GT_0}$
11	A_5/D_5	30	NC
12	A_4/D_4	29	NC
13	A_3/D_3	28	$\overline{S_2}$
14	A_2/D_2	27	$\overline{S_1}$
15	A_1/D_1	26	$\overline{S_0}$
16	A_0/D_0	25	QS_0
17	NC	24	QS_1
18	NC	23	BUSY
19	CLK	22	READY
20	V_{SS}	21	RESET

Fig. 8.3(*a*) Pinouts of the 8087 coprocessor

Software

The Intel mnemonics (instructions) for the 8087 are written to begin with the letter 'F' (no normal 8086 mnemonics begin with 'F'). For example, the mnemonic ADD specifies an 8086-based integer addition, while the mnemonic FADD selects an 8087-based floating-point addition.

The 8087 coprocessor for the 8086/8088 microprocessor is available in 40-pin DIP package using HMOS III technology. It provides additional 68-numeric processing instructions. It has 8-numbers of 80-bit register stack. It can be interfaced with 8086/8088 as well as with 80186/80188 microprocessors. It supports 7-types of data that includes 16, 32 and 64-bit integer 32, 64 and 80-bit floating-point and 18-digit BCD operands. It generates several arithmetic, trigonometric, exponential and logarithmic instructions. As the 8087 is interfaced with the 8086 externally, it provides additional register data types, instructions and control at the hardware level. Once connected, the host microprocessor and the coprocessor appears to be as single-unit to the programmers. The 8086/8088 microprocessor's status signal line $\overline{S_2} \Leftrightarrow \overline{S_0}$ and status queue lines $\overline{QS_1} \Leftrightarrow \overline{QS_0}$ enable the 8087 to monitor and decode the instructions in synchronism. For the 80186/ 80188 microprocessor system, the queue status lines of the 80186/80188 are synchronized with the 8087 by 8288 bus controller. The 8087 operates in parallel and also independently with the 8086/ 8088 and 80186/80188. A high on the BUSY pin of the 8087 indicates to the microprocessor that the 8087 is busy in executing the instruction and the microprocessor's WAIT instruction tests this signal to confirm that 8087 is now ready to execute the next instruction. The 8087 can interrupt the microprocessor when it detects an error. The 8087's interrupt register is connected to the microprocessor through the programmable interrupt controller 8259 in 8086/8088 system and INT for the 80186/80188 systems.

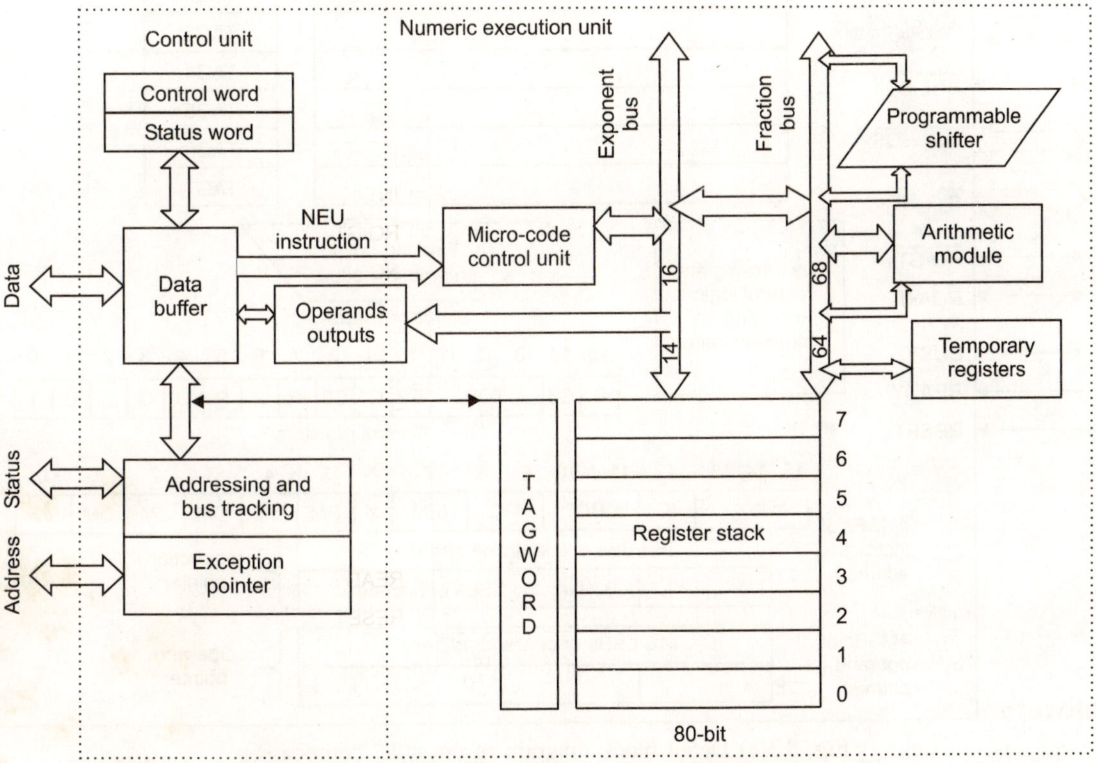

Fig. 8.3(*b*) Simplified block diagram of the 8087 coprocessor

The 8087 operates on memory operands of 7-different data types

- word,
- integer,
- short integer,
- packed BCD,
- short real,

- long real and
- temporary real.

The number of bytes, format and approximate range of each of these data types are given in Table 8.3(*a*). In the memory, the least significant byte of a number is always stored in the lowest address.

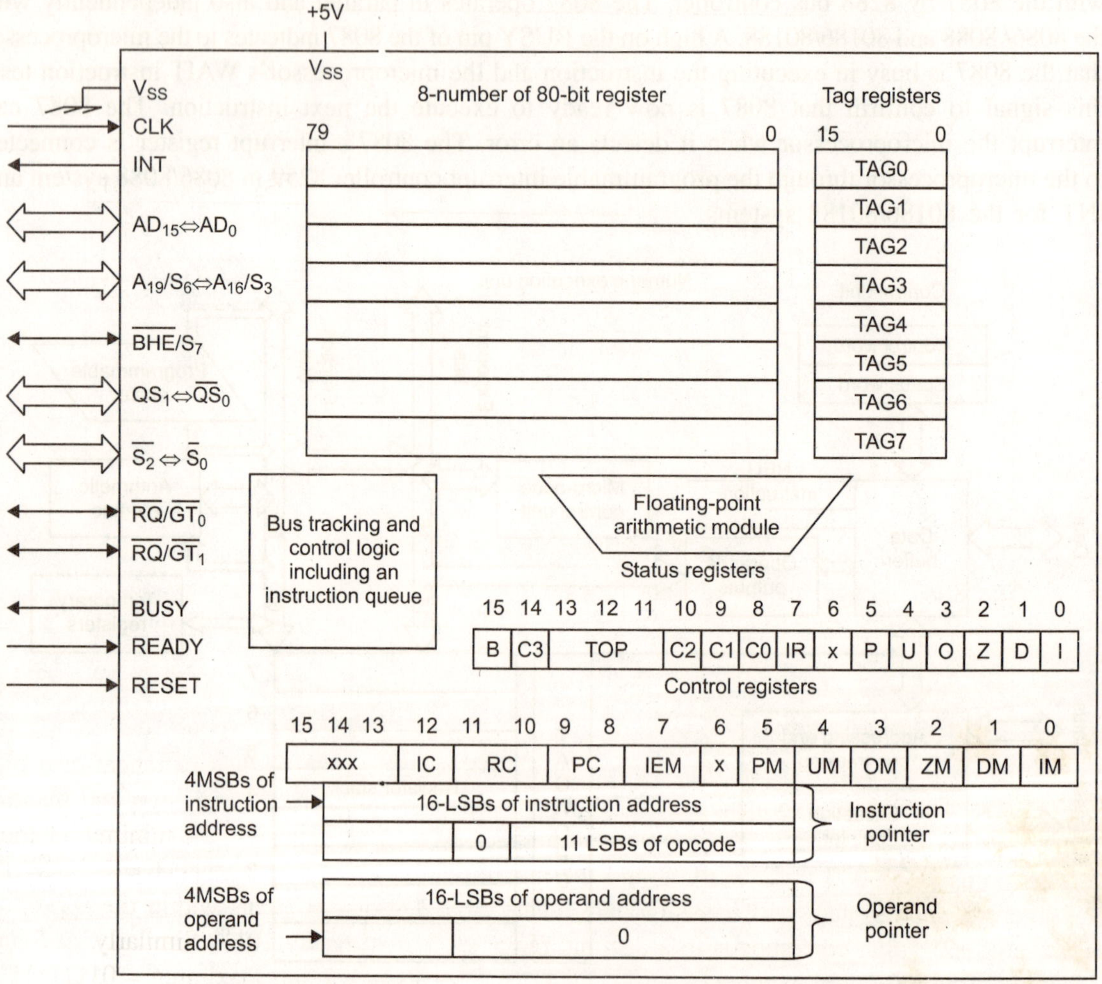

Fig. 8.3(*c*) Detail block diagram of the 8087 coprocessor

Table 8.3(*a*) Characteristics of data types

Data Type	*Bits/Bytes*	*Significant Digits (decimal)*	*Approx. Range (decimal)*
Word length	16/2	4	$-32768 \leq x \leq 32767$
Short integer	32/4	9	$-1*10^9 \leq x \leq 2*10^9$
Long integer	64/8	18	$-9*10^{18} \leq x \leq 9*10^{18}$

(Contd....)

Packed decimal	80/8	18	$-9999.... \leq x \leq 9999....$
Short Real (single precision)	32/4	6-7	$1*10^{-38}$ to $3*10^{38}$
Long Real (double precision)	64/8	15-16	$1*10^{-308}$ to 10^{308}
Temporary real	80/10	19	10^{-4932} to 10^{4932}

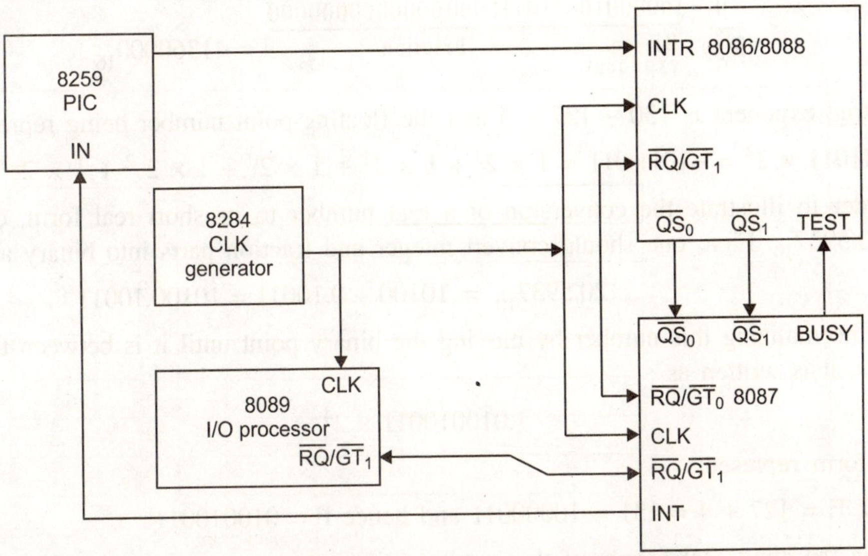

Fig. 8.3(*d*) Interfacing of 8086 with 8087

The 8087 recognizes 3-real data types

- short real,
- long real and
- temporary real.

The exponent of a binary floating-point number may be either **signed** exponent or a **biased** exponent. The signed exponent may be represented in 2's complement or in sign and magnitude notation. As an example, in 2's complement notation of an 8-bit exponent, the minimum exponent is given as $10000000 = -2^7 = -128$ and the maximum exponent is $= 01111111 = 2^7 - 1 = +127$. Therefore, such system will accommodate binary floating-point numbers within the approximate range of $\pm 2^{127}$ that corresponds to a decimal range of approximately $10^{\pm 38}$. Similarly, in S and M notation, the minimum exponent is $= 11111111 = -2^7 - 1 = -127$ and maximum $= 01111111 = 2^7 - 1 = 127$. This approximately results in the same range.

The standard notation is, however, the **biased** exponent. Some appropriate integer constant, called the biased exponent is added to all exponents in order to keep them always positive. The single precision floating-point, the IEEE standard suggests that a normalized nonzero number X is expressed as

$$(-1)^S \times 2^{e-127} \times \lambda F$$

where,

S = sign bit.

e = 8-bit exponent biased by 127.

F = 23-bit fraction that together with an implicit leading λ, yields the significand part λ.F.

Because 1 appearing before the fraction is always present, it is implied and is not physically stored. For, example, suppose that a short real number is stored as

$$\underbrace{0}_{\text{Sign}}\ \underbrace{10000010}_{\text{Biased exponent}}\underbrace{011011000000000000000000}_{\text{Fraction}} = 41360000_{16}$$

The true exponent is 130 – 127 = 3 and the floating-point number being represented as $+1.011011 \times 2^3 = +1011.011 = 1 \times 2^3 + 1 \times 2^1 + 1 \times 2^0 + 1 \times 2^{-2} + 1 \times 2^{-3} = 11.375_{10}$

In order to illustrate the conversion of a real number to its short real form, consider the number 20.5937_{10}. First, one should convert integer and fraction parts into binary as

$$20.5937_{10} = 10100 + 0.10011 = 10100.10011$$

After normalizing this number by moving the binary point until it is between the first and second bits, it is written as

$$1.010010011 \times 2^4$$

This form represents

S = 0, E = 127 + 4 = 131 = 10000011 and hence F = 010010011

And in the short real format of the number is

$$\underbrace{0}_{\text{Sign}}\ \underbrace{10000011}_{\text{Biased exponent}}\underbrace{010010011000000000000000}_{\text{Fraction}} = 41A4C000_{16}$$

For the short real data type, the valid range of the biased exponent is $0 < E < 255$. Consequently, the numbers that can be represented are from $\pm 2^{-126}$ to $\pm 2^{128}$, approximately $\pm 1\times10^{-38}$ to $\pm 3\times10^{38}$.

The long real format has 11 exponent bits and 152 fraction bits as the short real format uses 15-bits for exponent and 64-bits for the mantissa. Unlike the short and long real formats, the most significant bit in the mantissa is actually stored. Because of the extended precision (19 to 20 decimal digits), integer and packed BCD operands can be operated on internally using floating-point arithmetic and still yield exact results. A primary reason for using the temporary real format for internal data storage is to reduce the chances for overflows and underflows during a series of calculations that produce a final result that is within the required range. For example, consider the calculation

$$D \leftarrow (A * B) / C$$

where A, B and C are in the short real format. The multiply operation may yield a result that is too large to be represented in the short real format, but yet the final result, after the division by C, may still be within the valid range. The 15-bit exponent of the temporary real format extends the range of valid numbers to approximately $\pm10^{\pm4932}$. Therefore, for most applications the user need not worry about overflows and underflows in the intermediate calculations.

The 8086/8088 can operate on integers of 16-bits ranging from -2^{15} to $2^{15} - 1$. In addition to word integer format, the 8087 allows an integer operands to be represented using 32-bits (short integer) and 64-bit (long integer). Negative integers are coded in 2's complement form. The greater number of bits significantly extends the ranges of integers to -2^{-31} to $2^{31}-1$ for short integer format and -2^{-63} to $2^{63}-1$ for long range integer format. This is roughly $\pm 2 \times 10^9$ and $\pm 9 \times 10^{18}$ respectively.

In the packed BCD format, a decimal number is stored in 10-bytes. The entire most significant bit of this byte indicates whether a decimal number is positive (0) or negative (1). The remaining 9-bytes represent the magnitude with 2-BCD digits packed into each byte. Thus, the valid range of values in the packed BCD format is $-10^{18} + 1$ to $10^{18} - 1$.

The real data types (floating-point) can represent operands that varies from extremely small to extremely large values and retain a constant number of significant digits during calculation. As discussed, a real data format is divided into 3-fields : ***sign, exponent*** and ***mantissa*** i.e.,

$$X = \pm 2^{e} \times \text{mantissa}$$

The exponent adjusts the position of the binary point in the mantissa. Decreasing the exponent by 1 moves the binary point to the right by 1 position. Hence, using a negative exponent without loosing any position can represent very small number. However, except for numbers whose mantissa falls within the range of the format, a number may not be exactly representable, thus causing a roundoff error. If leading 0's are allowed, a given number may have more than 1 representation in a real number format. But, since mantissa has fixed number of bits, the leading 0's increase the roundoff error. Therefore, to minimize the roundoff, after each calculation the 8087 detects the leading 0's by properly adjusting the exponent. A nonzero real number is said to be normalized when its mantissa is in the form of 1.F, where F represents a fraction.

Normally, a bias value is added to the true exponent so that the true exponent is actually the number in the exponent field minus the bias value. Using bias exponent allows 2-normalized real numbers of the same sign to be compared by simply comparing the bytes from left to right as if they were integers.

FPU registers

The 80x87 FPUs add 13-registers to the 80386 and later processors:

- 8-floating-point data registers,
- a control register,
- a status register,
- a tag register,
- an instruction pointer and
- a data pointer.

The data registers are similar to the 80x86's general-purpose register set in so far as all floating-point calculations take place in these registers. The control register contains bits that allow us to decide how the 80x87 handles certain degenerate cases like rounding of inaccurate computations, control precision and so on. The status register is similar to the 80x86's flags register; it contains the condition code bits and several other floating-point flags that describe the state of the 80x87 chip. The tag register contains several groups of bits that determine the state of the value in each of the 8-general-purpose registers. The instruction and data pointer registers contain certain state information about the last floating-point instruction executed. We will not consider the last 3-registers in this text, see the Intel documentation for more details.

FPU data registers

The simplified programming model set of register of 8086/8087 is shown in Fig. 8.3(*e*). The monitor and control logic maintains a 6-byte instruction queue (4-byte for 8088) and tracks the instruction execution sequence of the host. If the instruction is an ESC type, the 8087 decodes the external opcode to perform the specified operation and also capture the operand address. Instructions other than ESC instruction are simply ignored by the 8087.

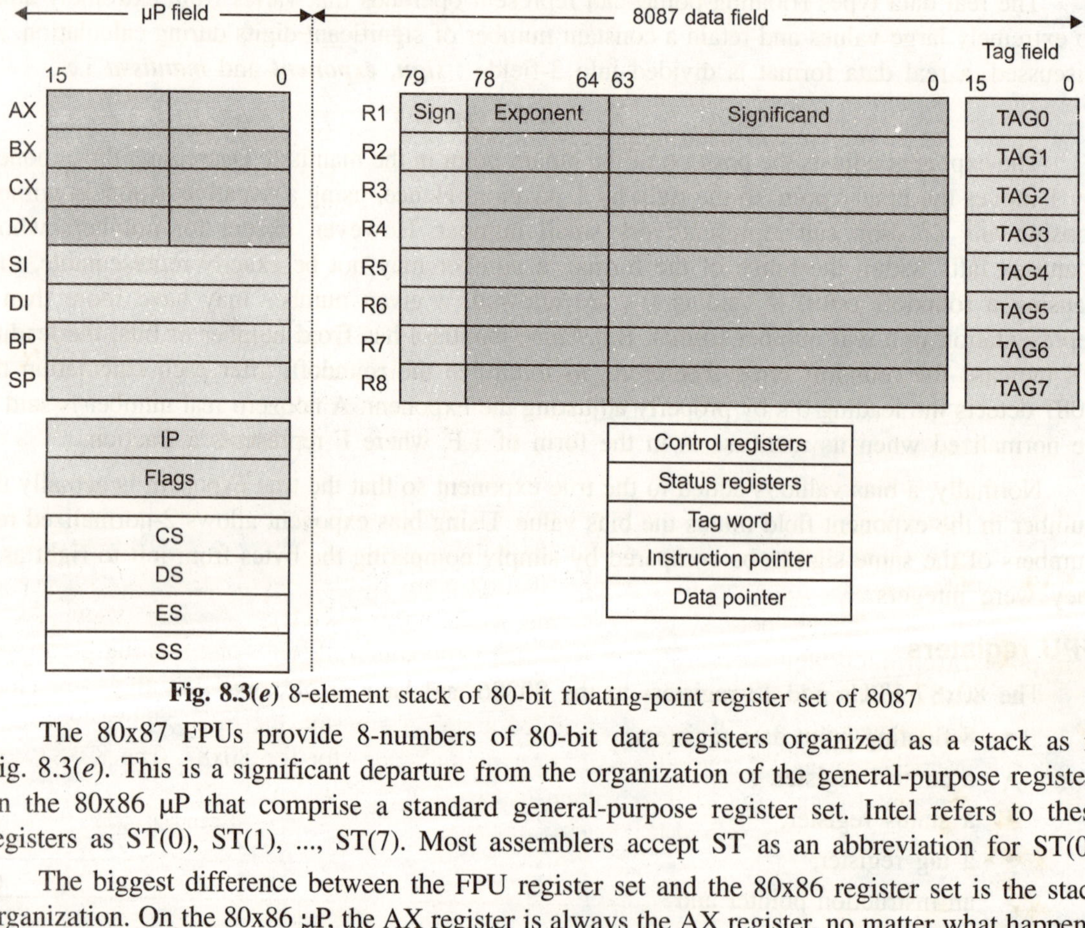

Fig. 8.3(*e*) 8-element stack of 80-bit floating-point register set of 8087

The 80x87 FPUs provide 8-numbers of 80-bit data registers organized as a stack as in Fig. 8.3(*e*). This is a significant departure from the organization of the general-purpose registers on the 80x86 μP that comprise a standard general-purpose register set. Intel refers to these registers as ST(0), ST(1), ..., ST(7). Most assemblers accept ST as an abbreviation for ST(0).

The biggest difference between the FPU register set and the 80x86 register set is the stack organization. On the 80x86 μP, the AX register is always the AX register, no matter what happens. On the 80x87, however, the register set is an 8-element stack of 80-bit floating-point values as in Fig. 8.3(*e*). ST(0) refers to the item on the top of the stack, ST(1) refers to the next item on the stack and so on. Many floating-point instructions push and pop items on the stack; therefore, ST(1) will refer to the previous contents of ST(0) after we push something onto the stack. It will take some thought and practice to get used to the fact that the registers are changing under us, but this is an easy problem to overcome.

Using an index to the stack pointer may reference each register. This is called relative stack addressing as indicated in Fig. 8.3(*f*). Any register can be at the top of the stack at any time. The top of the stack is indicated by ST(0). The push operation loads the data at the top of the stack from the memory. The data is loaded now sits atop the stack, with each of the other registers pushed down by one position.

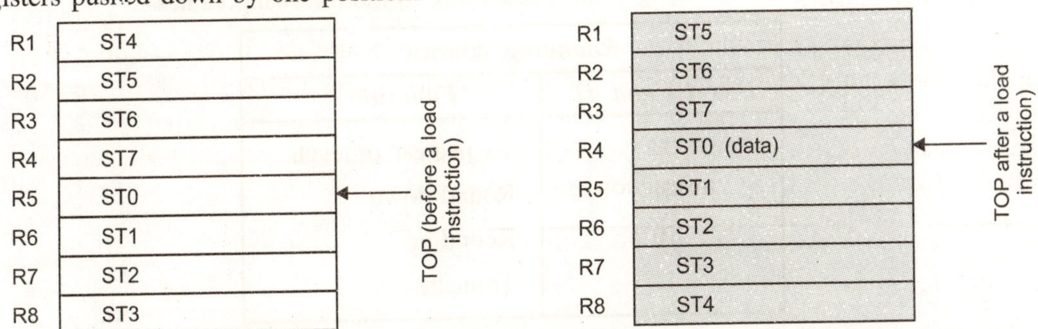

Fig. 8.3(*f*) 8-data register of the 8087 arranged for pushed down

The data popped from the stack causes the contents of ST(0) to be transferred to memory and all registers are moved up by one position. That is, ST(1) becomes ST(0) and ST(2) becomes ST(1) and so on.

FPU control register

When Intel designed the 80x87 (and essentially, the IEEE floating-point standard), there were no standards in floating-point hardware. Different (mainframe and mini) computer manufacturers all had different and incompatible floating-point formats. Unfortunately, many application software had been written taking into account the idiosyncrasies of these different floating-point formats. Intel wanted to designed an FPU that could work with the majority of the software out there (keep in mind, the IBM PC was 3-to-4 years away when Intel began designing the 8087), they could not rely on that 'mountain' of software available for the PC to make their chip popular. Unfortunately, many of the features found in these older floating-point formats were mutually exclusive. For example, in some floating-point systems rounding would occur when there was insufficient precision; in others, truncation would occur. Some applications would work with one floating-point system but not with the other. Intel wanted as many applications as possible to work with as few changes as possible on their 80x87 FPUs, so they added a special register, the FPU control register, that lets the user choose one of several possible operating modes for the 80x87. The 80x87 control register contains 16-bits organized as shown in Fig. 8.3(*g*).

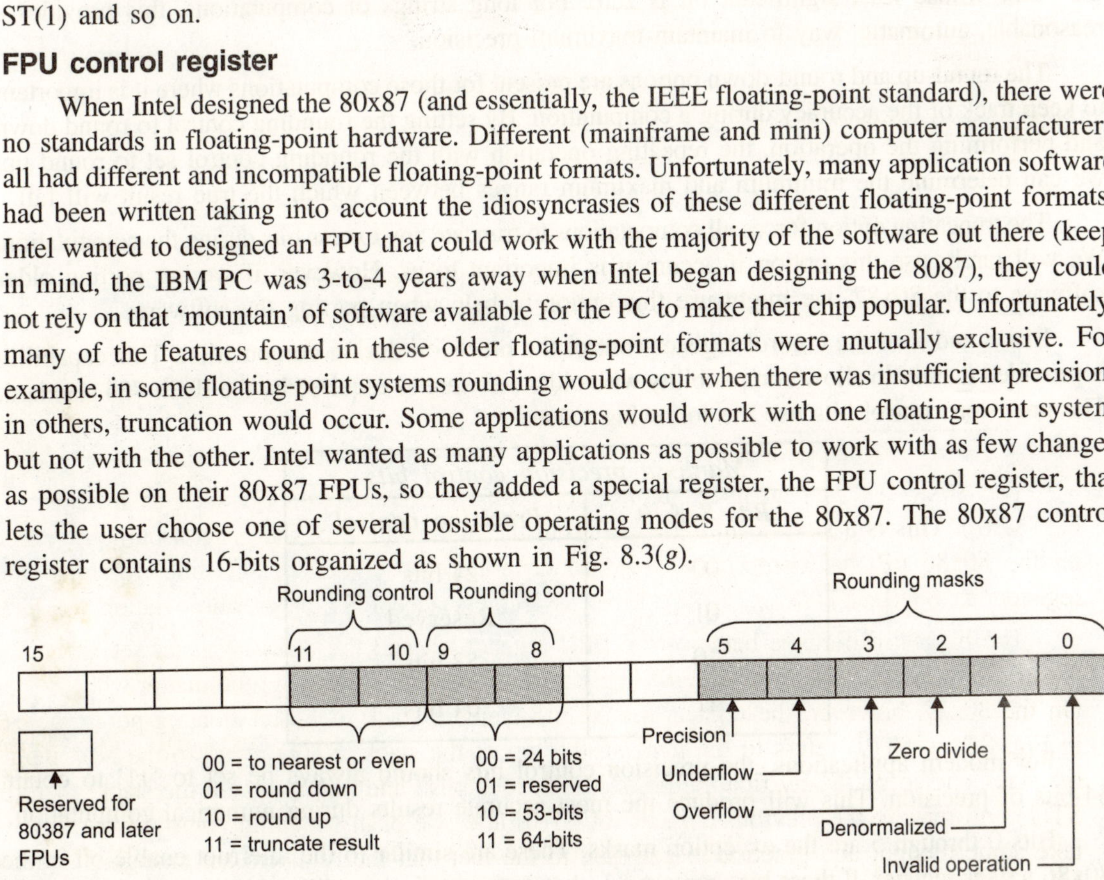

Fig. 8.3(*g*) Control register of 8087

Bit-12 of the control register is only present on the 8087 and 80287 chips. It controls how the 80x87 responds to infinity. The 80387 and later chips always use a form of infinitely known and a fine closure because this is the only form supported by the IEEE 754/854 standards. As such, we will ignore any further use of this bit and assume that it is always programmed with a 1.

Bits-10 and 11 provide rounding control according to the following values

Rounding control	
Bits 10 and 11	*Function*
00	To nearest or even
01	Round down
10	Round up
11	Truncate

The '00' setting is the default. The 80x87 rounds values above one-half of the least significant bit up. It rounds values below one-half of the least significant bit down. If the value below the least significant bit is exactly one-half the least significant bit, the 80x87 rounds the value towards the value whose least significant bit is zero. For long strings of computations, this provides a reasonable, automatic, way to maintain maximum precision.

The round-up and round-down options are present for those computations where it is important to keep track of the accuracy during a computation. By setting the rounding control to round down and performing the operation, the repeating operation with the rounding control set to round up, we can determine the minimum and maximum ranges between which the true result will fall.

The truncating option forces all computations to truncate any excess bits during the computation. We will rarely use this option if accuracy is important to us. However, if we are porting older software to the 80x87, we might use this option to help when porting the software.

Bits-8 and 9 of the control register control the precision during computation. This capability is provided mainly to allow compatibility with older software as required by the IEEE 754 standard. The precision control bits use the following values

Mantissa precision control bits	
Bits 8 & 9	*Precision control*
00	24 bits
01	Reserved
10	53 bits
11	64 bits

For modern applications, the precision control bits should always be set to '11' to obtain 64-bits of precision. This will produce the most accurate results during numerical computation.

Bits-0 through 5 are the exception masks. These are similar to the interrupt enable bit in the 80x86's flags register. If these bits contain a 1, the corresponding condition is ignored by the 80x87

FPU. However, if any bit contains 0 and the corresponding condition occurs, then the FPU immediately generates an interrupt so the program can handle the degenerate condition.

Bit-0 corresponds to an invalid operation error. This generally occurs as the result of a programming error. Problem which raise the invalid operation exception include pushing more than 8-items onto the stack or attempting to pop an item off an empty stack, taking the square root of a negative number or loading a non-empty register.

Bit-1 masks the denormalized interrupt that occurs whenever we try to manipulate denormalized values. Denormalized values generally occur when we load arbitrary extended precision values into the FPU or work with very small numbers just beyond the range of the FPU's capabilities. Normally, we would probably not enable this exception.

Bit-2 masks the 0 divide exception. If this bit contains 0, the FPU will generate an interrupt if we attempt to divide a nonzero value by 0. If we do not enable the 0 division exception, the FPU will produce NaN (not a number) whenever we perform a 0 division.

Bit-3 masks the overflow exception. The FPU will raise the overflow exception if a calculation overflows or if we attempt to store a value that is too large to fit into a destination operand (e.g., storing a large extended precision value into a single precision variable).

Bit-4, if set, masks the underflow exception. Underflow occurs when the result is too small to fit in the destination operand. Like overflow, this exception can occur whenever we store a small-extended precision value into a smaller variable (single or double precision) or when the result of a computation is too small for extended precision.

Bit-5 controls whether the precision exception occurs. A precision exception occurs whenever the FPU produces an imprecise result, generally the result of an internal rounding operation. Although many operations will produce an exact result, many more will not. For example, dividing one by ten will produce an inexact result. Therefore, this bit is usually one since inexact results are very common.

Bits-6 and 13-through-15 in the control register are currently undefined and reserved for future use. Bit-7 is the interrupt enable mask, but it is only active on the 8087 FPU; a 0 in these bit enables 8087 interrupts and a 1-disables FPU interrupts.

The 80x87 provides 2-instructions, FLDCW (load control word) and FSTCW (store control word), that allows us to load and store the contents of the control register. The single operand to these instructions must be a 16-bit memory location. The FLDCW instruction loads the control register from the specified memory location, FSTCW stores the control register into the specified memory location.

FPU status register

The FPU status register provides the status of the coprocessor at the instant we read it. The FSTSW instruction stores the16-bit floating-point status registers into the mod/reg/rm operand. The status register is a 16-bit register. Its layout is shown in Fig. 8.3(*h*).

Bits-0 through 5 are the exception flags. These bits are appearing in the same order as the exception masks in the control register. If the corresponding condition exists, then the bit is set. These bits are independent of the exception masks in the control register. The 80x87 sets and clears these bits regardless of the corresponding mask setting.

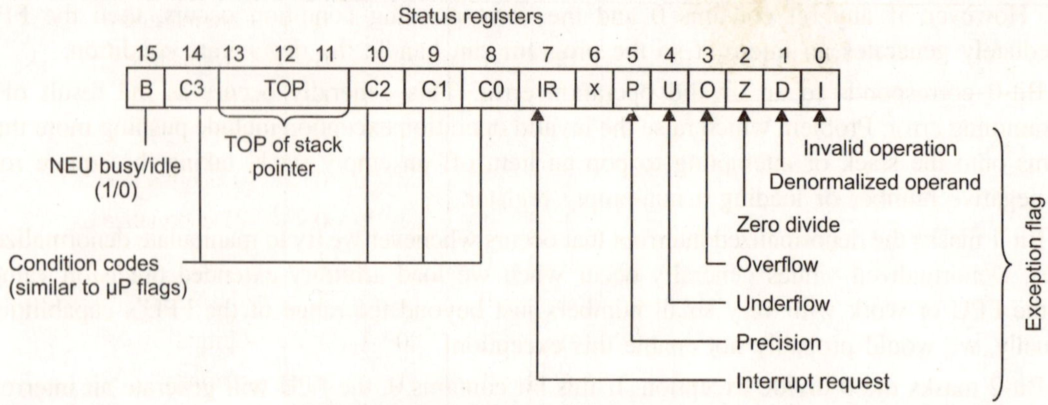

Fig. 8.3(*h*) FPU status register

Bit-6 (active only on 80386 and later processors) indicates a stack fault. A stack fault occurs whenever there is a stack overflow or underflow. When this bit is set, the C1 condition code bit determines whether there was a stack overflow (C1 = 1) or stack underflow (C1 = 0) condition.

Bit-7 of the status register is set if any error condition bit is set. It is the logical OR of bits-0 through-5. A program can test this bit to quickly determine if an error condition exists.

Bits 8, 9, 10 and 14 are the coprocessor condition code bits. Various instructions set the condition code bits as shown in Table 8.3(*b*).

Bits 11-13 of the FPU status register provide the register number of the top of stack. During computations, the 80x87 adds (modulo 8) the logical register numbers supplied by the programmer to these 3-bits to determine the physical register number at run time.

Bit-15 of the status register is the busy bit. It is set whenever the FPU is busy. Most programs will have little reason to access this bit.

Table 8.3(*b*) FPU condition code bits

Instruction	*Condition code bits*				
	C3	*C2*	*C1*	*C0*	
FCOM, FCOMP, FCOMPP, FICOM, FICOMP	0	0	x	0	ST > source
	0	0	x	1	ST < source
	1	0	x	0	ST = source
	1	1	x	1	ST or source undefined
FTST	0	0	x	0	ST is positive
	0	0	x	1	ST is negative
	1	0	x	0	ST is zero (+ or –)
	1	1	x	1	ST is incomparable
FXAM	0	0	0	0	+ Unnormalized
	0	0	1	0	– Unnormalized

(*Contd....*)

	0	1	0	0	+ Normalized
	0	1	1	0	– Normalized
	1	0	0	0	+ 0
	1	0	1	0	– 0
	1	1	0	0	+ Denormalized
	1	1	1	0	— Denormalized
	0	0	0	1	+ NaN
	0	0	1	1	– NaN
	0	1	0	1	+ Infinity
	0	1	1	1	– Infinity
	1	x	x	1	Empty register
FUCOM, FUCOMP, FUCOMPP	0	0	x	0	ST > source
	0	0	x	1	ST < source
	1	0	x	1	ST = source
	1	1	x	1	Unorder
	X = Do not care				

Condition code interpretation

Instruction(s)	*C0*	*C3*	*C2*	*C1*
FCOM, FCOMP, FCMPP, FTST, FUCOM, FUCOMP, FUCOMPP, FICOM, FICOMP	Result of comparison. See table above.	Result of comparison. See table above.	Operand is not comparable.	Result of comparison (see table above) or stack overflow/underflow (if stack exception bit is set).
FXAM	See previous table.	See previous table.	See previous table.	Sign of result, or stack overflow/underflow (if stack exception bit is set).
FPREM, FPREM1	Bit 2 of remainder	Bit 0 of remainder	0-reduction done. 1- reduction incomplete.	Bit 1 of remainder or stack overflow/underflow (if stack exception bit is set).
FIST, FBSTP, FRNDINT, FST, FSTP, FADD, FMUL, FDIV, FDIVR, FSUB, FSUBR, FSCALE, FSQRT, FPATAN, F2XM1, FYL2X, FYL2XP1	Undefined	Undefined	Undefined	Round up occurred or stacks overflow/underflow (if stack exception bit is set).
FPTAN, FSIN, FCOS, FSINCOS	Undefined	Undefined	0-reduction done. 1- reduction incomplete.	Round up occurred or stacks overflow/underflow (if stack exception bit is set).

(*Contd....*)

FCHS, FABS, FXCH, FINCSTP, FDECSTP, CONSTANT LOADS, FXTRACT, FLD, FILD, FBLD, FSTP (80 BIT)	Undefined	Undefined	Undefined	Zero result or stack over flow/underflow (if stack exception bit is set).
FLDENV, FSTOR	Restored from memory operand.	Restored from memory operand.	Restored from memory operand.	Restored from memory operand.
FLDCW, FSTENV, FSTCW, FSTSW, FCLEX	Undefined	Undefined	Undefined	Undefined
FINIT, FSAVE	Cleared to zero.	Cleared to zero.	Cleared to zero.	Cleared to zero.

Bits 11-13 of the FPU status register provide the register number of the top of stack. During computations, the 80x87 adds (modulo 8) the logical register numbers supplied by the programmer to these 3-bits to determine the physical register number at run time.

Bit-15 of the status register is the busy bit. It is set whenever the FPU is busy. Most programs will have little reason to access this bit.

FPU data types

The 80x87 FPU supports 7-different data types

- 3-integer types,
- a packed decimal type and
- 3-floating-point types.

Since the 80x86 μPs already support integer data types, these are few reasons why we would want to use the 80x87 integer types. The packed decimal type provides a 17-digit signed decimal (BCD) integer. However, we are avoiding BCD arithmetic in this text, so we will ignore this data type in the 80x87 FPU. The remaining 3-data types are the 32-bit, 64-bit and 80-bit floating-point data types we have looked at so far. The 80x87 data types appear in the following Fig. 8.3(*i*).

The 80x87 FPU generally stores values in a normalized format. When a floating-point number is normalized, the H.O. bit is always one. In the 32 and 64-bit floating-point formats, the 80x87 does not actually store this bit, the 80x87 always assumes that it is 1. Therefore, 32 and 64-bit floating-point numbers are always normalized. In the extended precision 80-bit floating-point format, the 80x87 does not assume that the H.O. bit of the mantissa is 1, the H.O. bit of the number appears as part of the string of bits.

Normalized values provide the greatest precision for a given number of bits. However, there are a large number of non-normalized values that we can represent with the 80-bit format. These values are very close to 0 and represent the set of values whose mantissa H.O. bit is not zero. The 80x87 FPUs support a special form of 80-bit known as denormalized values. Denormalized values allow the 80x87 to encode very small values it cannot encode using normalized values, but at a price. Denormalized values offer less bits of precision than normalized values. Therefore, using denormalized values in a computation may introduce some slight inaccuracy into a computation. Of course, this is always better than underflowing the denormalized value to 0 (which could make the computation even less accurate), but we must keep in mind that if we work with very small values we may lose some accuracy in our computations. Note that the 80x87 status register contains a bit we can use to detect when the FPU uses a denormalized value in a computation.

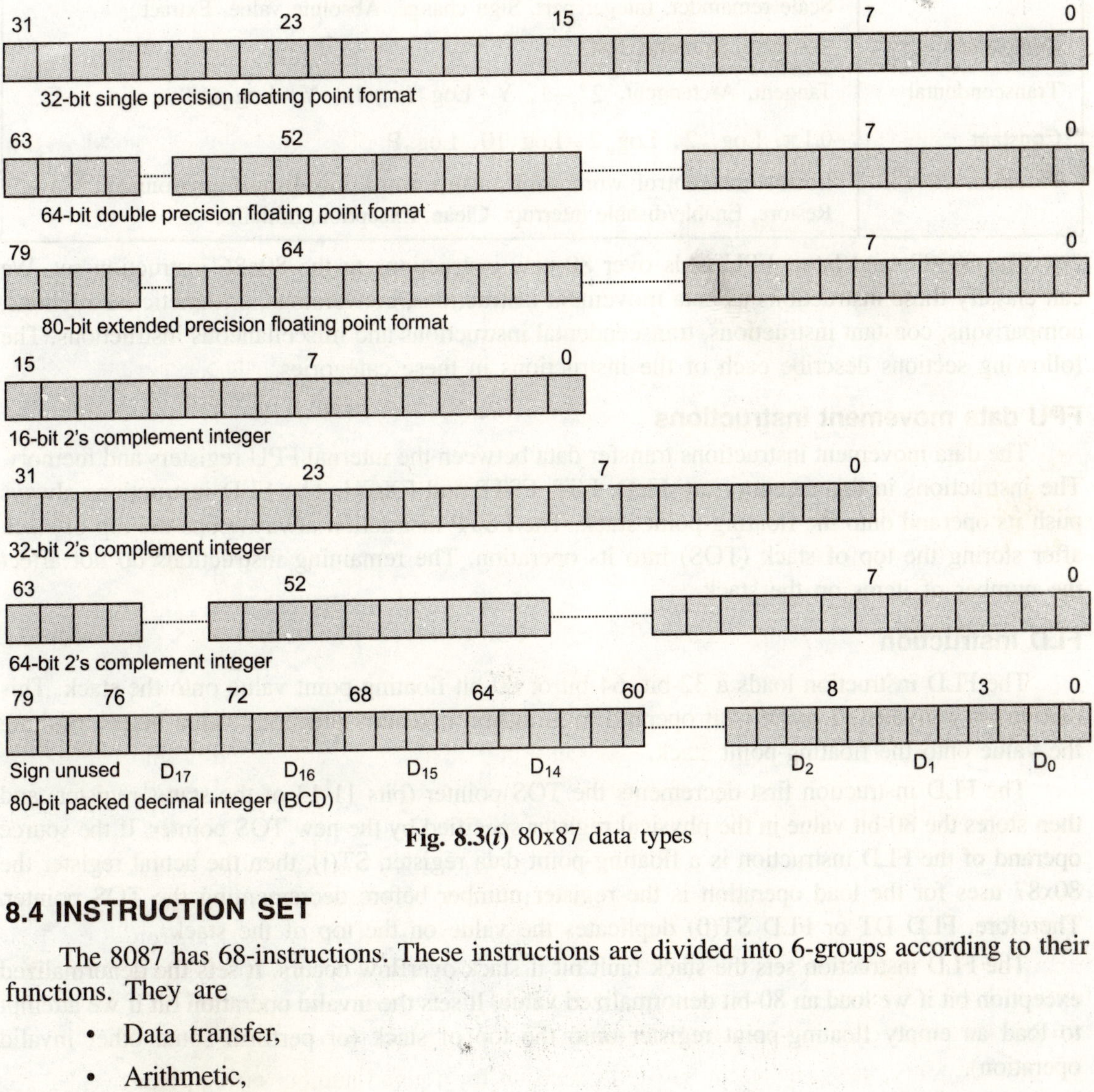

Fig. 8.3(*i*) 80x87 data types

8.4 INSTRUCTION SET

The 8087 has 68-instructions. These instructions are divided into 6-groups according to their functions. They are

- Data transfer,
- Arithmetic,
- Comparison,
- Transcendental,
- Constant,
- Processor control.

These groups are categorized as in Table 8.4(*a*).

Table 8.4(*a*) 8087 instruction type

Data transfer	Load/store (all types), Exchange
Arithmetic	Add, Subtract, Multiply, Divide, Subtract reversed, divide reversed, Square root, Scale remainder, Integer part, Sign change, Absolute value, Extract
Comparison	Compare, Examine, Test.
Transcendental	Tangent, Arctangent, $2^x - 1$, $Y * Log_2(x+1)$, $Y * Log_2(x)$
Constant	$0, 1, \pi$, $Log_{10} 2$, $Log_e 2$, $Log_2 10$, $Log_2 P$
Processor control	Load/store control word, store status word, Load/store environment, Save, Restore, Enable/disable interrupt, Clean, Exception, Initialize

The 80387 (and later) FPU adds over 80-new instructions to the 80x86 instruction set. We can classify these instructions as data movement instructions, conversions, arithmetic instructions, comparisons, constant instructions, transcendental instructions and miscellaneous instructions. The following sections describe each of the instructions in these categories.

FPU data movement instructions

The data movement instructions transfer data between the internal FPU registers and memory. The instructions in this category are FLD, FST, FSTP and FXCH. The FLD instructions always push its operand onto the floating-point stack. The FSTP instruction always pops the top of stack after storing the top of stack (TOS) into its operation. The remaining instructions do not affect the number of items on the stack.

FLD instruction

The FLD instruction loads a 32-bit, 64-bit or 80-bit floating-point value onto the stack. This instruction converts 32 and 64-bit operand to an 80-bit extended precision value before pushing the value onto the floating-point stack.

The FLD instruction first decrements the TOS pointer (bits 11-13 of the status register) and then stores the 80-bit value in the physical register specified by the new TOS pointer. If the source operand of the FLD instruction is a floating-point data register, ST(i), then the actual register the 80x87 uses for the load operation is the register number before decrementing the TOS pointer. Therefore, FLD DT or FLD ST(0) duplicates the value on the top of the stack.

The FLD instruction sets the stack fault bit if stack overflow occurs. It sets the denormalized exception bit if we load an 80-bit denormalized value. It sets the invalid operation bit if we attempt to load an empty floating-point register onto the top of stack (or perform some other invalid operation).

Examples

```
FLD    ST(1)
FLD    mem_32
FLD    MyRealVar
FLD    mem_64[bx]
```

FST and FSTP instructions

The FST and FSTP instructions copy the value on the top of the floating-point register stack to another floating-point register or to a 32, 64 or 80-bit memory variable. When copying data to a 32 or 64-bit memory variable, the 80-bit extended precision value on the top of stack is rounded to the smaller format as specified by the rounding control bits in the FPU control register.

The FSTP instruction pops the value off the top of stack when moving it to the destination location. It does this by incrementing the top of stack pointer in the status register after accessing the data in ST(0). If the destination operand is a floating-point register, the FPU stores the value at the specified register number before popping the data off the top of the stack.

Executing an FSTP ST(0) instruction effectively pops the data off the top of stack with no data transfer.

Examples

```
FST     mem_32
FSTP    mem_64
FSTP    mem_64[EBX*8]
FST     mem_80
FST     ST(2)
FSTP    ST(1)
```

The above example effectively pops ST(1) while leaving ST(0) on the top of the stack.

The FST and FSTP instructions will set the stack exception bit if a stack underflow occurs (attempting to store a value from an empty register stack). They will set the precision bit if there is a loss of precision during the store operation (this will occur, for example, when storing an 80-bit extended precision value into a 32 or 64-bit memory variable and there are some bits lost during conversion). They will set the underflow exception bit when storing an 80-bit value into a 32 or 64-bit memory variable, but the value is too small to fit into the destination operand. Likewise, these instructions will set the overflow exception bit if the value on the top of stack is too big to fit into a 32 or 64-bit memory variable. The FST and FSTP instructions set the denormalized flag when we try to store a denormalized value into an 80-bit register or variable. They set the invalid operation flag if an invalid operation (such as storing into an empty register) occurs. Finally, these instructions set the C1 condition bit if rounding occurs during the store operation (this only occurs when storing into a 32 or 64-bit memory variable and we have to round the mantissa to fit into the destination).

FXCH instruction

The FXCH instruction exchanges the value on the top of stack with one of the other FPU registers. This instruction takes 2-forms

- with a single FPU register as an operand and
- without any operands.

The 1st form exchanges the top of stack with the specified register. The 2nd form of FXCH swaps the top of stack with ST(1).

Many FPU instructions, e.g., FSQRT, operate only on the top of the register stack. If we want to perform such an operation on a value that is not on the top of stack, we can use the FXCH instruction to swap that register with TOS, perform the desired operation and then use the FXCH to swap the TOS with the original register. The following example takes the square root of ST(2)

```
FXCH    ST(2)
FSQRT
FXCH    ST(2)
```

The FXCH instruction sets the stack exception bit if the stack is empty. It sets the invalid operation bit if we specify an empty register as the operand. This instruction always clears the C1 condition code bit.

Conversions

The 80x87 chip performs all arithmetic operations on 80-bit real quantities. In a sense, the FLD and FST/FSTP instructions are conversion instructions as well as data movement instructions because they automatically convert between the internal 80-bit real format and the 32 and 64-bit memory formats. Nonetheless, we will simply classify them as data movement operations, rather than conversions, because they are moving real values to and from memory. The 80x87 FPU provides 5-routines that convert to or from integer or binary coded decimal (BCD) format when moving data. These instructions are FILD, FIST, FISTP, FBLD and FBSTP.

FILD instruction

The FILD (integer load) instruction converts a 16, 32 or 64-bit 2's complement integer to the 80-bit extended precision format and pushes the result onto the stack. This instruction always expects a single operand. This operand must be the address of a word, double word, or quad word integer variable. Although the instruction format for FILD uses the familiar mod/rm fields, the operand must be a memory variable, even for 16 and 32-bit integers. We cannot specify one of the 80386's 16 or 32-bit general-purpose registers. If we want to push an 80x86 general-purpose register onto the FPU stack, we must first store it into a memory variable and then use FILD to push that value of that memory variable.

The FILD instruction sets the stack exception bit and C1 (accordingly) if stack overflow occurs while pushing the converted value.

Examples

```
FILD    mem_16
FILD    mem_32[ECX*4]
FILD    mem_64[EBX+ECX*8]
```

FIST and FISTP instructions

The FIST and FISTP instructions convert the 80-bit extended precision variable on the top of stack to a 16, 32, or 64-bit integer and store the result away into the memory variable specified by the single operand. These instructions convert the value on TOS to an integer according to the rounding setting in the FPU control register (bits-10 and 11). As for the FILD instruction, the FIST and FISTP instructions will not let us to specify one of the 80x86's general purpose 16 or 32-bit registers as the destination operand.

The FIST instruction converts the value on the top of stack to an integer and then stores the result; it does not otherwise affect the floating-point register stack. The FISTP instruction pops the value off the floating-point register stack after storing the converted value.

These instructions set the stack exception bit if the floating-point register stack is empty (this will also clear C1). They set the precision (imprecise operation) and C1 bits if rounding occurs (that is, if there is any fractional component to the value in ST(0)). These instructions set the underflow exception bit if the result is too small (i.e., less than 1 but greater than 0 or less than 0 but greater than –1).

Examples

```
FIST    mem_16[BX]
FIST    mem_64
FISTP   mem_32
```

Do not forget that these instructions use the rounding control settings to determine how they will convert the floating-point data to an integer during the store operation. By default, the rounding control is usually set to 'round' mode; yet most programmers expect FIST/ FISTP to truncate the decimal portion during conversion. If we want FIST/ FISTP to truncate floating-point values when converting them to an integer, we will need to set the rounding control bits appropriately in the floating-point control register.

FBLD and FBSTP instructions

The FBLD and FBSTP instructions load and store 80-bit BCD values. The FBLD instruction converts a BCD value to its 80-bit extended precision equivalent and pushes the result onto the stack. The FBSTP instruction pops the extended precision real value on TOS, converts it to an 80-bit BCD value (rounding according to the bits in the floating-point control register) and stores the converted result at the address specified by the destination memory operand. Note that there is no FBST instruction that stores the value on TOS without popping it.

The FBLD instruction sets the stack exception bit and C1 if stack overflow occurs. It sets the invalid operation bit if we attempt to load an invalid BCD value. The FBSTP instruction sets the stack exception bit and clears C1 if stack underflow occurs (the stack is empty). It sets the underflow flag under the same conditions as FIST and FISTP.

Examples

; Assuming fewer than 8-items on the stack, the following code sequence is equivalent to an FBST instruction

```
FLD     ST(0)       ; Duplicate value on TOS.
FBSTP   mem_80
```

; The following example easily converts an 80-bit BCD value to a 64-bit integer:

```
FBLD    BCD_80    ; get BCD value to convert.
FIST    mem_64    ; store as an integer.
```

Arithmetic instructions

The arithmetic instructions make up a small but important, subset of the 80x87's instruction set. These instructions fall into 2-general categories

- those that operate on real values and
- those that operate on a real and an integer value.

FADD and FADDP instructions

These 2-instructions take the following forms

```
FADD
FADDP
FADD     ST(i), ST(0)
FADD     ST(0), ST(i)
FADDP    ST(i), ST(0)
FADD     mem
```

The first 2-forms are equivalent. They pop the 2-values on the top of stack, add them and push their sum back onto the stack.

The next 2-forms of the FADD instruction, those with 2-FPU register operands, behave like the 80x86's add instruction. They add the value in the 2nd register operand to the value in the 1st register operand. Note that one of the register operands must be ST(0).

The FADDP instruction with 2-operands adds ST (0) (which must always be the 2nd operand) to the destination (1st) operand and then pops ST(0). The destination operand must be one of the other registers of FPU.

The last form above, FADD with a memory operand, adds a 32 or 64-bit floating-point variable to the value in ST(0). This instruction will convert the 32 or 64-bit operands to an 80-bit extended precision value before performing the addition. Note that this instruction does not allow an 80-bit memory operand.

These instructions can raise the stack, precision, underflow, overflow, denormalization and illegal operation exceptions, as appropriate. If a stack fault exception occurs, C1 denotes stack overflow or underflow.

FSUB, FSUBP, FSUBR and FSUBRP instructions

These 4-instructions take the following forms

```
FSUB
FSUBP
FSUBR
FSUBRP
```

FSUB ST(i), ST(0)
FSUB ST(0), ST(i)
FSUBP ST(i), ST(0)
FSUB mem

FSUBR ST(i), ST(0)
FSUBR ST(0), ST(i)
FSUBRP ST(i), ST(0)
FSUBR mem

With no operands, the FSUB and FSUBP instructions operate identically. They pop ST(0) and ST(1) from the register stack, compute ST(0)-ST(1) and then push the difference back onto the stack. The FSUBR and FSUBRP instructions (reverse subtraction) operate in an almost identical fashion except they compute ST(1)-ST(0) and push that difference.

With 2-register operands (destination, source) the FSUB instruction computes destination = destination – source. One of the 2-registers must be ST(0). With 2-registers as operands, the FSUBP also computes destination = destination – source and then it pops ST(0) off the stack after computing the difference. For the FSUBP instruction, the source operand must be ST(0).

With 2-register operands, the FSUBR and FSUBRP instruction work in a similar fashion to FSUB and FSUBP, except they compute destination = (source – destination).

The FSUB mem and FSUBR mem instructions accept a 32 or 64-bit memory operand. They convert the memory operand to an 80-bit extended precision value and subtract this from ST(0) (FSUB) or subtract ST(0) from this value (FSUBR) and store the result back into ST(0).

These instructions can raise the stack, precision, underflow, overflow, denormalization and illegal operation exceptions, as appropriate. If a stack fault exception occurs, C1 denotes stack overflow or underflow.

FMUL and FMULP instructions

The FMUL and FMULP instructions multiply 2-floating-point values. These instructions allow the following forms

FMUL
FMULP

FMUL ST(0), ST(i)
FMUL ST(i), ST(0)
FMUL mem

FMULP ST(i), ST(0)

With no operands, FMUL and FMULP both do the same thing - they pop ST(0) and ST(1), multiply these values and push their product back onto the stack. The FMUL instructions with 2-register operands compute destination = destination*source. One of the registers (source or destination) must be ST(0).

The FMULP ST(i), ST(0) instruction computes ST(i) = ST(i)*ST(0) and then pops ST(0). This instruction uses the value for 'i' before popping ST(0). The FMUL mem instruction requires a 32 or 64-bit memory operand. It converts the specified memory variable to an 80-bit extended precision value and multiplies ST(0) by this value.

These instructions can raise the stack, precision, underflow, overflow, denormalization and illegal operation exceptions, as appropriate. If rounding occurs during the computation, these instructions set the C1 condition code bit. If a stack fault exception occurs, C1 denotes stack overflow or underflow.

FDIV, FDIVP, FDIVR and FDIVRP instructions

These 4-instructions allow the following forms

```
FDIV
FDIVP
FDIVR
FDIVRP

FDIV     ST(0), ST(i)
FDIV     ST(i), ST(0)
FDIVP    ST(i), ST(0)

FDIVR    ST(0), ST(i)
FDIVR    ST(i), ST(0)
FDIVRP   ST(i), ST(0)

FDIV     mem
FDIVR    mem
```

With zero operands, the FDIV and FDIVP instructions pop ST(0) and ST(1), compute ST(0)/ST(1) and push the result back onto the stack. The FDIVR and FDIVRP instructions also pop ST(0) and ST(1) but compute ST(1)/ST(0) before pushing the quotient onto the stack.

With two register operands, these instructions compute the following quotients:

```
FDIV     ST(0), ST(i)     ;ST(0) = ST(0)/ST(i)
FDIV     ST(i), ST(0)     ;ST(i) = ST(i)/ST(0)
FDIVP    ST(i), ST(0)     ;ST(i) = ST(i)/ST(0)
FDIVR    ST(i), ST(i)     ;ST(0) = ST(0)/ST(i)
FDIVRP   ST(i), ST(0)     ;ST(i) = ST(0)/ST(i)
```

The FDIVP and FDIVRP instructions also pop ST(0) after performing the division operation. The value for 'i' in this 2-instructions is computed before popping ST(0).

These instructions can raise the stack precision, underflow, overflow, denormalization, zero divide and illegal operation exceptions, as appropriate. If rounding occurs during the computation, these instructions set the C1 condition code bit. If a stack fault exception occurs, C1 denotes stack overflow or underflow.

FSQRT instruction

The FSQRT routine does not allow any operands. It computes the square root of the value on TOS and replaces ST(0) with this result. The value on TOS must be zero or positive, otherwise FSQRT will generate an invalid operation exception.

This instruction can raise the stack, precision, denormalization and invalid operation exceptions, as appropriate. If rounding occurs during the computation, FSQRT sets the C1 condition code bit. If a stack fault exception occurs, C1 denotes stack overflow or underflow.

Example

```
; Compute Z : = SQRT(x^2 + y^2);
        FLD     x       ; Load x.
        FLD     ST(0)   ; Duplicate x on TOS.
        FMUL            ; Compute x^2.
        FLD     y       ; Load y.
        FLD     ST(0)   ; Duplicate y on TOS.
        FMUL            ; Compute y^2.
        FADD            ; Compute x^2 + y^2.
        FSQRT           ; Compute SQRT(x^2 + y^2).
        FST     z       ; Store away result in z.
```

FSCALE instruction

The FSCALE instruction pops 2-values off the stack. It multiplies ST(0) by 2ST(1) and pushes the result back onto the stack. If the value in ST(1) is not an integer, FSCALE truncates it towards zero before performing the operation.

This instruction raises the stack exception if there are not 2-items currently on the stack (this will also clear C1 since stack underflow occurs). It raises the precision exception if there is a loss of precision due to this operation (this occurs when ST(1) contains a large negative value). Likewise, this instruction sets the underflow or overflow exception bits if we multiply ST(0) by a very large positive or negative power of 2. If the result of the multiplication is very small, FSCALE could set the denormalized bit. Also, this instruction could set the invalid operation bit if we attempt to FSCALE illegal values. FSCALE sets C1 if rounding occurs in an otherwise correct computation.

Example

```
        FILD    Sixteen     ; push 16 onto the stack.
        FLD     x           ; compute x * 2^16.
        FSCALE
        .
        .
        .
        Sixteen     word    16
```

FPREM and FPREM1 instructions

The FPREM and FPREM1 instructions compute a partial remainder. Intel designed the FPREM instruction before the IEEE finalized their floating-point standard. In the final draft of the IEEE floating-point standard, the definition of FPREM was a little different than Intel's original design. Unfortunately, Intel needed to maintain compatibility with the existing software that used the FPREM instruction, so they designed a new version to handle the IEEE partial remainder operation, FPREM1. We should always use FPREM1 in new software we write, therefore we will only discuss FPREM1 here, although we use FPREM in an identical fashion.

FPREM1 computes the partial remainder of ST(0)/ST(1). If the difference between the exponents of ST(0) and ST(1) is less than 64, FPREM1 can compute the exact remainder in one operation. Otherwise, we will have to execute the FPREM1 two or more times to get the correct remainder value. The C2 condition code bit determines when the computation is complete. Note that FPREM1 does not pop the 2-operands off the stack; it leaves the partial remainder in ST(0) and the original divisor in ST(1) in case we need to compute another partial product to complete the result.

The FPREM1 instruction sets the stack exception flag if there are not 2-values on the top of stack. It sets the underflow and denormal exception bits if the result is too small. It sets the invalid operation bit if the values on TOS are inappropriate for this operation. It sets the C2 condition code bit if the partial remainder operation is not complete. Finally it loads C3, C1 and C0 with bits zero, one and two of the quotient respectively.

Example

```
; Compute z = x mod y
              FLD      y
              FLD      x
PartialLp :   FPREM1
              FSTSW    AX          ; get condition bits in AX.
              TEST     AH,100B     ; see if C2 is set.
              JNZ      PartialLp   ; repeat if not done yet.
              FSTP     z           ; store remainder away.
              FSTP     ST(0)       ; pop old y value.
```

FRNDINT instruction

The FRNDINT instruction rounds the value on TOS to the nearest integer using the rounding algorithm specified in the control register.

This instruction sets the stack exception flag if there is no value on the TOS (it will also clear C1 in this case). It sets the precision and denormal exception bits if there was a loss of precision. It sets the invalid operation flag if the value on the TOS is not a valid number.

FXTRACT instruction

The FXTRACT instruction is the complement to the FSCALE instruction. It pops the value off the top of the stack and pushes a value which is the integer equivalent of the exponent (in 80-bit real form) and then pushes the mantissa with an exponent of zero (3FFFH in biased form).

This instruction raises the stack exception if there is a stack underflow when popping the original value or a stack overflow when pushing the 2-results (C1 determines whether stack overflow or underflow occurs). If the original top of stack was zero, FXTRACT sets the zero division exception flag. The denormalized flag is set if the result warrants it and the invalid operation flag is set if there are illegal input values when we execute FXTRACT.

Example

```
; The following example extracts the binary exponent of X and stores this into the 16-bit integer
;   variable XPONENT.
        FLD         x
        FXTRACT
        FSTP        ST(0)
        FISTP       XPONENT
```

FABS instruction

FABS computes the absolute value of ST(0) by clearing the sign bit of ST(0). It sets the stack exception bit and invalid operation bits if the stack is empty.

Example

```
; Compute x = SQRT(ABS( x ))
        FLD         x
        FABS
        FSQRT
        FSTP        x
```

FCHS instruction

FCHS changes the sign of ST(0)'s value by inverting its sign bit. It sets the stack exception bit and invalid operation bits if the stack is empty.

Example

```
; Compute x = –x if x is positive, x = x if x is negative.
        FLD         x
        FABS
        FCHS
        FSTP        x
```

Comparison instructions

The 80x87 provides several instructions for comparing real values. The FCOM, FCOMP, FCOMPP, FUCOM, FUCOMP and FUCOMPP instructions compare the 2-values on the top of stack and set the condition codes appropriately. The FTST instruction compares the value on the top of stack with zero. The FXAM instruction checks the value on TOS and reports sign, normalization and tag information.

Generally most programs test the condition code bits immediately after a comparison. Unfortunately, there are no conditional jump instructions that branch based on the FPU condition codes. Instead, we can use the FSTSW instruction to copy the floating-point status register (see FPU Status Register) into the AX register; then we can use the SAHF instruction to copy the AH register into the 80x86's condition code bits. After doing this, we can use the conditional jump instructions to test some condition. This technique copies C0 into the carry flag, C2 into the parity flag and C3 into the zero flag. The SAHF instruction does not copy C1 into any of the 80x86's flag bits.

Since the SAHF instruction does not copy any 80x87 processor status bits into the sign or overflow flags, we cannot use the JG, JL, JGE, or JLE instructions. Instead, use the JA, JAE, JB, JBE, JE and JZ instructions when testing the results of a floating-point comparison. Yes, these conditional jumps normally test unsigned values and floating-point numbers are signed values. However, use the unsigned conditional branches anyway, the FSTSW and SAHF instructions set the 80x86 flags register to use the unsigned jumps.

FCOM, FCOMP and FCOMPP instructions

The FCOM, FCOMP and FCOMPP instructions compare ST(0) to the specified operand and set the corresponding 80x87 condition code bits based on the result of the comparison. The legal forms for these instructions are

```
        FCOM
        FCOMP
        FCOMPP

        FCOM    ST(i)
        FCOMP   ST(i)

        FCOM    mem
        FCOMP   mem
```

With no operands FCOM, FCOMP and FCOMPP compare ST(0) against ST(1) and set the processor flags accordingly. In addition, FCOMP pops ST(0) off the stack and FCOMPP pops both ST(0) and ST(1) off the stack.

With a single register operand FCOM and FCOMP compare ST(0) against the specified register. FCOMP also pops ST(0) after the comparison.

With a 32 or 64-bit memory operand the FCOM and FCOMP instructions convert the memory variable to an 80-bit extended precision value and then compare ST(0) against this value, setting the condition code bits accordingly. FCOM also pops ST(0) after the comparison.

These instructions set C2 (which winds up in the parity flag), if the 2-operands are not comparable (e.g., NaN). If it is possible for an illegal floating-point value to wind up in a comparison, we should check the parity flag for an error before checking the desired condition.

These instructions set the stack fault bit if there are not 2-items on the top of the register stack. They set the denormalized exception bit if either or both operands are denormalized. They set the invalid operation flag if either or both operands are quite NaNs. These instructions always clear the C1 condition code.

FUCOM, FUCOMP and FUCOMPP instructions

These instructions are similar to the FCOM, FCOMP and FCOMPP instructions, although they only allow the following forms

```
FUCOM
FUCOMP
FUCOMPP
FUCOM     ST(i)
FUCOMP    ST(i)
```

The difference between FCOM/FCOMP/FCOMPP and FUCOM/FUCOMP/FUCOMPP is relatively minor. The FCOM/FCOMP/FCOMPP instructions set the invalid operation exception bit if we compare 2-NaNs. The FUCOM/FUCOMP/FUCOMPP instructions do not do this. In all other cases, these 2-sets of instructions behave identically.

FTST instruction

The FTST instruction compares the value in ST(0) against 0.0. It behaves just like the FCOM instruction would if ST(1) contained 0.0. Note that this instruction does not differentiate – 0.0 from +0.0. If the value in ST(0) is either of these values, FTST will set C3 to denote equality. If we need to differentiate – 0.0 from +0.0, we use the FXAM instruction. Note that this instruction does not pop ST(0) off the stack.

FXAM instruction

The FXAM instruction examines the value in ST(0) and reports the results in the condition code bits (see FPU Status Register for details on how FXAM sets these bits). This instruction does not pop ST(0) off the stack.

Constant instructions

The 80x87 FPU provides several instructions that let we load commonly used constants onto the FPU's register stack. These instructions set the stack fault, invalid operation and C1 flags if a stack overflow occurs; they do not otherwise affect the FPU flags. The specific instructions in this category include:

```
FLDZ      ;pushes +0.0.
FLD1      ;pushes +1.0.
FLDP      ;pushes p.
FLDL2T    ;pushes log2(10).
FLDL2E    ;pushes log2(e).
FLDLG2    ;pushes log10(2).
FLDLN2    ;pushes ln(2).
```

Transcendental instructions

The 80387 and later FPUs provide 8-transcendental (log and trigonometric) instructions to compute a partial tangent, partial arctangent, 2x–1, $y * \log_2(x)$ and $y * \log_2(x+1)$. Using various algebraic identities, it is easy to compute most of the other common transcendental functions using these instructions.

F2XM1 instruction

F2XM1 computes 2ST(0) –1. The value in ST(0) must be in the range –1.0 ST(0) +1.0. If ST(0) is out of range F2XM1 generates an undefined result but raises no exceptions. The computed value replaces the value in ST(0).

Example

```
; Compute 10x using the identity: 10x = 2xlg(10)(lg = log2).
FLD x
FLDL2T
FMUL
F2XM1
FLD1
FADD
```

Note that F2XM1 computes $2^x - 1$, which is why the code above adds 1.0 to the result at the end of the computation.

FSIN, FCOS and FSINCOS instructions

These instructions pop the value off the top of the register stack and compute the sine, cosine or both and push the result(s) back onto the stack. The FSINCOS pushes the sine followed by the cosine of the original operand, hence it leaves COS(ST(0)) in ST(0) and SIN(ST(0)) in ST(1).

These instructions assume ST(0) specifies an angle in radians and this angle must be in the range $-2^{63} < ST(0) < +2^{63}$. If the original operand is out of range, these instructions set the C2 flag and leave ST(0) unchanged. We can use the FPREM1 instruction, with a divisor of 2*pi* to reduce the operand to a reasonable range.

These instructions set the stack fault/C1, precision, underflow, denormalization and invalid operation flags according to the result of the computation.

FPTAN instruction

FPTAN computes the tangent of ST(0) and pushes this value and then it pushes 1.0 onto the stack. Like the FSIN and FCOS instructions, the value of ST(0) is assumed to be in radians and must be in the range $-2^{63} < ST(0) < +2^{63}$. If the value is outside this range, FPTAN sets C2 to indicate that the conversion did not take place. As with the FSIN, FCOS and FSINCOS instructions, we can use the FPREM1 instruction to reduce this operand to a reasonable range using a divisor of 2pi.

If the argument is invalid (i.e., 0 or *pi* radians, which causes a division by 0) the result is undefined and this instruction raises no exceptions. FPTAN will set the stack fault, precision, underflow, denormal, invalid operation, C2 and C1 bits as required by the operation.

FPATAN instruction

This instruction expects 2-values on the top of stack. It pops them and computes the following:

$$ST(0) = \tan^{-1}(ST(1)/ST(0))$$

The resulting value is the arctangent of the ratio on the stack expressed in radians. If we have a value we wish to compute the tangent of, we use FLD1 to create the appropriate ratio and then execute the FPTAN instruction.

This instruction affects the stack fault/ C1, precision, underflow, denormal and invalid operation bits if an problem occurs during the computation. It sets the C1 condition code bit if it has to round the result.

FYL2X and FYL2XP1 instructions

The **FYL2X** and **FYL2XP1** instructions compute $ST(1) \times \log_2(ST(0))$ and $ST(1) \times \log_2(ST(0) + 1)$, respectively. **FYL2X** requires that ST(0) be greater than zero, **FYL2XP1** requires ST(0) to be in the range

$$\left\langle -1-\left(\frac{\sqrt{2}}{2}\right)\right\rangle < ST(0) < \left\langle 1-\left(\frac{\sqrt{2}}{2}\right)\right\rangle$$

FYL2X is useful for computing logs to bases other than 2; FYL2XP1 is useful for computing compound interest, maintaining the maximum precision during computation.

FYL2X can affect all the exception flags. C1 denotes rounding if there is not other error, stack overflow/underflow if the stack fault bit is set.

The FYL2XP1 instruction does not affect the overflow or zero divide exception flags. These exceptions occur when ST(0) is very small or zero. Since FYL2XP1 adds one to ST(0) before computing the function, this condition never holds. FYL2XP1 affects the other flags in a manner identical to FYL2X.

Miscellaneous instructions

The 80x87 FPU includes several additional instructions that control the FPU, synchronize operations and let we test or set various status bits. These instructions include FINIT/FNINIT, FDISI/FNDISI, FENI/FNENI, FLDCW, FSTCW/FNSTCW, FCLEX/FNCLEX, FSAVE/FNSAVE, FRSTOR, FRSTPM, FSTSW/FNSTSW, FSTENV/FNSTENV, FLDENV, FINCSTP, FDECSTP, FWAIT, FNOP and FFREE. The FDISI/FNDISI, FENI/FNENI and FRSTPM are active only on FPUs earlier than the 80387, so we will not consider them here.

Many of these instructions have 2-forms. The 1st form is Fxxxx and the 2nd form is FNxxxx. The version without the 'N' emits an FWAIT instruction prior to opcode (which is standard for most coprocessor instructions). The version with the 'N' does not emit the FWAIT opcode ('N' stands for no wait).

FINIT and FNINIT instructions

The FINIT instruction initializes the FPU for proper operation. Our applications should execute this instruction before executing any other FPU instructions. This instruction initializes the control register to 37FH (see FPU Control Register), the status register to 0 (see FPU Status Register) and the tag word to 0FFFFH. The other registers are unaffected.

FWAIT instruction

The FWAIT instruction pauses the system until any currently executing FPU instruction completes. This is required because the FPU on the 80486SX and earlier μP/FPU combinations can execute instructions in parallel with the μP. Therefore, any FPU instruction that reads or writes memory could suffer from a data hazard if the main μP accesses that same memory location before the FPU reads or writes that location. The FWAIT instruction lets us synchronize the operation of the FPU by waiting until the completion of the current FPU instruction. This resolves the data hazard by, effectively, inserting an explict 'stall' into the execution stream.

FLDCW and FSTCW instructions

The FLDCW and FSTCW instructions require a single 16-bit memory operand:

```
FLDCW   mem_16
FSTCW   mem_16
```

These 2-instructions load the control register (see FPU Control Register) from a memory location (FLDCW) or store the control word to a 16-bit memory location (FSTCW).

When using the FLDCW instruction to turn on one of the exceptions, if the corresponding exception flag is set when we enable that exception, the FPU will generate an immediate interrupt before the μP executes the next instruction. Therefore, we should use the FCLEX instruction to clear any pending interrupts before changing the FPU exception enable bits.

FCLEX and FNCLEX instructions

The FCLEX and FNCLEX instructions clear all exception bits the stack fault bit and the busy flag in the FPU status register (see FPU Status Register).

FLDENV, FSTENV and FNSTENV instructions

```
FSTENV    mem_14B
FNSTENV   mem_14B
FLDENV    mem_14B
```

The FSTENV/FNSTENV instructions store a 14-byte FPU environment record to the memory operand specified. When operating in real mode (the only mode this text considers), the environment record takes the form:

			Offset
Data Ptr bits 19-16		Unused bits (set to zero)	12
Data Ptr (bits 15-0)			10
Instr. Ptr. bits 19-16	0	Instruction opcode (11 bits)	8
Instruction Ptr. (bit 15-0)			6
Tag word			4
Status word			2
Control word			0

We must execute the FSTENV and FNSTENV instructions with the μP interrupts disabled. Furthermore, we should always ensure that the FPU is not busy before executing this instruction. This is easily accomplished by using the following code

```
PUSHF                ; preserve I flag.
CLI                  ; disable interrupts.
FSTENV mem_14B       ; implicit wait for not busy.
FWAIT                ; wait for operation to finish.
POPF                 ; restore I flag.
```

The FLDENV instruction loads the FPU environment from the specified memory operand. Note that this instruction lets us load the status word. There is no explicit instruction like FLDCW to accomplish this.

FSAVE, FNSAVE and FRSTOR instructions

```
FSAVE     mem_94B
FNSAVE    mem_94B
FRSTOR    mem_94B
```

These instructions save and restore the state of the FPU. This includes saving all the internal control, status and data registers. The destination location for FSAVE/ FNSAVE (source location for FRSTOR) must be 94-bytes long. The first 14-bytes correspond to the environment record the FLDENV and FSTENV instructions use the remaining 80-bytes hold the data from the FPU register stack written out as ST(0) through ST(7). FRSTOR reloads the environment record and floating-point registers from the specified memory operand.

The FSAVE/ FNSAVE and FRSTOR instructions are mainly intended for task switching. We can also use FSAVE/ FNSAVE and FRSTOR as a 'push all' and 'pop all' sequence to preserve the state of the FPU.

Like the FSTENV and FLDENV instructions, interrupts should be disabled while saving or restoring the FPU state. Otherwise another interrupt service routine could manipulate the FPU registers and invalidate the operation of the FSAVE/FNSAVE or FRESTORE operation. The following code properly protects the environment data while saving and restore the FPU status.

```
; Preserve the FPU state, assume DI points at the environment record in memory.
        PUSHF
        CLI
        FSAVE   [SI]
        FWAIT
        POPF
        .
        .
        .
        PUSHF
        CLI
        FRSTOR  [SI]
        FWAIT
        POPF
```

FSTSW and FNSTSW instructions

```
FSTSW     AX
FNSTSW    AX
FSTSW     mem_16
FNSTSW    mem_16
```

These instructions store the FPU status register (see FPU Status Register) into a 16-bit memory location or the AX register. These instructions are unusual in the sense that they can copy an FPU value into one of the 80x86 general-purpose registers. Of course, the whole purpose behind allowing the transfer of the status register into ax is to allow the μP to easily test the condition code register with the SAHF instruction.

FINCSTP and FDECSTP instructions

The FINCSTP and FDECSTP instructions do not take any operands. They simply increment and decrement the stack pointer bits (mod 8) in the FPU status register. These 2-instructions clear the C1 flag, but do not affect the condition code bits in the FPU status register.

FNOP instruction

The FNOP instruction is simply an alias for FST ST, ST(0). It performs no other operation on the FPU.

FFREE instruction

FFREE ST(i)

This instruction modifies the tag bits for register 'i' in the tags register to mark the specified register as EMTPY. The value is unaffected by this instruction, but the FPU will no longer be able to access that data (without resetting the appropriate tag bits).

Integer operations

The 80x87 FPUs provide special instructions that combine integer to extended precision conversion along with various arithmetic and comparison operations. These instructions are the following

```
FIADD    INT
FISUB    INT
FISUBR   INT
FIMUL    INT
FIDIV    INT
FIDIVR   INT

FICOM    INT
FICOMP   INT
```

These instructions convert their 16 or 32-bit integer operands to an 80-bit extended precision floating-point value and then use this value as the source operand for the specified operation. These instructions use ST(0) as the destination operand.

8.5 80287

The 80287 coprocessor is enhanced version of its predecessor the 8087. It is interfaced with the 80286 microprocessor for providing floating-point, extended integer and BCD data types. The 80287 provides more than 50 additional instructions to the instruction set of the 80286. Similar to the 8087. It is designed using HMOS technology housed in 40-pin package called CERDIP. The block diagram and pinouts of the 80287 are shown in Figs. 8.5(*a*) and (*b*).

The large number of calculation that must be performed are no longer difficult for the current 32-bit and 64-bit microprocessor that perform multiplication and divisions in single instruction at the rate of MIPS, GIGA FLOPS, TERA FLOPS.

Like the host processor, it can also operate in either real mode or protected virtual address mode. After reset, it starts in the real mode and can switch to PVAM using FSETPM. The FSTSW AX instruction allows the 80287 to store its control word directly into the AX register of its host.

The data type of the 8087 are 32-bit, 64-bit and 80-bit floating-point, 32-bit, 64-bit integer, 18-bit BCD. The 8087 depends on its host processor to fetch and/ or store memory operands. The 80287 outputs a request at the PEREQ pin to write a word to the memory, the 80286 reads this requested data from the 80287, returns an acknowledgement through the $\overline{\text{PEACK}}$ pin and write the data word to the memory. The numeric processor selects NPS2 and $\overline{\text{NPS1}}$ and command CMD1 and CMD0 pins are used to indicate that the CPU is performing an ESC instruction and to direct the operation of the 80287. The 80287 can operate in parallel with the 80286. As soon as the 80287 detects an exception, it indicates the 80286 by asserting the ERROR signal. The 80286 and 80287 are programmed as a single processor. All memory addressing modes, physical memory and virtual memory of the 80286 are available in the 80287.

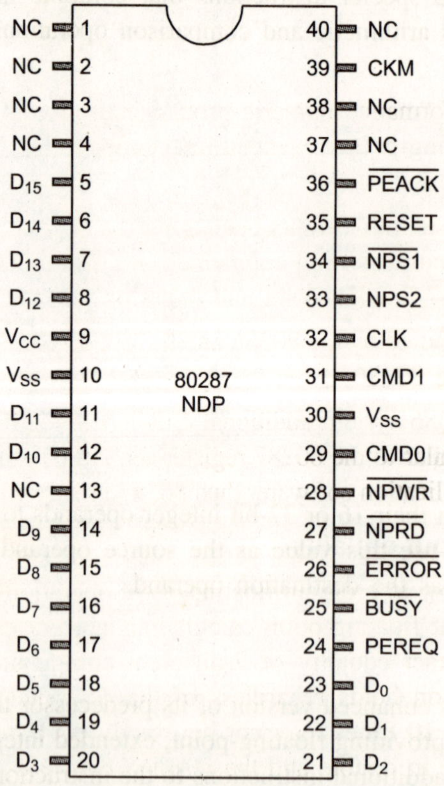

Fig. 8.5 (*a*) Pinouts of 80287

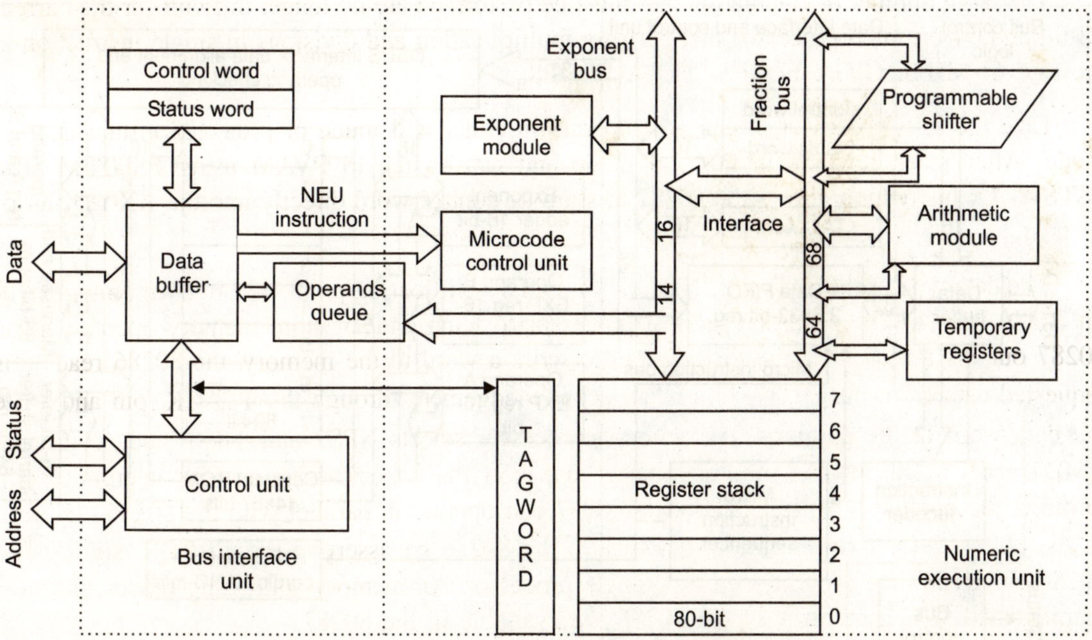

Fig. 8.5(*b*) Block diagram of 80287

8.6 80387

The 80387 is a high performance numeric processor used with the 80386. This extends the 80386's architecture with floating-point, extended integer and BCD data types. The block diagram of the 80387 is shown in Fig. 8.6(*a*).

The 80386 can be divided into 3-functional units that operates in parallel to increase the performance. The 80386 can be transferring commands and data to the 80387 control logic for the next instruction while the 80387 floating-point unit is performing the current numeric instruction.

All memory access are handled by the 80386 and the 80387 merely operates on the instructions and values passed to it by the 80386.

The 80387 adds to an 80386 system additional data types, register, interrupts and instructions. The register set of 80387 is similar to the 80287 register set. All communications between the 80386 and 80387 is transparent to application software, thus 80387 greatly enhances the 80386 capabilities.

A final word of caution about equality

That is all we need to know to program floating-point operations in (this) assembly language. However, let us remember that floating-point calculations are not exact and therefore it is not a really good idea to expect exact equality of 2-different computations, even if they should be theoretically equal. This caution holds regardless of the computer language we are using. For instance, adding 1/10 to itself 10 times may not be exactly equal to 1.0! It is more likely to be equal if the additions are done in double and the result is converted to single for comparing, but even this is not foolproof.

(x-1.0) is more safely written as abs(x-1.0) < 0.000001

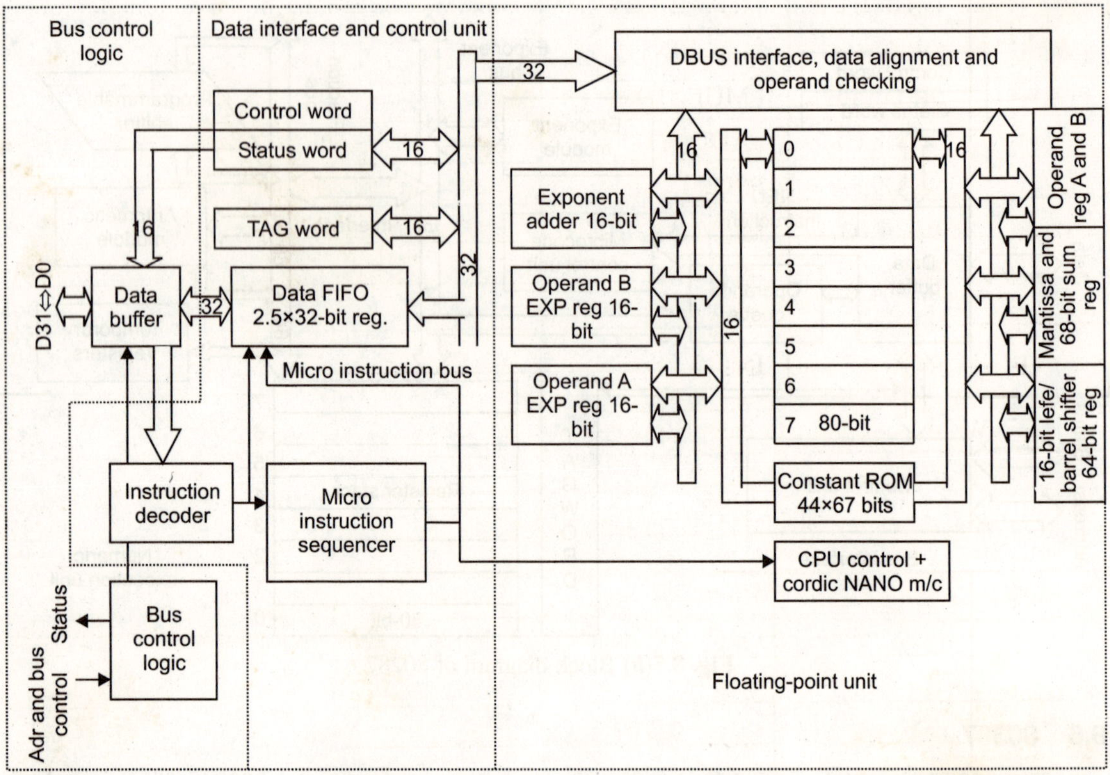

Fig. 8.6(*a*) Block diagram of the 80387

Moves to and from coprocessor

Since conversions are done by the coprocessor, it is necessary to move integer data back and forth between processor general registers and coprocessor registers. *This movement does not change the bit patterns, so the data must be converted before being used for operations.* The programmer must keep track of the type of data in each coprocessor register. It is also possible to have single floating data in the general registers, but not very useful.

8.1 Write a program to obtain square root of sum of square of two data 'a' and 'b' i.e., $\sqrt{a^2 + b^2}$. The data 'a' is stored at 0300H and 'b' is stored at 0304H. Store the result at 0308H.

0300H : 09H,00H,00H,00H and 0306H : 04H,00H,00H,00H

Location	*Opcode*	*Mnemonics*	*Comments*
0200	BB,00,03	MOV BX,0300H	; BX points to data a
0203	DB,07	FLD [BX]	; Load a
0205	9B,DA,0F	FMUL (BX)	; Multiply a with a = a^2
0208	9B,DD,D9	FSTP ST(1)	; Load a^2 in ST(1)

(*Contd....*)

020B	BB,04,03	MOV BX,0304H	; BX points to data b
020E	DB,07	FLD [BX]	; load b
0210	9B,DA,0F	FMUL (BX)	; multiply b with b = b^2
0213	9B,D8,C1	FADD ST(1)	; add $a^2 + b^2$ with result on top of stack
0216	9B,D9,FA	FSQRT	; takes square root of $a^2 + b^2$
0219	BB,08,03	MOV BX,0308H	; BX points at result location
021C	DB,17	FST [BX]	; store the result at top of stack at 0308H
021E	9B	FWAIT	; wait
021F	CC	INT 3	; break

8.2 Write a program to calculate Sin z where z is in degree.

Sin z can be calculated using tangent function of the 8087. $\tan\frac{z}{2} = \frac{2xy}{x^2+y^2}$

Location	*Opcode*	*Mnemonics*	*Comments*
0200	B4,00	MOV AH,00H	; clear AH
0202	D1,E8	SHR AX	; z = z/2
0204	A3,00,03	MOV WARG1,AX	; load AX in Arg.1
0207	DF,06,00,03,	FILD ARG1	; load z/2 on top of stack
020B	C7,06,02,03, B4,00	MOV WARG2,00B4	; (Arg.2) = B4 (180° in hex)
0211	9B,DF,06,02,03	FILD ARG2	; insert Fwait and load ARG2
0216	9B,DE,F9	FDIVRP	; wait and divide ST(0) = z/180°
0219	9B,D9,EB	FILDP	; ST(0) = pi
021C	9B,DE,C9	FMUL RP	; ST(0) = $z \times pi / 180$ = angle in radians
021F	9B,D9,F2	FTAN	; calculate tanz/2 = y/x
0222	9B,D9,F0	FILD ST(0)	; duplicate ST(0) on top of stack
0225	9B,D8,C9	FMUL ST,ST(1)	; ST(0) = x × x
0228	BB,D9,C2	FILD ST(2)	; ST(0) = y
022B	9B,D8,CB	FMUL ST(0),ST(3)	; ST(0) = y × y
022E	9B,DE,CI	FADDP ST(1),ST(0)	; y × y popped of and top of stack ST(0) = $x^2 + y^2$
0231	9B,DE,F9	FDIVRP	; ST(0) = $x/(x^2+y^2)$

(*Contd....*)

0234	9B,DE,C9	FMULRP	; ST(0) = $xy/(x^2+y^2)$
0237	9B,D9,C0	FLD ST(0)	; duplicate top of stack
023A	9E.DE,C1	FADD ST(1),ST(0)	; ST(0) = $2xy/(x^2+y^2)$ = sin z
023D	C7,06,04,03,10.27	MOV WARG3,2710	; load ARG3 with 10,000 value
0243	9B,DF,06,04,03	FILD ARG3	;
0248	9B,DE,C9	FMULRP	; ST(0) = 10,000 × sin z
024B	9B,D9,FC	FRND INT	; round the integer
024E	9B,DF,36,06,03	FSTP ARG4	; store in BCD format
0253	9B	FWAIT	
0254	A1,06,03	MOV AX,ARG4	; result stored at 0306 and AX
0257	CC	INT	; break

In order to execute the program, we have to enter the value of angle in AL i.e., 1EH for calculating the Sin for angle of 30°. To execute the program, we will find AX = 5000H. This output is in packed BCD form. A decimal point is to be assumed before the 1st BCD digit. For practice, we execute the values of Sin z.

Input (AL)	*Output (AX)*	*Sin z*
10	2756	sin 16°
2E	7193	sin 46°
3C	8660	sin 60°
56	9976	sin 86°

Stack operation after each instruction

(1) FILD ARG1 ST(0) → [ARG1] z

(2) FILD ARG2
ST(0) → [ARG2] 180°
ST(1) → [ARG1] z

(3) FDIVRP ST(0) → [z/180°]

(4) FLDPI
ST(0) → [pi]
ST(1) → [z/180°]

(5) FMULRP
ST(0) → [pi*z/180°]
ST(1) [z/180°]

(6) FTAN

ST(0) →	x
ST(1)	y
ST(2)	z/180°

(7) FLD

ST(0) →	x
ST(1)	x
ST(2)	y
ST(3)	z/180°

(8) FMUL ST(0),ST(1)

ST(0) →	x^2
ST(1)	x
ST(2)	y
ST(3)	z/180°

(9) FLD ST(2)

ST(0) →	y
ST(1)	x^2
ST(2)	x
ST(3)	y
ST(4)	z/180°

(10) FLD ST(0),ST(3)

ST(0) →	y^2
ST(1)	x^2
ST(2)	x
ST(3)	y
ST(4)	z/180°

(11) FADDP ST(1),ST(0)

ST(0) →	$x^2 + y^2$
ST(1)	x
ST(2)	y
ST(3)	z/180°

(12) FDIVRP

ST(0) →	$x/(x^2 + y^2)$
ST(1)	y
ST(2)	z/180°

(13) FMULRP

ST(0) →	$xy/(x^2 + y^2)$
ST(1)	z/180°

(14) FLD ST(0)

ST(0) →	$xy/(x^2 + y^2)$
ST(1)	$xy/(x^2 + y^2)$
ST(2)	z/180°

(15) FADDP ST(1),ST(0)   ST(0) → | $2xy/(x^2 + y^2)$ = sin z |

(16) FILD ARG3   ST(0) → | 10,000 | ; ST(1) | sin z |

(17) FMURP   ST(0) → | 10,000* sin z |

(18) FRND INT   ST(0) → | Rounding integer |

(19) FSTP ARG4

Argument has the result

8.3 Obtain the area of a circle using 8087.

Input : Radius in AX (byte integer)

Output : Area in BX : AX (short integer)

```
PROC        NEAR
MOV         Radius,AH         ; Load radius in memory
FILD        Radius            ; ST(0) = r
FMUL        ST(0) × ST(0)     ; ST(0) = r×r = r^2
FLDPI                         ; ST(0) = pi, ST(1) = r^2
FMUL        ST(0),ST(1)       ; ST(0) = pi×r^2
FIST        Area              ; store rounded result
FWAIT                         ; Be sure data is stored
MOV AX,Area                   ; Fetch low 16-bits
MOV BX,Area                   ; Fetch high 16-bits
RET                           ; Done
ENDP
ENDS
END
```

REVIEW QUESTIONS AND ANSWERS

Q.1. What are 2-main sections of the 8087?

Ans. **Control** unit and **numeric execution** unit.

Q.2. Which instruction is used by the 8087 to fetch in parallel with the host processor?

Ans. ESCAPE.

Q.3. Is 8087 compatible with both 8086 and 8088? How does it interface with these processors?

Ans. Yes, it monitors pin 34 $\overline{BHE}$ for 16-bit µP and $\overline{SS_0}$ for 8-bit µP.

Q.4. Normally 8087 access the memory using 8086 addressing modes. Does it have address calculating capability?

Ans. No.

Q.5. Which instruction would require DMA cycle via $\overline{RQ}/\overline{GT}$ control line of the 8087?

Ans. FSAVE writes 94-bytes.

Q.6. Which instruction ensures the storing of data in memory?

Ans. FWAIT.

Q.7. Is it possible to transfer data directly between the 8087 and 8086?

Ans. No, all data must pass through the memory.

Q.8. When a memory variable is automatically converted into temporary real format?

Ans. When pushed on the stack of the 8087.

Q.9. When the 8087 fix up an error or exception during processing?

Ans. When the exception masks of the control register are set.

Q.10. What does FADD ST(2), ST(3) instruction do?

Ans. $ST(2) \leftarrow (ST(2) + ST(3))$.

Q.11. Determine the decimal equivalent of the following hexadecimal numbers.

(*a*) CA5B2000 (*b*) 3C630000 (*c*) 4C341200

Ans. (*a*)

$$\underbrace{1}_{\text{Sign}}\ \underbrace{10010100}_{\text{Biased exponent}}\underbrace{1011011001\,0000000000000}_{\text{Fraction}}$$

Sign = negative,

Biased exponent = 10010100

Fraction = 10110110010000000000000 = 1011011001

Decimal = $1.1011011001 \times 2^{148-127} = 1.1011011001 \times 2^{21} = -1.7123046 \times 2^{21}$

= −35909642(D)

(*b*)

$$\underbrace{0}_{\text{Sign}}\ \underbrace{01111000}_{\text{Biased exponent}}\underbrace{1100011000\,0000000000000}_{\text{Fraction}}$$

Sign = positive,

Biased exponent = 01111000

Fraction = 11000110000000000000000 = 1100011

Decimal = $0.1100011 \times 2^{120-127} = 0.1100011 \times 2^{-7} = 0.0078125 \times 2^{-7}$ = 0.0138579(D)

(*c*)

$$\underbrace{0}_{\text{Sign}}\ \underbrace{10011000}_{\text{Biased exponent}}\underbrace{01101000001001000000000}_{\text{Fraction}}$$

Sign = positive,

Biased exponent = 10011000

Fraction = $0.01101000001001 \times 2^{152-127} = 0.01101000001001 \times 2^{25}$ = 47221424

Q.12. Convert the following decimal numbers to short real numbers in hexadecimal format.

(*a*) 1 −12345 (*b*) -148.4375×10^{-3} (*c*) 1.2345625×10^{4}

Ans. (*a*) −12345(D)

2|12345| 1
2| 6172| 0
2| 3086| 0
2| 1543| 1
2| 771| 1
2| 385| 1
2| 192| 0
2| 96| 0
2| 48| 0
2| 24| 0
2| 12| 0
2| 6| 0
2| 3| 1
1

2|140| 0
2| 70| 0
2| 35| 1
2| 17| 1
2| 8| 0
2| 4| 0
2| 2| 0 = 10001100
1

$= -11000000111001 = -1.1000000111001\times2^{13}$

S = 1, e = 127+13 = 140 = 10001100

Hence, number = – 11000110010000000111001 = – 1100 0110 0100 0000 1110 0100
= C640E400H

(*b*) $-148.4375\times10^{-3} = -0.1484375$

$0.1484375\times2 = 0.2968750|\ 0$

$0.2968750\times2 = 0.5937500|\ 0$

$0.5937500\times2 = 1.1875000|\ 1$ $= -0.0010011$

$0.1875000\times2 = 0.3750000|\ 0$ $= -1.0011\times2^{-3}$

$0.3750000\times2 = 0.7500000|\ 0$ = s = 1

$0.7500000\times2 = 1.5000000|\ 1$ $e = 127 - 3 = 124$

$0.5000000\times2 = 1.0000000|\ 1$

2|124| 0
2| 62| 0
2| 31| 1
2| 15| 1
2| 7| 1
2| 3| 1 = 1111100 = 01111100
1

Hence, number = 1011111000011 = 1011 1110 0001 1000 = BE180000H

(*c*) $1.2345625\times10^{4} = 12345.625$

The binary of 12345 = 11000000111001

$0.625\times2 = 1.250|\ 1$

$0.250\times2 = 0.500|\ 0$ = 0.101

$0.500\times2 = 1.000|\ 1$

2|140| 0
2| 70| 0
2| 35| 1
2| 17| 1
2| 8| 0
2| 4| 0
2| 2| 0 = 10001100
1

Hence, binary equivalent of 12345.625 = 11000000111001.101
$= 1.1000000111001101\times2^{13}$

Hence, e = 127+13 = 140 = 10001100

Hence, number = 01000110010000000111001101 = 0100 0110 0100 0000 1110 0110 1000
= 4640E680H

Instruction Set Summary of 89C51 *Appendix A*

MSN →	0	1	2	3	4	5	6	7
0 LSN ↓	NOP	JBC bit, rel [3B, 2C]	JB bit, rel [3B, 2C]	JNB bit, rel [3B, 2C]	JC rel [2B, 2C]	JNC rel [2B, 2C]	JZ rel [2B, 2C]	JNZ rel [2B, 2C]
1	AJMP (P0) [2B, 2C]	ACALL (P0) [2B, 2C]	AJMP (P1) [2B, 2C]	ACALL (P1) [2B, 2C]	AJMP (P2) [2B, 2C]	ACALL (P2) [2B, 2C]	AJMP (P3) [2B, 2C]	ACALL (P3) [2B, 2C]
2	LJMP addr16 [3B, 2C]	LCALL addr16 [3B, 2C]	RET [2C]	RETI [2C]	ORL dir, A [2B]	ANL dir, A [2B]	XRL dir, a [2B]	ORL C, bit [2B, 2C]
3	RR A	RRC A	RL A	RLC A	ORL dir, #data [3B, 2C]	ANL dir, #data [3B, 2C]	XRL dir, #data [3B, 2C]	JMP @A + DPTR [2C]
4	INC A	DEC A	ADD A, #data [2B]	ADDC A, #data [2B]	ORL A, #data [2B]	ANL A, #data [2B]	XRL A, #data [2B]	MOV A, #data [2B]
5	INC dir [2B]	DEC dir [2B]	ADD A, dir [2B]	ADDC A, dir [2B]	ORL A, dir [2B]	ANL A, dir [2B]	XRL A, dir [2B]	MOV dir, #data [3B, 2C]
6	INC @R0	DEC @R0	ADD A, @R0	ADDC A, @R0	ORL A, @R0	ANL A, @R0	XRL A, @R0	MOV @R0, @data [2B]
7	INC @R1	DEC @R1	ADD A, @R1	ADDC A, @R1	ORL A, @R1	ANL A, @R1	XRL A, @R1	MOV @R1, #data [2B]
8	INC R0	DEC R0	ADD A, R0	ADDC A, R0	ORL A, R0	ANL A, R0	XRL A, R0	MOV R0, #data [2B]
9	INC R1	DEC R1	ADD A, R1	ADDC A, R1	ORL A, R1	ANL A, R1	XRL A, R1	MOV R1, #data [2B]
A	INC R2	DEC R2	ADD A, R2	ADDC A, R2	ORL A, R2	ANL A, R2	XRL A, R2	MOV R2, #data [2B]
B	INC R3	DEC R3	ADD A, R3	ADDC A, R3	ORL A, R3	ANL A, R3	XRL A, R3	MOV R3, #data [2B]
C	INC R4	DEC R4	ADD A, R4	ADDC A, R4	ORL A, R4	ANL A, R4	XRL A, R4	MOV R4, #data [2B]
D	INC R5	DEC R5	ADD A, R5	ADDC A, R5	ORL A, R5	ANL A, R5	XRL A, R5	MOV R5, #data [2B]
E	INC R6	DEC R6	ADD A, R6	ADDC A, R6	ORL A, R6	ANL A, R6	XRL A, R6	MOV R6, #data [2B]
F	INC R7	DEC R7	ADD A, R7	ADDC A, R7	ORL A, R7	ANL A, R7	XRL A, R7	MOV R7, #data [2B]

Instruction Set Summary (Continued)

	8	9	A	B	C	D	E	F
0	SJMP REL [2B, 2C]	MOV DPTR,# data 16 [3B, 2C]	ORL C, /bit [2B, 2C]	ANL C, /bit [2B, 2C]	PUSH dir [2B, 2C]	POP dir [2B, 2C]	MOVX A, @DPTR [2C]	MOVX @DPTR, A [2C]
1	AJMP (P4) [2B, 2C]	ACALL (P4) [2B, 2C]	AJMP (P5) [2B, 2C]	ACALL (P5) [2B, 2C]	AJMP (P6) [2B, 2C]	ACALL (P6) [2B, 2C]	AJMP (P7) [2B, 2C]	ACALL (P7) [2B, 2C]
2	ANL C, bit [2B, 2C]	MOV bit, C [2B, 2C]	MOV C, bit [2B]	CPL bit [2B]	CLR bit [2B]	SETB bit [2B]	MOVX A, @R0 [2C]	MOVX wR0, A [2C]
3	MOVC A, @A + PC [2C]	MOVC A, @A + DPTR [2C]	INC DPTR [2C]	CPL C	CLR C	SETB C	MOVX A, @RI [2C]	MOVX @RI, A [2C]
4	DIV AB [2B, 4C]	SUBB A, #data [2B]	MUL AB [4C]	CJNE A, #data, rel [3B, 2C]	SWAP A	DA A	CLR A	CPL A
5	MOV dir, dir [3B, 2C]	SUBB A, dir [2B]	—	CJNE A, dir, rel [3B, 2C]	XCH A, dir [2B]	DJNZ dir, rel [3B, 2C]	MOV A, dir [2B]	MOV dir, A [2B]
6	MOV dir, @R0 [2B, 2C]	SUBB A, @R0	MOV @R0, dir [2B, 2C]	CJNE @R0, #data, rel [3B, 2C]	XCH A, @R0	XCHD A, @R0	MOV A, @R0	MOV @R0, A
7	MOV dir, @R1 [2B, 2C]	SUBB A, @R1	MOV @R1, dir [2B, 2C]	CJNE @R1, #data, rel [3B, 2C]	XCH A, @R1	XCHD A, @R1	MOV A, @R1	MOV @R1, A
8	MOV dir, R0 [2B, 2C]	SUBB A, R0	MOV R0, dir [2B, 2C]	CJNE R0, #data, rel [3B, 2C]	XCH A, R0	DJNZ R0, rel [2B, 2C]	MOV A, R0	MOV R0, A
9	MOV dir, R1 [2B, 2C]	SUBB A, R1	MOV R1, dir [2B, 2C]	CJNE R1, #data, rel [3B, 2C]	XCH A, R1	DJNZ R1, rel [2B, 2C]	MOV A, R1	MOV R1, A
A	MOV dir, R2 [2B, 2C]	SUBB A, R2	MOV R2, dir [2B, 2C]	CNJE R2, #data, rel [3B, 2C]	XCH A, R2	DJNZ R2, rel [2B, 2C]	MOV A, R2	MOV R2, A
B	MOV dir, R3 [2B, 2C]	SUBB A, R3	MOV R3, dir [2B, 2C]	CJNE R3, #data, rel [3B, 2C]	XCH A, R3	DJNZ R3, rel [2B, 2C]	MOV A, R3	MOV R3, A
C	MOV dir, R4 [2B, 2C]	SUBB A, R4	MOV R4, dir [2B, 2C]	CJNE R4, #data, rel [3B, 2C]	XCH A, R4	DJNZ R4, rel [2B, 2C]	MOV A, R4	MOV R4, A
D	MOV dir, R5 [2B, 2C]	SUBB A, R5	MOV R5, dir [2B, 2C]	CJNE R5, #data, rel [3B, 2C]	XCH A, R5	DJNZ R5, rel [2B, 2C]	MOV A, R5	MOV R5, A
E	MOV dir, R6 [2B, 2C]	SUBB A, R6	MOV R6, dir [2B, 2C]	CJNE R6, #data, rel [3B, 2C]	XCH A, R6	DJNZ R6, rel [2B, 2C]	MOV A, R6	MOV R6, A
F	MOV dir, R7 [2B, 2C]	SUBB A, R7	MOV R7, dir [2B, 2C]	CJNE R7, #data, rel [3B, 2C]	XCH A, R7	DJNZ R7, rel [2B, 2C]	MOV A, R7	MOV R7, A

Note: Key: [2B] = 2 Byte, [3B] = 3 Byte, [2C] = 2 Cycle, [4C] = 4 Cycle, Blank = 1 byte/1 cycle

Table 1. Instruction Set Summary of 89C51[1] ***Appendix B***

Mnemonic		Description	Byte	Oscillator Period
ARITHMETIC OPERATIONS				
ADD	A,R_n	Add register to Accumulator	1	12
ADD	A, direct	Add direct byte to Accumulator	2	12
ADD	A, @R_i	Add indirect RAM to Accumulator	1	12
ADD	A, #data	Add immediate data to Accumulator	2	12
ADDC	A, R_n	Add register to Accumulator with Carry	1	12
ADDC	A, direct	Add direct byte to Accumulator with Carry	2	12
ADDC	A, @R_i	Add indirect RAM to Accumulator with Carry	1	12
ADDC	A, #data	Add immediate data to Acc with Carry	2	12
SUBB	A, R_n	Subtract Register from Acc with borrow	1	12
SUBB	A, direct	Subtract direct byte from Acc with borrow	2	12
SUBB	A, @R_i	Subtract indirect RAM from ACC with borrow	1	12
SUBB	A, #data	Subtract immediate data from Acc with borrow	2	12
INC	A	Increment Accumulator	1	12
INC	R_n	Increment register	1	12
INC	direct	Increment direct byte	2	12
INC	@R_i	Increment direct RAM	1	12
DEC	A	Decrement Accumulator	1	12
DEC	R_n	Decrement Register	1	12
DEC	direct	Decrement direct byte	2	12
DEC	@R_i	Decrement indirect RAM	1	12
INC	DPTR	Increment Data Pointer	1	24
MUL	AB	Multiply A & B	1	48
DIV	AB	Divide A by B	1	48
DA	A	Decimal Adjust Accumulator	1	12

Note : 1. All mnemonics copyrighted © Intel Crop. 1980.

Mnemonic		Description	Byte	Oscillator Period
LOGICAL OPERATIONS				
ANL	A,R_n	AND Register to Accumulator	1	12
ANL	A, direct	AND direct byte to Accumulator	2	12
ANL	A, @R_i	AND indirect RAM to Accumulator	1	12
ANL	A, #data	AND immediate data to Accumulator	2	12
ANL	direct, A	AND Accumulator to direct byte	2	12
ANL	direct, #data	AND immediate data to direct byte	3	24
ORL	A, R_n	OR register to Accumulator	1	12
ORL	A, direct	OR direct byte to Accumulator	2	12
ORL	A, @R_i	OR indirect RAM to Accumulator	1	12
ORL	A, #data	OR immediate data to Accumulator	2	12
ORL	direct, A	OR Accumulator to direct byte	2	12
ORL	direct, #data	OR immediate data to direct byte	3	24
XRL	A, R_n	Exclusive-OR register to Accumulator	1	12
XRL	A, direct	Exclusive-OR direct byte to Accumulator	2	12
XRL	A, @R_i	Exclusive-OR indirect RAM to Accumulator	1	12
XRL	A, #data	Exclusive-OR immediate data to Accumulator	2	12
XRL	direct, A	Exclusive-OR Accumulator to direct byte	2	12
XRL	direct, #data	Exclusive-OR immediate data to direct byte	3	24
CLR	A	Clear Accumulator	1	12
CPL	A	Complement Accumulator	1	12
RL	A	Rotate Accumulator Left	1	12
RLC	A	Rotate Accumulator Left through the Carry	1	12
RR	A	Rotate Accumulator Right	1	12
RRC	A	Rotate Accumulator Right through the Carry	1	12
SWAP	A	Swap nibbles within the Accumulator	1	12

Mnemonic		Description	Byte	Oscillator Period
DATA TRANSFER				
MOV	A, R_n	Move register to Accumulator	1	12
MOV	A, direct	Move direct byte to Accumulator	2	12
MOV	A, @R_i	Move indirect RAM to Accumulator	1	12
MOV	A, #data	Move immediate data to Accumulator	2	12
MOV	R_n, A	Move Accumulator to register	1	12
MOV	R_n, direct	Move direct byte to register	2	24
MOV	R_n, #data	Move immediate data to register	2	12
MOV	direct, A	Move Accumulator to direct byte	2	12
MOV	direct, R_n	Move register to direct byte	2	24
MOV	direct, direct	Move direct byte to direct	3	24
MOV	direct, @R_i	Move indirect RAM to direct byte	2	24
MOV	direct, #data	Move immediate data to direct byte	3	24
MOV	@R_i, A	Move Accumulator to indirect RAM	1	12
MOV	@R_i, direct	Move direct byte to indirect RAM	2	24
MOV	@R_i, #data	Move immediate data to indirect RAM	2	12
MOV	DPTR, #data 16	Load Data Pointer with a 16-bit constant	3	24
MOVC	A, @A + DPTR	Move Code byte relative to DPTR to Acc	1	24
MOVC	A, @A + PC	Move Code byte relative to PC to Acc	1	24
MOVX	A, @R_i	Move External RAM (8-bit addr) to Acc	1	24
MOVX	A, @DPTR	Move External RAM (16-bit addr) to Acc	1	24
MOVX	@R_i,A	Move Acc to External RAM (8-bit addr)	1	24
MOVX	@DPTR, A	Move Acc to external RAM (16-bit addr)	1	24
PUSH	direct	Push direct byte onto stack	2	24
POP	direct	Pop direct byte from stack	2	24
XCH	A, R_n	Exchange register with Accumulator	1	12
XCH	A, direct	Exchange direct byte with Accumulator	2	12
XCH	A, @R_i	Exchange indirect RAM with Accumulator	1	12
XCHD	A, @R_i	Exchange low-order Digit indirect RAM with Acc	1	12

Mnemonic		Description	Byte	Oscillator Period
BOOLEAN VARIABLE MANIPULATION				
CLR	C	Clear Carry	1	12
CLR	bit	Clear direct bit	2	12
SETB	C	Set Carry	1	12
SETB	bit	Set direct bit	2	12
CPL	C	Complement Carry	1	12
CPL	bit	Complement direct bit	2	12
ANL	C, bit	AND direct bit to CARRY	2	24
ANL	C,/bit	AND complement of direct bit to Carry	2	24
ORL	C, bit	OR direct bit to Carry	2	24
ORL	C,/bit	OR complement of direct bit to Carry	2	24
MOV	C, bit`	Move direct bit to Carry	2	12
MOV	bit, C	Move Carry to direct bit	2	24
JC	rel	Jump if Carry is set	2	24
JNC	rel	Jump if carry not is set	2	24
JB	bit, rel	Jump if direct Bit is set	3	24
JNB	bit, rel	Jump if direct Bit is Not set	3	24
JBC	bit, rel	Jump if direct Bit is set & clear bit	3	24
PROGRAM BRANCHING				
ACALL	addr11	Absolute Subroutine Call	2	24
LCALL	addr16	Long Subroutine Call	3	24
RET		Return from Subroutine	1	24
RETI		Return from interrupt	1	24
AJMP	addr11	Absolute Jump	2	24
LJMP	addr16	Long Jump	3	24
SJMP	rel	Short Jump (relative addr)	2	24
JMP	@A + DPTR	Jump indirect relative to the DPTR	1	24
JZ	rel	Jump if Accumulator is Zero	2	24
JNZ	rel	Jump if Accumulator is Not Zero	2	24
PROGRAM BRANCHING (continued)				

Mnemonic		Description	Byte	Oscillator Period
CJNE	A, direct, rel	Compare direct byte to Acc and Jump if Not Equal	3	24
CJNE	A, #data, rel	Compare immediate to Acc and Jump if Not Equal	3	24
CJNE	R_n, #data, rel	Compare immediate to register and Jump if Not Equal	3	24
CNJE	@R_i, #data, rel	Compare immediate to indirect and Jump if Not Equal	3	24
DJNZ	R_n, rel	Decrement register and Jump if Not Zero	2	24
DJNZ	direct, rel	Decrement direct byte and Jump if Not Zero	3	24
NOP		No Operation	1	12

Table 2. Instruction Opcodes in Hexadecimal Order

Hex Code	Number of Bytes	Mnemonic	Operands
00	1	NOP	
01	2	AJMP	code addr
02	3	LJMP	code addr
03	1	RR	A
04	1	INC	A
05	2	INC	data addr
06	1	INC	@R0
07	1	INC	@R1
08	1	INC	R0
09	1	INC	R1
0A	1	INC	R2
0B	1	INC	R3
0C	1	INC	R4
0D	1	INC	R5
0E	1	INC	R6
0F	1	INC	R7
10	3	JBC	bit, addr, code addr
11	2	ACALL	code addr
12	3	LCALL	code addr
13	1	RRC	A
14	1	DEC	A
15	2	DEC	data addr
16	1	DEC	@R0
17	1	DEC	@R1
18	1	DEC	R0
19	1	DEC	R1
1A	1	DEC	R2
1B	1	DEC	R3
1C	1	DEC	R4
1D	1	DEC	R5
1E	1	DEC	R6
1F	1	DEC	R7
20	3	JB	bit addr, code addr
21	2	AJMP	code addr
22	1	RET	
23	1	RL	A
24	2	ADD	A,#data
25	2	ADD	A, data addr
26	1	ADD	A, @R0
27	1	ADD	A, @R1
28	1	ADD	A, R0
29	1	ADD	A, R1
2A	1	ADD	A, R2

Hex Code	Number of Bytes	Mnemonic	Operands
2B	1	ADD	A, R3
2C	1	ADD	A, R4
2D	1	ADD	A, R5
2E	1	ADD	A, R6
2F	1	ADD	A, R7
30	3	JNB	bit, addr, code addr
31	2	ACALL	code addr
32	1	RETI	
33	1	RLC	A
34	2	ADDC	A, #data
35	2	ADDC	A, data addr
36	1	ADDC	A, @R0
37	1	ADDC	A, @R1
38	1	ADDC	A, R0
39	1	ADDC	A, R1
3A	1	ADDC	A, R2
3B	1	ADDC	A, R3
3C	1	ADDC	A, R4
3D	1	ADDC	A, R5
3E	1	ADDC	A, R6
3F	1	ADDC	A, R7
40	2	JC	code addr
41	2	AJMP	code addr
42	2	ORL	data addr, A
43	3	ORL	data, addr, #data
44	2	ORL	A, #data
45	2	ORL	A, data addr
46	1	ORL	A, @R0
47	1	ORL	A, @R1
48	1	ORL	A, R0
49	1	ORL	A, R1
4A	1	ORL	A, R2
4B	1	ORL	A, R3
4C	1	ORL	A, R4
4D	1	ORL	A, R5
4E	1	ORL	A, R6
4F	1	ORL	A, R7
50	2	JNC	code addr
51	2	ACALL	code addr
52	2	ANL	data addr, A
53	3	ANL	data addr, #data
54	2	ANL	A, #data
55	2	ANL	A, data addr
56	1	ANL	A, @R0
57	1	ANL	A, @R1
58	1	ANL	A, R0
59	1	ANL	A, R1
5A	1	ANL	A, R2
5B	1	ANL	A, R3
5C	1	ANL	A, R4
5D	1	ANL	A, R5
5E	1	ANL	A, R6
5F	1	ANL	A, R7
60	2	JZ	code addr
61	2	AJMP	code addr
62	2	XRL	data addr, A
63	3	XRL	data addr, #data
64	2	XRL	A, #data
65	2	XRL	A, data addr
66	1	XRL	A, @R0
67	1	XRL	A, @R1
68	1	XRL	A, R0
69	1	XRL	A, R1
6A	1	XRL	A, R2
6B	1	XRL	A, R3
6C	1	XRL	A, R4
6D	1	XRL	A, R5
6E	1	XRL	A, R6
6F	1	XRL	A, R7
70	2	JNZ	code addr
71	2	ACALL	code addr
72	2	ORL	C, bit addr

Hex Code	Number of Bytes	Mnemonic	Operands
73	1	JMP	@A + DPTR
74	2	MOV	A, #data
75	3	MOV	data addr, #data
76	2	MOV	@R0, #data
77	2	MOV	@R1, #data
78	2	MOV	R0, #data
79	2	MOV	R1, #data
7A	2	MOV	R2, #data
7B	2	MOV	R3, #data
7C	2	MOV	R4, #data
7D	2	MOV	R5, #data
7E	2	MOV	R6, #data
7F	2	MOV	R7, #data
80	2	SJMP	code addr
81	2	AJMP	code addr
82	2	ANL	C, bit addr
83	1	MOVC	A, @A + PC
84	1	DIV	AB
85	3	MOV	data addr, data addr
86	2	MOV	data addr, @R0
87	2	MOV	data addr, @R1
88	2	MOV	data addr, R0
89	2	MOV	data addr, R1
8A	2	MOV	data addr, R2
8B	2	MOV	data addr, R3
8C	2	MOV	data addr, R4
8D	2	MOV	data addr, R5
8E	2	MOV	data addr, R6
8F	2	MOV	data addr, R7
90	3	MOV	DPTR, #data
91	2	ACALL	code addr
92	2	MOV	bit addr, C
93	1	MOVC	A, @A + DPTR
94	2	SUBB	A, #data
95	2	SUBB	A, data addr
96	1	SUBB	A, @R0

Hex Code	Number of Bytes	Mnemonic	Operands
97	1	SUBB	A, @R1
98	1	SUBB	A, R0
99	1	SUBB	A, R1
9A	1	SUBB	A, R2
9B	1	SUBB	A, R3
9C	1	SUBB	A, R4
9D	1	SUBB	A, R5
9E	1	SUBB	A, R6
9F	1	SUBB	A, R7
A0	2	ORL	C,/bit addr
A1	2	AJMP	code addr
A2	2	MOV	C, bit addr
A3	1	INC	DPTR
A4	1	MUL	AB
A5		reserved	
A6	2	MOV	@R0, data addr
A7	2	MOV	@R1, data addr
A8	2	MOV	R0, data addr
A9	2	MOV	R1, data addr
AA	2	MOV	R2, data addr
AB	2	MOV	R3, data addr
AC	2	MOV	R4, data addr
AD	2	MOV	R5, data addr
AE	2	MOV	R6, data addr
AF	2	MOV	R7, data addr
B0	2	ANL	C,/bit addr
B1	2	ACALL	code addr
B2	2	CPL	bit addr
B3	1	CPL	C
B4	3	CJNE	A,#data, code addr
B5	3	CJNE	A, data addr, code addr
B6	3	CJNE	@R0,#data, code addr
B7	3	CJNE	@R1,#data, code addr
B8	3	CJNE	R0,#data, code addr

Hex Code	Number of Bytes	Mnemonic	Operands
B9	3	CJNE	R1,#data, code addr
BA	3	CJNE	R2,#data, code addr
BB	3	CJNE	R3,#data, code addr
BC	3	CJNE	R4,#data, code addr
BD	3	CJNE	R5,#data, code addr
BE	3	CJNE	R6,#data, code addr
BF	3	CJNE	R7,#data, code addr
C0	2	PUSH	data addr
C1	2	AJMP	code addr
C2	2	CLR	bit addr
C3	1	CLR	C
C4	1	SWAP	A
C5	2	XCH	A, data addr
C6	1	XCH	A, @R0
C7	1	XCH	A, @R1
C8	1	XCH	A, R0
C9	1	XCH	A, R1
CA	1	XCH	A, R2
CB	1	XCH	A, R3
CC	1	XCH	A, R4
CD	1	XCH	A, R5
CE	1	XCH	A, R6
CF	1	XCH	A, R7
D0	2	POP	data addr
D1	2	ACALL	code addr
D2	2	SETB	bit addr
D3	1	SETB	C
D4	1	DA	A
D5	3	DJNZ	data addr, code addr
D6	1	XCHD	A, @R0
D7	1	XCHD	A, @R1
D8	2	DJNZ	R0, code addr
D9	2	DJNZ	R1, code addr
DA	2	DJNZ	R2, code addr
DB	2	DJNZ	R3, code addr
DC	2	DJNZ	R4, code addr
DD	2	DJNZ	R5, code addr
DE	2	DJNZ	R6, code addr
DF	2	DJNZ	R7, code addr
E0	1	MOVX	A, @DPTR
E1	2	AJMP	code addr
E2	1	MOVX	A, @R0
E3	1	MOVX	A, @R1
E4	1	CLR	A
E5	2	MOV	A, data addr
E6	1	MOV	A, @R0
E7	1	MOV	A, @R1
E8	1	MOV	A, R0
E9	1	MOV	A, R1
EA	1	MOV	A, R2
EB	1	MOV	A, R3
EC	1	MOV	A, R4
ED	1	MOV	A, R5
EE	1	MOV	A, R6
EF	1	MOV	A, R7
F0	1	MOVX	@DPTR,A
F1	2	ACALL	code addr
F2	1	MOVX	@R0, A
F3	1	MOVX	@R1, A
F4	1	CPL	A
F5	2	MOV	data addr, A
F6	1	MOV	@R0, A
F7	1	MOV	@R1, A
F8	1	MOV	R0, A
F9	1	MOV	R1, A
FA	1	MOV	R2, A
FB	1	MOV	R3, A
FC	1	MOV	R4, A
FD	1	MOV	R5, A
FE	1	MOV	R6, A
FF	1	MOV	R7, A

Instruction Definitions

Appendix C

ACALL addr11

Function : Absolute Call

Description : ACALL unconditionally calls a subroutine located at the indicated address. The instruction increments the PC twice to obtain the address of the following instruction, then pushes the 16-bit result onto the stack (low-order byte first) and increments the Stack Pointer twice. The destination address is obtained by successively concatenating the five high-order bits of the incremented PC, opcode bits 7 through 5, and the second byte of the instruction. The subroutine called must therefore start within the same 2 K block of the program memory as the first byte of the instruction following ACALL. No flags are affected.

Example : Initially SP equals 07H. The label SUBRTN is at program memory location 0345 H. After excecuting the following instruction,

ACALL SUBRTN

at location 0123H, SP contains 09H, internal RAM locations 08H and 09H will contain 25H and 01H, respectively, and the PC contains 0345H.

Bytes : 2

Cycles : 2

Encoding :

a10 a9 a8 1	0 0 0 1		a7 a6 a5 a4	a3 a2 a1 a0

Operation : ACALL

$(PC) \leftarrow (PC) + 2$

$(SP) \leftarrow (SP) + 1$

$((SP)) \leftarrow (PC_{7-0})$

$(SP) \leftarrow (SP) + 1$

$((SP)) \leftarrow (PC_{15-8})$

$(PC_{10-0}) \leftarrow$ page address

ADD A,< src-byte>

Function : Add

Description : ADD adds the byte variable indicated to the Accumulator, leaving the result in the Accumulator. The carry and auxiliary-carry flags are set, respectively, if there is a carry-out from bit 7 or bit 3, and cleared otherwise. When adding unsigned integers, the carry flag indicates an overflow occurred.

OV is set if there is a carry-out of bit 6 but not out of bit 7, or a carry-out of bit 7 but not bit 6; otherwise, OV is cleared. When adding signed integers, OV indicates a negative number produced as the sum of two positive operands, or a positive sum from two negative operands.

Four source operand addressing modes are allowed: register, direct, register-indirect, or immediate.

Example : The Accumulator holds 0C3H (11000011B), and register 0 holds 0AAH (10101010B). The following instruction,

ADD A, R0

leaves 6EH (01101101B) in the Accumulator with the AC flag cleared and both the carry flag and OV set to 1.

ADD A, R_n

Bytes : 1

Cycles : 1

Encoding :

0 0 1 0	1 r r r

Operation : ADD

$(A) \leftarrow (A) + (R_n)$

ADD A, direct

Bytes : 2

Cycles : 1

Encoding :

0 0 1 0	0 1 0 1	Direct address

Operation : ADD

$(A) \leftarrow (A) + (\text{direct})$

ADD, A, @R_i

Bytes : 1

Cycles : 1

Encoding :

0 0 1 0	0 1 1 i

Operation : ADD

$(A) \leftarrow (A) + ((R_i))$

ADD A,#data

Bytes : 2

Cycles : 1

Encoding :

0 0 1 0	0 1 0 0	immediate data

Operation : ADD

$(A) \leftarrow (A) + \#data$

ADDC A, <src-byte>

Function : Add with Carry

Description : ADDC simultaneously adds the byte variable indicated, the carry flag and the Accumulator contents, leaving the result in the Accumulator. The carry and auxiliary-carry flags are set respectively, if there is a carry-out from bit 7 or bit 3, and cleared otherwise. When adding unsigned integers, the carry flag indicates an overflow occurred.

OV is set if there is a carry-out of bit 6 but not of bit 7, or a carry-out of bit 7 but not out of bit 6; otherwise OV is cleared. When adding signed integers, OV indicates a negative number produced as the sum of two positive operands or a positive sum from two negative operands.

Four source operand addressing modes are allowed : register, direct, register-indirect, or immediate.

Example : The Accumulator holds 0C3H (11000011B) and register 0 holds 0AAH (10101010B) with the carry flag set. The following instruction,

ADDC A,R0

leaves 6EH (01101110B) in the Accumulator with AC cleared and both the Carry flag and OV set to 1.

ADDC A,R_n

Bytes : 1

Cycles : 1

Encoding :

0	0	1	1	1	r	r	r

Operation : ADDC

$(A) \leftarrow (A) + (C) + (R_n)$

ADDC A, direct

Bytes : 2

Cycles : 1

Encoding :

0	0	1	1	0	1	0	1	Direct address

Operation : ADDC

$(A) \leftarrow (A) + (C) + (\text{direct})$

ADDC A, @R_i

Bytes : 1

Cycles : 1

Encoding :

0	0	1	1	0	1	1	i

Operation : ADDC

$(A) \leftarrow (A) + (C) + ((R_i))$

ADDC A, #data

Bytes : 2

Cycles : 1

Encoding :

0	0	1	1	0	1	0	0		immediate data

Operation : ADDC

(A) ← (A) + (C) + #data

AJMP addr11

Function : Absolute Jump

Description : AJMP transfers program execution to the indicated address, which is formed at run-time by concatenating the high-order five bits of the PC (after incrementing the PC twice), opcode bits 7 through 5, and the second byte of the instruction. The destination must therefore be within the same 2 K block of program memory as the first byte of the instruction following AJMP.

Example : The label JMPADR is at program memory location 0123H. The following instruction,

AJMP JMPADR

is at location 0345H and loads the PC with 0123H.

Bytes : 2

Cycles : 2

Encoding :

a10	a9	a8	0	0	0	0	1		a7	a6	a5	a4	a3	a2	a1	a0

Operation : AJMP

(PC) ← (PC) + 2

(PC_{10-0}) ← page address

ANL <dest-byte>,<src-byte>

Function : Logical-AND for byte variables

Description : ANL performs the bitwise logical-AND operation between the variables indicated and stores the results in the destination variable. No flags are affected.

The two operands allow six addressing mode combinations. When the destination is the Accumulator, the source can use register, direct, register-indirect, or immediate addressing; when the destination is a direct address, the source can be the Accumulator or immediate data.

Note: When this instruction is used to modify an output port, the value used as the original port data will be read from the output data latch, *not* the input pins.

Example : If the Accumulator holds 0C3H (11000011B), and register 0 holds 55H (01010101B), then the following instruction,

ANL A,R0

leaves 41H (01000001 B) in the Accumulator.

When the destination is a directly addressed byte, this instruction clears combinations of bits in any RAM location or hardware register. The mask byte determining the pattern of bits to be cleared would either be a constant contained in the instruction or a value computed in the Accumulator at run-time. The following instruction,

ANL P1,#01110011B

clears bits 7, 3, and 2 of output port 1.

ANL A,R_n

Bytes : 1

Cycles : 1

Encoding :

0	1	0	1	1	r	r	r

Operation : ANL

$(A) \leftarrow (A) \wedge (R_n)$

ANL A,direct

Bytes : 2

Cycles : 1

Encoding :

0	1	0	1	0	1	0	1	direct address

Operation : ANL

$(A) \leftarrow (A) \wedge (direct)$

ANL A,@R_i

Bytes : 1

Cycles : 1

Encoding :

0	1	0	1	0	1	1	i

Operation : ANL

$(A) \leftarrow (A) \wedge ((R_i))$

ANL A,#data

Bytes : 2

Cycles : 1

Encoding :

0	1	0	1	0	1	0	0	immediate data

Operation : ANL

$(A) \leftarrow (A) \wedge$ #data

ANL direct,A

Bytes : 2

Cycles : 1

Encoding :

0	1	0	1	0	0	1	0	direct address

Operation : ANL

(direct) ← (direct) ∧ (A)

ANL direct,#data

Bytes : 3

Cycles : 2

Encoding :

0	1	0	1	0	0	1	1	direct address	immediate data

Operation : ANL

(direct) ← (direct) ∧ #data

ANL C,<src-bit>

Function : Logical-AND for bit variables

Description : If the Boolean value of the source bit is a logical 0, then ANL C clears the carry flag; otherwise, this instruction leaves the carry flag in its current state. A slash (/) preceding the operand in the assembly language indicates that the logical complement of the addressed bit is used as the source value, *but the source bit itself is not affected.* No other flags are affected.

Only direct addressing is allowed for the source operand.

Example : Set the carry flag if, and only if, P1.0 = 1, ACC.7 = 1, and OV = 0:

```
MOV   C,P1.0    ; LOAD CARRY WITH INPUT PIN STATE
ANL   C,ACC.7   ; AND CARRY WITH ACCUM. BIT 7
ANL   C,/OV     ; AND WITH INVERSE OF OVERFLOW FLAG
```

ANL C, bit

Bytes : 2

Cycles : 2

Encoding :

1	0	0	0	0	0	1	0	bit address

Operation : ANL

(C) ← (C) ∧ (bit)

ANL C,/bit

Bytes : 2

Cycles : 2

Encoding :

1	0	1	1	0	0	0	0	bit address

Operation : ANL

$(C) \leftarrow (C) \wedge \neg (bit)$

CJNE <dest-byte>,<src-byte>, rel

Function : Compare and Jump if Not Equal.

Description : CJNE compares the magnitudes of the first two operands and branches if their values are not equal. The branch destination is computed by adding the signed relative-displacement in the last instruction byte to the PC, after incrementing the PC to the start of the next instruction. The carry flag is set if the unsigned integer value of <dest-byte> is less than the unsigned integer value of <src-byte>; otherwise, the carry is cleared. Neither operand is affected.

The first two operands allow four addressing mode combinations: the Accumulator may be compared with any directly addressed byte or immediate data, and any indirect RAM location or working register can be compared with an immediate constant.

Example : The Accumulator contains 34H. Register 7 contains 56H. The first instruction in the sequence,

```
            CJNE    R7, # 60H, NOT_EQ
;           ...     .....          ;R7 = 60H.
NOT_EQ:     JC      REQ_LOW        ;IF R7 < 60H.
;           ...     .....          ;R7 > 60H.
```

sets the carry flag and branches to the instruction at label NOT_EQ. By testing the carry flag, this instruction determines whether R7 is greater or less than 60H.

If the data being presented to Port 1 is also 34H, then the following instruction,

```
WAIT: CJNE   A, P1,WAIT
```

clears the carry flag and continues with the next instruction in sequence, since the Accumulator does equal the data read from P1. (If some other value was being input on P1, the program loops at this point until the P1 data changes to 34H.)

CJNE A,direct,rel

Bytes : 3

Cycles : 2

Encoding :

1	0	1	1	0	1	0	1	direct address	rel. address

Operation : $(PC) \leftarrow (PC) + 3$

IF (A) < > (*direct*)
THEN
(PC) ← (PC) + *relative offset*
IF (A) < (*direct*)
THEN
(C) ← 1
ELSE
(C) ← 0

CJNE A,#data,rel

Bytes : 3

Cycles : 2

Encoding :

	0	1	1	0	1	0	0	immediate data	rel. address

Operation : (PC) ← (PC) + 3
IF (A) < > *data*
THEN
(PC) ← (PC) + *relative offset*
IF (A) < *data*
THEN
(C) ← 1
ELSE
(C) ← 0

CJNE R_n,#data,rel

Bytes : 3

Cycles : 2

Encoding :

1	0	1	1	1	r	r	r	immediate data	rel. address

Operation : (PC) ← (PC) + 3
IF (R_n) < > *data*
THEN
(PC) ← (PC) + *relative offset*
IF (R_n) < data
THEN
(C) ← 1
ELSE
(C) ← 0

CJNE @R_i,data,rel

Bytes : 3

Cycles : 2

Encoding :

1	0	1	1	0	1	1	i	immediate data	rel. address

Operation : (PC) ← (PC) + 3

IF $((R_i) <> \textit{data}$

THEN

(PC) ← (PC) + *relative offset*

IF $((R_i)) < \textit{data}$

THEN

(C) ← 1

ELSE

(C) ← 0

CLR A

Function : Clear Accumulator

Description : CLR A clears the Accumulator (all bits set to 0). No flags are affected.

Example : The Accumulator contains 5CH (01011100B). The following instruction,CLR A leaves the Accumulator set to 00H (00000000B).

Bytes : 1

Cycles : 1

Encoding :

1	1	1	0	0	1	0	0

Operation : CLR

(A) ← 0

CLR bit

Function : Clear bit

Description : CLR bit clears the indicated bit (reset to 0). No other flags are affected. CLR can operate on the carry flag or any directly addressable bit.

Example : Port 1 has previously been written with 5DH (01011101B). The following instruction, CLR P1.2 leaves the port set to 59H (01011001B).

CLR C

Bytes : 1

Cycles : 1

Encoding :

1	1	0	0	0	0	1	1

Operation : CLR

(C) ← 0

CLR bit

Bytes : 2

Cycles : 1

Encoding :

1	1	0	0	0	0	1	0	bit address

Operation : CLR

(bit) ← 0

CPL A

Function : Complement Accumulator

Description : CPLA logically complements each bit of the Accumulator (one's complement). Bits which previously contained a 1 are changed to a 0 and vice versa. No flags are affected.

Example : The Accumulator contains 5CH (01011100B). The following instruction,

CPL A

leaves the Accumulator set to 0A3H (10100011B).

Bytes : 1

Cycles : 1

Encoding :

1	1	1	1	0	1	0	0

Operation : CPL

(A) ← ¬ (A)

CPL bit

Function : Complement bit

Description : CPL bit complements the bit variable specified. A bit that had been a 1 is changed to 0 and vice-versa. No other flags are affected. CLR can operate on the carry or any directly addressable bit.

Note: When this instruction is used to modify an output pin, the value used as the original data is read from the output data latch, *not* the input pin.

Example : Port 1 has previously been written with 5BH (01011011B). The following instruction sequence, CPL P1.1CPL P1.2 leaves the port set to 5BH (01011011B).

CPL C

Bytes : 1

Cycles : 1

Encoding :

1	0	1	1	0	0	1	1

Operation : CPL

(C) ← ¬ (C)

CPL bit

Bytes : 2

Cycles : 1

Encoding :

1	0	1	1	0	0	1	0	bit address

Operation : CPL

(bit) ← ¬ (bit)

DA A

Function : Decimal-adjust Accumulator for Addition

Description : DA A adjusts the eight-bit value in the Accumulator resulting from the earlier addition of two variables (each in packed-BCD format), producing two four-bit digits. Any ADD or ADDC instruction may have been used to perform the addition.

If Accumulator bits 3 through 0 are greater than nine (xxxx1010-xxxx1111), or if the AC flag is one, six is added to the Accumulator producing the proper BCD digit in the low-order nibble. This internal addition sets the carry flag if a carry-out of the low-order four-bit field propagates through all high-order bits, but it does not clear the carry flag otherwise.

If the carry flag is now set, or if the four high-order bits now exceed nine (1010xxxx-1111xxxx), these high-order bits are incremented by six, producing the proper BCD digit in the high-order nibble. Again, this sets the carry flag if there is a carry-out of the high-order bits, but does not clear the carry. The carry flag thus indicates if the sum of the original two BCD variables is greater than 100, allowing multiple precision decimal addition. OV is not affected.

All of this occurs during the one instruction cycle. Essentially, this instruction performs the decimal conversion by adding 00H, 06H, 60H, or 66H to the Accumulator, depending on initial Accumulator and PSW conditions.

Note : DA A *cannot* simply convert a hexadecimal number in the Accumulator to BCD notation, nor does DAA apply to decimal subtraction.

Example : The Accumulator holds the value 56H (01010110B), representing the packed BCD digits of the decimal number 56. Register 3 contains the value 67H (01100111B), representing the packed BCD digits of the decimal number 67. The carry flag is set. The following instruction sequence

ADDC A,R3

DA A

first performs a standard two's-complement binary addition, resulting in the value 0BEH (10111110) in the Accumulator. The carry and auxiliary carry flags are cleared.

The Decimal Adjust instruction then alters the Accumulator to the value 24H (00100100B), indicating the packed BCD digits of the decimal number 24, the low-order two digits of the decimal sum of 56, 67, and the carry-in. The carry flag is set by the Decimal Adjust instruction, indicating that a decimal overflow occurred. The true sum of 56, 67, and 1 is 124.

BCD variables can be incremented or decremented by adding 01H or 99H. If the Accumulator initially holds 30H (representing the digits of 30 decimal), then the following instruction sequence,

ADD A, # 99H

DA A

leaves the carry set and 29H in the Accumulator, since 30 + 99 = 129. The low-order byte of the sum can be interpreted to mean 30 – 1 = 29.

Bytes : 1

Cycles : 1

Encoding :

1	1	0	1	0	1	0	0

Operation : DA

-contents of Accumulator are BCD

IF $[[(A_{3\text{-}0}) > 9] \vee [(AC) = 1]]$

THEN $(A_{3\text{-}0}) \leftarrow (A_{3\text{-}0}) + 6$

AND

IF $[[(A_{7\text{-}4}) > 9] \vee [(C) = 1]]$

THEN $(A_{7\text{-}4}) \leftarrow (A_{7\text{-}4}) + 6$

DEC byte

Function : Decrement

Description : DEC byte decrements the variable indicated by 1. An original value of 00H underflows to 0FFH. No flags are affected. Four operand addressing modes are allowed: accumulator, register, direct, or register-indirect.

Note: When this instruction is used to modify an output port, the value used as the original port data will be read from the output data latch, *not* the input pins.

Example : Register 0 contains 7FH (01111111B). Internal RAM locations 7EH and 7FH contain 00H and 40H, respectively. The following instruction sequence,

DEC @R0

DEC R0

DEC @R0

leaves register 0 set to 7EH and internal RAM locations 7EH and 7FH set to 0FFH and 3FH.

DEC A

Bytes : 1

Cycles : 1

Encoding :

0	0	0	1	0	1	0	0

Operation : DEC

(A) ← (A) − 1

DEC R_n

Bytes : 1

Cycles : 1

Encoding :

0	0	0	1	1	r	r	r

Operation : DEC

$(R_n) \leftarrow (R_n) - 1$

DEC direct

Bytes : 2

Cycles : 1

Encoding :

0	0	0	1	0	1	0	1	direct address

Operation : DEC

(direct) ← (direct) − 1

DEC @R_i

Bytes : 1

Cycles : 1

Encoding :

0	0	0	1	0	1	1	i

Operation : DEC

$((R_i)) \leftarrow ((R_i)) - 1$

DIV AB

Function : Divide

Description : DIV AB divides the unsigned eight-bit integer in the Accumulator by the unsigned eight-bit integer in register B. The Accumulator receives the integer part of the quotient; register B receives the integer remainder. The carry and OV flags are cleared.

Exception: if B had originally contained 00H, the values returned in the Accumulator and B-register are undefined and the overflow flag are set. The carry flag is cleared in any case.

Example : The Accumulator contains 251 (0FBH or 11111011B) and B contains 18 (12H or 00010010B). The following instruction,

DIV AB

leaves 13 in the Accumulator (0DH or 00001101B) and the value 17 (11H or 00010001B) in B, since 251 = (13 × 18) + 17. Carry and OV are both cleared.

Bytes : 1

Cycles : 4

Encoding :

1 0 0 0	0 1 0 0

Operation : DIV

$(A)_{15-8} \leftarrow (A)/(B)$

$(B)_{7-0}$

DJNZ <byte>,<rel-addr>

Function : Decrement and Jump if Not Zero

Description : DJNZ decrements the location indicated by 1, and branches to the address indicated by the second operand if the resulting value is not zero. An original value of 00H underflows to 0FFH. No flags are affected. The branch destination is computed by adding the signed relative-displacement value in the last instruction byte to the PC, after incrementing the PC to the first byte of the following instruction.

The location decremented may be a register or directly addressed byte.

Note: When this instruction is used to modify an output port, the value used as the original port data will be read from the output data latch, *not* the input pins.

Example : Internal RAM locations 40H, 50H, and 60H contain the values 01H, 70H, and 15H, respectively. The following instruction sequence,

```
DJNZ    40H,LABEL_1
DJNZ    50H,LABEL_2
DJNZ    60H,LABEL_3
```

causes a jump to the instruction at label LABEL_2 with the values 00H, 6FH, and 15H in the three RAM locations. The first jump was *not* taken because the result was zero.

This instruction provides a simple way to execute a program loop a given number of times or for adding a moderate time delay (from 2 to 512 machine cycles) with a single instruction. The following instruction sequence,

```
          MOV     R2, #8
TOGGLE:   CPL     P1.7
          DJNZ    R2,TOGGLE
```

toggles P1.7 eight times, causing four output pulses to appear at bit 7 of output Port 1. Each pulse lasts three machine cycles; two for DJNZ and one to alter the pin.

DJNZ R_n,rel

Bytes : 2

Cycles : 2

Encoding :

1	1	0	1	1	r	r	r		rel. address

Operation : DJNZ

(PC) ← (PC) + 2

$(R_n) \leftarrow (R_n) - 1$

IF $(R_n) > 0$ or $(R_n) < 0$

THEN

(PC) ← (PC) + rel

DJNZ direct,rel

Bytes : 3

Cycles : 2

Encoding :

1	1	0	1	0	1	0	1		direct address		rel. address

Operation : DJNZ

(PC) ← (PC) + 2

(direct) ← (direct) –1

IF (direct) > 0 or (direct) < 0

THEN

(PC) ← (PC) + rel

INC <byte>

Function : Increment

Description : INC increments the indicated variable by 1. An original value of 0FFH overflows to 00H. No flags are affected. Three addressing modes are allowed: register, direct, or register-indirect.

Note: When this instruction is used to modify an output port, the value used as the original port data will be read from the output data latch, *not* the input pins.

Example : Register 0 contains 7EH (01111110B). Internal RAM locations 7EH and 7FH contain 0FFH and 40H, respectively. The following instruction sequence,

```
INC     @R0
INC     R0
INC     @R0
```

leaves register 0 set to 7FH and internal RAM locations 7EH and 7FH holding 00H and 41H, respectively.

INC A

Bytes : 1

Cycles : 1

Encoding :

0	0	0	0	0	1	0	0

Operation : INC

$(A) \leftarrow (A) + 1$

INC R_n

Bytes : 1

Cycles : 1

Encoding :

0	0	0	0	1	r	r	r

Operation : INC

$(R_n) \leftarrow (R_n) + 1$

INC direct

Bytes : 2

Cycles : 1

Encoding :

0	0	0	0	0	1	0	1	direct address

Operation : INC

$(\text{direct}) \leftarrow (\text{direct}) + 1$

INC $@R_i$

Bytes : 1

Cycles : 1

Encoding :

0	0	0	0	0	1	1	i

Operation : INC

$((R_i)) \leftarrow ((R_i)) + 1$

INC DPTR

Function : Increment Data Pointer

Description : INC DPTR increments the 16-bit data pointer by 1. A 16-bit increment (modulo 2^{16}) is performed, and an overflow of the low-order byte of the data pointer (DPL) from 0FFH to 00H increments the high-order byte (DPH). No flags are affected.

This is the only 16-bit register which can be incremented.

Example : Registers DPH and DPL contain 12H and 0FEH, respectively. The following instruction sequence,

```
INC     DPTR
INC     DPTR
INC     DPTR
```

changes DPH and DPL to 13H and 01H.

Bytes : 1

Cycles : 2

Encoding :

1 0 1 0	0 0 1 1

Operation : INC

(DPTR) ← (DPTR) + 1

JB bit,rel

Function : Jump if Bit set

Description : If the indicated bit is a one, JB jump to the address indicated; otherwise, it proceeds with the next instruction. The branch destination is computed by adding the signed relative-displacement in the third instruction byte to the PC, after incrementing the PC to the first byte of the next instruction. The bit tested is not modified. No flags are affected.

Example : The data present at input port 1 is 11001010B. The Accumulator holds 56 (01010110B). The following instruction sequence,

```
JB      P1.2,LABEL1
JB      ACC.2,LABEL2
```

causes program execution to branch to the instruction at label LABEL2.

Bytes : 3

Cycles : 2

Encoding :

0 0 1 0	0 0 0 0	bit address	rel. address

Operation : JB

(PC) ← (PC) + 3

IF (bit) = 1

THEN

(PC) ← (PC) + rel

JBC bit,rel

Function : Jump if Bit is set and Clear bit

Description : If the indicated bit is one, JBC branches to the address indicated; otherwise, it proceeds with the next instruction. *The bit will not be cleared if it is already a zero.* The branch destination is computed by adding the signed relative-displacement in the third instruction byte to the PC, after incrementing the PC to the first byte of the next instruction. No flags are affected.

Note: When this instruction is used to test an output pin, the value used as the original data will be read from the output data latch, *not* the input pin.

Example : The Accumulator holds 56H (01010110B). The following instruction sequence,

```
JBC     ACC.3,LABEL1
JBC     ACC.2,LABEL2
```

causes program execution to continue at the instruction identified by the label LABEL2, with the Accumulator modified to 52H (01010010B).

Bytes : 3

Cycles : 2

Encoding :

0 0 0 1	0 0 0 0	bit address	rel. address

Operation : JBC

(PC) ← (PC) + 3

IF (bit) = 1

THEN

(bit) ← 0

(PC) ← (PC) + rel

JC rel

Function : Jump if Carry is set

Description : If the carry flag is set, JC branches to the address indicated; otherwise, it proceeds with the next instruction. The branch destination is computed by adding the signed relative-displacement is the second instruction byte to the PC, after incrementing the PC twice. No flags are affected.

Example : The carry fiag is cleared. The following instruction sequence,

```
JC      LABEL1
CPL     C
JC      LABEL2
```

sets the carry and causes program execution to continue at the instruction identified by the lable LABEL2.

Bytes : 2

Cycles : 2

Encoding :

0	1	0	0	0	0	0	0	rel. address

Operation : JC

(PC) ← (PC) + 2

IF (C) = 1

THEN

(PC) ← (PC) + rel

JMP @A+DPTR

Function : Jump indirect

Description : JMP @A+DPTR adds the eight-bit unsigned contents of the Accumulator with the 16-bit data pointer and loads the resulting sum to the program counter. This is the address for subsequent instruction fetches. Sixteen-bit addition is performed (modulo 2^{16}): a carry-out from the low-order eight bits propagates through the higher-order bits. Neither the Accumulator nor the Data Pointer is altered. No flags are affected.

Example : An even number from 0 to 6 is in the Accumulator. The following sequence of instructions branches to one of four AJMP instructions in a jump table starting at JMP_TBL.

```
          MOV     DPTR, # JMP_TBL
          JMP     @A + DPTR
JMP_TBL:  AJMP    LABEL0
          AJMP    LABEL1
          AJMP    LABEL2
          AJMP    LABEL3
```

If the Accumulator equals 04H when starting this sequence, execution jumps to label LABEL2. Because AJMP is a 2-byte instruction, the jump instructions start at every other address.

Bytes : 1

Cycles : 2

Encoding :

0 1 1 1	0 0 1 1

Operation : JMP

(PC) ← (A) + (DPTR)

JNB bit,rel

Function : Jump if Bit Not set

Description : If the indicated bit is a 0, JNB branches to the indicated address; otherwise, it proceeds with the next instruction. The branch destination is computed by adding the signed relative-displacement in the third instruction byte to the PC, after incrementing the PC to the first byte of the next instruction. *The bit tested is not modified.* No flags are affected.

Example : The data present at input port 1 is 11001010B. The Accumulator holds 56H (01010110B). The following instruction sequence,

```
JNB     P1.3,LABEL1
JNB     ACC.3,LABEL2
```

causes program execution to continue at the instruction at label LABEL2.

Bytes : 3

Cycles : 2

Encoding :

0 0 1 1	0 0 0 0	bit address	rel. address

Operation : JNB

(PC) ← (PC) + 3

IF (bit) = 0

THEN (PC) ← (PC) + rel

JNC rel

Function : Jump if Carry not set

Description : If the carry flag is a 0, JNC branches to the address indicated; otherwise, it proceeds with the next instruction. The branch destination is computed by adding the signal relative-displacement in the second instruction byte to the PC, after incrementing the PC twice to point to the next instruction. The carry flag is not modified.

Example : The carry flag is set. The following instruction sequence,

```
JNC     LABEL1
CPL     C
JNC     LABEL2
```

clears the carry and causes program execution to continue at the instruction identified by the label LABEL2.

Bytes : 2

Cycles : 2

Encoding :

0	1	0	1	0	0	0	0

rel. address

Operation : JNC

(PC) ← (PC) + 2

IF (C) = 0

THEN (PC) ← (PC) + rel

JNZ rel

Function : Jump if Accumulator Not Zero

Description : If any bit of the Accumulator is a one, JNZ branches to the indicated address; otherwise, it proceeds with the next instruction. The branch destination is computed by adding the signed relative-displacement in the second instruction byte to the PC, after incrementing the PC twice. The Accumulator is not modified. No flags are affected.

Example : The Accumulator originally holds 00H. The following instruction sequence,

```
JNZ     LABEL1
INC     A
JNZ     LABEL2
```

sets the Accumulator to 01H and continues at label LABEL2.

Bytes : 2

Cycles : 2

Encoding :

0	1	1	1	0	0	0	0

rel. address

Operation : JNZ

(PC) ← (PC) + 2

IF (A) ≠ 0

THEN (PC) ← (PC) + rel

JZ rel

Function : Jump if Accumulator Zero

Description : If all bits of the Accumulator are 0, JZ branches to the address indicated; otherwise, it proceeds with the next instruction. The branch destination is computed by adding the signed relative-displacement in the second instruction byte to the PC, after incrementing the PC twice. The Accumulator is not modified. No flags are affected.

Example : The Accumulator originally contains 01H. The following instruction sequence,

```
JZ      LABEL1
DEC     A
JZ      LABEL2
```

changes the Accumulator to 00H and causes program execution to continue at the instruction identified by the label LABEL2.

Bytes : 2

Cycles : 2

Encoding :

0	1	1	0	0	0	0	0	rel. address

Operation : JZ

(PC) ← (PC) + 2

IF (A) = 0

THEN (PC) ← (PC) + rel

LCALL addr16

Function : Long call

Description : LCALL calls a subroutine located at the indicated address. The instruction adds three to the program counter to generate the address of the next instruction and then pushes the 16-bit result onto the stack (low byte first), incrementing the Stack Pointer by two. The high-order and low-order bytes of the PC are then loaded, respectively, with the second and third bytes of the LCALL instruction. Program execution continues with the instruction at this address. The subroutine may therefore begin anywhere in the full 64K byte program memory address space. No flags are affected.

Example : Initially the Stack Pointer equals 07H. The label SUBRTN is assigned to program memory location 1234H. After executing the instruction,

```
LCALL       SUBRTN
```

at location 0123H, the Stack Pointer will contain 09H, internal RAM locations 08H and 09H will contain 26H and 01H, and the PC will contain 1234H.

Bytes : 3

Cycles : 2

Encoding :

0	0	0	1	0	0	1	0	addr15-addr8	addr7-addr0

Operation : LCALL

(PC) ← (PC) + 3

(SP) ← (SP) + 1

$((SP)) \leftarrow (PC_{7-0})$

$(SP) \leftarrow (SP) + 1$

$((SP)) \leftarrow (PC_{15-8})$

$(PC) \leftarrow addr_{15-0}$

LJMP addr16

Function : Long Jump

Description : LJMP causes an unconditional branch to the indicated address, by loading the high-order and low-order bytes of the PC (respectively) with the second and third instruction bytes. The destination may therefore be anywhere in the full 64K program memory address space. No flags are affected.

Example : The label JMPADR is assigned to the instruction at program memory location 1234H. The instruction,

LJMP JMPADR

at location 0123H will load the program counter with 1234H.

Bytes : 3

Cycles : 2

Encoding :

0 0 0 0	0 0 1 0	addr15-addr8	addr7-addr0

Operation : LJMP

$(PC) \leftarrow addr_{15-0}$

MOV <dest-byte>,<src-byte>

Function : Move byte variable

Description : The byte variable indicated by the second operand is copied into the location specified by the first operand. The source byte is not affected. No other register or flag is affected.

This is by far the most flexible operation. Fifteen combinations of source and destination addressing modes are allowed.

Example : Internal RAM location 30H holds 40H. The value of RAM location 40H is 10H. The data present at input port 1 is 11001010B(0CAH).

```
MOV     R0,#30H     ;R0 < = 30H
MOV     A,@R0       ;A < = 40H
MOV     R1,A        ;R1< = 40H
MOV     B,@R1       ;B < = 10H
MOV     @R1 ,P1     ;RAM (40H) < = OCAH
MOV     P2,P1       ;P2 #0CAH
```

leaves the value 30H in register 0, 40H in both the Accumulator and register 1, 10H in register B, and 0CAH (11001010B) both in RAM location 40H and output on port 2.

MOV A,R_n

Bytes : 1

Cycles : 1

Encoding :

1	1	1	0	1	r	r	r

Operation : MOV

$(A) \leftarrow (R_n)$

***MOV A,direct**

Bytes : 2

Cycles : 1

Encoding :

1	1	1	0	0	1	0	1	direct address

Operation : MOV

(A) ← (direct)

***MOV A,ACC is not a valid Instruction.**

MOV A,@R_i

Bytes : 1

Cycles : 1

Encoding :

1	1	1	0	0	1	1	i

Operation : MOV

$(A) \leftarrow ((R_i))$

MOV A,#data

Bytes : 2

Cycles : 1

Encoding :

0	1	1	1	0	1	0	0	Immediate data

Operation: MOV

(A) ← #data

MOV R_n,A

Bytes : 1

Cycles : 1

Encoding :

1	1	1	1	1	r	r	r

Operation : MOV

$(R_n) \leftarrow (A)$

MOV R_n,direct

Bytes : 2

Cycles : 2

Encoding :

1 0 1 0	1 r r r	direct addr.

Operation : MOV

$(R_n) \leftarrow$ (direct)

MOV R_n,#data

Bytes : 2

Cycles : 1

Encoding :

0 1 1 1	1 r r r	immediate data

Operation : MOV

$(R_n) \leftarrow$ #data

MOV direct,A

Bytes : 2

Cycles : 1

Encoding :

1 1 1 1	0 1 0 1	direct address

Operation : MOV

(direct) $\leftarrow$ (A)

MOV direct, R_n

Bytes : 2

Cycles : 2

Encoding :

1 0 0 0	1 r r r	direct address

Operation : MOV

(direct) $\leftarrow (R_n)$

MOV direct,direct

Bytes : 3

Cycles : 2

Encoding :

1 0 0 0	0 1 0 1	dir. addr. (scr)	dir. addr. (dest)

Operation : MOV

(direct) $\leftarrow$ (direct)

MOV direct,@R_i

Bytes : 2

Cycles : 2

Encoding :

1	0	0	0	0	1	1	1	direct addr.

Operation : MOV

(direct) ← $((R_i))$

MOV direct,#data

Bytes : 3

Cycles : 2

Encoding :

0	1	1	1	0	1	0	1	direct address	immediate data

Operation : MOV

(direct) ← #data

MOV @R_i,A

Bytes : 1

Cycles : 1

Encoding :

1	1	1	1	0	1	1	i

Operation : MOV

$((R_i))$ ← (A)

MOV @R_i,direct

Bytes : 2

Cycles : 2

Encoding :

1	0	1	0	0	1	1	i	direct addr.

Operation : MOV

((R)) ← (direct)

MOV @R_i,#data

Bytes : 2

Cycles : 1

Encoding :

0	1	1	1	0	1	1	i	immediate data

Operation : MOV

$((R_i))$ ← #data

MOV <dest-bit>,<src-bit>

Function : Move bit data

Description : MOV <dest-bit>,<src-bit> copies the Boolean variable indicated by the second operand into the location specified by the first operand. One of the operands must be the carry flag; the other may be any directly addressable bit. No other register or flag is affected.

Example : The carry flag is originally set. The data present at input Port 3 is 11000101B. The data previously written to output Port 1 is 35H (00110101B).

```
MOV    P1.3,C
MOV    C,P3.3
MOV    P1.2,C
```

leaves the carry cleared and changes Port 1 to 39H (00111001B).

MOV C,bit

Bytes : 2

Cycles : 1

Encoding :

1 0 1 0	0 0 1 0	bit address

Operation : MOV

(C) ← (bit)

MOV bit,C

Bytes : 2

Cycles : 2

Encoding :

1 0 0 1	0 0 1 0	bit address

Operation : MOV

(bit) ← (C)

MOV DPTR,#data16

Function : Load Data Pointer with a 16-bit constant

Description : MOV DPTR,#data16 loads the Data Pointer with the 16-bit constant indicated. The 16-bit constant is loaded into the second and third bytes of the instruction. The second byte (DPH) is the high-order byte, while the third byte (DPL) holds the lower-order byte. No flags are affected.

This is the only instruction which moves 16 bits of data at once.

Example: The instruction,

```
MOV    DPTR, #1234H
```

loads the value 1234H into the Data Pointer: DPH holds 12H, and DPL holds 34H.

Bytes : 3

Cycles : 2

Encoding :

1	0	0	1	0	0	0	0	immed. data15-8	immed. data7-0

Operation : MOV

(DPTR) ← #$data_{15-0}$

DPH ← DPL ← #$data_{15-8}$ ← #$data_{7-0}$

MOVC A,@A+ <base-reg>

Function : Move Code byte

Description : The MOVC instructions load the Accumulator with a code byte or constant from program memory. The address of the byte fetched is the sum of the original unsigned 8-bit Accumulator contents and the contents of a 16-bit base register, which may be either the Data Pointer or the PC. In the latter case, the PC is incremented to the address of the following instruction before being added with the Accumulator; otherwise the base register is not altered. Sixteen-bit addition is performed so a carry-out from the low-order eight bits may propagate through higher-order bits. No flags are affected.

Example : A value between 0 and 3 is in the Accumulator. The following instructions will translate the value in the Accumulator to one of four values defined by the DB (define byte) directive.

```
REL_PC:  INC    A
         MOVC   A,@A+PC
         RET
         DB     66H
         DB     77H
         DB     88H
         DB     99H
```

If the subroutine is called with the Accumulator equal to 01H, it returns with 77H in the Accumulator. The INC A before the MOVC instruction is needed to "get around" the RET instruction above the table. If several bytes of code separate the MOVC from the table, the corresponding number is added to the Accumulator instead.

MOVC A,@A+DPTR

Bytes : 1

Cycles : 2

Encoding :

1	0	0	1	0	0	1	1

Operation : MOVC

(A) ← ((A) + (DPTR))

MOVC A,@A+PC

Bytes : 1

Cycles : 2

Encoding :

1	0	0	0	0	0	1	1

Operation : MOVC

(PC) ← (PC) + 1

(A) ← ((A) + (PC))

MOVX <dest-byte>,<src-byte>

Function : Move External

Description : The MOVX instructions transfer data between the Accumulator and a byte of external data memory, which is why "X" is appended to MOV. There are two types of instructions, differing in whether they provide an 8-bit or 16-bit indirect address to the external data RAM.

In the first type, the contents of R0 or R1 in the current register bank provide an 8-bit address multiplexed with data on P0. Eight bits are sufficient for external I/O expansion decoding or for a relatively small RAM array. For somewhat larger arrays, any output port pins can be used to output higher-order address bits. These pins are controlled by an output instruction preceding the MOVX.

In the second type of MOVX instruction, the Data Pointer generates *a* 16-bit address. P2 outputs the high-order eight address bits (the contents of DPH), while P0 multiplexes the low-order eight bits (DPL) with data. The P2 Special Function Register retains its previous contents, while the P2 output buffers emit the contents of DPH. This form of MOVX is faster and more efficient when accessing very large data arrays (up to 64K bytes), since no additional instructions are needed to set-up the output ports.

It is possible to use both MOVX types in some situations. A large RAM array with its high-order address lines driven by P2 can be addressed via the Data Pointer, or with code to output high-order address bits to P2, followed by a MOVX instruction using R0 or R1.

Example : An external 256 byte RAM using multiplexed address/data lines is connected to the 8051 Port 0. Port 3 provides control lines for the external RAM. Ports 1 and 2 are used for normal I/O. Registers 0 and 1 contain 12H and 34H. Location 34H of the external RAM holds the value 56H. The instruction sequence,

MOVX A,@R1

MOVX @R0,A

copies the value 56H into both the Accumulator and external RAM location 12H.

MOVX A,@R_i

Bytes : 1

Cycles : 2

Encoding :

1	1	1	0	0	0	1	i

Operation : MOVX

(A) ← ((R_i))

MOVX A,@DPTR

Bytes : 1

Cycles : 2

Encoding :

1	1	1	0	0	0	0	0

Operation : MOVX

(A) ← ((DPTR))

MOVX @R_i,A

Bytes : 1

Cycles : 2

Encoding :

1	1	1	1	0	0	1	i

Operation : MOVX

((R_i)) ← (A)

MOVX @DPTR,A

Bytes : 1

Cycles : 2

Encoding :

1	1	1	1	0	0	0	0

Operation : MOVX

(DPTR) ← (A)

MUL AB

Function : Multiply

Description : MUL AB multiplies the unsigned 8-bit integers in the Accumulator and register B. The low-order byte of the 16-bit product is left in the Accumulator, and the high-order byte in B. If the product is greater than 255 (0FFH), the overflow flag is set; otherwise it is cleared. The carry flag is always cleared.

Example : Originally the Accumulator holds the value 80 (50H). Register B holds the value 160 (0A0H). The instruction,

MUL AB

will give the product 12,800 (3200H), so B is changed to 32H (00110010B) and the Accumulator is cleared. The overflow flag is set, carry is cleared.

Bytes : 1

Cycles : 4

Encoding :

1	0	1	0	0	1	0	0

Operation : MUL

$(A)_{7-0} \leftarrow (A) \times (B)$

$(B)_{15-8}$

NOP

Function : No Operation

Description : Execution continues at the following instruction. Other than the PC, no registers or flags are affected.

Example : A low-going output pulse on bit 7 of Port 2 must last exactly 5 cycles. A simple SETB/CLR sequence generates a one-cycle pulse, so four additional cycles must be inserted. This may be done (assuming no interrupts are enabled) with the following instruction sequence,

```
CLR     P2.7
NOP
NOP
NOP
NOP
SETB    P2.7
```

Bytes : 1

Cycles : 1

Encoding :

0	0	0	0	0	0	0	0

Operation : NOP

$(PC) \leftarrow (PC) + 1$

ORL <dest-byte> <src-byte>

Function : Logical-OR for byte variables

Description : ORL performs the bitwise logical-OR operation between the indicated variables, storing the results in the destination byte. No flags are affected.

The two operands allow six addressing mode combinations. When the destination is the Accumulator, the source can use register, direct, register-indirect, or immediate addressing; when the destination is a direct address, the source can be the Accumulator or immediate data.

Note: When this instruction is used to modify an output port, the value used as the original port data is read from the output data latch, *not* the input pins.

Example : If the Accumulator holds 0C3H (11000011B) and R0 holds 55H (01010101B) then the following instruction,

ORL A,R0

leaves the Accumulator holding the value 0D7H (1101011IB).When the destination is a directly addressed byte, the instruction can set combinations of bits in any RAM location or hardware register. The pattern of bits to be set is determined by a mask byte, which may be either a constant data value in the instruction or a variable computed in the Accumulator at run-time. The instruction,

ORL P1,#00110010B

sets bits 5, 4, and 1 of output Port 1.

ORL A,R_n

Bytes : 1

Cycles : 1

Encoding :

0	1	0	0	1	r	r	r

Operation : ORL

$(A) \leftarrow (A) \vee (R_n)$

ORL A,direct

Bytes : 2

Cycles : 1

Encoding :

0	1	0	0	0	1	0	1	direct address

Operation : ORL

$(A) \leftarrow (A) \vee (\text{direct})$

ORL A,@R_i

Bytes : 1

Cycles : 1

Encoding :

0	1	0	0	0	1	1	i

Operation : ORL

$(A) \leftarrow (A) \vee ((R_i))$

ORL A,#data

Bytes : 2

Cycles : 1

Encoding :

0	1	0	0	0	1	0	0	immediate data

Operation : ORL

$(A) \leftarrow (A) \vee \#data$

ORL direct,A

Bytes : 2

Cycles : 1

Encoding :

0	1	0	0	0	0	1	0	direct address

Operation : ORL

$(direct) \leftarrow (direct) \vee (A)$

ORL direct,#data

Bytes : 3

Cycles : 2

Encoding :

0	1	0	0	0	0	1	1	direct addr.	immediate data

Operation : ORL

$(direct) \leftarrow (direct) \vee \#data$

ORL C,<src-bit>

Function : Logical-OR for bit variables

Description : Set the carry flag if the Boolean value is a logical 1; leave the carry in its current state otherwise. A slash (/) preceding the operand in the assembly language indicates that the logical complement of the addressed bit is used as the source value, but the source bit itself is not affected. No other flags are affected.

Example : Set the carry flag if and only if P1.0 = 1, ACC. 7 = 1, or OV = 0:

```
MOV    C, P1.0     ;LOAD CARRY WITH INPUT PIN P10
ORL    C,ACC.7     ;OR CARRY WITH THE ACC. BIT 7
ORL    C,/OV       ;OR CARRY WITH THE INVERSE OF OV
```

ORL C,bit

Bytes : 2

Cycles : 2

Encoding :

0	1	1	1	0	0	1	0	bit address

Operation : ORL

(C) ← (C) ∨ (bit)

ORL C/bit

Bytes : 2

Cycles : 2

Encoding :

1	0	1	0	0	0	0	0	bit address

Operation : ORL

(C) ← (C) ∨ $(\overline{\text{bit}})$

POP direct

Function : Pop from stack.

Description : The contents of the internal RAM location addressed by the Stack Pointer is read, and the Stack Pointer is decremented by one. The value read is then transferred to the directly addressed byte indicated. No flags are affected.

Example : The Stack Pointer originally contains the value 32H, and internal RAM locations 30H through 32H contain the values 20H, 23H, and 01H, respectively. The following instruction sequence,

```
POP    DPH
POP    DPL
```

leaves the Stack Pointer equal to the value 30H and sets the Data Pointer to 0123H. At this point, the following instruction,

```
POP    SP
```

leaves the Stack Pointer set to 20H. In this special case, the Stack Pointer was decremented to 2FH before being loaded with the value popped (20H).

Bytes : 2

Cycles : 2

Encoding :

1	1	0	1	0	0	0	0	direct address

Operation : POP

(direct) ← ((SP))

(SP) ← (SP) − 1

PUSH direct

Function : Push onto stack

Description : The Stack Pointer is incremented by one. The contents of the indicated variable is then copied into the internal RAM location addressed by the Stack Pointer. Otherwise no flags are affected.

Example : On entering an interrupt routine, the Stack Pointer contains 09H. The Data Pointer holds the value 0123H. The following instruction sequence,

PUSH DPL

PUSH DPH

leaves the Stack Pointer set to 0BH and stores 23H and 01H in internal RAM locations 0AH and 0BH, respectively.

Bytes : 2

Cycles : 2

Encoding :

1	1	0	0	0	0	0	0	direct address

Operation : PUSH

$(SP) \leftarrow (SP) + 1$

$((SP)) \leftarrow (direct)$

RET

Function : Return from subroutine

Description : RET pops the high- and low-order bytes of the PC successively from the stack, decrementing the Stack Pointer by two. Program execution continues at the resulting address, generally the instruction immediately following an ACALL or LCALL. No flags are affected.

Example : The Stack Pointer originally contains the value 0BH. Internal RAM locations 0AH and 0BH contain the values 23H and 01H, respectively. The following instruction,

RET

leaves the Stack Pointer equal to the value 09H. Program execution continues at location 0123H.

Bytes : 1

Cycles : 2

Encoding :

0	0	1	0	0	0	1	0

Operation : RET

$(PC_{15-8}) \leftarrow ((SP))$

$(SP) \leftarrow (SP) - 1$

$(PC_{7-0}) \leftarrow ((SP))$

$(SP) \leftarrow (SP) - 1$

RETI

Function : Return from interrupt

Description : RETI pops the high- and low-order bytes of the PC successively from the stack and restores the interrupt logic to accept additional interrupts at the same priority level as the one just processed. The Stack Pointer is left decremented by two. No other registers are affected; the PSW is *not* automatically restored to its pre-interrupt status. Program execution continues at the resulting address, which is generally the instruction immediately after the point at which the interrupt request was detected. If a lower- or same-level interrupt was pending when the RETI instruction is executed, that one instruction is executed before the pending interrupt is processed.

Example : The Stack Pointer originally contains the value 0BH. An interrupt was detected during the instruction ending at location 0122H. Internal RAM locations 0AH and 0BH contain the values 23H and 01H, respectively. The following instruction,

RETI

leaves the Stack Pointer equal to 09H and returns program execution to location 0123H.

Bytes : 1

Cycles : 2

Encoding :

0	0	1	1	0	0	1	0

Operation : RETI

$(PC_{15-8}) \leftarrow ((SP))$

$(SP) \leftarrow (SP) - 1$

$(PC_{7-0}) \leftarrow ((SP))$

$(SP) \leftarrow (SP) - 1$

RL A

Function : Rotate Accumulator Left

Description : The eight bits in the Accumulator are rotated one bit to the left. Bit 7 is rotated into the bit 0 position. No flags are affected.

Example : The Accumulator holds the value 0C5H (11000101B). The following instruction,

RL A

leaves the Accumulator holding the value 8BH (10001011B) with the carry unaffected.

Bytes : 1

Cycles : 1

Encoding :

0	0	1	0	0	0	1	1

Operation : RL

$(A_n + 1) \leftarrow (A_n)$ n = 0 – 6

$(A_0) \leftarrow (A_7)$

RLC A

Function : Rotate Accumulator Left through the Carry flag

Description : The eight bits in the Accumulator and the carry flag are together rotated one bit to the left. Bit 7 moves into the carry flag; the original state of the carry flag moves into the bit 0 position. No other flags are affected.

Example : The Accumulator holds the value 0C5H (11000101B), and the carry is zero. The following instruction,

RLC A

leaves the Accumulator holding the value 8AH (10001010B) with the carry set.

Bytes : 1

Cycles : 1

Encoding :

0	0	1	1	0	0	1	1

Operation : RLC

$(A_n + 1) \leftarrow (A_n)$ n = 0 – 6

$(A_0) \leftarrow (C)$

$(C) \leftarrow (A_7)$

RR A

Function : Rotate Accumulator Right

Description : The eight bits in the Accumulator are rotated one bit to the right. Bit 0 is rotated into the bit 7 position. No flags are affected.

Example : The Accumulator holds the value 0C5H (11000101B). The following instruction,

RR A

leaves the Accumulator holding the value 0E2H (11100010B) with the carry unaffected.

Bytes : 1

Cycles : 1

Encoding :

0	0	0	0	0	0	1	1

Operation : RR

$(A_n) \leftarrow (A_n + 1)$ n = 0 – 6

$(A_7) \leftarrow (A_0)$

RRC A

Function : Rotate Accumulator Right through Carry flag

Description : The eight bits in the Accumulator and the carry flag are together rotated one bit to the right. Bit 0 moves into the carry flag; the original value of the carry flag moves into the bit 7 position. No other flags are affected.

Example : The Accumulator holds the value 0C5H (11000101B), the carry is zero. The following instruction,

RRC A

leaves the Accumulator holding the value 62 (01100010B) with the carry set.

Bytes : 1

Cycles : 1

Encoding :

0	0	0	1	0	0	1	1

Operation : RRC

$(A_n) \leftarrow (A_n + 1)$ n = 0 – 6

$(A_7) \leftarrow (C)$

$(C) \leftarrow (A_0)$

SETB <bit>

Function : Set Bit

Description : SETB sets the indicated bit to one. SETB can operate on the carry flag or any directly addressable bit. No other flags are affected.

Example : The carry flag is cleared. Output Port 1 has been written with the value 34H (00110100B). The following instructions,

SETB C

SETB P1.0

sets the carry flag to 1 and changes the data output on Port 1 to 35H (00110101B).

SETB C

Bytes : 1

Cycles : 1

Encoding :

1	1	0	1	0	0	1	1

Operation : SETB
(C) ← 1

SETB bit

Bytes : 2

Cycles : 1

Encoding :

1	1	0	1	0	0	1	0	bit address

Operation : SETB
(bit) ← 1

SJMP rel

Function : Short Jump

Description : Program control branches unconditionally to the address indicated. The branch destination is computed by adding the signed displacement in the second instruction byte to the PC, after incrementing the PC twice. Therefore, the range of destinations allowed is from 128 bytes preceding this instruction 127 bytes following it.

Example : The label RELADR is assigned to an instruction at program memory location 0123H. The following instruction,

SJMP RELADR

assembles into location 0100H. After the instruction is executed, the PC contains the value 0123H.

Note: Under the above conditions the instruction following SJMP is at 102H. Therefore, the displacement byte of the instruction is the relative offset (0123H-0102H) = 21H. Put another way, an SJMP with a displacement of 0FEH is a one-instruction infinite loop.

Bytes : 2

Cycles : 2

Encoding :

1	0	0	0	0	0	0	0	rel. address

Operation : SJMP
(PC) ← (PC) + 2
(PC) ← (PC) + rel

SUBB A,<src-byte>

Function : Subtract with borrow

Description : SUBB subtracts the indicated variable and the carry flag together from the Accumulator, leaving the result in the Accumulator. SUBB sets the carry (borrow) flag if a borrow is needed for bit 7 and clears C otherwise. (If C was set *before* executing a SUBB instruction, this indicates that a borrow was needed for the previous step in a multiple-precision subtraction, so the carry is subtracted from the Accumulator along with the source operand.) AC is set if a borrow is needed for bit 3 and cleared otherwise. OV is set if a borrow is needed into bit 6, but not into bit 7, or into bit 7, but not bit 6.

When subtracting signed integers, OV indicates a negative number produced when a negative value is subtracted from a positive value, or a positive result when a positive number is subtracted from a negative number.

The source operand allows four addressing modes: register, direct, register-indirect, or immediate.

Example : The Accumulator holds 0C9H (11001001B), register 2 holds 54H (01010100B), and the carry flag is set. The instruction,

SUBB A,R2

will leave the value 74H (01110100B) in the accumulator, with the carry flag and AC cleared but OV set.

Notice that 0C9H minus 54H is 75H. The difference between this and the above result is due to the carry (borrow) flag being set before the operation. If the state of the carry is not known before starting a single or multiple-precision subtraction, it should be explicitly cleared by CLR C instruction.

SUBB A,R_n

Bytes : 1

Cycles : 1

Encoding :

1	0	0	1	1	r	r	r

Operation : SUBB

$(A) \leftarrow (A) - (C) - (R_n)$

SUBB A,direct

Bytes : 2

Cycles : 1

Encoding :

1	0	0	1	0	1	0	1	direct address

Operation : SUBB

$(A) \leftarrow (A) - (C) - (\text{direct})$

SUBB A,@R_i

Bytes : 1

Cycles : 1

Encoding :

1	0	0	1	0	1	1	i

Operation : SUBB

$(A) \leftarrow (A) - (C) - ((R_i))$

SUBB A,#data

Bytes : 2

Cycles : 1

Encoding :

1	0	0	1	0	1	0	0		immediate data

Operation : SUBB

$(A) \leftarrow (A) - (C) - \#data$

SWAP A

Function : Swap nibbles within the Accumulator

Description : SWAP A interchanges the low- and high-order nibbles (four-bit fields) of the Accumulator (bits 3 through 0 and bits 7 through 4). The operation can also be thought of as a 4-bit rotate instruction. No flags are affected.

Example : The Accumulator holds the value 0C5H (11000101B). The instruction,

```
SWAP        A
```

leaves the Accumulator holding the value 5CH (01011100B).

Bytes : 1

Cycles : 1

Encoding :

1	1	0	0	0	1	0	0

Operation : SWAP

(A_{3-0}) D (A_{7-4})

XCH A,<byte>

Function : Exchange Accumulator with byte variable

Description : XCH loads the Accumulator with the contents of the indicated variable, at the same time writing the original Accumulator contents to the indicated variable. The source/destination operand can use register, direct, or register-indirect addressing.

Example : R0 contains the address 20H. The Accumulator holds the value 3FH (00111111B). Internal RAM location 20H holds the value 75H (01110101B). The following instruction,

```
XCH         A,@R0
```

leaves RAM location 20H holding the values 3FH (00111111B) and 75H (01110101B) in the accumulator.

XCH A,R_n

Bytes : 1

Cycles : 1

Encoding :

1	1	0	0	1	r	r	r

Operation : XCH

(A) D (R_n)

XCH A,direct

Bytes : 2

Cycles : 1

Encoding :

1	1	0	0	0	1	0	1	direct address

Operation : XCH

(A) D (direct)

XCH A,@R_i

Bytes : 1

Cycles : 1

Encoding :

1	1	0	0	0	1	1	i

Operation : XCH

(A) D ((R_i))

XCHD A, @ R_i

Function : Exchange Digit

Description : XCHD exchanges the low-order nibble of the Accumulator (bits 3 through 0), generally representing a hexadecimal or BCD digit, with that of the internal RAM location indirectly addressed by the specified register. The high-order nibbles (bits 7-4) of each register are not affected. No flags are affected.

Example : R0 contains the address 20H. The Accumulator holds the value 36H (00110110B). Internal RAM location 20H holds the value 75H (01110101B). The following instruction,

XCHD A,@R0

leaves RAM location 20H holding the value 76H (01110110B) and 35H (00110101B) in the Accumulator.

Bytes : 1

Cycles : 1

Encoding :

1	1	0	1	0	1	1	i

Operation : XCHD

(A_{3-0}) D $((R_{i3-0}))$

XRL <dest-byte>,<src-byte>

Function : Logical Exclusive-OR for byte variables

Description : XRL performs the bitwise logical Exclusive-OR operation between the indicated variables, storing the results in the destination. No flags are affected.

The two operands allow six addressing mode combinations. When the destination is the Accumulator, the source can use register, direct, register-indirect, or immediate addressing; when the destination is a direct address, the source can be the Accumulator or immediate data.

Note: When this instruction is used to modify an output port, the value used as the original port data is read from the output data latch, *not* the input pins.

Example : If the Accumulator holds 0C3H (11000011B) and register 0 holds 0AAH (10101010B) then the instruction,

XRL A,R0

leaves the Accumulator holding the value 69H (01101001B).

When the destination is a directly addressed byte, this instruction can complement combinations of bits in any RAM location or hardware register. The pattern of bits to be complemented is then determined by a mask byte, either a constant contained in the instruction or a variable computed in the Accumulator at run-time. The following instruction,

XRL P1,#00110001B

complements bits 5, 4, and 0 of output Port 1.

XRL A,R_n

Bytes : 1

Cycles : 1

Encoding :

0	1	1	0	1	r	r	r

Operation : XRL

$(A) \leftarrow (A) \forall (R_n)$

XRL A,direct

Bytes : 2

Cycles : 1

Encoding :

0	1	1	0	0	1	0	1	direct address

Operation : XRL

$(A) \leftarrow (A) \forall (direct)$

XRL A,@R$_i$

Bytes : 1

Cycles : 1

Encoding :

0	1	1	0	0	1	1	i

Operation : XRL

$(A) \leftarrow (A) \forall ((R_i))$

XRL A,#data

Bytes : 2

Cycles : 1

Encoding :

0	1	1	0	0	1	0	0	immediate data

Operation : XRL

$(A) \leftarrow (A) \forall$ #data

XRL direct,A

Bytes : 2

Cycles : 1

Encoding :

0	1	1	0	0	0	1	0	direct address

Operation : XRL

(direct) $\leftarrow$ (direct) $\forall$ (A)

XRL direct,#data

Bytes : 3

Cycles : 2

Encoding :

0	1	1	0	0	0	1	1	direct address	immediate data

Operation : XRL

(direct) $\leftarrow$ (direct) $\forall$ #data

INDEX

SYMBOLS

C

J

L

M

N

O

R

S